Properties of Commonly used Figures

Geometric Area	Area Properties
(rectangle with dimensions $\bar{x}$, $\bar{y}$, h, b)	$A = bh$ $\bar{x} = b/2$ $\bar{y} = h/2$
(right triangle with dimensions $\bar{x}$, $\bar{y}$, h, b)	$A = bh/2$ $\bar{x} = b/3$ $\bar{y} = h/3$
(Vertex, 2nd order curve, dimensions $\bar{x}$, $\bar{y}$, h, b)	$A = 2bh/3$ $\bar{x} = 3b/8$ $\bar{y} = 2h/5$
(2nd order curve, Vertex, dimensions $\bar{x}$, $\bar{y}$, h, b)	$A = bh/3$ $\bar{x} = b/4$ $\bar{y} = 3h/10$

STRUCTURAL ANALYSIS

BICENTENNIAL
BICENTENNIAL
1807
WILEY
2007
BICENTENNIAL
BICENTENNIAL

THE WILEY BICENTENNIAL—KNOWLEDGE FOR GENERATIONS

*E*ach generation has its unique needs and aspirations. When Charles Wiley first opened his small printing shop in lower Manhattan in 1807, it was a generation of boundless potential searching for an identity. And we were there, helping to define a new American literary tradition. Over half a century later, in the midst of the Second Industrial Revolution, it was a generation focused on building the future. Once again, we were there, supplying the critical scientific, technical, and engineering knowledge that helped frame the world. Throughout the 20th Century, and into the new millennium, nations began to reach out beyond their own borders and a new international community was born. Wiley was there, expanding its operations around the world to enable a global exchange of ideas, opinions, and know-how.

For 200 years, Wiley has been an integral part of each generation's journey, enabling the flow of information and understanding necessary to meet their needs and fulfill their aspirations. Today, bold new technologies are changing the way we live and learn. Wiley will be there, providing you the must-have knowledge you need to imagine new worlds, new possibilities, and new opportunities.

Generations come and go, but you can always count on Wiley to provide you the knowledge you need, when and where you need it!

William J. Pesce

PRESIDENT AND CHIEF EXECUTIVE OFFICER

Peter Booth Wiley

CHAIRMAN OF THE BOARD

STRUCTURAL ANALYSIS

STRUCTURAL ANALYSIS
Using Classical and Matrix Methods

Fourth Edition

Jack C. McCormac
Clemson University

John Wiley and Sons, Inc.

ACQUISTIONS EDITOR Jennifer Welter
ASSOCIATE PUBLISHER Dan Sayre
MARKETING MANAGER Phyllis Cerys
SENIOR PRODUCTION EDITOR Sandra Dumas
DESIGNER Michael St. Martine
MEDIA EDITOR Stefanie Liebman
PRODUCTION MANAGEMENT SERVICES mb editorial services
COVER PHOTO Courtesy of American Institute of Timber Construction

This book was typeset in 10/12 Times Roman by Thomson Digital Limited and printed and bound by Hamilton Printing Company. The cover was printed by Phoenix Color Corporation.

The paper in this book was manufactured by a mill whose forest management programs include sustained yield harvesting of its timberlands. Sustained yield harvesting principles ensure that the number of trees cut each year does not exceed the amount of new growth.

ISBN 13 978-0470-03608-2
ISBN 10 0-470-03608-7

Printed in the United States of America.

10 9 8 7 6 5 4 3 2 1

This Book Is Dedicated to My Wife Mary
And to My Daughters Mary Christine and Becky

About the Author

Jack C. McCormac is Alumni Distinguished Professor of Civil Engineering, Emeritus at Clemson University. He holds a BS in civil engineering from the Citadel, an MS in civil engineering from Massachusetts Institute of Technology, and a Doctor of Letters from Clemson University. His contributions to engineering education and the engineering profession have been recognized by many, including the American Society for Engineering Education, the American Institute of Steel Construction, and the American Concrete Institute. Professor McCormac was included in the *International Who's Who in Engineering*, and was named by the *Engineering News-Record* as one of the top 125 engineers or architects in the world in the last 125 years for his contributions to the construction industry. He was one of only two educators living in the world today to receive this honor.

Professor McCormac belongs to the American Society of Civil Engineers and served as the principal civil engineering grader for the National Council of Examiners for Engineering and Surveying for many years.

Other Wiley Books by Jack C. McCormac

Design of Reinforced Concrete
Seventh Edition
ACI 318-05 Code Edition
0-471-76132-X

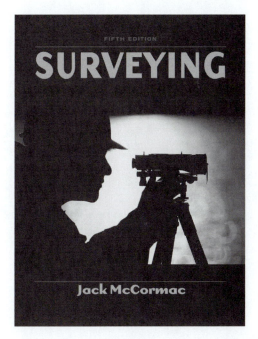

Surveying
Fifth Edition
0-471-23758-2

www.willey.com/college/mccormac

Preface

INTENDED AUDIENCE

This textbook presents an introduction to elementary structural analysis. It is primarily written for an undergraduate course for civil engineering, architecture, architectural engineering, and construction students. However, sufficient information is included for an additional course at the senior or graduate level.

NEW TO THIS EDITION

- The load sections of Chapters 2 and 3 have been revised to conform to ASCE Standard 7-02 as well as to the 2003 International Building Code.
- The matrix chapters (22–25) have been completely revised and substantially expanded.
- Most of the homework problems have been revised and their number substantially increased.
- Classical methods (influence lines, conjugate-beam analysis for deflections, and approximate methods and moment distribution for statically indeterminate structures) are included in this edition.

CLASSICAL METHODS

Matrix methods and computer applications have in effect made many of the older "classical" methods of structural analysis redundant. Matrix methods and structural analysis software such as SAP2000 are the tools that most engineers use in industry today.

I have elected to include several of the older "classical" methods in this text because they give the student an excellent knowledge of the behavior of structures subject to varying loading. Such knowledge or "feel" may very well not be developed if the student studies only matrix methods and computer applications. These methods also give students an alternative resource for confirming the validity of computer-generated data.

The following classical methods are included:

- influence lines
- conjugate-beam analysis for deflections
- approximate methods and slope deflection for statically indeterminate structures

STUDENT RESOURCES

The Web site for this text is located at *www.wiley.com/college/mccormac*. Available on the Web site are two software programs that may be used to solve most of the homework problems in the text:

SAP2000—Student version of a commercial structural analysis program frequently used in industry.

SABLE32—This program was developed by Jack McCormac and Jim Nelson to introduce students to software for structural analysis and to provide a tool to obtain a better understanding of structural behavior.

INSTRUCTOR RESOURCES

The Web site for this text is located at *www.wiley.com/college/mccormac*. All resources included on the student section of the site are also available to instructors.

Also available to instructors:

- Solutions Manual—Complete solutions for all of the homework problems in the text.
- Image Gallery—All figures and tables from the text, appropriate for use in PowerPoint presentations.

These resources are password-protected, and are only available to instructors who adopt the text for their course. Visit the instructor section of the book Web site to register for a password to access these resources.

ACKNOWLEDGMENTS

The author wishes to thank the following persons who reviewed this edition of the text. Their comments and suggestions are very much appreciated.

Robert W. Barnes, Auburn University

Robert Bruce, Tulane University

Shen-En Chen, University of North Carolina at Charlotte

Joel P. Conte, University of California, San Diego

Major Scott R. Hamilton

Paul Heyliger, Colorado State University

Sukhvarsh Jerath, University of North Dakota

Donald D. Liou, University of North Carolina at Charlotte

Lance Manuel, University of Texas at Austin

Andrzej S. Nowak, University of Nebraska

Brent Phares, Iowa State University

Kevin Rens, University of Colorado at Denver

Avi Singhal, Arizona State University

William A. Wood, Youngstown State University

Special thanks are extended to Dr. James K. Nelson of the University of Texas at Tyler for his preparation of the SABLE32 software. The author is also particularly grateful to Dr. Scott Schiff of Clemson University for his assistance in preparing the wind and seismic load material in Chapter 2 and for completely reviewing in great detail the page proof for the entire text.

Finally, the author thanks the reviewers and users of earlier editions for their suggestions and criticisms. He is always very grateful to anyone who takes the time to contact him concerning any part of the book.

Jack C. McCormac

Brief Table of Contents

APPENDICES

Table of Contents

PART TWO:
STATICALLY INDETERMINATE STRUCTURES
Classical Methods

PART THREE:
STATICALLY INDETERMINATE STRUCTURES
Common Methods in Current Practice

CHAPTER 21
Moment Distribution for Frames   **433**

CHAPTER 22
Introduction to Matrix Methods   **461**

CHAPTER 23
Fundamentals of the Displacement or Stiffness Method   **470**

CHAPTER 24
Stiffness Matrices for Inclined Members   **494**

STRUCTURAL ANALYSIS

PART ONE

STATICALLY DETERMINATE STRUCTURES

Chapter 1

Introduction

1.1 STRUCTURAL ANALYSIS AND DESIGN

The application of loads to a structure causes the structure to deform. Due to the deformation, various forces are produced in the components that comprise the structure. Calculating the magnitude of these forces, and the deformations that caused them is referred to as *structural analysis*, which is an extremely important topic to society. Indeed, almost every branch of technology becomes involved at some time or another with questions concerning the strength and deformation of structural systems.

Structural design includes the arrangement and proportioning of structures and their parts so they will satisfactorily support the loads to which they may be subjected. More specifically, structural design involves the following: the general layout of the structural system; studies of alternative structural configurations that may provide feasible solutions; consideration of loading conditions; preliminary structural analyses and design of the possible solutions; the selection of a solution; and the final structural analysis and design of the structure. Structural design also includes the preparation of design drawings.

This book is devoted to structural analysis, with only occasional remarks concerning the other phases of structural design. Structural analysis can be so interesting to engineers that they become completely attached to it and have the feeling that they want to become 100% involved in the subject. Although analyzing and predicting the behavior of structures and their parts is an extremely important part of structural design, it is only one of several important and interrelated steps. Consequently, it is rather unusual for an engineer to be employed solely as a structural analyst. An engineer, in almost all probability, will be involved in several or all phases of structural design.

It is said that Robert Louis Stevenson studied structural engineering for a time, but he apparently found the "science of stresses and strains" too dull for his lively imagination. He went on to study law for a while before devoting the rest of his life to writing prose and poetry.[1] Most of us who have read *Treasure Island, Kidnapped*, or his other works would agree that the world is a better place because of his decision. Nevertheless, there are a great number of us who regard structural analysis and design as extremely interesting topics. In fact some of us have found it so interesting that we have

---

[1]Proceedings of the First United States Conference on Prestressed Concrete (Cambridge, Mass.: Massachusetts Institute of Technology, 1951), 1.

White Bird Canyon Bridge, White Bird, Idaho (Courtesy of the
American Institute of Steel Construction, Inc.)

gone on to practice in the field of structural engineering. The author hopes that this book
will inspire more engineers to do the same.

1.2 HISTORY OF STRUCTURAL ANALYSIS

Structural analysis as we know it today evolved over several thousand years. During this
time many types of structures such as beams, arches, trusses, and frames were used in
construction for hundreds or even thousands of years before satisfactory methods of
analysis were developed for them. Though ancient engineers showed some understanding
of structural behavior (as evidenced by their successful construction of great bridges,
cathedrals, sailing vessels, and so on), real progress with the theory of structural analysis
occurred only in the last 175 years.

The Egyptians and other ancient builders surely had some kinds of empirical rules
drawn from previous experiences for determining sizes of structural members. There is,
however, no evidence that they had developed any theory of structural analysis. The
Egyptian Imhotep who built the great step pyramid of Sakkara in about 3000 B.C.E.
sometimes is referred to as the world's first structural engineer.

Although the Greeks built some magnificent structures, their contributions to
structural theory were few and far between. Pythagoras (about 582–500 B.C.E.), who is
said to have originated the word *mathematics*, is famous for the right angle theorem that
bears his name. This theorem actually was known by the Sumerians in about 2000 B.C.E.
Further, Archimedes (287–212 B.C.E.) developed some fundamental principles of statics
and introduced the term *center of gravity*.

The Romans were excellent builders and very competent in using certain structural
forms such as semicircular masonry arches. But, as did the Greeks, they too had little
knowledge of structural analysis and made even less scientific progress in structural
theory. They probably designed most of their beautiful buildings from an artistic
viewpoint. Perhaps their great bridges and aqueducts were proportioned with some
rules of thumb; however if these methods of design resulted in proportions that were
insufficient, the structures collapsed and no historical records were kept. Only their
successes endured.

One of the greatest and most noteworthy contributions to structural analysis, as
well as to all other scientific fields, was the development of the Hindu–Arabic

system of numbers. Unknown Hindu mathematicians in the 2nd or 3rd century B.C.E. originated a numbering system of one to nine. In about 600 C.E. the Hindus invented the symbol *sunya* (meaning empty), which we call zero. The Mayan Indians of Central America, however, had apparently developed the concept of zero about 300 years earlier.[2]

In the 8th century C.E. the Arabs learned this numbering system from the scientific writings of the Hindus. In the following century, a Persian mathematician wrote a book that included the system. His book later was translated into Latin and brought to Europe.[3] In around 1000 C.E., Pope Sylvester II decreed that the Hindu–Arabic numbers were to be used by Christians.

Before real advances could be made with structural analysis, it was necessary for the science of mechanics of materials to be developed. By the middle of the 19th century, much progress had been made in this area. A French physicist Charles Augustin de Coloumb (1736–1806) and a French engineer–mathematician Claude Louis Marie Henri Navier (1785–1836), building upon the work of numerous other investigations over hundreds of years, are said to have founded the science of mechanics of materials. Of particular significance was a textbook published by Navier in 1826, in which he discussed the strengths and deflections of beams, columns, arches, suspension bridges, and other structures.

Andrea Palladio (1508–1580), an Italian architect, is thought to have been the first person to use modern trusses. He may have revived some ancient types of Roman structures and their empirical rules for proportioning them. It was actually 1847, however, before the first rational method of analyzing jointed trusses was introduced by Squire Whipple (1804–1888). His was the first significant American contribution to structural theory. Whipple's analysis of trusses often is said to have signalled the beginning of modern structural analysis. Since that time there has been an almost continuous series of important developments in the subject.

Several excellent methods for calculating deflections were published in the 1860s and 1870s, which further accelerated the development of structural analysis. Among the important investigators and their accomplishments were James Clerk Maxwell (1831–1879) of Scotland, for the reciprocal deflection theorem in 1864; Otto Mohr (1835–1918) of Germany, for the method of elastic weights presented in 1870; Carlo Alberto Castigliano (1847–1884) of Italy, for the least-work theorem in 1873; and Charles E. Greene (1842–1903) of the United States, for the moment-area theorems in 1873.

The advent of railroads gave a great deal of impetus to the development of structural analysis. It was suddenly necessary to build long-span bridges capable of carrying very heavy moving loads. As a result, the computation of stresses and strains became increasingly important as did the need to analyze statically indeterminate structures.

One method for analyzing continuous statically indeterminate beams—the three-moment theorem—was introduced in 1857 by the Frenchman B. P. E. Clapeyron (1799–1864), and was used for analyzing many railroad bridges. In the decades that followed, many other advances were made in indeterminate structural analysis that were based upon the recently developed deflection methods.

Otto Mohr, who worked with railroads, is said to have reworked into practical, usable forms many of the theoretical developments up to his time. Particularly notable in

[2]*The World Book Encyclopedia* (Chicago, IL, 1993, Book N–O), pg. 617.
[3]*Ibid.*

this regard was his 1874 publication of the method of consistent distortions for analyzing statically indeterminate structures.

In the United States, two great developments in statically indeterminate structure analysis were made by G. A. Maney (1888–1947) and Hardy Cross (1885–1959). In 1915 Maney presented the slope deflection method, whereas Cross introduced moment distribution in 1924.

In the first half of the 20th century, many complex structural problems were expressed in mathematical form, but sufficient computing power was not available for practically solving the resulting equations. This situation continued in the 1940s, when much work was done with matrices for analyzing aircraft structures. Fortunately, the development of digital computers made the use of equations practical for these and many other types of structures, including high-rise buildings.

Pacific Gas and Electric Company headquarters, San Francisco (Courtesy of Bethlehem Steel Corporation)

Some particularly important historical references on the development of structural analysis include those by Kinney,[4] Timoshenko,[5] and Westergaard.[6] They document the slow but steady development of the fundamental principles involved. It seems ironic that the college student of today can learn in a few months the theories and principles of structural analysis that took many scholars several thousand years to develop.

[4]J. S. Kinney, *Indeterminate Structural Analysis* (Reading, Mass.: Addison-Wesley, 1957), 1–16.

[5]S. P. Timoshenko, *History of Strength of Materials* (New York: McGraw-Hill, 1953), 1–439.

[6]H. M. Westergaard, "One Hundred Fifty Years Advance in Structural Analysis," (*ASCE-94*, 1930), 226–240.

Cold-storage warehouse, Grand Junction, Colorado
(Courtesy of the American Institute of Steel Construction, Inc.)

1.3 BASIC PRINCIPLES OF STRUCTURAL ANALYSIS

Structural engineering embraces an extensive variety of structural systems. When speaking of structures, people typically think of buildings and bridges. There are, however, many other types of systems with which structural engineers deal, including sports and entertainment stadiums, radio and television towers, arches, storage tanks, aircraft and space structures, concrete pavements, and fabric air-filled structures. These structures can vary in size from a single member as is the case of a light pole to buildings or bridges of tremendous size. The Sears Tower in Chicago is over 1450 ft tall while the Taipei 101 building in Taiwan has a height of 1670 ft. Among the world's great bridges are the Humber Estuary Bridge in England, which has a suspended span of over 4626 ft, and the Akashi-Kaikyo bridge in Japan with its main suspended clear span of 6530 ft. Plans are now underway to build a bridge connecting Sicily to mainland Italy during the next decade. Its suspended main span is projected to be an almost unbelievable 2.05 miles.

To be able to analyze this wide range of sizes and types of structures, a structural engineer must have a solid understanding of the basic principles that apply to all structural systems. It is unwise to learn how to analyze a particular structure, or even a few different types of structures. Rather, it is more important to learn the fundamental principles that apply to all structural systems, regardless of their type or use. One never knows what types of problems the future holds or what type of structural system may be conceived for a particular application, but a firm understanding of basic principles will help us to analyze new structures with confidence.

The fundamental principles used in structural analysis are Sir Isaac Newton's laws of inertia and motion, which are:

1. A body will exist in a state of rest or in a state of uniform motion in a straight line unless it is forced to change that state by forces imposed on it.
2. The rate of change of momentum of a body is equal to the net applied force.
3. For every action there is an equal and opposite reaction.

These laws of motion can be expressed by the equation

$$\sum F = ma$$

In this equation, $\sum F$ is the summation of all the forces that are acting on the body, m is the mass of the body, and a is its acceleration.

In this textbook, we will be dealing with a particular type of equilibrium called *static equilibrium*, in which the system is not accelerating. The equation of equilibrium thus becomes

$$\sum F = 0$$

These structures either are not moving, as is the case for most civil engineering structures, or are moving with constant velocity, such as space vehicles in orbit. Using the principle of static equilibrium, we will study the forces that act on structures and methods to determine the response of structures to these forces. By response, the author means the displacement of the system and the forces that occur in each component of the system. This emphasis should provide readers with a solid foundation for advanced study, and hopefully convince them that structural theory is not difficult and that it is not necessary to memorize special cases.

1.4 STRUCTURAL COMPONENTS AND SYSTEMS

All structural systems are composed of components. The following are considered to be the primary components in a structure:

- *Ties:* those members that are subjected to axial tension forces only. Load is applied to ties only at the ends. Ties cannot resist flexural forces.
- *Struts:* those members that are subjected to axial compression forces only. Like ties, struts can be loaded only at their ends and cannot resist flexural forces.
- *Beams* and *Girders:* those members that are primarily subjected to flexural forces. They usually are thought of as being horizontal members that are primarily subjected to gravity forces; but there are frequent exceptions (e.g., inclined rafters).
- *Columns:* those members that are primarily subjected to axial compression forces. A column may be subjected to flexural forces also. Columns usually are thought of as being vertical members, but they may be inclined.
- *Diaphragms:* structural components that are flat plates. Diaphragms generally have very high in-plane stiffness. They are commonly used for floors and shear-resisting walls. Diaphragms usually span between beams or columns. They may be stiffened with ribs to better resist out-of-plane forces.

Structural components are assembled to form structural systems. In this textbook, we will be dealing with typical framed structures. A building frame is shown in Figure 1.1. In this figure, a girder is considered to be a large beam with smaller beams framing into it.

A *truss* is a special type of structural frame. It is composed entirely of struts and ties. That is to say, all of its components are connected in such a manner that they are subjected only to axial forces. All of the external loads acting on trusses are assumed to act at the joints and not directly on the components, where they might cause bending in the truss members. An older type of bridge structure consisting of two trusses is shown in Figure 1.2. In this figure, the *top and bottom chords* and the *diagonals* are the primary load carrying components of trusses. *Floor beams* are used to support the roadway. They are placed under the roadway and perpendicular to the trusses.

There are other types of structural systems. These include fabric structures (e.g., tents and outdoor arenas) and curved shell structures (e.g., dams or sports arenas). The analysis of these types of structures requires advanced principles of structural mechanics and is beyond the scope of this book.

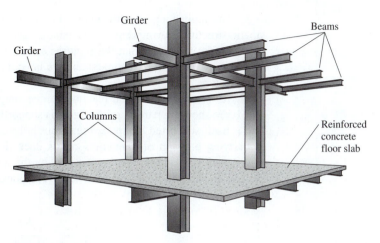

Figure 1.1   A typical building frame

Figure 1.2   Some components of a railroad bridge truss

1.5 STRUCTURAL FORCES

A structural system is acted upon by forces. Under the influence of these forces, the entire structure is assumed herein to be in a state of static equilibrium and, as a consequence, each component of the structure also is in a state of static equilibrium. The forces that act on a structure include the applied loads and the resulting reaction forces.

Las Vegas Convention Center (Courtesy of Bethlehem Steel Corporation)

The applied loads are the known loads that act on a structure. They can be the result of the structure's own weight, occupancy loads, environmental loads, and so on. The reactions are the forces that the supports exert on a structure. They are considered to be part of the external forces applied and are in equilibrium with the other external loads on the structure.

To introduce loads and reactions, three simple structures are shown in Figure 1.3. The beam shown in part (a) of the figure is supporting a uniformly distributed gravity load and is itself supported by upward reactions at its ends. The barge in part (b) of the figure is carrying a group of containers on its deck. It is in turn supported by a uniformly distributed hydrostatic pressure provided by the water beneath. Part (c) shows a building frame subjected to a lateral wind load. This load tends to overturn the structure, thus

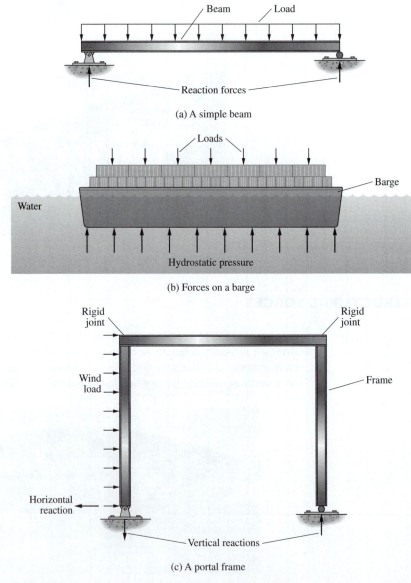

(a) A simple beam

(b) Forces on a barge

(c) A portal frame

Figure 1.3   Loads and reactions for three simple structures

requiring an upward reaction at the right-hand support and a downward one at the left-hand support. These forces create a couple that offsets the effect of the wind force. A detailed discussion of reactions and their computation is presented in Chapter 4.

1.6 STRUCTURAL IDEALIZATION (LINE DIAGRAMS)

To calculate the forces in the various parts of a structure with reasonable simplicity and accuracy, it is necessary to represent the structure in a simple manner that is conducive to analysis. Structural components have width and thickness. Concentrated forces rarely act at a single point; rather, they are distributed over small areas. If these characteristics are taken into consideration in detail, however, an analysis of the structure will be very difficult, if not impossible to perform.

The process of replacing an actual structure with a simple system conducive to analysis is called *structural idealization*. Most often, lines that are located along the centerlines of the components represent the structural components. The sketch of a structure idealized in this manner usually is called a *line diagram*.

The preparation of line diagrams is shown in Figure 1.4. In part (a) of the figure, the wood beam shown supports several floor joists and in turn is supported by three concrete-block walls. The actual distribution of the forces acting on the beam is shown in part (b) of the figure. For purposes of analysis, though, we can conservatively represent the beam and its loads and reactions with the line diagram of part (c). The loaded spans are longer with the result that shears and moments are higher than actually occur.

Another line diagram is presented in Figure 1.5 for the floor system of a steel frame building. Various other line diagrams are presented throughout the text as needed.

Sometimes the idealization of a structure involves assumptions about the behavior of the structure. As an example, the bolted steel roof truss of Figure 1.6(a) is considered. The joints in trusses often are made with large connection or gusset plates and, as such, can transfer moments to the ends of the members. However, experience has shown that the stresses caused by the axial forces in the members greatly exceed the stresses caused by flexural forces. As a result, for purposes of analysis we can assume that the truss consists of a set of pin-connected lines, as shown in Figure 1.6(b).

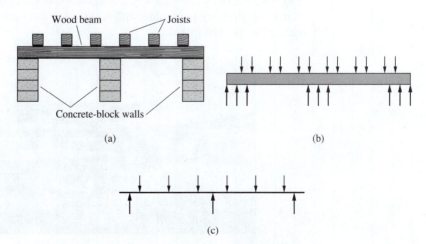

Figure 1.4   Replacing a structure and its forces with a line diagram

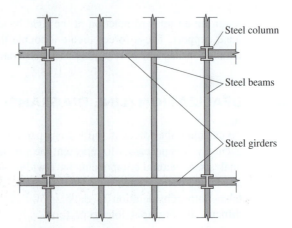

Figure 1.5   Line diagram for part of the floor system of a steel frame building

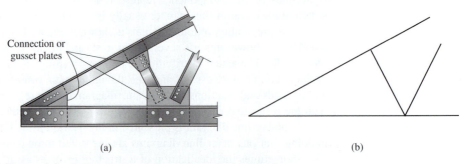

(a)   (b)

Figure 1.6   A line diagram for a portion of a steel roof truss

Although the use of simple line diagrams for analyzing structures will not result in perfect analyses, the results usually are quite acceptable. Sometimes, though, there may be some doubt in the mind of the analyst as to the exact line diagram or model to be used for analyzing a particular structure. For instance, should beam lengths be clear spans between supports, or should they equal the distances center to center of those supports? Should it be assumed that the supports are free to rotate under loads, are fixed against rotation, or do they fall somewhere in between? Because of many questions such as these, it may be necessary to consider different models and perform the analysis for each one to determine the worst cases.

Access Bridge, Renton, Washington (Courtesy of Bethlehem Steel Corporation)

1.7 CALCULATION ACCURACY

A most important point that many students with their superb pocket calculators and personal computers have difficulty understanding is that structural analysis is not an exact science for which answers can confidently be calculated to eight or more significant digits. Computations to only three places probably are far more accurate than the estimates of material strengths and magnitudes of loads used for structural analysis and design. The common materials dealt with in structures (wood, steel, concrete, and a few others) have ultimate strengths that can only be estimated. The loads applied to structures may be known within a few hundred pounds or no better than a few thousand pounds. It therefore seems inconsistent to require force computations to more than three or four significant figures.

Hungry Horse Dam and Reservoir, Rocky Mountains, in northwest Montana (Courtesy of the Montana Travel Promotion Division)

Several partly true assumptions will be made about the construction of trusses such as: truss members are connected with frictionless pins, the deformation of truss members under load is so slight as to cause no effect on member forces, and so on. These deviations from actual conditions emphasize that it is of little advantage to carry the results of structural analysis to many significant figures. Furthermore, calculations to more than three or four significant figures may be misleading in that they may give you a false sense of precision.

1.8 CHECKS ON PROBLEMS

A definite advantage of structural analysis is the possibility of making either mathematical checks on the analysis by some method other than the one initially used, or by the same method from some other position on the structure. You should be able in nearly every situation to determine if your work has been done correctly.

All of us, unfortunately, have the weakness of making exasperating mistakes, and the best that can be done is to keep them to the absolute minimum. The application of the simple arithmetical checks suggested in the following chapters will eliminate many of these costly blunders. The best structural designer is not necessarily the one who makes the fewest mistakes initially, but probably is the one who discovers the largest percentage of his or her mistakes and corrects them.

Oxford Valley Mall, Langehorne, Pennsylvania (Courtesy of Bethlehem Steel Corporation)

1.9 IMPACT OF COMPUTERS ON STRUCTURAL ANALYSIS

The availability of personal computers has drastically changed the way in which structures are analyzed and designed. In nearly every engineering school and office, computers are used to address structural problems. It is interesting to note, however, that up to the present time the feeling at most engineering schools has been that the best way to teach structural analysis is with chalk and blackboard, perhaps supplemented with some computer examples.

A rather large percentage of structural engineering professors feel that students should first learn the theories involved in structural analysis and the solution of problems with their pocket calculators before computer applications are introduced. As a result the author has placed computer applications at the ends of chapters so they can either be used at that time, skipped, or temporarily bypassed until a later date as the professor might prefer. The reader should realize that no theory is presented in the computer coverage contained herein which is not included in other sections of the book.

Two computer programs are provided for this book. These are SABLE32 (Structural Analysis and Behavior for Learning Engineering) and SAP2000. Both programs are available for download from the book's website at www.wiley.com/college/mccormac.

The author had quite a difficult time in deciding whether to include one of these programs or both of them. SABLE32 was specifically prepared to handle structural analysis problems of the types included in this text as well as the kinds of problems encountered in an elementary text dealing with reinforced concrete design. Sometime after the preparation of SABLE32, the author was granted access to the student version of the far more comprehensive structural program SAP2000.

A person not familiar with either of these programs can learn to use SABLE32 in very short order whereas the use of SAP2000 will require a considerable amount of study.

SAP2000 herein is a student version of a widely used commercial program. Its full version is used extensively in engineering practice not only in the United States but in many other countries as well. Though it will take students appreciably more time and effort to to learn SAP2000, they will be amply rewarded for their efforts. It is the kind of program they will use after graduation if they work for an engineering firm. Perhaps such knowledge will give them a head start on their early jobs. Very little information is contained herein on the application of SAP2000 as it is felt that the HELP sections of the program provide a sufficient set of directions. As a result, most of the examples provided herein and in the solutions manual available to professors are handled with the simple, direct program SABLE32. Several example problems that make use of SAP2000 are presented in Appendix D of this text.

Chapter 2

Structural Loads

2.1 INTRODUCTION

On rather frequent occasions structures professors ask their former students if they feel they were adequately prepared by their structures courses for their initial jobs. So often the answer is: Yes in most areas but probably not in the area of estimating design loads. This is a rather disturbing answer because the accurate estimation of the magnitude and character of the loads structures will have to support during their lives is probably the designer's most important task.

A very large number of different types of structures (beams, frames, trusses, and so forth), subjected to all sorts of loads, are introduced in this text. The student may as a result wonder "Where in the world did the author get all of these loads?" This very important question is addressed in this chapter and the next.

Structural engineers today generally use computer software in their work. Although the typical software will enable them to quickly analyze and design structures after the loads are established, it will provide little help in selecting the loads.

In this chapter, various types of loads are introduced and specifications are presented with which the individual magnitudes of the loads may be estimated. Our objective is to be able to answer questions such as the following: How heavy could the snow load be on a school in Minneapolis? What maximum wind force might be expected on a hotel in Miami? How large a rain load is probable for a flat roof in Houston?

The methods used for estimating loads are constantly being refined and may involve some very complicated formulas. Do not be concerned about committing such expressions to memory. Rather, learn the types of loads that can be applied to a structure and where to obtain information for estimating the magnitude of these loads.

The author is quite concerned that the enclosed sections on wind and seismic loads may be a little overwhelming for students just beginning the study of structures. He has included the material primarily to be used as an introduction and as a reference for future study and not as an essential part of an elementary course in structural analysis. The estimation of the magnitudes of wind and seismic loads are so involved that they each are often the subject of entire college courses. When design courses, such as those in structural steel and reinforced concrete, are taken in later semesters the student will learn more about wind and earthquake forces. It is to be realized that the procedures for estimating wind and seismic loads are constantly changing through the years as a result of continuing research in those topics.

Once the author opened the door to these loads, he found it very difficult to find a reasonable stopping place. It is probable that instructors of elementary structural analysis classes, for which this book was prepared, will not require students to learn in detail the information presented. The author's purpose here is to give the student an idea of the items involved in estimating the magnitude of wind and earthquake loads and to serve as a starting point and reference for further study when it becomes necessary in later work.

2.2 STRUCTURAL SAFETY

A structure must be adequate to support all of the loads to which it may foreseeably be subjected during its lifetime. Not only must it safely support these loads but it must do so in such a manner that deflections and vibrations are not so great as to frighten the occupants or to cause unsightly cracks.

The reader might think that all a structural designer has to do is to look at some structures similar to the one he or she is preparing to design, estimate the loads those structures are supporting, and then design his or her structure to be sufficiently strong to support those loads and a little more for safety's sake. However, its not quite that simple because there are so many uncertainties in design. Some of these uncertainties follow:

1. Material strengths may vary appreciably from their assumed values and they will vary more with time due to creep, corrosion, and fatigue.
2. The methods of analysis of structures are often subject to appreciable errors.
3. The so-called "beggaries of nature" or acts of God (earthquakes, hurricanes, rain and snow storms) cause loads that are extremely difficult to predict.
4. There are technological changes that cause increased loads to occur, such as larger trucks, trains, or army tanks crossing our bridges.
5. Loads occurring during construction operations can be severe and their magnitudes are difficult to predict.
6. Among other uncertainties that structures face are variations in member sizes, residual stresses, and stress concentrations.

Many years of design experience, both favorable and unfavorable, have led to detailed specifications and building codes. This chapter is devoted to some of the load requirements of those specifications. Ultimately the safety of the public is the major issue in this topic of selecting the magnitudes of loads for design.

2.3 SPECIFICATIONS AND BUILDING CODES

The design of most structures is controlled by specifications. Even if they are not so controlled, the designer probably will refer to them as a guide. No matter how many structures a person has designed, it is impossible for him or her to have encountered every situation. By referring to specifications, an engineer is making use of the best available material on the subject. Engineering specifications that are developed by various organizations present the best opinion of those organizations as to what represents good practice.

Municipal and state governments concerned with the safety of the public have established building codes with which they control the construction of various structures within their jurisdiction. These codes, which actually are laws or ordinances, specify design loads, design stresses, construction types, material quality, and other factors. They can vary considerably from city to city, which can cause some confusion among architects and engineers.

A structure housing the world's largest radio telescope, Green Bank, West Virginia (Courtesy of Lincoln Electric Company)

The determination of the magnitude of loads is only a part of determining the structural loads. The structural engineer must be able to determine which loads can be reasonably expected to act concurrently on a structure. For example, would a highway bridge completely covered with ice and snow be simultaneously subjected to fast moving lines of heavily loaded trucks in every lane and a 90-mile-per-hour lateral wind? Instead, is some lesser combination of these loads more reasonable and realistic? The topic of concurrent loads is addressed initially in Chapter 3 along with a related problem; the placement of loads on a structure so as to create the most severe conditions.

Quite a few organizations publish recommended practices for regional or national use. Their specifications are not legally enforceable, however, unless they are embodied in local building codes or made part of a particular contract. Among these organizations are the American Society of Civil Engineers (ASCE),[1] the American Association of State Highway and Transportation Officials (AASHTO),[2] and the American Railway Engineering Association (AREA).[3]

Recently, the International Code Council has developed the *International Building Code®*.[4] This code was developed to meet the need for a modern building code for building systems that emphasize performance. The *International Building Code* (IBC-2003) is intended to serve as a set of model code regulations to safeguard the public in all communities.

Readers should note that logical and clearly written codes are quite helpful to structural engineers. Furthermore, far fewer structural failures occur in areas with good

[1] American Society of Civil Engineers, *ASCE 7-02, Minimum Design Loads for Buildings and Other Structures*, Reston, Virginia 20191-4400, 2002.

[2] American Association of State Highway and Transportation Officials, *AASHTO LRFD Bridge Design Specifications*, (Washington, D.C. 2002).

[3] American Railway Engineering Association, *Specifications for Steel Railway Bridges*, (Washington, D.C. 2003).

[4] *2003 International Building Code*, (Falls Church, Virginia 22041-3401, International Code Council, Inc., 2003).

codes that are strictly enforced. The specifications published by the organizations mentioned are frequently used to estimate the maximum loads to which buildings, bridges, and some other structures may be subjected during their estimated lifetimes.

Some people feel that specifications prevent engineers from thinking for themselves—and there may be some basis for the criticism. The pundits say that the ancient engineers who built the great pyramids, the Parthenon, and the great Roman bridges were controlled by few specifications. This statement is certainly true. On the other hand, only a few score of these great projects were built over many centuries, and they were apparently built without regard to the cost of material, labor, or human life. They probably were built by intuition and by certain rules-of-thumb developed by observing the minimum size or strength of members that would fail under given conditions. Quite likely, there were numerous failures not recorded in history. Only the successes endured.

Today, at any one time hundreds of projects are being constructed in the United States that rival in importance and magnitude the famous structures of the past. Building codes and specifications are prepared by experts with knowledge in particular topics to provide guidance to engineers and a minimum standard of acceptable practice for design in a particular region. The result is that there are fewer disastrous failures and the public is better protected. *The important thing to remember about specifications is that their purpose is not to restrict engineers. Rather, their purpose is to protect the public.* Yet, no matter how many specifications are written, they cannot address every possible situation. Consequently, no matter which code or specification is or is not being used, the ultimate responsibility for the design of a safe structure lies with the structural engineer.

Specifications on many occasions clearly prescribe the minimum loads for which structures are to be designed. Despite the availability of this information, however, the designer's ingenuity and knowledge of the situation often are needed to predict the loads a particular structure will have to support in years to come. For example, over the past several decades, insufficient estimates of future traffic loads by bridge designers have resulted in a great amount of replacement with wider and stronger structures.

This chapter introduces the basic types of loads with which the structural engineer needs to be familiar. Its purpose is to help the reader develop an understanding of structural loads and their behavior, and to provide a foundation for estimating their magnitudes. It should not be regarded, however, as a complete essay on the subject of the loads that might be applied to any and every type of structure the engineer may design.

Since building loads are the most common type encounted by designers, they are the loads most frequently referred to in this text. The basic document currently being used by a large number of structural designers for estimating the loads to be applied to buildings is the ASCE 7 specification. It is often referred to in this chapter and has been incorporated into many model building codes. This specification was originally prepared and published by the American National Standards Institute and was referred to as ANSI 58.1 Standard. It has gone through several revisions. In 1988, it was taken over by ASCE and renamed ANSI/ASCE 7. Much information in this book is based on the 2002 edition of this specification, which now is called ASCE 7-02. Reference is also made to the *International Building Code* (IBC-2003).

When studying the information provided in this chapter, or when reviewing any standard providing design loads, the reader is cautioned that minimum design load standards are presented. An engineer should always view minimum design standards with some skepticism. The design standards are excellent and well prepared for most situations. However, there may be a building configuration, or a building use, for which the specified design loads are not adequate. A structural engineer should evaluate the minimum specified design loads to determine whether they are adequate for the structural system being designed.

South Fork Feather River Bridge in Northern California, being erected by use of a 1626-ft-long cableway strung from 210-ft-high masts anchored on each side of the canyon (Courtesy Bethlehem Steel Corporation)

2.4 TYPES OF STRUCTURAL LOADS

Structural loads usually are categorized by means of their character and duration. Loads commonly applied to buildings are categorized as follows:

- *Dead loads:* those loads of constant magnitude that remain in one position. They include the weight of the structure under consideration, as well as any fixtures that are permanently attached to it.
- *Live loads:* those loads that can change in magnitude and position. They include occupancy loads, warehouse materials, construction loads, overhead service cranes, and equipment operating loads. In general, live loads are caused by gravity.
- *Environmental loads:* those loads caused by the environment in which the structure is located. For buildings, the environmental loads are caused by rain, snow, ice, wind, temperature, and earthquakes. Strictly speaking, these are also live loads, but they are the result of the environment in which the structure is located.

2.5 DEAD LOADS

The dead loads that must be supported by a particular structure include all of the loads that are permanently attached to that structure. They include the weight of the structural frame and also the weight of the walls, roofs, ceilings, stairways, and so on.

Permanently attached equipment, described as "fixed service equipment" in ASCE 7-02, also is included in the dead load applied to the building. This equipment will include ventilating and air-conditioning systems, plumbing fixtures, electrical cables, support racks, and so forth. Depending upon the use of the structure, kitchen equipment such as ovens and dishwashers, laundry equipment such as washers and dryers, or suspended walkways could be included in the dead load.

The dead loads acting on the structure are determined by reviewing the architectural, mechanical, and electrical drawings for the building. From these drawings, the structural engineer can estimate the size of the frame necessary for the building layout and the equipment and finish details indicated. Standard handbooks and manufacturers' specifications can be used to determine the weight of floor and ceiling finishes, equipment, and fixtures. The approximate weights of some common materials used for walls, floors, and ceilings are shown in Table 2.1.

TABLE 2.1   WEIGHT OF SOME COMMON BUILDING MATERIALS

Building Material	Unit Weight	Building Material	Unit Weight
Reinforced Concrete	150 pcf	2 × 12 @ 16-in. Double Wood Floor	7 psf
Acoustical Ceiling Tile	1 psf	Linoleum or/Asphalt Tile	1 psf
Suspended Ceiling	2 psf	Hardwood Flooring ($\frac{7}{8}$-in.)	4 psf
Plaster on Concrete	5 psf	1-in. Cement on Stone-Concrete Fill	32 psf
Asphalt Shingles	2 psf	Movable Steel Partitions	4 psf
3-Ply Ready Roofing	1 psf	Wood Studs w/$\frac{1}{2}$-in. gypsum	8 psf
Mechanical Duct Allowance	4 psf	Clay Brick, 4 in. wythe	39 psf

The estimates of building weight or other structural dead loads may have to be revised one or more times during the analysis–design process. Before a structure can be designed, it must be analyzed. Among the loads used for the first analysis are the estimates of the weights of the components of the frame, which then are designed using those values. The component weights may then be recomputed with the sizes just calculated and compared with the initially estimated values. If there are significant differences, the structure should be reanalyzed using the revised weight estimates. The cycle is repeated as many times as necessary.

2.6 LIVE LOADS

Live loads are those loads that can vary in magnitude and position with time. They are caused by the building being occupied, used, and maintained. Most of the loads applied to a building that are not dead loads are live loads. Environmental loads, which are actually live loads by our usual definition, are listed separately in ASCE 7-02 and IBC-2003. Although environmental loads do vary with time, they are not all caused by gravity or operating conditions, as is typical with other live loads.

Some typical live loads that act on building structures are presented in Table 2.2. The loads shown in the table were taken from Table 4-1 in ASCE 7-02 and Table 1607.1 in IBC-2003. They are acting downward and are distributed uniformly over the entire floor or roof.

Many building specifications provide concentrated loads to be considered in design. This is the situation in Section 4.3 of the ASCE 7-02 and Section 1607.4 of IBC-2003. These specifications state that the designer must consider the effect of certain concentrated

TABLE 2.2   SOME TYPICAL UNIFORMLY DISTRIBUTED LIVE LOADS

Area Utilization	Live Load	Area Utilization	Live Load
Lobbies of Assembly Areas	100 psf	Classrooms in Schools	40 psf
Dance Halls and Ballrooms	100 psf	Upper Floor Corridors in Schools	80 psf
Library Reading Rooms	60 psf	Stairs and Exitways	100 psf
Library Stack Rooms	150 psf	Heavy Storage Warehouses	250 psf
Light Manufacturing Buildings	125 psf	Retail Stores—First Floor	100 psf
Offices in Office Buildings	50 psf	Retail Stores—Upper Floors	75 psf
Residential Dwelling Areas	40 psf	Walkways and Elevated Platforms	60 psf

loads as an alternative to the previously discussed uniform loads. The intent, of course, is that the loading used for design be the one that causes the most severe stresses.

Presented in Table 4-1 of ASCE 7-02 and Table 1607.1 of IBC-2003 are the minimum concentrated loads to be considered. Some typical values from these tables are shown in Table 2.3. The appropriate loads are to be positioned on a particular floor or roof so as to cause the greatest stresses (a topic to be discussed in more detail in Chapters 3 and 10). Unless otherwise specified, each of the concentrated loads is assumed to be uniformly distributed over a square area 2.5 ft $\times$ 2.5 ft (6.25 ft^2).

TABLE 2.3   TYPICAL CONCENTRATED LIVE LOADS

Area or Structural Component	Concentrated Live Load
Elevator Machine Room Grating on 4-in.^{2}	300 lbs
Office Floors	2000 lbs
Center of Stair Tread on 4-in.^{2}	300 lbs
Sidewalks	8000 lbs
Accessible Ceilings	200 lbs

When estimating the magnitudes of the live loads that may be applied to a particular structure during its lifetime, engineers need to consider the future utilization of that structure. For example, modern office buildings often are constructed with large open spaces that may later be divided into offices and other work areas by means of partitions. These partitions may be moved, removed, or added to during the life of the structure. Building codes typically require that partition loads be considered if the floor live load is less than 80 psf, even if partitions are not shown on the drawings. A rather common practice of structural designers is to increase the otherwise specified floor design live loads of office buildings by 20 psf to estimate the effect of the impossible-to-predict partition configurations of the future. This is the minimum partition load specified in Section 1607.5 of IBC-2003.

The method used to establish the magnitude of the ASCE 7-02 live loads is a rather complicated process that is described in the Commentary of that specification. Among the factors contributing to a particular specified value are the mean expected load, its variation over time, the magnitude of short duration transient loads, and the reference time period, which is typically assumed to be 50 years.

To convince the reader that the specified loads are reasonable, a brief examination of one of the specified values is considered. The example used here is the 100-psf live load specified by ASCE 7-02 for the lobbies of theaters and for assembly areas. Determine if such a load is reasonable for a crowd of people standing quite close together. Assume that the area in question is full of average adult males each weighing

TABLE 2.4   LIVE LOAD IMPACT FACTORS

Equipment or Component	Impact Factor
Elevator Machinery	100%
Motor-Driven Machinery	20%
Reciprocating Machinery	50%
Hangers for Floors and Balconies	33%

165 pounds and each occupying an area 20 in. by 12 in., or 1.67 sq ft. The average load applied equals $165/1.67 = 98.8$ psf. As such, the 100-psf live load specified seems reasonable. It actually is on the conservative side, as it would be rather difficult to have men standing that close together over a floor area that is either small or large.

2.7 LIVE LOAD IMPACT FACTORS

Impact loads are caused by the vibration and sudden stopping or dropping of moving or movable loads. It is obvious that a crate dropped on the floor of a warehouse or a truck bouncing on the uneven pavement of a bridge cause greater forces than would occur if the loads were applied gently and gradually. Impact loads are equal to the difference between the magnitude of the loads actually caused and the magnitude of the loads had they been dead loads. In other words, impact loads result from the dynamic effects of a load as it is applied to a structure. For static loads, these effects are short lived and do not necessitate a dynamic structural analysis. They do, however, cause an increase in stress in the structure that must be considered. Impact loads usually are specified as percentage increases of the basic live load. Table 2.4 shows the impact percentages for buildings given in Section 4.7.2 of ASCE 7-02 and Section 1607.8.2 of IBC-2003.

2.8 LIVE LOADS ON ROOFS

The live loads that act on roofs are handled in most building codes in a little different manner than are the other building live loads. The pitch of the roof (the ratio of the rise of the roof to its span) affects the amount of load that realistically can be placed upon it. As the pitch increases, the amount of load that can be placed on the roof before it begins to slide off decreases. Furthermore, as the area of the roof that contributes to the load acting on a supporting component increases it is less likely that the entire area will be loaded at any one time to its maximum live-load value.

The largest roof live loads usually are caused by repair and maintenance operations that probably do not occur simultaneously over the entire roof. This is not true of the environmental rain and snow loads, however, which are considered in Sections 2.9 and 2.15 of this chapter.

In the equations presented in this section, the term *tributary area* is used. This term, discussed in detail in Chapter 3, is defined as the loaded area of a structure that directly contributes to the load applied to a particular member. The tributary area for a member is assumed to extend from the member in question halfway to the adjacent members in each direction. When a building is being analyzed, it is customary for the analyst to assume that the load supported by a member is the load that is applied to its tributary area. The tributary area for a column is shown in Figure 2.1.

The basic minimum roof live load to be used in design is 20 psf. This value is specified in Section 4.9 of ASCE 7-02 and Section 1607.11.2.1 of IBC 2003. Depending

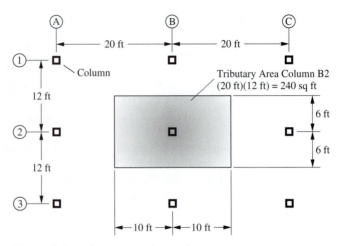

Figure 2.1   Tributary area for a column

on the size of the tributary area and the rise of the roof, this value may be reduced. The actual value to be used is determined with the expression:

$$L_r = 20R_1R_2$$

$$12 < L_r \leq 20$$

The term L_r represents the roof live load in psf of horizontal projection, whereas R_1 and R_2 are reduction factors. R_1 is used to account for the size of the tributary area A_t and R_2 is included to estimate the effect of the rise of the roof. The greater the tributary area (or the greater the rise of the roof) the larger will be the applicable reduction factor and the smaller the roof live load. The maximum roof live load is 20 psf and the minimum is 12 psf. Expressions for computing R_1 and R_2 are

$$R_1 = \begin{cases} 1.0 & A_t \leq 200 \, \text{ft}^2 \\ 1.2 - 0.001A_t & 200 \, \text{ft}^2 < A_t < 600 \, \text{ft}^2 \\ 0.6 & A_t \geq 600 \, \text{ft}^2 \end{cases}$$

$$R_2 = \begin{cases} 1.0 & F \leq 4 \\ 1.2 - 0.05F & 4 < F < 12 \\ 0.6 & F \geq 12 \end{cases}$$

The term F represents the number of inches of rise of the roof per foot of span. If the roof is a dome or an arch, the term F is the rise-to-span ratio of the structure multiplied by 32.

2.9 RAIN LOADS

It has been claimed that almost 50% of the lawsuits faced by building designers are concerned with roofing systems.[5] *Ponding*, a problem with many flat roofs, is one of the common subjects of such litigation. If water accumulates more rapidly on a roof than it runs off, ponding results because the increased load causes the roof to deflect into a dish

[5]Gary Van Ryzin, "Roof Design: Avoid Ponding by Sloping to Drain," *Civil Engineering* (New York: ASCE, January 1980), 77–81.

shape that can hold more water, which causes greater deflections, and so on. This process continues until equilibrium is reached or collapse occurs. Through proper selection of loads and good design providing adequate roof stiffness, the designer tries to avoid the latter situation. Many useful references on the subject of ponding are available.[6,7,8,9]

During a rainstorm, water accumulates on a roof for two reasons. First, when rain falls, time is required for the rain to run off the roof. Therefore, some water will accumulate. Second, roof drains may not be even with the roof surface and they may become clogged. Generally, roofs with slopes of 0.25 in. per ft or greater are not susceptible to ponding, unless the roof drains become clogged, thus enabling deep ponds to form.

In addition to ponding, another problem may occur for very large flat roofs (with perhaps an acre or more of surface area). During heavy rainstorms strong winds frequently occur. If there is a great deal of water on the roof, a strong wind may very well push a large quantity of the water toward one end. The result can be a dangerous water depth influencing the load on that end of the roof. For such situations, *scuppers* are sometimes used. Scuppers are large holes or tubes in walls or parapets that enable water above a certain depth to quickly drain off the roof.

Generally, two different drainage systems are provided for roofs. These normally are referred to as the primary and secondary drains. Usually the primary system will collect the rainwater through surface drains on the roof and direct it to storm sewers. The secondary system consists of scuppers or other openings or pipes through the walls that permit the rainwater to run over the sides of the building. The inlets of the secondary drains are normally located at elevations above the inlets to the primary drains.

The secondary drainage system is used to provide adequate drainage of the roof in the event that the primary system becomes clogged or disabled in some manner. The rainwater design load therefore is based on the amount of water that can accumulate before the secondary drainage system becomes effective.

The determination of the water that can accumulate on a roof before runoff during a rainstorm will depend upon local conditions and the elevation of secondary drains. Section 8 of ASCE 7-02 specifies that the rain load (in psf) on an undeflected roof can be computed from

$$R = 5.2(d_s + d_h)$$

This is the same equation as that in IBC-2003. The term d_s is the depth of water (in inches) on the undeflected roof up to the inlet of the secondary drainage system when the primary drainage system is blocked. This is the static head, which can be determined from the drawings of the roof system. The term d_h is the additional depth of water on the undeflected roof above the inlet of the secondary drainage system at its design flow. This is the hydraulic head. It is dependent upon the capacity of the drains installed and the rate at which rain falls.

From Section 8.3 of the ASCE 7-02 Commentary, the flow rate (in gallons per minute) that a particular drain must accommodate can be computed from

$$Q = 0.0104Ai$$

[6]F. J. Marino, "Ponding of Two-Way Roof System," *Engineering Journal*, AISC, 3rd quarter, no. 3 (1966), 93–100.

[7]L. B. Burgett, "Fast Check for Ponding," *Engineering Journal*, AISC, 10, 1st quarter, no. 1 (1973), 26–28.

[8]J. Chinn, "Failure of Simply-Supported Flat Roofs by Ponding of Rain," *Engineering Journal*, AISC, 2nd quarter, no. 2 (1965), 38–41.

[9]J. L. Ruddy, "Ponding of Concrete Deck Floors," *Engineering Journal*, AISC, 23, 3rd quarter, no. 2 (1986), 107–115.

The term A is the area of the roof (in square feet) that is served by a particular drain, and i is the rainfall intensity (in inches per hour). The rainfall intensity is specified by the code that has jurisdiction in a particular area. After the flow quantity is determined, the hydraulic head can be determined from Table 2.5 (ASCE 7-02, Table C8-1) for the type of drainage system being used. If the secondary drainage system is simply runoff over the edge of the roof, the hydraulic head will equal zero.

TABLE 2.5 FLOW RATE, Q, IN GALLONS PER MINUTE OF VARIOUS DRAINAGE SYSTEMS AND HYDRAULIC HEADS

	Hydraulic Head d_h (in.)									
Drainage System	1	2	2.5	3	3.5	4	4.5	5	7	8
4-in. diameter drain	80	170	180							
6-in. diameter drain	100	190	270	380	540					
8-in. diameter drain	125	230	340	560	850	1100	1170			
6-in. wide channel scupper	18	50		90		140		194	321	393
24-in. wide channel scupper	72	200		360		560		776	1284	1572
6-in. wide, 4-in. high, closed scupper	18	50		90		140		177	231	253
24-in. wide, 4-in. high, closed scupper	72	200		360		560		708	924	1012
6-in. wide, 6-in. high, closed scupper	18	50		90		140		194	303	343
24-in. wide, 6-in. high, closed scupper	72	200		360		560		776	1212	1372

Note: Interpolation is appropriate, including between scupper widths. Closed scuppers are four sided and channel scuppers are open topped.

Example 2.1 illustrates the calculation of the design rainwater load for a roof with scuppers using the ASCE 7-02 specification.

EXAMPLE 2.1

A roof measuring 240 feet by 160 feet has 6-in. wide channel-shaped scuppers serving as secondary drains. The scuppers are 4 inches above the roof surface and are spaced 20 feet apart along the two long sides of the building. The design rainfall for this location is 3 inches per hour. What is the design roof rain load?

Solution. The area served by each scupper is

$$A = (20\,\text{ft})(80\,\text{ft}) = 1600\,\text{ft}^2$$

The runoff quantity for each scupper is

$$Q = 0.0104Ai = 0.0104(1600\,\text{ft}^2)(3\,\text{in./hr}) = 49.92\,\text{gpm}$$

Referring to Table 2.5, observe that the hydraulic head at this flow rate for the scupper used is 2 in. The design roof rain load, then, is

$$R = 5.2(d_s + d_h) = 5.2(4\,\text{in.} + 2\,\text{in.}) = 31.2\,\text{psf} \quad \blacksquare$$

2.10 WIND LOADS

A survey of engineering literature for the past 150 years reveals many references to structural failures caused by wind. Perhaps the most infamous of these have been bridge failures such as those of the Tay Bridge in Scotland in 1879 (which caused the deaths of 75 persons) and the Tacoma Narrows Bridge (Tacoma, Washington) in 1940. However, there were some disastrous building failures due to wind during the same period, such as the Union Carbide Building in Toronto in 1958. It is important to realize that a large percentage of building failures due to wind have occurred during their construction.[10]

Considerable research has been conducted in recent decades on the subject of wind loads. Nevertheless a great deal more study is needed, as the estimation of wind forces can by no means be classified as an exact science.

The average structural designer would love to have a simple rule with which he or she could compute the magnitude of design wind loads, such as: The wind pressure is to be 20 psf for all parts of structures 100 ft or less above the ground and 30 psf for parts that are more than 100 ft above the ground. However, a simple specification such as this one, though of the type used for many years, has never been satisfactory. If we are to prevent future, perhaps catastrophic, mishaps we must do better.

Jacobs Field, the home of the Cleveland Indians (Courtesy of
The Lincoln Electric Company)

Wind forces act as pressures on vertical windward surfaces, pressures or suctions on sloping windward surfaces (depending on slope), and suction or uplift on flat surfaces and on leeward vertical and sloping surfaces (due to the creation of negative pressures or vacuums). The student may have noticed this definite suction effect where shingles or other roof coverings have been lifted from the leeward roof surfaces of buildings. Suction or uplift can easily be demonstrated by holding a piece of paper horizontally at two of its diagonally opposite corners and blowing above it. You will see that the far end of the paper moves upwards. For some common structures, uplift loads may be as large as 20 or 30 psf or even more.

---

[10]Wind Forces on Structures, Task Committee on Wind Forces. Committee on Loads and Stresses, Structural Division, ASCE, Final Report, *Transactions ASCE* 126, Part II (1961): 1124–1125.

The ASCE 7-02 standard provides equations with which wind pressures can be estimated for various parts of buildings. Though the use of these equations is complicated, the work is somewhat simplified with the tables and charts presented in the specification. Several of these charts and tables are shown in Appendix C of this book with the permission of the ASCE. The reader should particularly note that the information provided is for buildings of regular shape. Should domes, A-frames, or buildings with roofs sloped at angles greater than 45 degrees or buildings with unusual floor plans as H or Y shapes or others be encountered, it will be desirable to conduct wind tunnel studies. Guidelines for making such studies are presented in Section 6.6 of the aforementioned specification.

The values obtained using these specifications are not satisfactory for tornadic winds although it has been estimated that buildings designed for the wind forces obtained with ASCE 7-02 are able to resist with little damage about one-half of recorded tornadoes.

The introductory discussion of wind forces presented in this section provides only a brief introduction to the topic. Furthermore, only wind forces applied to the main wind force resisting systems are discussed and then only for low-rise buildings with roof slopes of less than 10 degrees.

There are many factors which affect wind pressures. Among them are wind speed, the exposures of structures, characteristics of surrounding terrain, presence of nearby structures, and so on. In preparing designs for wind, another factor that much be considered is the relative importance of the structure. Is it to be a hospital or a school or an agricultural barn? The factors mentioned here are briefly discussed in the paragraphs to follow.

Design Wind Speed, V

The basic wind speed to be used in design for the locality involved may be estimated from Appendix Figure C1. The values provided in this figure are not applicable to mountainous areas, gorges, and other regions where unusual wind conditions may exist. For such areas special studies will have to be made. The velocities obtained are the estimated worst 3-second gust speeds in miles per hour (mph) that would occur at 33 feet above the ground surface during a 50-year period.

Importance Factor, I

The importance factor is intended to bring into the calculation of wind forces a measure of the consequences of failure. Critical buildings, such as schools and hospitals, will have a higher importance factor and therefore higher design wind forces. Buildings whose failure will have little consequence on human life, such as farm buildings, will have a lower importance factor and therefore lower design wind forces.

Buildings are classified by the ASCE as falling into categories I, II, III, or IV as shown in Appendix Table C.1 of this text. These classifications are applicable not only to wind loads but also to flood, snow, earthquake, and ice loadings. After the category is selected for the building in question, its importance factor is determined from Appendix Table C.4.

Surface Roughness Categories

In ASCE 7-02 Section 6.5.6.2, ground surfaces around structures are classified as to roughness as being B, C, or D. These classifications, which are shown in Appendix Table C.2, run from urban and suburban areas with numerous closely spaced obstructions all the way to smooth mud or salt flats or unbroken ice. Originally there was a Class A roughness category but it is no longer listed by the ASCE.

Two methods are presented by ASCE 7-02 for estimating wind loads. These are the Simplified Procedure (ASCE Section 6.4) and the Analytical Procedure (ASCE Section 6.5). The first procedure is applicable to a limited group of structures while the second procedure has a much wider range of application. Actually, as previously mentioned, there is another procedure for very complicated situations that involves wind tunnel tests. Only the first of these methods is described in detail herein.

2.11 SIMPLIFIED ASCE PROCEDURE FOR ESTIMATING WIND LOADS

Section 6.4 of the ASCE Standard presents a simplified method for estimating wind loads. However, the procedure is only satisfactory for buildings with certain limitations. These limitations, listed in detail in Section 6.4.1.1 of the ASCE Standard, include the following:

1. The building must be low rise, enclosed, regular shape rigid, nearly symmetrical, and have a simple diaphragm. (For flat roof or gable/hip roofs with slope $\leq 45°$)
2. Its roof height must not exceed 60 ft.
3. There may not be any expansion joints or separations in the building structure.
4. There are also some requirements concerning wind-borne debris and response characteristics. (No topographical effects included.)

The wind pressures for such buildings may be estimated with the following expression

$$p_s = \lambda I p_{S30}$$

To use this expression the following values need to be determined:

V = estimated wind pressure, psf
λ = adjustment factor for building height and exposure
I = importance factor
p_{S30} = simplified design wind pressure for exposure B at 30-ft height and for $I = 1.0$.

Example 2.2 presents sample calculations for estimating wind pressures with the simplified procedure.

EXAMPLE 2.2

The building shown in Figure 2.2 is to be constructed in a suburban area with numerous closely spaced small buildings in Galveston on the Texas gulf coast. Its primary use will be for hotel rooms and there will be no areas in which more than 300 people can congregate. Compute the estimated wind pressures acting on the various areas of this

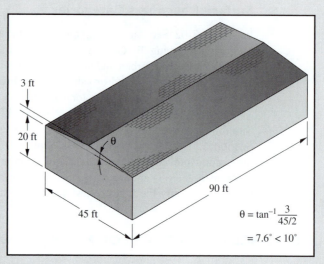

Figure 2.2

enclosed rigid frame simple diaphragm structure using the ASCE 7-02 simplified procedure.

Solution.

(1) Wind speed $= 140$ mph for Galveston (Appendix Figure C.1)

(2) Building will be in Category II because of occupancy (Appendix Table C.1)

(3) Surface roughness category will be C (Appendix Table C.2)

(4) $\lambda = 1.29$ (by interpolation from Appendix Figure C.2 for 20 ft since $\theta < 10°$)

Estimated pressures or suctions are represented by p_{S30} in Appendix Figure C.2. The applicable values are selected for the building in question and are recorded in Table 2.6, which follows. Then these values are multiplied by λ I to adjust them for our particular conditions for which $\lambda = 1.29$ and I $= 1.0$. The values so obtained are shown in the table for the various zones A to H that are shown in Figure 2.3. Note that negative signs represent suctions.

TABLE 2.6   ESTIMATED WIND PRESSURES FOR BUILDING IN EXAMPLE 2.2

| Zone | Horizontal Pressure, p_{S30} with I $= 1.0$ | | | Adjusted pressure, psf, $= p_s = \lambda I p_{S30}$ |
	If $\theta = 5°$	If $\theta = 10°$	If $\theta = 7.6°$	
A	31.1	35.1	33.2	42.8
B	-16.1	-14.5	-15.3	-19.7
C	20.6	23.3	22.0	28.3
D	-9.6	-8.5	-9.0	-11.6
E	-37.3	-37.3	-37.3	-48.1
F	-21.2	-22.8	-22.0	-28.3
G	-26.0	-26.0	-26.0	-32.5
H	-16.4	-17.5	-17.0	-21.9

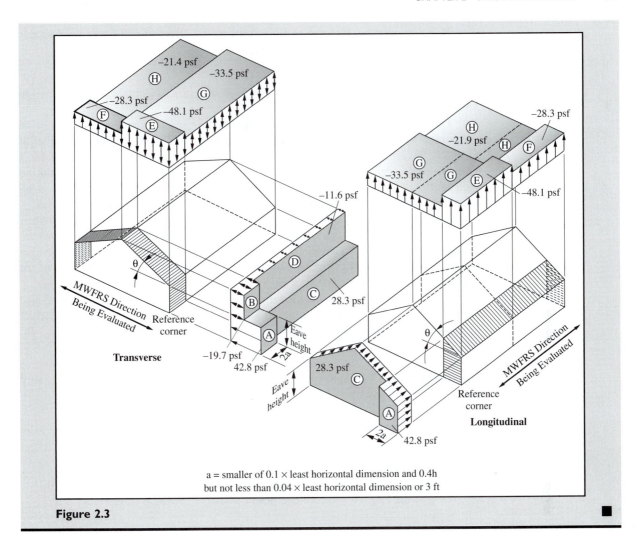

a = smaller of 0.1 × least horizontal dimension and 0.4h
but not less than 0.04 × least horizontal dimension or 3 ft

Figure 2.3

2.12 DETAILED ASCE PROCEDURE FOR ESTIMATING WIND LOADS

The ASCE 7-02 standard presents a more detailed procedure for estimating wind pressures for buildings and other structures. Due to the length of its application it is only briefly introduced in this section.

In its Section 6.5.10, ASCE 7-02 provides the following expression for estimating the velocity pressure at a height z above the ground surface

$$q_z = 0.00256 \, K_z K_{zt} K_d V^2 I$$

As a part of the simplified procedure described in the preceding section, the determination of wind velocities (V) and importance factors (I) were described. These terms are included in the equation for q_z as are a few other terms.

These others are:

K_z, a velocity pressure exposure coefficient that is dependent upon the structure height and the characteristics of the nearby terrain

K_{zt}, a topographic factor used for estimating the effect of increasing wind speeds caused by sudden changes in topography

K_d, a wind directionality factor

Various tables and figures are available for estimating the magnitude of these terms in ASCE 7-02.

Once the value of q_z is determined, it may be used to compute estimated wind pressures at various parts of the building being considered. ASCE 7-02 provides a set of equations for estimating such pressures. The choice of the equation to be used for a particular case is dependent upon the flexibility and height of the structure in question and upon whether the main wind resisting system of the building or its secondary wind resistance provided by the components of the building and its cladding are involved.

The reader can easily understand that since buildings are not completely airtight their interiors are subject to pressures just as are their exteriors. These internal pressures are caused by wall openings such as doors, windows, and so on. The reader needs to remember that windows can break or operable windows or doors may be left open. Section 6.2 of ASCE 7-02 provides a description of enclosed, partially enclosed, and open buildings.

The design wind pressure for low-rise (h < 60 ft) rigid buildings can be determined from the expression to follow in which p is the design wind pressure, p_e is the external wind pressure, and p_i is the internal wind pressure.

$$p = p_e - p_i$$

This expression can be expressed in more detail as follows for windward and leeward walls:

$$p = q_h[(GC_{pf}) - (GC_{pi})] \qquad \text{ASCE 7-02 Eq. 6-18 for leeward and side walls}$$

In which

q_h is the velocity pressure calculated for height h
GC_{pf} is the external pressure coefficient available in ASCE 7-02 Figure 6.10
GC_{pi} is the internal pressure coefficient available in ASCE 7-02 Figure 6.5

2.13 SEISMIC LOADS

Many areas of the world fall into "earthquake territory," and in those areas it is necessary to consider seismic forces in the design of all types of structures. Through the centuries there have been catastrophic failures of buildings, bridges, and other structures during earthquakes. It has been estimated that as many as 50,000 people died in the 1988 earthquake in Armenia.[11] Even more people were killed in the 2005 Kashmir earthquake. The 1989 Loma Prieta and 1994 Northridge earthquakes in California caused many billions of dollars of property damage as well as considerable loss of life.

The earth's outer crust is composed of hard plates as large or larger than entire continents. These plates float on the soft molten rock beneath. Their movement is very slow, perhaps only a few inches per year. It has been noted that this is slower than your fingernails grow. When the plates encounter each other, significant horizontal and vertical motion of the ground surface may be caused. These motions may cause very large inertia

---

[11]V. Fairweather, "The Next Earthquake," *Civil Engineering* (New York: ASCE, March 1990), 54–57.

forces in structures. The distribution and characteristics of the earth materials in a particular area greatly affect the magnitudes of the ground motions caused.

Recent earthquakes have clearly shown that the average building or bridge that has not been designed for earthquake forces can be destroyed by a relatively moderate earthquake. Most structures can be economically designed and constructed to withstand the forces caused during most earthquakes. On the other hand, the cost of providing seismic resistance to existing structures (called retrofitting) can be extremely high.

Some engineers seem to think that the seismic loads needed in design are merely percentage increases of the wind loads. This assumption is incorrect, however. Seismic loads are different in their action and are not proportional to the exposed area of the building, but rather are proportional to the distribution of the mass of the building above the particular level being considered.

Building codes around the United States where earthquakes are most likely to occur require that some type of seismic design be used. Even in areas less prone to earthquakes, seismic loading should be seriously considered for high-rise buildings, hospitals, nuclear plants, and other important structures.

The Interagency Committee on Seismic Safety in Construction has prepared an order (signed by the president) stating that all federally owned or leased buildings as well as all federally regulated and assisted construction should be constructed to mitigate seismic hazards.

To assess the importance of designing for earthquake forces, a seismic ground acceleration map such as the one shown in Appendix Figure C.3 should be examined. This particular map presents what are thought to be the estimated peak ground accelerations that might very well occur in the various parts of the United States during a 50-year period. High-risk areas for damaging earthquakes, such as coastal California, are quite obvious on this map while low-risk areas are also clearly shown in Florida and parts of Texas.

Earthquakes apply loads to structures in an indirect fashion. The ground is displaced and since the structures are connected to the ground, they are subject to sudden movements. These movements generate acclerations in the building leading to differential movement of the building levels. These deformations cause horizontal shears to be produced. From this information it is clear that no external forces are actually applied to buildings above ground by earthquakes.

The usual procedure for analyzing structures for seismic forces is to represent them with sets of supposedly equivalent loads. The magnitudes of the loads selected are based upon the distribution of mass in the various structures, the accelerations of the ground, and the dynamic characteristics of the systems. Another factor to be considered in seismic design is the soil condition. Almost all of the structural damage and loss of life in the Loma Prieta earthquake occurred in areas having soft clay soils. Apparently, these soils amplified the motions of the underlying rock.

Sections 9.5.3 through 9.5.8 of ASCE 7-02 present six different seismic analytical procedures for estimating seismic forces to be used in structural analysis and design. The situations where each of the methods may be used are listed in ASCE 7-02 Table 9.5.2.5.1. The first of the methods is a static analysis referred to as "index force analysis." The method is considered to be satisfactory only for certain small structures in a few areas of the country where seismic risks are low. The following expression is given to compute static lateral forces that are to be applied at each level of the structure.

$$F_x = 0.01 w_x$$

Where

F_x = the design lateral force applied at each story

w_x = the portion of the total gravity load W assigned to level x from above

W = the effective seismic weight of the structure, which includes the dead loads plus some portion of the live loads (such as at least 25% of the floor live load in storage buildings).

Notice that the total horizontal shear at a particular level is equal to the sum of all the F_x loads above.

The second method listed by the ASCE so-called "simplified analysis" is another static load analysis but one which involves increased forces. As such, it may be used for a little wider range of structures. Proceeding through the other methods, we move into the more complex methods of analysis. The third method listed by ASCE 7-02 is entitled the "Equivalent Lateral Force Procedure." It is the method with which today's designer needs to be the most familiar as it is the most commonly used method in practice in the United States. Consequently, it is the only one of the six methods illustrated in this text and that is done in the next section.

The fourth method called "Modal Analysis Procedure" is generally used instead of the third method when some floors of a building are quite different from the other floors as to stiffness, weight, and so forth. The final two methods "Linear Response History Analysis" and "Nonlinear Response History Analysis" are rarely used in design practice today. They are primarily used in various research studies.

As the reader will see in the next chapter, the calculated loadings on structures and their components due to seismic effects are combined with other loads. In making the combinations, the reader should understand that the methods just mentioned are based on a strength limit state beyond first yielding of the structures. The necessity for understanding this fact will become clear after one studies load combinations in the next chapter.

2.14 EQUIVALENT LATERAL FORCE PROCEDURE FOR ESTIMATING SEISMIC LOADS

In this section the author briefly introduces the equivalent lateral force procedure for estimating seismic forces. As the student studies this method, the author wants him or her to give a great deal of thought to the accuracies obtained. With this procedure equivalent, static loads are computed to estimate the effect of dynamic seismic forces. In addition the structures considered are assumed to resist the design loads in an elastic manner—but their resistance is actually inelastic during a design event.

With the equivalent lateral force procedure, a total base shear is estimated based on properties of the structures and the ground motion expected at the building site. Empirical equations are presented to estimate the total lateral shear applied to the building and to apportion those shears to the various floor levels. These equations contain several terms, which are listed and defined here in the order in which they will be used in the calculations.

1. The *fundamental natural period of a building*, which is represented by the letter T, is the time required for the building to go through one complete cycle of motion. Its magnitude is dependent upon the mass of the structure and its stiffness and can be estimated with the equation given at the end of this paragraph. The application of this equation usually provides periods which are somewhat smaller than the real periods of the structures involved. Such a

situation causes the calculated shears to be a little on the high side and hopefully puts us a little on the safe side.

$$T_a = C_t h_n^x$$

In this expression C_t is the building period coefficient. Its value is 0.035 for moment resisting frames of structural steel and 0.02 for most other structures, such as braced frames. The term h_n is the height of the highest level of the building while x is 0.9 for reinforced concrete moment frames and 0.75 for other systems.

2. The *design spectral accelerations*, represented by the terms S_{D1} and S_{DS}, may be determined from seismic maps. Such maps provide the estimated intensities of design earthquakes with $T = 1$ second (S_{D1}) and with $T = 0.2$ seconds (S_{DS}). The numbers obtained represent proportions of g, the gravitational acceleration. For Salt Lake City, the values are respectively 0.5 g and 1.2 g. The values given are for structure foundations supported by moderately strong rock. If weaker foundation materials are involved, S_{D1} and S_{DS} will be larger.

3. The *response modification factor* is used to estimate the ability of a structure to resist seismic forces. Its value varies from 1.25 up to 8 with the high values applicable to ductile structures and the lower values applicable to brittle structures. For structures with reinforced concrete shear walls, R values of about 4 are used while for structural steel frames and for reinforced concrete frames with rigid joints, R will equal approximately 8. Other values are provided in ASCE 7-02. The larger the R value; the smaller will be the computed seismic design forces.

4. The *importance factor*, I, of a structure provides a measure of the consequences of failure. The higher the number, the more important the structure. For instance, ASCE 7-02 provides a value of 1.5 for hospitals, police stations, and other public buildings but only 1.0 for office buildings.

5. The *effective seismic weight* of a building, W, includes the total dead load of the structure plus applicable portions of other loads. For instance, a minimum of 25% of floor live loads must be included along with a 10 psf allowance for partitions if they are present. Furthermore, W must include the total weight of permanent equipment. Also where flat roof snow loads exceed 30 psf, 20 percent of the snow load is included in the seismic weight.

6. The ASCE 7-02 specification in its Section 9.5.4 provides the following expression for estimating the total static lateral base shear in a given direction for a building.

$$V = \frac{S_{D1}}{T(R/I)} W$$

However, the calculated V need not be greater than the value obtained with the following expression. (When very stiff structures with small T values are involved, the preceding equation yields unnecessarily high values.)

$$V_{max} = \frac{S_{DS} W}{R/I}$$

A practical minimum value of V to be used is given with the following equation.

$$V_{min} = 0.044 S_{DS} IW$$

7. The portion of the base shear V to be distributed to a particular floor is determined with the following equation:

$$F_x = \frac{w_x h_x^k}{\sum\limits_{i=1}^{n} w_i h_i^k} V$$

In which

F_x = lateral seismic force to be applied to level x

w_i and w_x = the weights assigned to levels i and x. The student will understand the values of i and x after Example 2.3 is studied.

h_i and h_x = height of levels i and x

n = floor level in question

k = a distribution exponent related to the fundamental natural period of the structure in question. If T is 0.5 seconds or less, k = 1.0. Should T be > 0.5 seconds and ≤ 2.5 seconds, k can be determined from the expression to follow. If T > 2.5 seconds, k = 2.0.

$$k = 1 + \frac{T - 0.5}{2}$$

EXAMPLE 2.3

Using the ASCE 7-02 equivalent lateral force procedure, compute the lateral force to be applied to the third floor of the proposed structural steel office building shown in Figure 2.4. The building, which is to be located in Salt Lake City, has special moment resisting connections. Values of $S_{DS} = 1.2$ g and $S_{D1} = 0.5$ g have been selected. The estimated weight, w, of each level is 500 k.

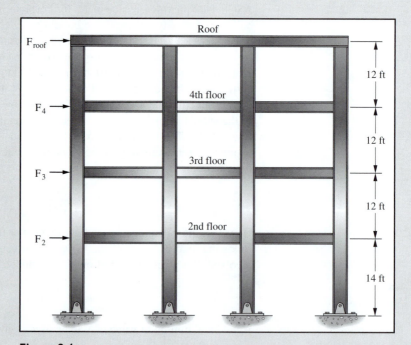

Figure 2.4

Solution.

(1) $T_a = C_t h_n^x = 0.028(50\,\text{ft})^{0.8} = 0.64\,\text{sec}$

(2) $S_{DS} = 1.2\,\text{g}$ and $S_{D1} = 0.5\,\text{g}$ (given)

(3) $R = 8$ (special moment resisting steel frame)

(4) Since the building has an Occupancy Category of II, the Seismic Use Group is assigned to I with an Importance Factor equal to 1.0.

(5) $W = (4)\,(500) = 2000\,\text{k}$

(6) $V = \dfrac{S_{D1}}{T(R/I)}\,W = \dfrac{0.5}{0.64\left(\dfrac{8}{1.0}\right)}(2000) = 195.3\,\text{k} \leftarrow$

but not more than

$$V_{max} = \frac{S_{DS}}{R/I}\,W = \left(\frac{1.2}{8/1.0}\right)(2000) = 300\,\text{k}$$

and not less than

$$V_{min} = 0.044 S_{DS}\,IW = 0.044(1.2)(1.0)(2000) = 105.6\,\text{k}$$

(7) Determining the lateral force applied to third floor

$$k = 1 + \frac{T - 0.5}{2} = 1 + \frac{0.64 - 0.5}{2} = 1.07$$

$$F_3 = \frac{w_3 h_3^k}{\displaystyle\sum_{i=1}^{m} w_i h_i^k}(V)$$

$$= \frac{500\,(26)^{1.07}}{500\,(14)^{1.07} + 500\,(26)^{1.07} + 500\,(38)^{1.07} + 500\,(50)^{1.07}}\,(195.3\,\text{k})$$

$$= 0.199(195.3\,\text{k}) = 38.8\,\text{k}$$

In a similar manner, the shears at the other floors are determined

$$F_{roof} = 78.2\,\text{k}$$
$$F_4 = 58.3\,\text{k}$$
$$F_2 = 20.0\,\text{k}$$

Note that the sum of the applied forces equals the design base shear of 195.3 k. ∎

2.15 SNOW LOADS

In the colder states, snow and ice loads are often quite important. One inch of snow is equivalent to approximately 0.5 psf, but it may be higher at lower elevations where snow is denser. For roof design, snow loads ranging from 10 to 40 psf are usually specified. The magnitude depends primarily on the slope of the roof, and to a lesser degree on the character of the roof surface. The larger values are used for flat roofs and the smaller values for sloped roofs. Snow tends to slide off sloped roofs, particularly those with metal or slate surfaces. A load of approximately 10 psf might be used for roofs with 45° slopes while a 40-psf load might be used for flat roofs. Studies of snowfall records in areas with severe winters may indicate the occurrence of snow loads much greater than 40 psf, with values as high as 100 psf in northern Maine.

Snow is a variable load that may cover an entire roof or only part of it. There may be drifts against walls or buildup in valleys or between parapets. Snow may slide off one roof onto a lower one. The snow may blow off one side of a sloping roof or it may crust over and remain in position even during very heavy winds.

The snow loads that are applied to a structure are dependent upon many factors, including geographic location, the pitch of the roof, sheltering, and the shape of the roof. The discussion that follows is intended to provide only an introduction to the determination of snow loads on buildings. When estimating these loads, consult ASCE 7-02 for information that is more complete.

According to Section 7.3 of ASCE 7-02 the basic snow load to be applied to structures with flat roofs in the contiguous United States can be obtained from the expression

$$p_f = 0.7 C_e C_t I p_g$$

This expression is for unobstructed flat roofs with slopes equal to or less than 5° (a 1 in./ft slope is equal to 4.76°). In the equation, C_e is the exposure index. It is intended to account for the snow that can be blown from the roof because of the surrounding locality. The exposure coefficient is lowest for highly exposed areas and is highest when there is considerable sheltering. Values of C_e are presented in Table 2.7 (Table 7-2 in ASCE 7-02).

TABLE 2.7   EXPOSURE COEFFICIENTS FOR SNOW LOADS

Terrain Category (See ASCE 7-02 Section 6.5.6.1)	Exposure of the Roof		
	Fully Exposed	Partially Exposed	Sheltered
A: Large city center	N/A	1.1	1.3
B: Urban and suburban areas	0.9	1.0	1.2
C: Open terrain with scattered obstructions	0.9	1.0	1.1
D: Unobstructed areas with wind over open water	0.8	0.9	1.0
Above the tree line in windswept mountainous areas	0.7	0.8	N/A
Alaska in areas with trees not within 2 miles of the site	0.7	0.8	N/A

The terrain category and roof exposure condition chosen must be representative of the anticipated conditions during the life of the structure. In this table the following definitions are used for the exposure of the roof:

- *Partially exposed:* All roofs except as described as fully exposed or sheltered.
- *Fully exposed:* Roofs exposed on all sides with no shelter afforded by terrain, higher structures, or trees. Roofs that contain several large pieces of mechanical equipment or other obstructions are not in this category.
- *Sheltered:* Roofs located tight in among conifers that qualify as obstructions.

Obstructions within a distance of $10h_o$ provide "shelter." The term h_o is the height of the obstruction above the roof level. If the only obstructions are a few deciduous trees that are leafless in winter, the "fully exposed" category should be used except for terrain category A. Please note that these are heights above the roof. The height used to establish the Terrain Category in Section 6.5.3 of ASCE 7-02 is the height above the ground.

The term C_t is the thermal index. Values of this coefficient are shown in Table 2.8 (Table 7-3 in ASCE 7-02). As shown in the table, the coefficient is equal to 1.0 for heated

TABLE 2.8   THERMAL FACTOR FOR SNOW LOADS

Representative Anticipated Winter Thermal Conditions	C_t
All structures except as indicated below	1.0
Structures kept just above freezing with cold ventilated roofs in which the thermal resistance between ventilated and heated space exceeds 25°Fhft²/BTU	1.1
Unheated structures and structures intentionally kept below freezing	1.2
Continuously heated greenhouses with a roof having a thermal resistance of less than 2.0°Fhft²/BTU.	0.85

structures, 1.1 for structures that are minimally heated to keep them from freezing, and 1.2 for unheated structures.

The values of the importance factors I for snow loads are shown in Table 2.9 (Table 7-4 in ASCE 7-02). The categories of building use are the same as those used for the computation of wind loads and are described in Table C.1 in Appendix C.

TABLE 2.9   IMPORTANCE FACTOR I FOR SNOW LOADS

Building Use Category	I
I	0.8
II	1.0
III	1.1
IV	1.2

The last term, p_g, is the ground snow load in pounds per square foot. Typical ground snow loads for the United States are shown in Figure C.4 in Appendix C. These values are dependent upon the climatic conditions at each site. If data are available that show local conditions are more severe than the values given in the figure, the local conditions should always be used.

The minimum value of p_f is $p_g(I)$ in areas where the ground snow load is less than or equal to 20 psf. In other areas, the minimum value of p_f is 20 psf (I).

Example 2.4, which follows, illustrates the calculation of the design snow load for a building in Chicago according to the ASCE 7-02 specification.

EXAMPLE 2.4

A shopping center is being designed for a location in Chicago. The building will be located in a residential area with minimal obstructions from surrounding buildings and the terrain. It will contain large department stores and enclosed public areas in which more that 300 people can congregate. The roof will be flat, but to provide for proper drainage it will have a slope equal to 0.5 in./ft. What is the roof snow load that should be used for design?

Solution.   Because the slope of the roof is less than 5°, it can be designed as a flat roof. From Figure C.6 the ground snow load, p_g, for Chicago is 25 psf. The exposure factor, C_e, can be taken to equal 0.9 because there are minimal obstructions, though not necessarily an absence of all obstructions. Furthermore, because the building will be located in a residential area, it is unlikely there will be any obstructions to wind blowing across the roof. The thermal factor C_t is 1.0 because this will have to be a heated structure. Lastly, the importance factor I to be used is 1.1 because more than 300 people can congregate in one area. The design snow load to be used, then, is

$$p_f = 0.7C_eC_tIp_g = 0.7(0.9)(1.0)(1.1)(25) = 17.3\,\text{psf}$$

or

$$p_f = 20\,\text{psf}(1.1) = 22\,\text{psf} \qquad \text{CONTROLS} \quad \blacksquare$$

Another expression is given in Section 7.4 of ASCE 7-02 for estimating the snow load on sloping roofs. It involves the multiplication of the snow load for flat roofs by a roof slope factor, C_s. Values of C_s and illustrations are given for warm roofs, cold roofs, and for other situations in Sections 7.4.1 through 7.4.4 of ASCE 7-02.

2.16 OTHER LOADS

There are quite a few other kinds of loads that the designer may occasionally face. These include the following

Traffic Loads on Bridges

Bridge structures are subject to a series of concentrated loads of varying magnitude caused by groups of truck or train wheels. Such loads are discussed in detail in Sections 10.9 and 10.10 of this text.

Ice Loads

Ice has the potential of causing the application of extraordinarily large forces to structural members. Ice can emanate from two sources: (1) surface ice on frozen lakes, rivers, and other bodies of water; and (2) atmospheric ice (freezing rain and sleet). The latter can form even in warmer climates.

In colder climates, ice loads often will greatly affect the design of marine structures. One such situation occurs when ice loads are applied to bridge piers. For such situations it is necessary to consider dynamic pressures caused by moving sheets of ice, pressures caused by ice jams, and uplift or vertical loads in water of varying levels causing the adherence of ice.

The breaking up of ice and its movement during spring floods can significantly affect the design of bridge piers. During the breakup, tremendous chunks of ice may be heaved upward, and when the jam breaks up, the chunks may rush downstream, striking and grinding against the piers. Furthermore, the wedging of pieces of ice between two piers can be extremely serious. It is thus necessary to either keep the piers out of the dangerous areas of the stream or to protect them in some way. Section 3 of the AASHTO specifications provides formulas for estimating the dynamic forces caused by moving ice.

Bridges and towers—any structure, for that matter—are sometimes covered with layers of ice from 1 to 2 in. thick. The weight of the ice runs up to about 10 psf. A factor that influences wind loads is the increased surface area of the ice-coated members.

Atmospheric icing is discussed in Section 10 of ASCE 7-02. Detailed information for estimating the thickness and weight of ice accumulations is provided. This ice typically accumulates on structural members as shown on Figure 10.1 in ASCE 7-02. The thickness to which the ice accumulates must be determined from historic data for the site. It can also be determined from a meteorological investigation of the conditions at the site. The extent of ice accumulation is such a localized phenomenon that general tables cannot be reasonably prepared.

Miscellaneous Loads

Some of the many other loads with which the structural designer will have to contend are

- *soil pressures:* such as the exertion of lateral earth pressures on walls or upward pressures on foundations;
- *hydrostatic pressures:* such as water pressure on dams, inertia forces of large bodies of water during earthquakes, and uplift pressures on tanks and basement structures;
- *flooding:* caused by heavy rain or melting snow and ice;
- *blast loads:* caused by explosions, sonic booms, and military weapons;
- *thermal forces:* due to changes in temperature resulting in structural deformations and resulting structural forces;
- *centrifugal forces:* such as those caused on curved bridges by trucks and trains, or similar effects on roller coasters, etc.;
- and *longitudinal loads:* caused by stopping trucks or trains on bridges, ships running into docks, and the movement of traveling cranes that are supported by building frames.

2.17 PROBLEMS FOR SOLUTION

Use the basic building layout shown in the figure for the solution of Problems 2.1 through 2.5. Determine the requested loads for your area of residence, or for an area specified by your instructor. Use the building code specified by your instructor. If no building code is specified, use ASCE 7-02. You do not need to determine the weight of the beams, girders, and columns for these problems. The author used ASCE 7-02 to obtain the answers provided. You are not expected to have a copy of this specification; sufficient information is provided herein for solving these problems.

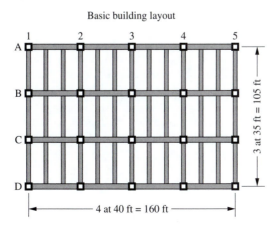

Basic building layout

2.1 The roof of the building is flat. It is composed of 4-ply felt and gravel on 3 in. of reinforced concrete. The ceiling beneath the roof is a suspended steel channel system. Determine
a. The roof dead load (*Ans*. 45 psf)
b. The live load in psf to be applied to Column B2 (*Ans*. 12 psf)

2.2 The roof of the building is flat. It is composed of 2 in. of reinforced concrete on 18-gauge metal decking that weighs 3 psf. A single-ply waterproof sheet 0.7 psf, will be used. The ceiling beneath the roof is unfinished, but allowance for mechanical ducts should be provided. Determine
a. The roof dead load
b. The live load in psf to be applied to Column A1.

2.3 Determine the dead load and live load for the second floor of a library in which any area can be used for stacks. Assume that there will be a steel channel ceiling system (2 psf) and asphalt tile on the floors. The floors are 6-in. reinforced concrete. Allowance should be provided for mechanical ducts. (*Ans*. $p_D = 82$ psf, $p_L = 150$ psf)

2.4 Determine the dead load and live load for a floor in a light manufacturing warehouse/office complex in which any area can be used for storage. Assume that there will be no ceiling or floor finish and that the floors are 4-in. reinforced concrete. Allowance for mechanical ducts should be provided.

2.5 Determine the dead load and live load for a typical upper floor in an office building with movable steel partition walls. The ceiling is a suspended steel channel system and the floors have a linoleum finish. The floors are 3-in. reinforced concrete. Allowance should be provided for mechanical ducts. (*Ans*. $p_D = 48.5$ psf, p_L in offices $= 50$ psf, p_L in hallways $= 80$ psf.)

2.6 Determine the loads on an upper floor in a school with steel stud walls (1/2-in. gypsum on each side), a steel channel ceiling system, and a 3-in. reinforced concrete floor with asphalt tile covering. Allowance should be made for mechanical ducts.

2.7 Determine the design rain load on the roof of a building that is 250 ft wide and 500 ft long. The architect has decided to use 6-in. diameter drains spaced uniformly around the perimeter of the building at 40-ft intervals for the secondary drainage system. The drains are located 1.5 in. above the roof surface. The rainfall intensity at the location of this building is 2.5 in. per hour. (*Ans*. 14.72 psf)

Chapter 3

System Loading and Behavior

3.1 INTRODUCTION

In Chapter 2 different types of loads that might be applied to structural systems were discussed. Methods for estimating the individual magnitudes of the loads were presented. In that discussion, however, we did not consider whether the loads acted at the same time or at different times, nor did we address how and where to place them on the structure to cause maximum system response.

System response is a catchall phrase that really refers to a particular quantity of structural behavior. The response could be the negative bending moment in a floor beam, the displacement at a particular location in the structure, or the force at one of the structural supports. The student probably knows very little about how to calculate these aspects of response at this time.

After the magnitudes of the loads have been computed, the next step in the analysis of a particular structure includes the placing of the loads on the structure and the calculation of its response to those loads. When placing the loads on a structure, two distinct tasks must be performed:

1. We must decide which loads can reasonably be expected to act concurrently in time. Because different loads act on the structure at different times, several different loading conditions must be evaluated. Each of these loading conditions will cause the structural system to respond in a different manner.
2. We also need to determine where to place those loads on the structure. After loads are placed on the structure, the response of the structure is computed. If the same loads are placed on the structure in different positions, the response of the system will be different. We need to determine where to place the loads to obtain maximum response. For example, would the *bending moments* in the floor beams be greater if we placed the floor live loads on every span or on every other span?

Placing the live loads to cause the worst effects on any member of a structure is the responsibility of the structural engineer. Theoretically, his or her calculations are subject to the review of the appropriate building officials, but seldom do such individuals have the time and/or the ability to make significant reviews. Consequently, these calculations remain the responsibility of the engineer.

Martin Towers, Bethlehem, Pennsylvania (Courtesy Bethlehem Steel Corporation)

3.2 TRIBUTARY AREAS

In Section 2.8, the term *tributary area* was briefly defined. In this section, this term is discussed at greater length. In the next section, a related term, *influence area*, is introduced.

The tributary area is the loaded area of a particular structure that directly contributes to the load applied to a particular member in the structure. It is best defined as the area that is bounded by lines located halfway to the next beam or to the next column. Tributary areas are shown for several beams and columns by the shaded areas in Figure 3.1 for a structure with one-way bending between the beams. The component that the tributary area serves is indicated in black.

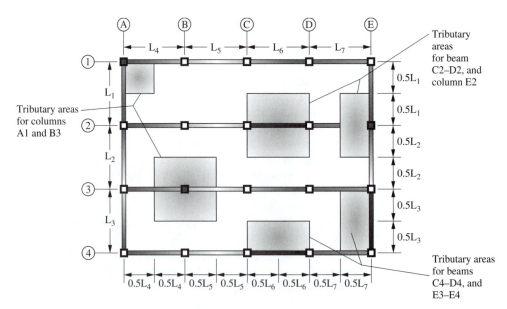

Figure 3.1   Tributary area for selected columns and beams

The tributary areas shown for the beams in Figure 3.1 are the tributary areas used in common practice for one-way or two-way bending. The theoretical tributary area for a typical interior beam and typical edge beam are shown in Figure 3.2 for a structure with a two-way floor system spanning between the beams.

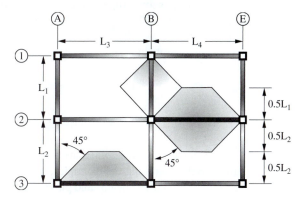

Figure 3.2   Theoretical tributary area for a beam with two-way bending

Beams are members that support transverse loads. They are usually thought of as being used in horizontal positions and subjected to gravity or vertical loads, but there are frequent exceptions—rafters, for instance. The term *girder* is used rather loosely, usually indicating a large beam and perhaps one into which smaller beams are framed. The loads are assumed to be applied by deck slabs.

We see that in the middle of the beam the tributary area extends halfway to the next beam in each direction. At the ends of the beam, however, load is supported partly by the beams in the perpendicular direction. Therefore, the theoretical boundary of the tributary area will fall halfway between the two, that is, at a 45° angle. The tributary areas for the beams shown in Figure 3.2 are rarely, if ever, used in practice because of the difficulty in dealing with the resulting trapezoidal load. Using the tributary areas shown on Figure 3.1 instead of those shown on Figure 3.2 is conservative because there will be more load acting on the member when it is analyzed than will actually occur.

Very often floor loads are supported by floor beams as shown in Figure 3.3(a). Floor beams extend from one girder to another. Normally the floor beams are connected to the

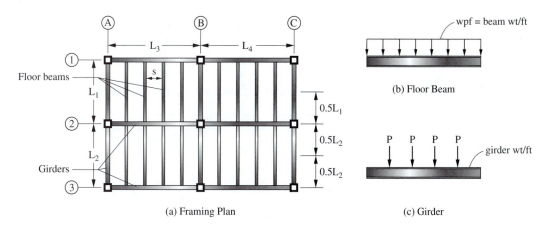

(a) Framing Plan

(b) Floor Beam

(c) Girder

Figure 3.3   A typical floor framing system

girders with connections that can be idealized as simple supports, that is, no moment is assumed to occur at their ends. When the floor beams are framed in this manner, the loads that act on them are shown in Figure 3-3(b). The term w represents the uniform load the beam must support per foot including its own weight.

The girders must support the reactions from the beams connected to them as well as their own weights. Therefore the total reactions applied to the interior and exterior girders of Figure 3.3 are as follows. Notice that in addition to the beam reactions, a girder must support its own weight as well

$$\text{For an interior girder} \qquad P = w\left(\frac{L_1 + L_2}{2}\right)s$$

$$\text{For an exterior girder} \qquad P = w\left(\frac{L_1 \text{ or } L_2}{2}\right)s$$

Examples 3.1 and 3.2 illustrate the computation of loads acting on columns, beams, and girders. Before working the examples, one additional comment should be made regarding structural framing. The beams and girders in a frame can be connected to the columns either as simple supports or in a manner that allows moment to exist at the ends of the members. If moment is resisted at the ends of the beams and girders, the frame is referred to as a *moment-resisting frame*. If simple nonmoment resisting connections are used, diagonal bracing or shear walls must be provided for lateral stability.

EXAMPLE 3.1

The building floor shown Figure 3.4 is to be designed to support a uniformly distributed load of 50 psf over its entire area. Neglecting member weights, determine the loads to be supported by (a) interior column B3, (b) edge column E2 and (c) corner column A1.

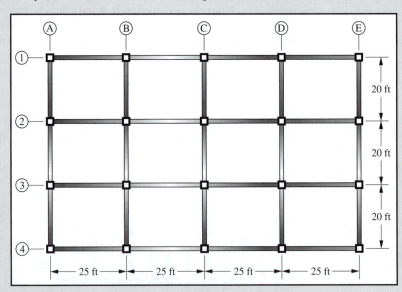

Figure 3.4

Solution.
Column B3

$$P = (50\,\text{psf})\left(\frac{20\,\text{ft} + 20\,\text{ft}}{2}\right)\left(\frac{25\,\text{ft} + 25\,\text{ft}}{2}\right) = 25{,}000\,\text{lbs}$$

Column E2

$$P = (50\,\text{psf})\left(\frac{20\,\text{ft} + 20\,\text{ft}}{2}\right)\left(\frac{25\,\text{ft}}{2}\right) = 12{,}500\,\text{lbs}$$

Column A1

$$P = (50\,\text{plf})\left(\frac{20\,\text{ft}}{2}\right)\left(\frac{25\,\text{ft}}{2}\right) = 6250\,\text{lbs} \quad \blacksquare$$

EXAMPLE 3.2

The building floor shown in Figure 3.5 is to be designed to support a uniformly distributed load of 50 psf over its entire area. Neglecting member weights, determine the following:

(a) The uniform load per foot to be supported by a typical interior beam
(b) The concentrated loads, or P loads, applied to the interior girder A2-B2
(c) The concentrated loads, or P loads, applied to the exterior girder A1-B1

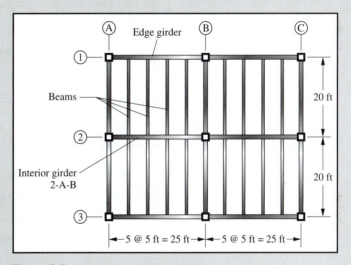

Figure 3.5

Solution.

(a) Interior Floor Beam

$$w = (50\,\text{psf})(5\,\text{ft}) = 250\,\text{plf}$$

(b) Interior girder A2-B2

$$P = (250\,\text{plf})\left(\frac{20\,\text{ft} + 20\,\text{ft}}{2}\right) = 5000\text{-lb}$$

concentrated loads 5 ft on center

(c) Exterior girder A1-B1

$$P = (250\,\text{plf})\left(\frac{20\,\text{ft}}{2}\right) = 2500\text{-lb}$$

concentrated loads 5 ft on center $\quad \blacksquare$

3.3 INFLUENCE AREAS

Influence areas are those areas that when loaded affect the design forces in a particular member of a structure. As we will see, these areas are different from the tributary areas previously described. Referring to the drawing of the building floor shown in Figure 3.6 it can be seen that a load placed anywhere in the upper left rectangle of the building floor will directly affect the force applied to the upper left column A1. This entire rectangle is referred to as the influence area for that particular column. The tributary area for this same column is only the upper left quarter of the same rectangle. We can see that the influence area for this column is four times its tributary area.

The influence areas for several different beams and columns in this same floor system are shown in Figure 3.6. In each case the member in question is shown darkened in the figure.

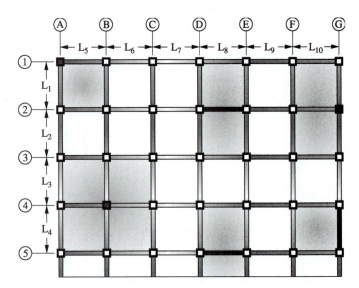

Figure 3.6   Influence areas for selected beams and columns

In both ASCE 7-02 and IBC-2003, the influence area is defined as:

$$A_I = K_{LL}A_T$$

The influence area is equal to the live load element factor, K_{LL}, times the tributary area, A_T. The live load element factor can be calculated from building geometry, or can be taken from Table 3.1.

From Figure 3.6 and the table you can see that the influence area for an interior column is four times as large as its tributary area, whereas that for an interior beam is twice as large as its tributary areas.

3.4 LIVE LOAD REDUCTION

Under some circumstances the code-specified live loads for a building can be reduced. For this discussion, it is assumed that the maximum specified live load for a particular building floor is 100 psf. If the influence area for a particular member is 1000 sq ft, the

TABLE 3.1   LIVE LOAD ELEMENT FACTOR, K_{LL}[*]

Type of Element	K_{LL}
Interior column,	
Exterior columns without cantilever slabs.	4
Edge columns with cantilever slabs.	3
Corner columns with cantilever slabs,	
Edge beams without cantilever slabs,	
Interior beams.	2
All other beams not identified including: Edge beams with cantilever slabs, cantilever beams, two-way slabs, and members without provisions for continuous shear transfer normal to their span.	1

[*]IBC2003 Table 1607.9.1

likelihood of having the maximum live load 100 psf applied to every single square foot of that area seems much less likely than if the area was 200 sq ft. Thus we say that as the influence area contributing to the load applied to a particular member increases, the possibility of having the full design live load applied to every square foot decreases. Consequently, building codes usually permit some reduction in the specified live loads when large areas are involved. In Section 4.8 of ASCE 7-02 and Section 1607.9.1 of IBC-2003 the following reduction factor is given:

$$L = L_0 \left(0.25 + \frac{15}{\sqrt{K_{LL} A_T}} \right)$$

In this equation, L is the reduced live load, L_0 is the code-specified live load, and the term in parentheses is the reduction factor. The terms K_{LL} and A_T were defined in the last section.

From this equation, we can show that live loads will be reduced only when the influence area is greater than 400 sq ft. There are limits, though, as to how much the live load can be reduced. If the structural member supports one floor only, the live load cannot be reduced by more than 50%. For members supporting more than one floor, the live load cannot be reduced by more than 60%. Usually, individual beams are involved in the support of only one floor. On the other hand, columns are often involved in the support of more than one floor, in fact, they support all of the floors above them. A plot of live-load reduction factors, as obtained from the preceding equation is presented in Figure 3.7 for different influence areas.

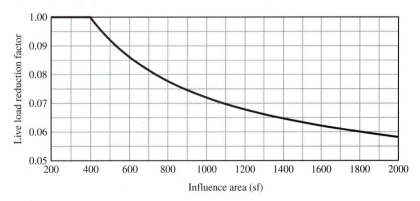

Figure 3.7   Live load reduction factor versus influence area

However, live-load reduction is not permitted in all cases by ASCE 7-02. If the unit live load is 100 psf or less or if the loaded area is used for a place of public assembly, the reduction cannot be taken. Also, the reduction cannot be taken if the loaded area is a roof or a one-way slab and the unit live load is less than or equal to 100 psf. A one-way slab is a slab that is primarily supported on two opposite edges only.

When the unit live load exceeds 100 psf, a maximum reduction of 20% can be taken for structural members supporting more than one floor. There is no permitted reduction for members that support only one floor when the unit live load is greater than 100 psf. The basis for this 20% reduction is that higher unit live loads tend to occur in buildings used for storage and warehousing. In this type of building, several adjacent spans may be loaded concurrently, but studies have indicated that rarely is an entire floor loaded to more than 80% of its rated design load.

The provision for live load reduction and the related limitations have two significant implications on structural analysis. First, the loads used to obtain column design forces and those used to obtain floor beam design forces may be different. This situation occurs because the live load reduction factors for each are likely to be different. Also, because roofs and floors are treated differently, it does not appear that the live-load reduction is always permitted when columns support a floor and the roof. Typical of such columns are those supporting the top story in a building. When a column supports a floor and the roof, that column should be considered to support a single floor for purposes of determining the permissible live-load reduction.

3.5 LOADING CONDITIONS FOR ALLOWABLE STRESS DESIGN

There are two general methods used for the design of structures. These are the allowable stress and the strength design procedures. This section is concerned with the allowable stress method while the next section is concerned with the strength method.

With the allowable stress method, the most severe loading conditions are estimated and elastic stresses in the members are computed. These stresses are limited to certain maximums that are appreciably below the ultimate stresses the materials can withstand.

To determine the most severe loadings that the structure must be able to safely support, it is necessary to consider which of the loads (dead, live, and environmental) can occur simultaneously. For this discussion the following nomenclature is used:

D = dead loads

E = seismic or earthquake load effects

F = loads due to the weight and pressure of fluids

H = loads due to weight and lateral earth pressure of soils, groundwater pressure, or pressure of bulk materials

L = live loads

L_r = roof live loads

R = rain loads

S = snow loads

T = total effects of temperature, creep, shrinkage, differential settlement, and shrinkage-compensating concrete

W = wind loads

In accordance with Section 2 of ASCE 7-02 and Section 1605 of IBC-2003, the following possible simultaneous load situations may occur and must be considered for determining the most severe situations:

1. $D + F$
2. $D + H + F + L + T$
3. $D + H + F + (L_r$ or S or R$)$
4. $D + H + F + 0.75(L + T) + 0.75(L_r$ or S or R$)$
5. $D + H + F + (W$ or $0.7E)$
6. $D + H + F + 0.75(W$ or $0.7E) + 0.75L + 0.75(L_r$ or S or R$)$
7. $0.6D + W + H$
8. $0.6D + 0.7E + H$

When impact effects need to be considered, they should be included with the live loads. Such situations occur when those loads are quickly applied, as they are for parking garages, elevators, loading docks, and others.

You will note that all of these loads, other than dead loads, will vary appreciably with time—there is not always snow on the structure and the wind is not always blowing, for example. In the second through sixth ASCE 7-02 loading conditions, the dead load has been combined with multiple variable loads, yet in the fourth and eighth loading conditions the dead load has been combined with only two variable loads. Observe also in the seventh and eighth loading combinations that the full dead load is not being considered. The two variable loads in these combinations, the wind and earthquake loads, generally have a lateral component. As such, they tend to cause the structure to overturn. A dead load, on the other hand, is a gravity load, which tends to keep the structure from overturning. Consequently, a more severe condition can occur if for some reason the full dead load is not acting.

Most likely when two or more loads are acting on a structure in addition to dead load, the loads other than dead load are not likely to achieve their absolute maximum values simultaneously. Load surveys seem to bear out this assumption. The ASCE and IBC codes permit the load effects in these loading conditions, except dead load, to be multiplied by 0.75 provided the result is not less than that produced by dead load and the load causing the greatest effect. *Remember that the code is listing the minimum conditions that must be considered.* If the design engineer feels that the maximum values of two variable loads (wind and rain, for example) may occur at the same time in his or her area, it is not required that the 0.75 value be used.

These load combinations are only the recommended minimum load combinations that need to be considered. As with the determination of the loads themselves, the engineer must evaluate the structure being analyzed and determine whether these load combinations comprise all of the possible combinations for a particular structure. Under some conditions, other loads and load combinations may be appropriate.

EXAMPLE 3.3

An observation deck at an airport has girders configured as shown in Figure 3.8.

Figure 3.8

These girders are spaced 15 ft on-center. Assume that all of the loads are uniformly distributed over the deck and have been found to be as follows

Dead load: 32 psf

Live load: 100 psf

Snow load: 24 psf

Rain load: 10 psf

Using the loading conditions from ASCE 7-02 and IBC-2003, what are the combined loads that can reasonably be expected to act on this girder?

Solution. Because the girders are 15 ft on-center, the tributary area for this beam has a width of 15 ft. The applicable loading combinations for this beam are

(1) $w = 15\,[32 + 0] = 480$ plf

(2) $w = 15\,[32 + 0 + 0 + 100 + 0] = 1980$ plf $\leftarrow$

(3) $w = 15\,[32 + 0 + 0 + 24] = 840$ plf

(4) $w = 15\,[32 + 0 + 0(0.75)(100 + 0) + (0.75)(24)] = 1875$ plf

(5) $w = 15\,[32 + 0 + 0 + 0] = 480$ plf

(6) $w = 15\,[32 + 0 + 0 + 0 + 0 + (0.75)(100) + (0.75)(24)] = 1875$ plf

(7) $w = 15\,[(0.6)(32) + 0 + 0] = 288$ plf

(8) $w = 15\,[(0.6)(32) + 0 + 0] = 288$ plf

The girders must be designed to support a maximum load of 1980 plf. ∎

3.6 LOADING CONDITIONS FOR STRENGTH DESIGN

A design philosophy that has become common in recent years is the strength-design procedure. With this method the estimated loads are multiplied by certain load factors that are almost always larger than 1.0 and the resulting ultimate or "factored" loads are then used for designing the structure. The structure is proportioned to have a design ultimate strength sufficient to support the ultimate loads.

The purpose of load factors is to increase the loads to account for the uncertainties involved in estimating the magnitudes of dead or live loads. For instance, how close in percent could you estimate the largest wind or snow loads that ever will be applied to the building that you are now occupying?

The load factors used for dead loads are smaller than those used for live loads because engineers can estimate more accurately the magnitudes of dead loads than they can the magnitudes of live loads. In this regard, notice that loads that remain in place for long periods will be less variable in magnitude, whereas those applied for brief periods, such as wind loads, will have larger variations.

When a structure is to be designed using strength procedures, other load combinations and load factors may apply. Although we are not directly concerned about design while performing analysis, we must be cognizant of the design method that will be used so that the analysis results will have meaning for the design engineer.

The recommended load combinations and load factors for strength design usually are presented in the specifications of the different code-writing bodies such as the International Council of Building Officials, the American Institute of Steel

Construction, or the American Concrete Institute. The load factors for use with strength design are determined statistically, and consideration is given to the type of structure upon which the loads are acting. As such, the load combinations and factors for a building will be different from those used for a bridge. Both of these will likely be different from those used for an offshore oil production platform. A structural analyst always should refer to the design guide or recommendation appropriate for the system being analyzed.

Following are the load combinations for building structures as recommended by ASCE 7-02 and IBC-2003.

1. $U = 1.4(D + F)$
2. $U = 1.2(D + F + T) + 1.6(L + H) + 0.5(L_r \text{ or } S \text{ or } R)$
3. $U = 1.2D + 1.6(L_r \text{ or } S \text{ or } R) + (1.0L \text{ or } 0.8W)$
4. $U = 1.2D + 1.6W + 1.0L + 0.5(L_r \text{ or } S \text{ or } R)$
5. $U = 1.2D + 1.0E + 1.0L + 0.2S$
6. $U = 0.9D + 1.6W + 1.6H$
7. $U = 0.9D + 1.0E + 1.6H$

The load combinations presented in the last two of these expressions contain a 0.9D value. This 0.9 factor accounts for cases where larger dead loads tend to reduce the effects of other loads. One obvious example of such a situation may occur in tall buildings that are subject to lateral wind and seismic forces where overturning may be a possibility. As a result, the dead loads are reduced by 10% to take into account situations where they may have been overestimated.

The reader must realize that the sizes of the load factors do not vary in relation to the seriousness of failure. You may think that larger load factors should be used for hospitals or high-rise buildings than for cattle barns, but such is not the case. The load factors were developed on the assumption that designers would consider the seriousness of possible failure in specifying the magnitude of their service loads. Furthermore, the ASCE load factors are minimum values, and designers are perfectly free to use larger factors as they desire.

Example 3.4 presents the calculation of factored loads for the girders of Example 3.3 using the strength load combinations. The largest value obtained is referred to as the critical or governing load combination and is the value to be used in design.

EXAMPLE 3.4

Repeat Example 3.3 but this time determine the load combinations for a structure to be designed using strength design procedures.

Solution.   The applicable load combinations are as follows:

(1) $U = 15[(1.4)(32 + 0)] = 672 \text{ plf}$
(2) $U = 15[(1.2)(32 + 0 + 0) + (1.6)(100 + 0) + (0.5)(24)] = 3156 \text{ plf} \leftarrow$
(3) $U = 15[(1.2)(32) + (1.6)(24) + (1.0)(100)] = 2652 \text{ plf}$
(4) $U = 15[(1.2)(32) + (1.6)(0) + (1.0)(100) + (0.5)(24)] = 2256 \text{ plf}$
(5) $U = 15[(1.2)(32) + (1.0)(0) + (1.0)(100) + (0.2)(24)] = 2148 \text{ plf}$
(6) $U = 15[(0.9)(32) + (1.6)(0) + (1.6)(0)] = 432 \text{ plf}$
(7) $U = 15[(0.9)(32) + (1.0)(0) + (1.6)(0)] = 432 \text{ plf}$ ∎

For some special situations, the codes permit reductions in the specified load factors. These situations are as follows:

(a) In the third through fifth equations, the load factors for live loads may be reduced to 0.5, except for garages, for areas used for public assembly, and for all areas where the live loads are larger than 100 psf.
(b) In the sixth and seventh equations, the load factor for H is zero if the structural action of H counteracts that due to W or E. If lateral earth pressure opposes actions from other forces, it should not be included in H but should be included in the design resistance.

Frequently, building codes and design load references provide seismic loads in strength-level values (that is, in effect they have already been multiplied by a load factor). This is the situation assumed in the fifth and seventh equations. If, however, service level seismic forces are specified, it will be necessary to use 1.4E in these two equations.

Example 3.5 presents the calculation of factored loads for a column using the strength combinations. The largest value obtained is referred to as the critical or governing load combination and is the value to be used in design. *Notice that the values of the wind and seismic loads can have two values depending on the direction of those forces, and it may be possible for the sign of those loads to be different (that is, compression or tension). As a result, we may have to apply the applicable equations two times each to take into account the different values.* This same situation can occur with the load combinations required for allowable stress design that were described in the last section. The substitution into all of these load combinations is a little bit tedious but can be easily handled with computer programs.

EXAMPLE 3.5

The axial loads for a building column have been estimated with the following results: $D = 150$ k, live load from roof $L_r = 60$ k, $L = 300$ k, compression wind $W = 70$ k, tensile wind $W = 60$ k, seismic compression load $= 50$ k, and tensile seismic load $= 40$ k. Determine the critical design load using the load factors given in this section.

Solution.

(1) $U = (1.4)(150 + 0) = 210$ k
(2) $U = (1.2)(150 + 0 + 0) + (1.6)(300 + 0) + (0.5)(60) = 690$ k ←
(3)(a) $U = (1.2)(150) + (1.6)(60) + (1.0)(300) = 576$ k
    (b) $U = (1.2)(150) + (1.6)(60) + (0.8)(70) = 332$ k
    (c) $U = (1.2)(150) + (1.6)(60) + (0.8)(-60) = 228$ k
(4)(a) $U = (1.2)(150) + (1.6)(70) + (1.0)(300) + (0.5)(60) = 622$ k
    (b) $U = (1.2)(150) + (1.6)(-60) + (1.0)(300) + (0.5)(60) = 414$ k
(5)(a) $U = (1.2)(150) + (1.0)(50) + (1.0)(300) + (0.2)(0) = 530$ k
    (b) $U = (1.2)(150) + (1.0)(-40) + (1.0)(300) + (0.2)(0) = 440$ k
(6)(a) $U = (0.9)(150) + (1.6)(70) + (1.6)(0) = 247$ k
    (b) $U = (0.9)(150) + (1.6)(-60) + (1.6)(0) = 39$ k
(7)(a) $U = (0.9)(150) + (1.0)(50) + (1.6)(0) = 185$ k
    (b) $U = (0.9)(150) + (1.0)(-40) + (1.6)(0) = 95$ k   ■

3.7 CONCEPT OF THE FORCE ENVELOPE

When loads are applied to a structure, the structure responds in reaction to those loads. The forces in a particular component in the system are caused by (1) the loads acting on the structure, and (2) where those loads are located. From analysis of the response to the various forces acting, we can determine the maximum and minimum forces that can occur in any component. This range of forces is called a *force envelope*.

Example 3.6 further illustrates the idea of force envelopes. A force envelope is prepared for the bending moment in a simple beam using the procedures learned in earlier courses.

EXAMPLE 3.6

During its lifetime the beam shown in Figure 3.9 will be loaded with a dead uniform load equal to 3 klf for its full length. It will also be loaded with a concentrated live load at its center line which can vary from 0 up to a maximum value of 25 k. Draw moment diagrams for the dead load only, for maximum live load only and then draw a diagram for the dead load plus the full live load.

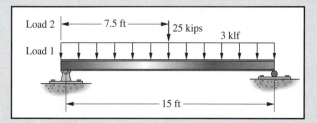

Figure 3.9

Solution. The moments are drawn using the principles of mechanics. The preparation of moment diagrams is discussed in great detail in Chapter 5 of this text. The moment diagram for the uniform 3 klf load is drawn as shown below.

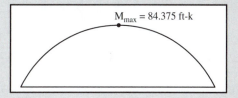

In a similar manner the moment diagram for the 25 k concentrated load is prepared.

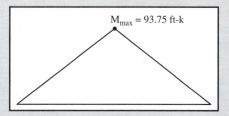

The dead load moment diagram and the dead load plus maximum live load moment diagram are shown in the following figure. The shaded area represents the moment envelope or the range of possible moment values for the given loads.

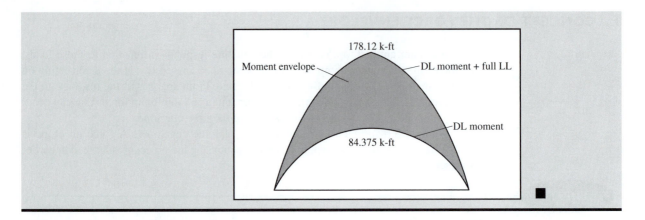

3.8 PROBLEMS FOR SOLUTION

For Problems 3.1 through 3.5 compute the values requested using the basic floor framing plan shown in the accompanying figure. Consider live-load reduction following the provisions of ASCE 7-02 as appropriate. For purposes of these problems assume that this is an upper story in a multistory office building, but it is not the top story.

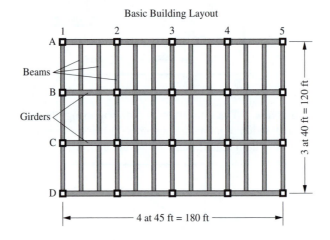

Basic Building Layout

3.1 Load on Column B3 contributed by this floor if the live load is 150 psf (*Ans.* 216,000 lbs)

3.2 Load on Column A3 contributed by this floor if the live load is 75 psf

3.3 Load on Column A1 contributed by this floor if the live load is 100 psf (*Ans.* 27,160 lbs)

3.4 Load on an interior floor beam if the live load is 75 psf

3.5 Load on Girder B2–B3 if the live load is 50 psf (*Ans.* 16,680 lbs)

For Problems 3.6 through 3.10, given the loads specified, compute the maximum combined load using the ASCE 7-02 load combinations for working stress design.

3.6 D = 50 psf, L_r = 75 psf, R = 8 psf, S = 20 psf

3.7 D = 45 psf, L = 60 psf (*Ans.* 105 psf)

3.8 D = 2750 lbs, L = 4500 lbs, L_r = 1500 lbs, R = 1250 lbs, S = 1000 lbs

3.9 D = 87 psf, L = 150 psf (*Ans.* 237 psf)

3.10 D = 75 psf, L_r = 35 psf, R = 12 psf

For Problems 3.11 to 3.15 repeat Problems 3.6 to 3.10 using the ASCE 7-02 strength design load combinations.

3.11 Repeat Problem 3.6 (*Ans.* 180 psf)

3.12 Repeat Problem 3.7

3.13 Repeat Problem 3.8 (*Ans.* 11,250 lbs)

3.14 Repeat Problem 3.9

3.15 Repeat Problem 3.10 (*Ans.* 146 psf)

Chapter 4

Reactions

4.1 EQUILIBRIUM

A body at rest is said to be in *static equilibrium*. The resultant of the external forces acting on the body—including the supporting forces, which are called reactions—is zero. Not only must the sum of all forces (or their components) acting in any possible direction be zero, but the sum of the moments of all forces about any axis also must be equal to zero.

If a structure, or part of a structure, is to be in equilibrium under the action of a system of loads, it must satisfy the six equations of static equilibrium. Using the Cartesian x, y, and z coordinate system, the equations of static equilibrium can be written as

$$\sum F_x = 0 \qquad \sum F_y = 0 \qquad \sum F_z = 0$$
$$\sum M_x = 0 \qquad \sum M_y = 0 \qquad \sum M_z = 0$$

For purposes of analysis and design, the large majority of structures can be considered as being planar structures without loss of accuracy. For these structures, which are usually assumed to be in the xy plane, the sum of the forces in the x and y directions and the sum of the moments about an axis perpendicular to the plane must be zero. The equations of equilibrium reduce to

$$\sum F_x = 0 \qquad \sum F_y = 0 \qquad \sum M_z = 0$$

These equations are commonly written as

$$\sum F_H = 0 \qquad \sum F_V = 0 \qquad \sum M = 0$$

These equations cannot be proved algebraically; they are merely statements of Sir Isaac Newton's observation that for every action on a body at rest there is an equal and opposite reaction. Whether the structure under consideration is a beam, a truss, a rigid frame, or some other type of assembly supported by various reactions, the equations of static equilibrium must apply if the body is to remain at rest.

The structures discussed in the first seven chapters of this textbook are called planar structures; each structure lies in a plane and their loads are applied in the same plane. Three-dimensional trusses, also called space trusses, are discussed in Chapter 8.

4.2 MOVING BODIES

The statement was made in the preceding section that a body at rest is in a state of static equilibrium. However, being at rest is not a necessary condition for static equilibrium.

A body that is moving at constant velocity also can be in a state of static equilibrium; the net force acting on the body is equal to zero. This concept can be proved with the impulse–momentum relationship:

$$F(\Delta T) = m(\Delta v)$$

In this equation, F is the net force acting on the body, ΔT is the time that the force acts, Δv is the change in velocity of the body, and m is the mass of the body. If the net force is equal to zero, the left side of the equations becomes zero, which implies that the change in velocity must be equal to zero since the mass of any real body cannot be equal to zero. When the change in velocity is equal to zero, the body is not accelerating—it is moving at constant velocity. There is nothing in the relationship to imply that the body is stationary.

When a body is accelerating, there are additional forces acting that must be included in equilibrium calculations. These additional forces are the inertial forces, which are caused by the mass of the body. When the inertial forces are included and the net force acting on the body, including the inertial forces, is equal to zero, the body is said to be in a state of dynamic equilibrium. We will not investigate dynamic equilibrium in this book.

4.3 CALCULATION OF UNKNOWNS

To completely define a force, three properties of that force must be defined. These properties are its magnitude, its line of action, and the direction in which it acts along the line of action. All these properties are generally known for each of the externally applied loads. However, when dealing with structural reactions, only the points at which the reaction forces act, and perhaps their directions, are known. The magnitude of the reaction forces, and sometimes the directions in which they act, are unknown quantities that must be determined.

YKK USA Building, Dublin, Georgia (Courtesy of Britt, Peters and Associates)

The number of unknowns that can be determined using the equations of static equilibrium is limited by the number of independent equations of static equilibrium available. For any structure lying in a plane, there are only three independent equations of static equilibrium. These three equations can be used to determine at most three unknown quantities for the structure. When there are more than three unknowns to evaluate, additional equations must be used in conjunction with the equations of static equilibrium. We will see that in some instances, because of special construction features, equations of condition are available in addition to the usual equations of static equilibrium. In later chapters, we will learn to write some other equations for use in analysis. These equations will pertain to the compatibity of displacements in the structure.

4.4 TYPES OF SUPPORT

Structural frames may be supported by hinges, rollers, fixed ends, or links. These supports are discussed in the following paragraphs.

A *hinge* or pin-type support (represented herein by the symbol ⏢ or ⏢) is assumed to be connected to the structure with a frictionless pin. This type of support prevents movement in a horizontal or vertical direction, but does not prevent slight rotation about the hinge. There are two unknown forces at a hinge: the magnitude of the force required to prevent horizontal movement and the magnitude of the force required to prevent vertical movement. The support supplied at a hinge may also be referred to as an inclined force, which is the resultant of the horizontal force and the vertical force at the support. Two unknowns remain: the magnitude and direction of the inclined resultant.

A roller type of support (represented herein by the symbol ⚬) is assumed to offer resistance to movement only in a direction perpendicular to the supporting surface beneath the roller. There is no resistance to slight rotation about the roller or to movement parallel to the supporting surface. The magnitude of the force required to prevent movement perpendicular to the supporting surface is the one unknown. *Rollers may be installed in such a manner that they can resist movement either toward or away from the supporting surface.* Should the temperature of the beam be raised, the beam will expand. However, since there is no longitudinal restraint supplied at the roller or expansion support, no stress will be developed in the beam or in the supporting walls or other structural members.

Hinged support for a bridge girder (Courtesy Bethlehem Steel Corporation)

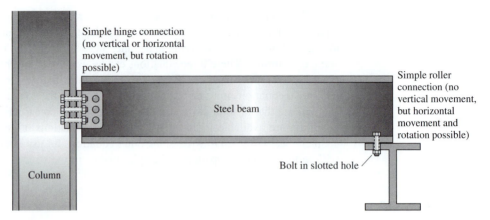

Figure 4.1   Simple connections for a steel beam

A *fixed-end* support (represented herein by the symbol ▐▬) is assumed to offer resistance to both rotation about the support and to movement vertically and horizontally. There are three unknowns: the magnitude of the force required to prevent horizontal movement, the magnitude of the force required to prevent vertical movement, and the magnitude of the force required to prevent rotation.

A *link* type of support (represented herein by the symbol ◖▬◗) is similar to the roller in its action because the pins at each end are assumed to be frictionless. The line of action of the supporting force must be in the direction of the link and through the two pins. One unknown is present: the magnitude of the force in the direction of the link.

Figure 4.1 shows hinge and expansion (or roller) type connections as they might be used for a steel beam.

A hinge or simple connection theoretically should be free for some rotation when loads are applied to the member. The ends of the structural steel beam shown in Figure 4.1 are fairly free to rotate. Two other types of end connections for steel members that have a similar rotation capacity are shown in Figure 4.2.

Two connections that provide considerable moment resistance and thus approach fixed ends are shown in Figure 4.3. Notice in each case some type of connection is

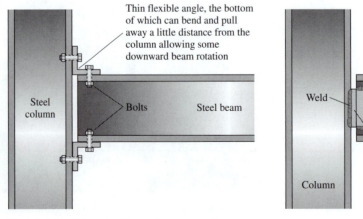

Figure 4.2   More simple or hinge connections

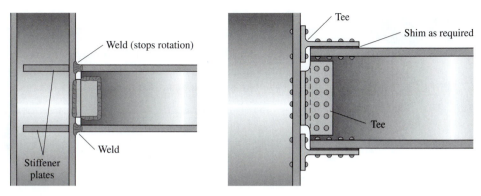

Figure 4.3   Moment resisting connections

provided at the top and bottom of the ends of the beams to prevent downward rotation. Other connections will be discussed as needed at various places in the text.

4.5 STABILITY, DETERMINACY, AND INDETERMINACY

The discussion of supports shows there are three unknown reaction components at a fixed end, two at a hinge, and one at a roller or link. If, for a particular structure, the total number of forces of reaction is equal to the number of equations of static equilibrium available, the unknowns may be determined and the structure is then said to be statically determinate externally. Should the number of unknowns be greater than the number of equations available, the structure is statically indeterminate externally; if less, it is unstable externally. From this discussion, note that stability, determinacy, and indeterminacy are dependent upon the configuration of the structure; they are not dependent upon the loads applied to the structure. The following examples will demonstrate the application of these concepts to structural systems.

EXAMPLE 4.1

Determine the statical classification of the simply supported beam shown in Figure 4.4. The unknown forces of reaction are illustrated. There are two unknown forces of reaction at the left support because it is a pinned support. There is one unknown force of reaction at the right support because it is a roller.

Figure 4.4

There are three applicable equations of static equilibrium: summation of forces vertically, summation of forces horizontally, and summation of moments. Because there are three unknown forces of reaction and three applicable equations of equilibrium, the beam is stable and statically determinate externally—the number of unknowns is equal to the number of equations of static equilibrium. ∎

EXAMPLE 4.2

Determine the statical classification of the simply supported beam shown in Figure 4.5. The unknown forces of reaction are illustrated. There is one unknown force of reaction at each of the two supports because the supports are rollers.

Figure 4.5

There are three applicable equations of static equilibrium: summation of forces vertically, summation of forces horizontally, and summation of moments. Because there are two unknown forces of reaction and three applicable equations of equilibrium, the beam is unstable—the number of equations of static equilibrium exceeds the number of unknowns.

A structure may be stable under one arrangement of loads, but if it is not stable under any other conceivable set of loads, it is unstable. This beam cannot hold its position when subjected to a horizontal load. It is unstable. ■

EXAMPLE 4.3

Determine the statical classification of the continuous beam shown in Figure 4.6. The unknown forces of reaction are illustrated. There are two unknown forces of reaction at the left support because that support is pinned. There is one unknown force of reaction at each of the other supports because they are rollers.

Figure 4.6

There are three applicable equations of static equilibrium. Because there are five unknown forces of reaction and three applicable equations of equilibrium, the beam is stable and statically indeterminate to the second degree externally—there are two more unknowns than there are equations of static equilibrium. ■

EXAMPLE 4.4

Determine the statical classification of the propped cantilever beam shown in Figure 4.7. The unknown forces of reaction are illustrated. There are three unknown forces of reaction at the left support because it is fixed. There is one unknown force of reaction at the right support because that support is a roller.

There are three applicable equations of static equilibrium. Because there are four unknown forces of reaction and three applicable equations of equilibrium, the beam is

Figure 4.7

stable and statically indeterminate to the first degree externally—there is one more unknown than there are equations of static equilibrium. ■

The internal arrangement of some structures is such that one or more equations of condition are available. This situation can occur when there are hinges or links in the structure. A special condition exists because the internal moment at the hinge, or at the ends of the link, must be zero regardless of the loading. The internal moment is zero because of the "frictionless" pin used to make the connection:—rotation cannot be transferred between the adjacent parts of the structure. A similar statement cannot be made for any continuous section of the beam. If the number of condition equations plus the three equations of static equilibrium is equal to the number of unknowns, the structure is statically determinate; if more, it is unstable; and if less, it is statically indeterminate.

EXAMPLE 4.5

Determine the statical classification of the propped cantilever beam shown in Figure 4.8. The unknown forces of reaction are illustrated. This beam is the same as in the last example except that now a link is used at the right support instead of a roller. There are three unknown forces of reaction at the left support because it is a fixed support. There is only one unknown force of reaction at the right side even though the link is pin-connected at the support. A link can only transmit force along its axis, therefore the direction in which the reaction force is acting is known.

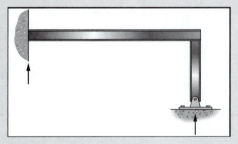

Figure 4.8

There are three applicable equations of static equilibrium. Because there are four unknown forces of reaction and three applicable equations of equilibrium, the beam is stable and statically indeterminate to the first degree externally—there is one more unknown force than there are equations of static equilibrium. ■

EXAMPLE 4.6

Determine the statical classification of the structural system shown in Figure 4.9.

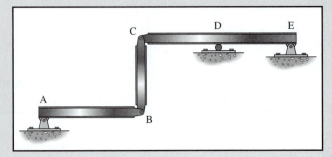

Figure 4.9

In this structural system, there are five unknown forces of reaction: 2 forces at A, 1 force at D, and 2 forces at E. On the surface, this structure appears to be statically indeterminate to the second degree because there are only three applicable equations of static equilibrium. However, because of the link connected to points B and C in the system, an equation of condition can be introduced. The pins at B and C are assumed frictionless so the moment at B and at C must be equal to zero. Given this condition, the two free-body diagrams shown here can be drawn.

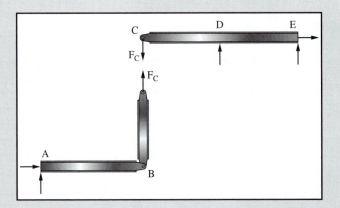

Observe that there are six unknown forces: the forces of reaction previously mentioned and the force in the pin at C. There are three applicable equations of static equilibrium for each of the free-bodies. As such, we have six equations of static equilibrium and six unknown forces. This structure is therefore stable and statically determinate. Note that the free bodies could be cut at B instead of C. ■

4.6 UNSTABLE EQUILIBRIUM AND GEOMETRIC INSTABILITY

The ability of a structure to adequately support the loads applied to it is dependent not only on the number of reaction components but also on the arrangement of those components. A structure can be unstable and yet be stable under a certain set of loads.

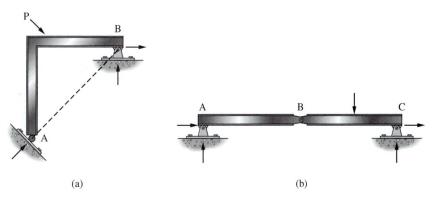

(a)   (b)

Figure 4.10   Geometrically unstable structures

The beam previously shown in Figure 4.5 is an example of such a structure. This beam is supported on its ends with rollers only and is unstable. The beam will slide laterally if any horizontal force is applied. However, the beam can support vertical loads and is stable if only vertical loads are applied. This condition is sometimes referred to as *unstable equilibrium.*

It is also possible for a structure to have as many or more reaction components than there are equations available and yet still be unstable. This condition is referred to as *geometric instability.* The frame of Figure 4.10(a) has three reaction components and three equations available for their solution. However, a study of the moment at B shows that the structure is unstable. The line of action of the reaction at A passes through the reaction at B. Unless the line of action of the force P passes through the same point, the sum of the moments about B cannot be equal to zero. There would be no resistance to rotation about B, and the frame would immediately begin to rotate. It might not collapse, but it would rotate until a stable situation was developed when the line of action of the reaction at A did not pass through B. Of prime importance to the engineer is for a structure to hold its position under load (even though it may deform). One that does not do so is unstable.

Another geometrically unstable structure is shown in Figure 4.10(b). Four equations are available to compute the four unknown forces of reaction—three equations of static equilibrium and one equation of condition. Nevertheless, rotation will instantaneously occur about the hinge at B. After a slight vertical deflection at B, the structure probably will become stable.

4.7 SIGN CONVENTION

The particular sign convention used for tension, compression, and so forth is of little consequence as long as a consistent system is used. However, the use of a standard sign convention makes it easier for engineers to communicate. The author uses the following signs in his computations.

1. For *tension* a positive sign is used, the thought being that pieces in tension become longer or have plus lengths.
2. A negative sign is used for pieces in *compression* because they are compressed or shortened and therefore have minus lengths.

3. The sign convention for internal forces (moments) in a beam is discussed in Chapter 5. A positive moment causes the top of a beam to be in compression and the bottom in tension.

4. On many occasions it is possible to determine the direction of a *reaction* by inspection, but where it is not possible a direction is assumed and the appropriate statics equation is written. If on solution of the equation the numerical value for the reaction is positive, the assumed direction was correct; if negative, the correct direction is opposite that assumed.

4.8 FREE-BODY DIAGRAMS

For a structure to be in equilibrium, every part of the structure also must be in equilibrium. The equations of static equilibrium are as applicable to each piece of a structure as they are to an entire structure. It is therefore possible to draw a diagram of any part of a structure, including all of the forces that are acting on that part of the structure, and apply the equations of static equilibrium to that part. Such a diagram is called a *free-body diagram*. The forces acting on the free-body are the external forces acting on that piece of the structure, including structural reactions, and the internal forces applied from the adjoining parts of the structure.

A simple beam is shown in Figure 4.11(a). This beam has two supports and is acted upon by two loads. A free-body diagram of the entire beam Figure 4.11(b) shows all of the forces of reaction. We also can cut the beam at point A and draw a free-body diagram for each of the two pieces. The result is shown in Figure 4.11(c). Observe that we now have included the internal forces at the location of the cut on the diagrams. The internal forces are the same on the two pieces, but the direction in which they act is reversed. In essence, the effects of the right side of the beam on the left side are shown on the left free-body and vice versa. For instance, the right-hand part of the beam tends to push the left-hand free body down while the left-hand part is trying to push the right-hand free body up.

Isolating certain sections of structures and considering the forces applied to those sections is the basis of all structural analysis. It is doubtful that this procedure can be

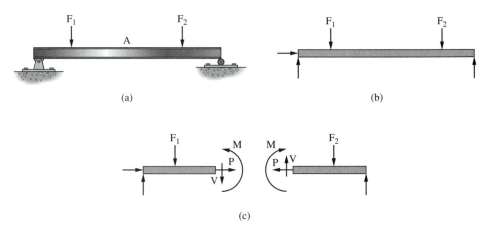

(a)   (b)

(c)

Figure 4.11   A beam and two free-body diagrams

overemphasized to the reader, who will, it is hoped, discover over and over that free-body diagrams open the way to the solution of structural problems.

4.9 HORIZONTAL AND VERTICAL COMPONENTS

It is good practice to compute the horizontal and vertical components of inclined forces for use in making calculations. If this practice is not followed, the perpendicular distances from the lines of action of inclined forces to the point where moment is being taken will have to be found. The calculation of these distances is often difficult, and the possibility of making mistakes in setting up the equations is greatly increased.

4.10 REACTIONS BY PROPORTIONS

The calculation of reactions is fundamentally a matter of proportions. To illustrate this point, reference is made to Figure 4.12(a). The load P is three-fourths of the distance from the left-hand support A to the right-hand support B. By proportions, the right-hand support will carry three-fourths of the load and the left-hand support will carry the remaining one-fourth of the load.

Similarly, for the beam of Figure 4.12(b), the 10-kip (k or kip is the abbreviation for kilopound) load is one-half of the distance from A to B and each support will carry half of it, or 5 kips. The 20-kip load is three-fourths of the distance from A to B. The B support will carry three-fourths of it, or 15 kips, and the A support will carry one-fourth or 5 kips. In this manner the total reaction at the A support was found to be 10 kips and the total reaction at the B support 20 kips.

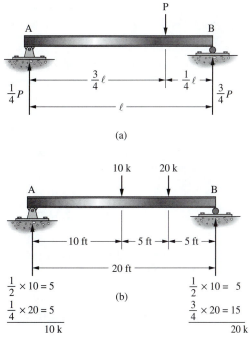

(a)

(b)

Figure 4.12

4.11 REACTIONS CALCULATED BY EQUATIONS OF STATICS

Reaction calculations by the equations of statics are illustrated by Examples 4.7 to 4.9. In applying the $\sum M = 0$ equation, a point may usually be selected as the center of moments so that the lines of action of all but one of the unknowns pass through the point. The unknown is determined from the moment equation, and the other reaction components are found by applying the $\sum F_H = 0$ and $\sum F_V = 0$ equations.

The beam of Example 4.7 has three unknown reaction components: vertical and horizontal ones at A and a vertical one at B. Moments are taken about A to find the value of the vertical component at B. All of the vertical forces are equated to zero, and the vertical reaction component at A is found. A similar equation is written for the horizontal forces applied to the structure, and the horizontal reaction component at A is found to be zero.

The solutions of reaction problems may be checked by taking moments about the other support, as illustrated in this example. For future examples space is not taken to show the checking calculations. *A problem, however, should be considered incomplete until a mathematical check of this nature is made.*

EXAMPLE 4.7

Compute the reaction components for the beam shown in Figure 4.13.

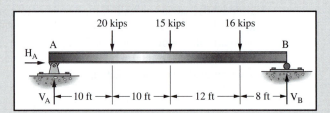

Figure 4.13

Solution. The forces of reaction and their assumed direction are shown on the figure. Begin the solution by summing forces horizontally to determine H_A.

$$\sum F_H = H_A = 0$$
$$\therefore H_A = 0$$

Next, sum moments clockwise about the left support. By doing so we obtain the equation

$$\sum M_A = 20(10) + 15(20) + 16(32) - V_B(40) = 0$$
$$\therefore V_B = 25.3 \, \text{kips} \uparrow$$

The result for V_B is positive so the assumed direction is correct; the reaction at B acts upward. Lastly, forces are summed vertically to compute the remaining reaction.

$$\sum F_V = V_A - 20 - 15 - 16 + V_B = 0$$
$$V_A - 20 - 15 - 16 + 25.3 = 0$$
$$\therefore V_A = 25.7 \, \text{kips} \uparrow$$

Again, the computed reaction at A is positive so the assumed direction is correct. We can sum moments about B to check our calculations.

$$\sum M_B = 25.7(40) - 20(30) - 15(20) - 16(8)$$
$$\sum M_B = 0$$

Because the summation of moments is equal to zero, the computed reactions are correct. ■

EXAMPLE 4.8

Find all of the reaction components for the structure shown in Figure 4.14.

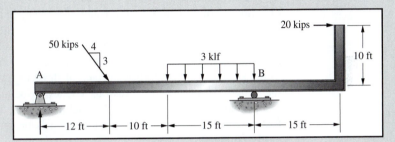

Figure 4.14

Solution.   The free-body diagram for this beam is shown here. Observe that the inclined force has been replaced with its horizontal and vertical components, and the distributed load has been replaced with an equivalent concentrated load acting at its centroid.

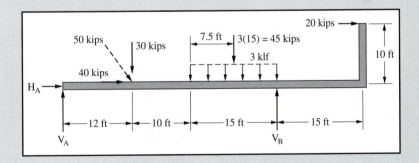

Begin by summing forces horizontally to compute the horizontal reaction at A.

$$\sum F_H = 40 + 20 + H_A = 0$$
$$\therefore H_A = -60 \, \text{kips} \; \leftarrow$$

Observe that the computed value is negative so the actual direction of the reaction is opposite that assumed. Next, moments are taken clockwise about A to determine the vertical reaction at B.

$$\sum M_A = 30(12) + (3 \times 15)(29.5) - V_B(37) + 20(10) = 0$$
$$\therefore V_B = 51 \, \text{kips} \uparrow$$

Because the computed sign on V_B is positive the assumed direction is correct. Forces are summed vertically to compute the vertical reaction at A.

$$\sum F_V = V_A - 30 - 45 + V_B = 0$$
$$V_A - 30 - 45 + 51 = 0$$
$$\therefore V_A = 24 \, \text{kips} \uparrow$$

Again, the assumed direction of V_A is correct. ■

The roller in the frame of Figure 4.15 is supported by an inclined surface. The direction of the reaction at the roller is known; it is perpendicular to the supporting surface. If the direction of the reaction is known, the relationship between the vertical component, the horizontal component, and the reaction itself is known. Here the reaction has a slope of four vertically to three horizontally (4:3), which is perpendicular to the slope of the supporting surface of three to four (3:4). Moments are taken about the left support, which gives an equation including the horizontal and vertical components of the reaction at the inclined roller. Both components are in terms of that reaction. Therefore only one unknown is present in the equation and its value is easily obtained.

EXAMPLE 4.9

Compute the reactions for the frame shown in Figure 4.15.

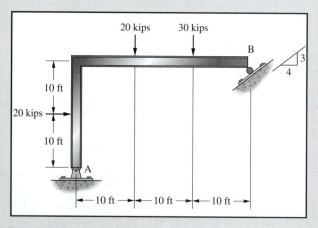

Figure 4.15

Solution. The free-body diagram for this frame is shown next. Observe that the inclined force of reaction at B has been replaced with its horizontal and vertical components. These components act at the same point as the reaction.

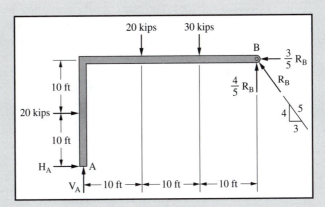

We will begin by summing moments about A to determine the reaction at B. In the calculation we will use the components of the reaction.

$$\sum M_A = 20(10) + 20(10) + 30(20) - \frac{4}{5}R_B(30) - \frac{3}{5}R_B(20) = 0$$

$$\therefore R_B = 27.8 \text{ kips} \ \diagdown$$

The sign on the computed reaction at B is positive so the assumed direction is correct. Next, forces are summed vertically to obtain the vertical reaction at A.

$$\sum F_V = V_A - 20 - 30 + \frac{4}{5}R_B = 0$$

$$V_A - 20 - 30 + \frac{4}{5}(27.8) = 0$$

$$\therefore V_A = 27.8 \ \uparrow$$

Again, the sign on the computed reaction is positive so the assumed direction is correct. By summing the horizontal forces, the horizontal reaction at A can be evaluated.

$$\sum F_H = H_A + 20 - \frac{3}{5}R_B = 0$$

$$H_A + 20 - \frac{3}{5}(27.8) = 0$$

$$\therefore H_A = -3.3 \text{ kips} \ \leftarrow$$

The sign on the computed reaction is negative so the force of reaction is actually acting to the left, opposite the direction indicated.   ■

4.12 PRINCIPLE OF SUPERPOSITION

As we proceed with our study of structural analysis, we will encounter structures subject to large numbers of forces and to different kinds of forces (concentrated, uniform, triangular, dead, live, impact, etc.). To assist in handling such situations there is available an extremely useful tool called the *principle of superposition*. The principle of superposition can be stated as follows:

If the structural behavior is linearly elastic, the forces acting on a structure may be separated or divided in any convenient fashion and the structure analyzed for the separate cases. The final results can then be obtained by adding algebraically the individual results. The author previously made use of this principle in Figure 4.12(b) when the reactions for the two loads were determined separately by proportions and then were added together to obtain the final values. The principle applies not only to reactions, but also to shears, moments, stresses, strains, and displacements. This concept is graphically represented in Figure 4.16.

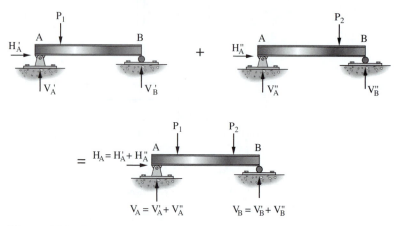

Figure 4.16   Principle of superposition

The principle of superposition is not valid in two important situations. The first occurs where the geometry of the structure is appreciably changed under the action of the loads. The second occurs where the structure consists of a material for which stresses are not directly proportional to strains. This latter situation can occur when the stress is beyond the elastic limit of the material. It can also occur when the material does not follow Hooke's law for any part of its stress-strain curve.

United Airlines hangar, San Francisco, California (Courtesy of the Lincoln Electric Company)

4.13 THE SIMPLE CANTILEVER

The simple cantilever in Example 4.10 has three unknown reaction components supporting it at the fixed end; they are the forces required to resist horizontal movement, vertical movement, and rotation. They may be determined with the equations of static equilibrium as illustrated.

EXAMPLE 4.10

Find all reaction components for the cantilever beam shown in Figure 4.17.

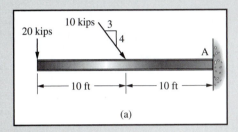

(a)

Figure 4.17

Solution. The free-body diagram used for the analysis is illustrated here.

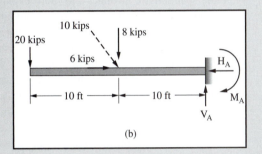

(b)

Summing forces vertically yields the vertical reaction at A

$$\sum F_V = -20 - 8 + V_A = 0$$
$$\therefore V_A = 28 \, \text{kips} \uparrow$$

Summing forces horizontally yields the horizontal reaction at A

$$\sum F_H = 6 - H_A = 0$$
$$\therefore H_A = 6 \, \text{kips} \leftarrow$$

Finally, summing moments clockwise about A yields the rotational component of the reaction.

$$\sum M_A = -20(20) - 8(10) + M_A = 0$$
$$\therefore M_A = 480 \, \text{ft-k} \; \circlearrowright \; \blacksquare$$

4.14 CANTILEVERED STRUCTURES

Moments in structures that are simply supported increase rapidly as their spans become longer. We will see that bending increases approximately in proportion to the square of

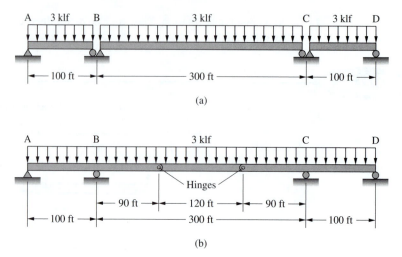

(a)

(b)

Figure 4.18

the span length. Stronger and more expensive structures are required to resist the greater moments. For very long spans, moments are so large that it becomes economical to introduce special types of structures that will reduce the moments. One of these types is cantilevered construction, as illustrated in Figure 4.18(b).

A cantilevered type of structure is substituted for the three simple beams of Figure 4.18(a) by making the beam continuous over the interior supports B and C and introducing hinges in the center span as indicated in (b). An equation of condition ($\Sigma M_{hinge} = 0$) is available at each of the hinges introduced, giving a total of five equations and five unknowns. The structure is statically determinate.

The moment advantage of cantilevered construction is illustrated in Figure 4.19. The diagrams shown give the variation of bending moment in each of the structures of Figure 4.18 due to a uniform load of 3 klf for the entire spans. The maximum moment for the cantilevered type is seen to be considerably less than that for the simple spans, and this permits lighter and more economical construction. The plotting of moment diagrams is fully explained in Chapter 5. For the cantilevered structure of Figure 4.18(b) it is possible to better balance the positive and negative moments by moving the unsupported hinges closer to supports B and C.

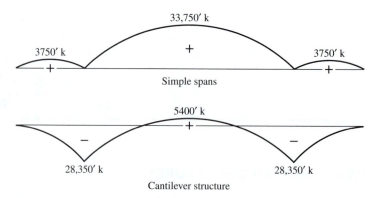

Figure 4.19

4.15 REACTION CALCULATIONS FOR CANTILEVERED STRUCTURES

Cantilevered construction consists essentially of two simple beams, each with an overhanging or cantilevered end as follows:

with another simple beam in between supported by the cantilevered ends:

The first step in determining the reactions for cantilevered construction is to isolate the center simple beam and compute the forces necessary to support it at each end. Second, these forces are applied as downward loads on the respective cantilevers, and as a final step the end-beam reactions are determined individually. Example 4.11 illustrates the entire process.

EXAMPLE 4.11

Calculate all reactions for the cantilevered structure illustrated in Figure 4.20.

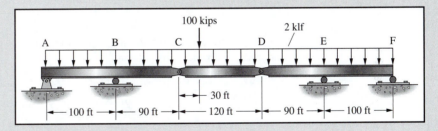

Figure 4.20

Solution. The free-body diagram

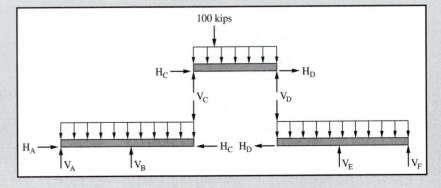

Because of the equations of condition that exist at C and D, we were able to isolate the center section from the two end sections with the forces acting at the hinge, as shown on the free-body. We can then use the free-body of the center section to compute the forces V_C and V_D,

$$\sum M_C = 100(30) + 2(120)\left(\frac{120}{2}\right) - V_D(120) = 0$$

$$\therefore V_D = 145 \text{ kips}$$

then by summing forces on the center section vertically,

$$\sum F_V = V_C - 100 - 2(120) + V_D = 0$$

$$\therefore V_C = 195 \text{ kips}$$

Finally, by summing forces horizontally on the center section a relationship is obtained between the horizontal forces.

$$\sum F_H = H_C + H_D = 0$$

$$\therefore H_C = -H_D$$

Now that we have computed the forces acting on the center section, we will turn to the right section. First, sum moments counterclockwise about the right end.

$$\sum M_F = -V_D(190) - 2(190)\left(\frac{190}{2}\right) + V_E(100) = 0$$

$$-(145)(190) - 2(190)\left(\frac{190}{2}\right) + V_E(100) = 0$$

$$\therefore V_E = 636.5 \text{ kips} \uparrow$$

We can then sum forces vertically and horizontally to obtain the other components of reaction on the right section.

$$\sum F_V = -V_D - 2(190) + V_E + V_F = 0$$

$$-145 - 2(190) + 636.5 - V_F = 0$$

$$\therefore V_F = -111.5 \text{ kips} \downarrow$$

The negative sign indicates that the reaction is acting downward, that is, opposite the direction assumed.

$$\sum F_H = H_D = 0$$

$$\therefore H_D = 0 = H_C$$

We will work with the left section to compute the remaining forces of reaction. Begin by summing moments clockwise about A.

$$\sum M_A = 2(190)\left(\frac{190}{2}\right) + V_C(190) - V_B(100) = 0$$

$$2(190)\left(\frac{190}{2}\right) + 195(190) - V_B(100) = 0$$

$$\therefore V_B = 731.5 \text{ kips} \uparrow$$

The remaining vertical reaction is found by summing forces vertically.

$$\sum F_V = V_A - 2(190) - V_C + V_B = 0$$
$$V_A - 2(190) - 195 + 731.5$$
$$\therefore V_A = -156.5 \, \text{kips} \; \downarrow$$

Again, the negative sign indicates that the force of reaction is acting opposite that assumed. Finally, the horizontal reaction is computed.

$$\sum F_H = H_A - H_C = 0 = H_A - 0$$
$$\therefore H_A = 0 \quad \blacksquare$$

An examination of the reactions obtained in the previous example for the left-end and right-end sections of this beam show why cantilevered bridges often are called "seesaw" bridges. These structures are primarily supported by the first interior supports on each end, where the reactions are quite large. The end supports may very well have to provide downward reaction components. Thus, an end section of a cantilevered structure seems to act as a seesaw over the first interior support with downward loads on both sides.

4.16 ARCHES

Historically, arches were the only feasible form that could be used for large structures made up of materials with negligible tensile strength, such as bricks and stones. Masonry arches of such materials have been used for thousands of years.

In effect, an arch takes vertical loads and turns them into thrusts that run around the arch and put the elements of the arch in compression, as shown in Figure 4.21. The parts of a stone arch are called *voussoirs*. As you can see in the figure, the voussoirs are pushed together in compression.

Arches are very rigid, stable structures that are not appreciably affected by movements of their foundations. It is rather interesting to note that excavations of ancient ruins show that arches are the structures that generally have survived best.

Theoretically, an arch can be designed for a single set of gravity loads so that only axial compression stresses are developed throughout. Unfortunately, however, in practical

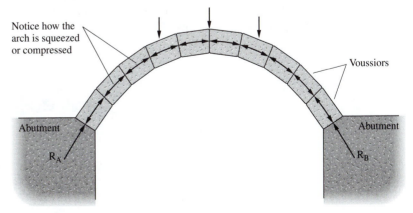

Figure 4.21   Stone arch

structures the loads change and move so that bending stresses are developed in the arch. Nevertheless, arches generally are designed so that their predominant loading primarily causes compression stresses.

Structural arch construction: U.S.A.F. hangar, Edwards Air Force Base (Courtesy Bethlehem Steel Corporation)

4.17 THREE-HINGED ARCHES

Arches may be constructed with two or three hinges; very rarely are they constructed with only one hinge. Quite often in reinforced concrete construction, arches are built without any hinges. The three-hinged arch is discussed at this time because it is the only statically determinate arch.

Examination of the three-hinged arch pictured in Figure 4.22 reveals two reaction components at each support, for a total of four. Three equations of static equilibrium and one condition equation are available to find the unknowns. The equation of condition is the summation of moments about the crown hinge from either the left or the right.

The arch in Example 4.12 is analyzed by taking moments about one of the supports to obtain the vertical reaction component at the other support. Because the supports are on the same level, the horizontal reaction component at the second support passes through the point where moments are being taken. The horizontal reaction components are obtained by taking moments at the crown hinge of the forces either to the left or to the right. The only unknown appearing in either equation is the horizontal reaction component on that side, and the equation is solved for its value. Once the reactions are determined, it is easy to compute the moment, shear, and axial force in the arch at any point using equations of static equilibrium.

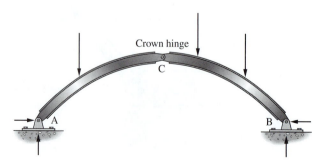

Figure 4.22 Three-hinged arch

Laminated timber arches, Maumee, Ohio (Courtesy of
Unit Structures, Inc.)

EXAMPLE 4.12

Find all reaction components for the three-hinged arch shown in Figure 4.23.

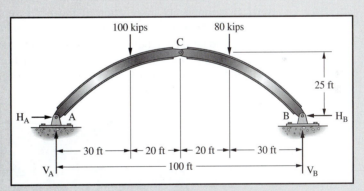

Figure 4.23

Solution. Summing moments about the left support.

$$\sum M_A = 100(30) + 80(70) - V_B(100) = 0$$
$$\therefore V_B = 86 \text{ kips} \uparrow$$

Then summing all forces vertically to compute the vertical component of reaction
at the left support.

$$\sum F_V = V_A - 100 - 80 + V_B = 0$$
$$V_A - 100 - 80 + 86 = 0$$
$$\therefore V_A = 94 \text{ kips} \uparrow$$

To compute the magnitude of the horizontal components of reactions, it is necessary to use the free-body shown in the following diagram and sum moments about C, the crown hinge. This free-body was obtained from the equation of condition that finds moment at C is equal to zero.

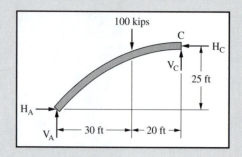

$$\sum M_C = V_A(50) - H_A(25) - 100(20) = 0$$
$$94(50) - H_A(25) - 100(20) = 0$$
$$\therefore H_A = 108\,\text{kips} = H_B \quad \blacksquare$$

The computation of reactions for the arch of Example 4.13 is slightly more complicated because the supports are not on the same level. Summing moments about one of the supports results in an equation involving both the horizontal and vertical components of reaction at the other support. Moments may then be taken about the crown hinge of the forces on the same side as those two unknowns. The resulting equation contains the same two unknowns. Solving the equations simultaneously yields the magnitude of those reactions and then, by summing forces vertically and horizontally, the magnitude of the remaining reaction components can be determined.

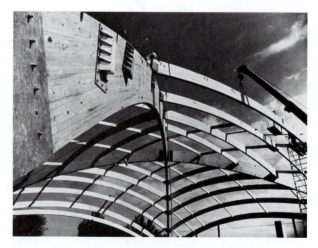

North Dakota State Teachers College fieldhouse, Valley City, North Dakota (Courtesy of the American Institute of Timber Construction)

EXAMPLE 4.13

Determine the components of the reactions for the structure shown in Figure 4.24.

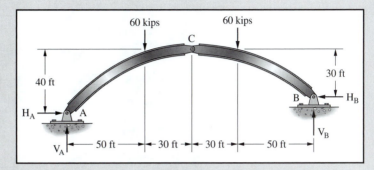

Figure 4.24

Solution.   Summing moments clockwise about A.

$$\sum M_A = 60(50) + 60(110) - H_B(10) - V_B(160) = 0$$
$$10H_B + 160V_B = 9600 \qquad (1)$$

We then utilize the equation of condition at C to obtain the free-body shown next. Using this free-body, we sum moments about C.

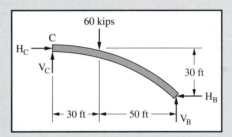

$$\sum M_C = 60(30) + H_B(30) - V_B(80) = 0$$
$$30H_B - 80V_B = -1800 \qquad (2)$$

Solving Equations (1) and (2) simultaneously gives

$$H_B = 85.7\,\text{kips} \quad \leftarrow$$
$$V_B = 54.6\,\text{kips} \quad \uparrow$$

By summing forces vertically and horizontally, the remaining components of reaction are computed.

$$\sum F_H = H_A - H_B = H_A - 85.7 = 0$$
$$\therefore H_A = 85.7\,\text{kips} \quad \rightarrow$$

$$\sum F_V = V_A - 60 - 60 + V_B = V_A - 60 - 60 + 54.6 = 0$$
$$\therefore V_A = 65.4\,\text{kips} \uparrow \quad \blacksquare$$

Glued laminated three-hinged arches, The Jai Alai Fronton, Riviera Beach, Florida (Courtesy of the Forest Products Association)

The reactions for the arch of Example 4.13 may be computed without using simultaneous equations. The horizontal and vertical axes (on the basis of which the $\Sigma H = 0$ and $\Sigma V = 0$ equations are written) are so rotated that the horizontal axis passes through the two supporting hinges (Figure 4.25). Components of forces can be computed parallel to the X′ and Y′ axes and the $\Sigma X' = 0$ and $\Sigma Y' = 0$ equations applied, but the computations are very cumbersome.

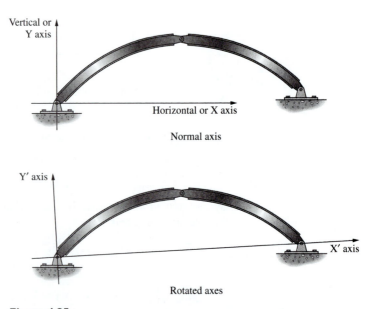

Figure 4.25

4.18 USES OF ARCHES AND CANTILEVERED STRUCTURES

Three-hinged steel arches are used for short and medium-length bridges with spans of up to approximately 600 ft. They are used for buildings when large clear spans are required underneath, as for hangars, field houses, and armories. Two-hinged steel arches generally are economical for bridges from 600 to 900 ft in length, with a few exceptional spans being over 1600 ft long. Reinforced-concrete arches without hinges are used for bridges with spans ranging from 100 to 400 ft. Cantilever-type bridges are used for spans from approximately 500 ft up to very long spans, such as the 1800-ft center span of the Quebec Bridge.

An arch is a structure that requires foundations capable of resisting the large horizontal reaction components, called thrusts, at the supports. In arches for buildings, it is possible to carry the thrusts by tying the supports together with steel rods, with steel sections, or even with specially designed floors. Arches so constructed are referred to as *tied arches*. For many locations with poor foundation conditions, and thereby the possibility of settlement, the three-hinged arch is selected over the statically indeterminate arches because the forces will not appreciably change in a three-hinged arch when settlement occurs. We will see in later chapters that foundation settlement can cause severe stress changes in statically indeterminate structures. Ease of erection is another advantage of three-hinged arches. It often is convenient to assemble and ship the two halves of a precast-concrete, structural steel, or laminated-timber arch separately and assemble them on the site.

The fact that cantilever-type construction reduces bending moments for long spans has been previously demonstrated. Arch-type construction also reduces moments, because the reactions at the supports tend to cause bending in the arch in a direction opposite to that caused by the downward loads. Because of this characteristic of having small bending moments, arches were admirably suited for masonry construction as practiced by the ancient builders.

4.19 CABLES

Cables provide perhaps the simplest means for supporting loads. They are used for supporting bridge and roof systems, as guys for derricks, radio towers, and similar type structures, as well as for many other applications. To the student, the most common use of cables may seem to be the cable car systems at hundreds of ski slopes around the world.

Steel cables are economically manufactured from high-strength steel wire, providing perhaps the lowest cost-to-strength ratio of any common structural members. Further, they are easily handled and positioned, even for very long spans. For the discussion to follow, the cable weight is neglected. When a cable of a given length is suspended between two supports, the shape it takes is determined by the applied loads.

The shape cables take in resisting loads is called a *funicular* curve. You may have noticed that the cable car systems in Europe often are called *funiculars*. Cables are quite flexible and support their loads in pure tension as shown in Figure 4.26. It can be seen in this figure that the load P must be balanced by the vertical components of tension in the cable; thus, the cable must have a vertical projection in order to support the load. The greater the vertical projection, the smaller

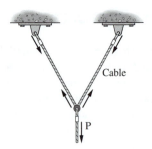

Figure 4.26   A simple cable structure

will be the cable tension. If the cable moves or if other loads are applied, the cable will change shape.

The resultant tension at any point can be obtained from the following equation:

$$T = \sqrt{H^2 + V^2}$$

In this equation, H and V are, respectively, the horizontal and vertical components of tensile force in the cable at that point. We can see from this equation that the tension varies with different slopes along the length of the cable. However, if only vertical loads are present the value of H will be constant throughout the cable.

Cables are assumed to be so flexible that they cannot resist bending or compression; they act only in direct tension. An equation of condition is available for analysis: the summation of moments to the left or right at any point along the cable is equal to zero. Should the position or sag of a cable at a particular point be known, the reactions at the cable ends and the sag at any other point in the cable can be determined with these equations. A numerical example follows. The weight of the cable is assumed to be negligible in this case.

Cable-stayed Sitka Harbor Bridge, Sitka, Alaska (Courtesy of the Alaska Department of Transportation)

EXAMPLE 4.14

Determine the reactions for the cable in Figure 4.27 and the sag at the 40-kip load.

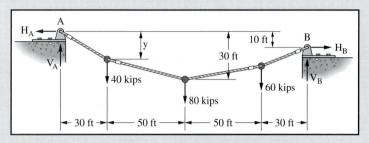

Figure 4.27

Solution.   Begin the solution by summing moments clockwise about the right reaction.

$$\sum M_B = V_A(160) - H_A(10) - 40(130) - 80(80) - 60(30) = 0$$
$$160V_A - 10H_A = 13,400 \qquad (1)$$

Next utilize the equation of condition and sum moments of the forces to the left of the 80-kip load.

$$\sum M_{80} = V_A(80) - H_A(30) - 40(50) = 0$$
$$80V_A - 30H_A = 2000 \qquad (2)$$

Solving equations (1) and (2) simultaneously gives

$$H_A = 188\,k \;\leftarrow$$
$$V_A = 95.5\,k \;\uparrow$$

The vertical reaction at the left support is equal to 95.5 kips and the horizontal reaction at that support is equal to 188 kips. To determine the other two reaction components we can sum forces vertically and horizontally.

$$\sum F_V = V_A - 40 - 80 - 60 + V_B = 0$$
$$95.5 - 40 - 80 - 60 + V_B = 0$$
$$\therefore V_B = 84.5\,\text{kips} \;\uparrow$$

$$\sum F_H = -H_A + H_B = 0$$
$$-188 + H_B = 0$$
$$\therefore H_B = 188\,\text{kips} \;\rightarrow$$

Lastly, we can use the equation of condition again and determine the sag at the 40-kip load by summing moments to the left of the 40-kip load.

$$\sum M_{40} = V_A(30) - H_A(y) = 0$$
$$95.5(30) - 188(y) = 0$$
$$\therefore y = 15.24\,\text{ft}$$

The sag, y, at the location of the 40-kip load is 15.24 ft.   ∎

Very often, the actual geometry of the cable structure will not be known initially. We may know only the total length of the cable and the locations at which the loads are to be applied. From this information, we must determine the geometry of the cable structure before we can determine the forces that exist in the cable. The following example demonstrates such a situation.

EXAMPLE 4.15

For the cable structure shown in Figure 4.28, determine the forces of reaction, the tension in each segment of the cable, and the sag in the cable. The total length of the cable is 65 ft.

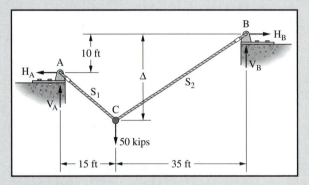

Figure 4.28

Solution.   Before the forces of reaction can be computed, the sag in the cable, Δ, must be computed. Using the Pythagorean theorem, equations for the sag can be written in terms of the left and right cable segments. These equations are

$$\Delta = 10 + \sqrt{S_1^2 - 15^2} \quad \text{in terms of the left segment}$$

$$\Delta = \sqrt{S_2^2 - 35^2} \qquad \text{in terms of the right segment}$$

Both of these equations can be expressed in terms of one of the segment lengths, S_1 or S_2. To do so, we must solve for the length of one segment in terms of the other. We will solve for S_1 in terms of S_2, which yields:

$$S = S_1 + S_2 = 65\text{ft} = \text{cable length}$$
$$\therefore S_1 = 65 - S_2$$

If we substitute this expression for S_1 into the equations for sag, we obtain the equations:

$$\Delta = 10 + \sqrt{(65 - S_2)^2 - 15^2}$$

$$\Delta = \sqrt{S_2^2 - 35^2}$$

We can iterate on S_2 until the sag we compute using both equations is the same. Upon doing so we find that

$$S_2 = 43.361\,\text{ft}$$
$$\Delta = 25.596\,\text{ft}$$

We can now solve for the components of reaction and the force in each segment of the cable. By summing moments clockwise about the left support, we obtain the equation

$$\sum M_A = H_B(10) - V_B(15+35) + 50(15) = 0$$
$$10H_B - 50V_B = -750$$

summing horizontal forces

$$\sum F_H = -H_A + H_B = 0$$
$$-H_A + H_B = 0$$

summing vertical forces

$$\sum F_V = V_A + V_B - 50 = 0$$
$$V_A + V_B = 50$$

Lastly, we can use the equation of condition that exists at C. At that location on the cable the bending moment is equal to zero. As such, if we sum moments in the right segment about C we obtain the equation

$$\sum M_{C(\text{Right})} = H_B(\Delta) - V_B(35) = 0$$
$$25.596H_B - 35V_B = 0$$

solving these equations simultaneously, we obtain the following values:

$$V_B = 20.65\,\text{k} \uparrow$$
$$H_B = 28.23\,\text{k} \rightarrow$$

then by $\sum F_H = 0$ and $\sum F_V = 0$

$$V_A = 29{,}35\,\text{k} \uparrow$$
$$H_A = 28{,}23\,\text{k} \leftarrow$$

Using the free-body diagram shown in the following figure we can compute the tension in the left cable segment.

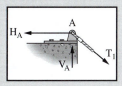

$$T_1 = \sqrt{H_A^2 + V_A^2} = \sqrt{28.23^2 + 29.35^2} = 40.73\,\text{kips}$$

Using a similar free-body diagram, we find the tension in the right cable segment to be

$$T_2 = \sqrt{H_B^2 + V_B^2} = \sqrt{28.23^2 + 20.65^2} = 34.98\,\text{kips} \quad\blacksquare$$

The distortion, or deflection, of most structures is assumed to be negligible when computing the forces produced in those structures. Such an assumption is not correct, however, for many cable structures, particularly the flat ones where a little sag can drastically affect cable tensions. This topic is not considered herein, but is described very well in a book by Firmage.[1] Flat cables cause very large horizontal reaction components and thus have very high tensile forces.

A cable supporting a load that is uniform along its length, such as a cable loaded only by its own weight, will take the form of a catenary. On many occasions, concentrated loads are applied to cables by hangers. If they are closely spaced, the loading will approach that of a uniform load along the horizontal projection of the cable. Cables supporting the roadway of suspension bridges usually are assumed to fall into this class. The analysis of such cables is presented in Appendix A.

4.20 PROBLEMS FOR SOLUTION

For Problem 4.1, determine which of the structures shown in the accompanying illustration are statically determinate, statically indeterminate (including the degree of indeterminacy), and unstable in regards to outer forces.

4.1

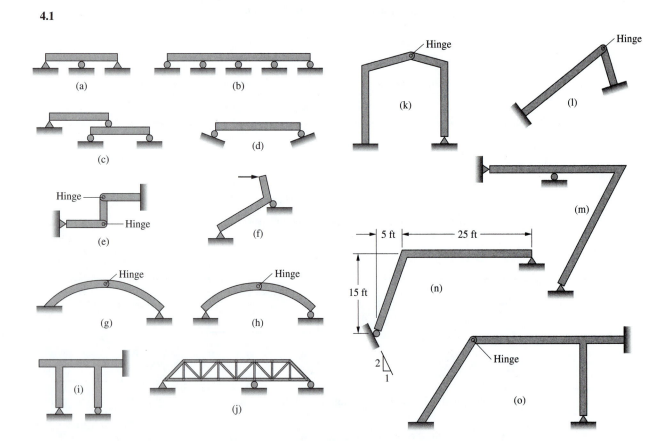

[1]D. A. Firmage, *Fundamental Theory of Structures* (New York: Wiley, 1963), 258–265.

For Problems 4.2 to 4.45 compute the reactions for the structures.

4.2

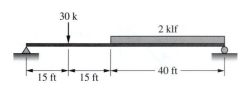

4.3   (*Ans.* $V_L = 115\,k \uparrow$, $V_R = 95\,k \uparrow$)

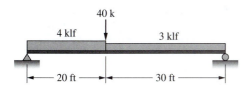

4.4

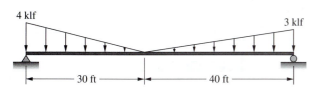

4.5   (*Ans.* $V_L = 115.83\,k \uparrow$, $V_R = 104.17\,k \uparrow$)

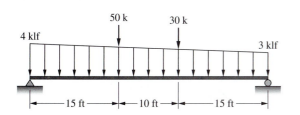

4.6

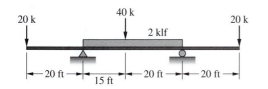

4.7   (*Ans.* $V_L = 62.5\,k \uparrow$, $H_R = 30\,k \rightarrow$)

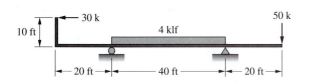

4.8

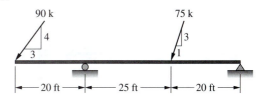

4.9   (*Ans.* $V_R = 103.33\,k \uparrow$, $H_L = 50\,k \leftarrow$)

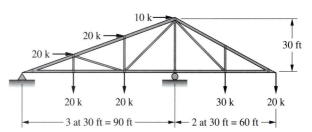

4.10

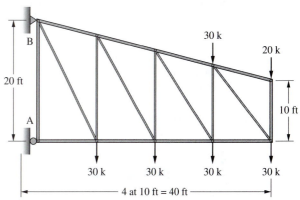

4.11   (*Ans.* $V_L = 112.96\,kN \uparrow$, $H_L = 19.40\,kN \leftarrow$)

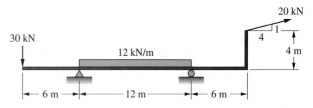

4.12

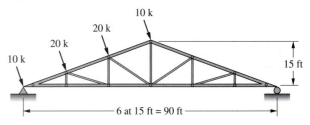

4.13 (*Ans.* $V_R = 81$ k ↑, $H_L = 50$ k ←)

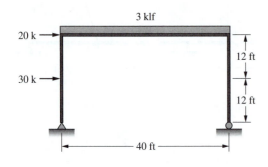

4.14

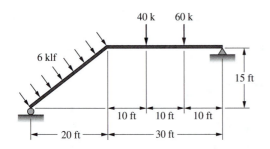

4.15 (*Ans.* $V_L = 36.25$ k ↑, $V_R = 98.75$ k)

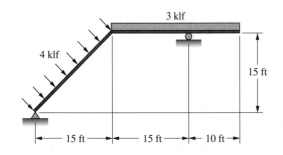

4.16

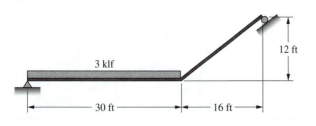

4.17 (*Ans.* $V_R = 50$ k ↑, $H_R = 30$ k ←)

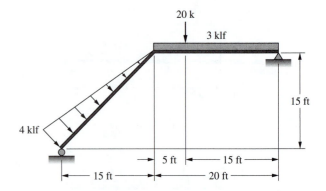

4.18

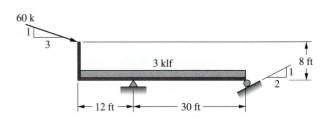

4.19 (*Ans.* $V_L = 106.41$ k ↑, $V_R = 45.59$ k ↑)

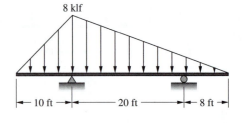

4.20

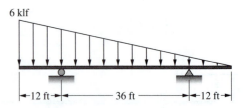

4.21 (*Ans.* $V_L = 59.22$ k ↑, $H_L = 31.16$ k →)

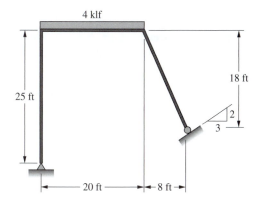

4.22

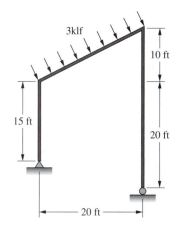

4.23 (*Ans.* $V_L = 130$ k, $H_L = 22.5$ k ←)

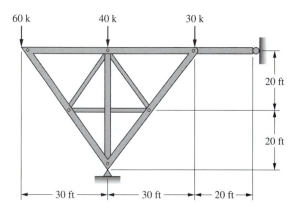

4.24

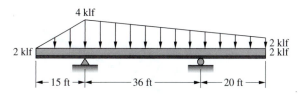

4.25 (*Ans.* $V_R = 73.36$ k ↑, $H_R = 48.93$ k ←)

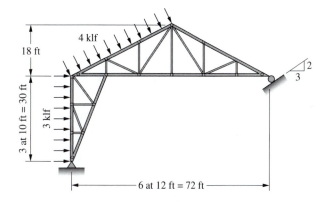

4.26 Consider only a 1 ft length of frame.

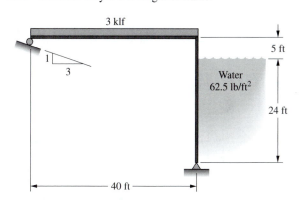

4.27 (*Ans.* $V_L = 13.09$ k ↑, $H_L = 18.82$ k ←, $V_R = 22.91$ k ↑)

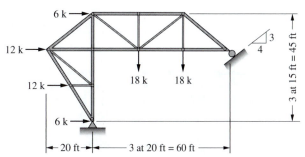

4.28

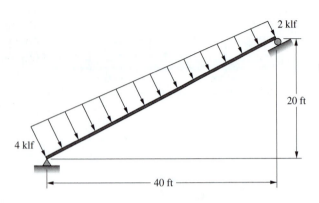

4.29 (*Ans.* $V_A = 68.96$ k ↑, $V_B = 157.79$ k ↑)

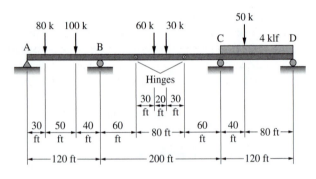

4.30

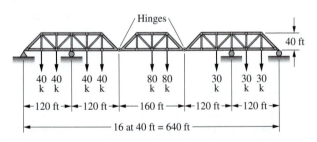

4.31 (*Ans.* $V_A = 122.5$ k ↓, $V_B = 575$ k ↑, $V_C = 67.5$ k ↑)

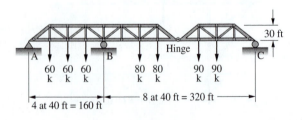

4.32

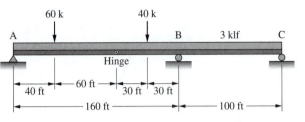

4.33 (*Ans.* $V_L = 85$ k ↑, $H_R = 63.33$ k ←)

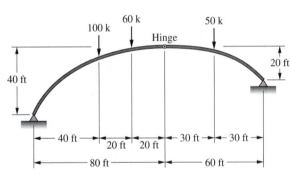

4.34

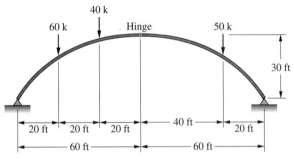

4.35 (*Ans.* $V_L = 89.27$ k ↑, $H_L = 98.54$ k →, $H_R =$ 31.57 k ←)

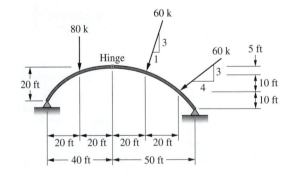

4.36

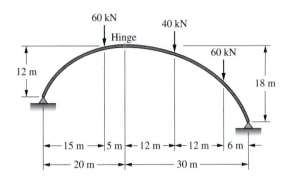

4.41   (*Ans.* $V_L = 97.07$ k, $V_R = 102.93$ k $\uparrow$, $H_R = 126.75$ k $\rightarrow$)

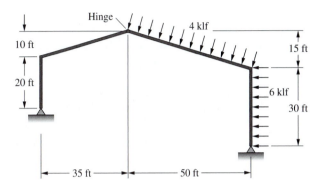

4.37   (*Ans.* $V_R = 126.67$ k $\uparrow$, $H_R = 166.67$ k $\leftarrow$)

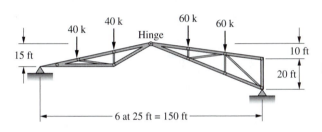

4.42

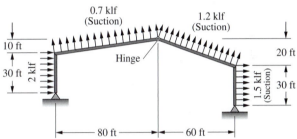

4.38

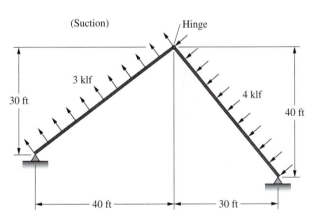

4.43   (*Ans.* $V_R = -4.23$k $\downarrow$, $H_L = 11.22$ k $\leftarrow$)

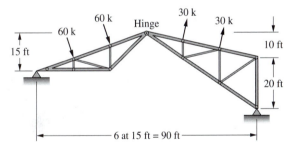

4.44

4.39   Repeat Problem 4.38 if the 3 klf load is increased to 4 klf and the 4 klf load is increased to 6 klf. (*Ans.* $V_R = 10$ k $\uparrow$, $H_R = 180$ k $\rightarrow$)

4.40   Repeat Problem 4.38 if the 4 klf load is removed.

4.45   (*Ans.* $V_A = -16.67$ k ↓, $V_B = 246.67$ k ↑, $V_C =$
50 k ↑)

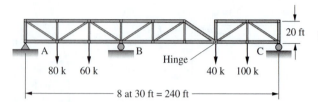

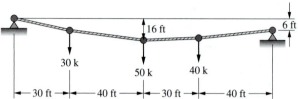

the cable. (*Ans.* $V_L = 69.89$ k ↑, $Y_{30} = 9.09$ ft, $Y_{40} =$
14.69 ft)

4.46   Determine the reaction components, cable sag at the
60 k load, and the maximum tensile force in the cable.

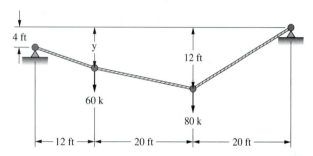

4.47   Determine the reaction components, cable sag at the
30- and 40-kip loads, and the maximum tensile force in

4.48   Determine the reaction components and the sag for the
cable shown. The cable is 120 ft long.

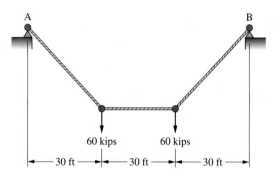

Chapter 5

Shearing Force and Bending Moment

5.1 INTRODUCTION

An important part of structural engineering, and indeed the understanding of structural behavior, is the understanding of the shearing forces and bending moments that exist within a structural system. Equations for shearing force and bending moment are needed to compute structural deflections. Very often the shearing force and bending moment are represented on diagrams to provide a visualization of structural response. These diagrams, from which the values of shearing force and bending moment at any point in a beam are immediately available, are very convenient in design since they visually provide the magnitude and location of maximum design forces. In this chapter we will examine methods for developing the equations for shearing force and bending moment in structural systems, as well as methods for constructing shearing force and bending moment diagrams. It is doubtful that there is any other topic for which careful study will give more reward in structural engineering knowledge.

To examine the internal conditions in a structure, a free-body is taken out and studied to see what forces must be present for the body to remain in a state of equilibrium. Shearing force and bending moments are two actions of the external loads on a structure that need to be understood to study properly the internal forces.

Shear is defined as the algebraic summation of the forces to the left *or* to the right of a section that are perpendicular to the axis of the member. *Bending moment* is the algebraic summation of the moments of all the forces to the left *or* to the right of a particular section—the moments being taken about an axis through the centroid of the section.

The selection of a sign convention for shear (V), moment (M) and normal or axial force (N) is actually a matter of personal choice. In practice, however the convention shown in Figure 5.1 is commonly used. In this figure the author has passed a section

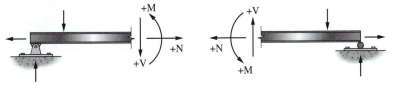

Free body to left of section   Free body to right of section

Figure 5.1   Commonly used sign convention for internal shear, moment, and axial forces.

through a beam and shown free bodies of that beam to the left and to the right of the section. For each of these free bodies, internal values for plus shear, plus moment, and plus axial or normal force are given using this convention.

Throughout this text the author commonly uses the sign convention shown in Figure 5.1. Actually though, he normally works with external forces. If he sums up the external shear forces to the left of the section and the result is up, the shear is positive. If he sums up the external shear forces to the right of the section and the result is down, the shear is considered to be positive. In a similar fashion, when the moment of the external forces to the left of the section is clockwise, the moment is considered to be positive. The reverse is true for moments to the right of the section.

The calculations for shear and bending moment at two sections in a simple beam are shown in Example 5.1.

EXAMPLE 5.1

Find the shear at sections a–a and b–b, Figure 5.2.

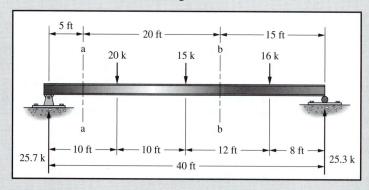

Figure 5.2

Solution.   Shear at section a–a:

$$V_{a-a} \text{ to left} = 25.7 \, k \uparrow, \text{ or } +25.7 \, k$$
$$V_{a-a} \text{ to right} = 20 + 15 + 16 - 25.3 = 25.7 \, k \downarrow, \text{ or } +25.7 \, k$$

Shear at section b–b:

$$V_{b-b} \text{ to left} = 25.7 - 20 - 15 = 9.3 \, k \downarrow = -9.3 \, k$$
$$V_{b-b} \text{ to right} = 16 - 25.3 = -9.3 \, k \uparrow = -9.3 \, k$$

The bending moments at sections a–a and b–b in the beam of Example 5.1 are computed as follows:

Moment at section a–a:

$$M_{a-a} \text{ to left} = (25.7)(5) = 128.5' \, k \circlearrowleft, \text{ or } +128.5' \, k$$
$$M_{a-a} \text{ to right} = (25.3)(35) - (16)(27) - (15)(15) - (20)(5)$$
$$= 128.5' \, k \circlearrowright, \text{ or } +128.5' \, k \quad \blacksquare$$

Moment at section b–b:

$$M_{b-b} \text{ to left} = (25.7)(25) - (20)(15) - (15)(5) = 267.5' \, k \circlearrowleft, \text{ or } +267.5' \, k$$
$$M_{b-b} \text{ to right} = (25.3)(15) - (16)(7) = 267.5' \, k \circlearrowright, \text{ or } +267.5' \, k$$

The Tridge, a triple-span pedestrian bridge using glued laminated timber, Midland, Michigan. (Courtesy of Unit Structures, Inc.)

Later in this textbook when methods are introduced for the analysis of rather complicated statically indeterminate structures, a more detailed sign convention is presented for axial forces, shears, and bending moments. This is particularly necessary when analysis is made using matrix methods. (See Chapters 22–25.)

5.2 SHEAR DIAGRAMS

Shear diagrams are quite simple to draw in most cases. The standard method is to start with the left end of the structure and work to the right. As each concentrated load or reaction is encountered, a vertical line is drawn to represent the quantity and direction of the force involved. Between the forces a horizontal line is drawn to indicate no change in shear.

Where uniform loads are encountered, the shear is changing at a constant rate per unit length and can be represented by a straight but inclined line on the diagram. When an ordinate on the shear diagram is above the line a positive shear is indicated because the sum of the forces to the left of that point is up. A shear diagram for a simple beam is drawn in Example 5.2.

EXAMPLE 5.2

Draw a shear diagram for the beam shown in Figure 5.3.

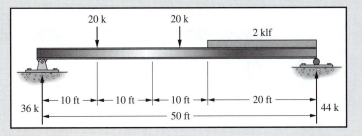

Figure 5.3

Solution.

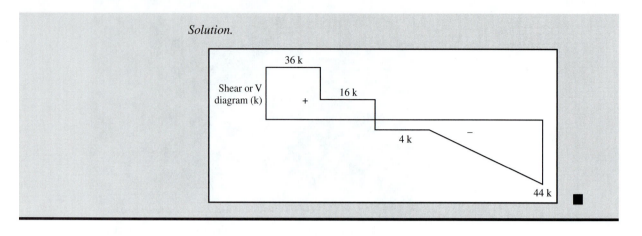

5.3 MOMENT DIAGRAMS

The moments at various points in a structure necessary for plotting a bending-moment diagram may be obtained algebraically by taking moments at those points, but the procedure is quite tedious if there are more than two or three loads applied to the structure. The method developed in the next section is much more practical.

5.4 RELATIONS AMONG LOADS, SHEARING FORCES, AND BENDING MOMENTS

There are significant mathematical relations among the loads, shears, and moments in a beam. These relations are discussed in the following paragraphs with reference to Figure 5.4.

For this discussion an element of a beam of length dx [Figure 5.4(a)] is considered. This particular element is loaded with a uniform load of magnitude w klf (it doesn't have to be uniform). The shear and moment at the left end of this element at section 1-1 may be written as follows:

$$V_{1-1} = R_A - P - wa$$

$$M_{1-1} = R_A x - P(a + b) - \frac{wa^2}{2}$$

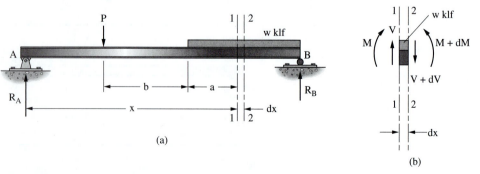

Figure 5.4

If we move a distance dx from section 1-1 to section 2-2 at the right end of the element, the new values of shear and moment can be written as at the end of this paragraph. The changes in these values may be expressed as dV and dM respectively.

$$V_{2-2} = V_{1-1} + dV = R_A - P - wa - w\,dx$$

$$M_{2-2} = M_{1-1} + dM = R_A x - P(a+b) - \frac{wa^2}{2} + V_{1-1}\,dx - \frac{w\,dx^2}{2}$$

From these expressions and with reference to part (b) of Figure 5.3, the changes in shear and moment in a dx distance are as follows:

$$\frac{dV}{dX} = -w$$

$$\frac{dM}{dX} = V \text{ neglecting the infinitesimal higher-order term involving } (dx)^2$$

From the preceding expressions the change in shear from one section to another can be written as follows (noting that some textbooks get rid of the minus sign by using a sign convention in which upward loads are given minus signs):

$$\Delta V = - \int_{1-1}^{2-2} w\,dx$$

And the change in moment in the same distance is

$$\Delta M = \int_{1-1}^{2-2} V\,dx$$

These two relationships are very useful to the structural designer. The first indicates that the rate of change of shear at any point equals the load per unit length at the point, meaning that the slope of the shear diagram at any point is equal to the load at that point. The second equation indicates that the rate of change of moment at any point equals the shear. This relation means that the slope of the bending-moment curve at any point equals the shear.

The procedure for drawing shear and moment diagrams, to be described in Section 5.5, is based on the above equations and is applicable to all structures regardless of loads or spans. Before the process is described, it may be well to examine the equations more carefully. A particular value of dV/dx or dM/dx is good only for the portion of the structure at which the function is continuous. For instance, in the beam of Example 5.4 the rate of change of shear from A to B equals the uniform load, 4 klf. At the 30-kip load, which is assumed to act at a point, the rate of change of shear and the slope of the shear diagram are infinite, and a vertical line is drawn on the shear diagram to represent a concentrated load. The rate of change of moment from A to B has been constant, but at B the shear changes decidedly, as does the rate of change of moment. In other words, an expression for shear or moment from A to B is not the same as the expression from B to C beyond the concentrated load. The equations of the diagrams are not continuous beyond such a point.

5.5 MOMENT DIAGRAMS DRAWN FROM SHEAR DIAGRAMS

The change in moment between two points on a structure has been shown to equal the shear between those points times the distance between them (dM = V dx); therefore, the change in moment equals the area of the shear diagram between the points.

The relationship between shear and moment greatly simplifies the drawing of moment diagrams. To determine the moment at a particular section, it is necessary only to compute the total area beneath the shear curve, either to the left or to the right of the section, taking into account the algebraic signs of the various segments of the shear curve. Shear and moment diagrams are self-checking. If they are initiated at one end of a structure, usually the left, and check out to the proper value on the other end, the work probably is correct.

The author usually finds it convenient to compute the area of each part of a shear diagram and record that value on the part in question. This procedure is followed for the examples in this text where the values enclosed in the shear diagrams are areas. These recorded values appreciably simplify the construction of the moment diagrams.

The rate of change of moment at a point has been shown to equal the shear at that point ($dM/dx = V$). Whenever the shear passes through zero, the rate of change of moment must be zero ($dM/dx = 0$), and the moment is at a maximum or a minimum. If the moment diagram is being drawn from left to right and the shear diagram changes from positive to negative, the moment will reach a positive maximum at that point. Beyond that point it begins to diminish as the negative shear area is added. If the shear diagram changes from negative to positive, the moment reaches a negative maximum and then begins to taper off as the positive shear area is added.

This theory, which indicates that maximum moment occurs where the shear is zero, is not always applicable. In some cases (at the end of a beam or at a point of discontinuity) the maximum moment can occur where the shear is not zero. Should a cantilevered beam be loaded with gravity loads, the maximum shear and the maximum moment both will occur at the fixed end.

When shear and moment diagrams are drawn for inclined members, the components of loads and reactions perpendicular to the centroidal axes of the members are used and the diagrams are drawn parallel to the members. Examples 5.3 to 5.5 illustrate the procedure for drawing shear and moment diagrams for ordinary beams. In studying these diagrams particular emphasis should be given to their shapes under uniform loads, between concentrated loads, and so on.

You will note that in the beam of Example 5.4 there is an axial tension force of 30 k for the entire length of the member. This force does not, however, affect the shear and moment in the beam.

EXAMPLE 5.3

Draw shear and moment diagrams for the beam shown in Figure 5.5.

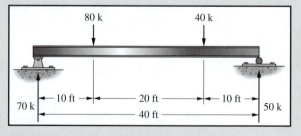

Figure 5.5

Solution.

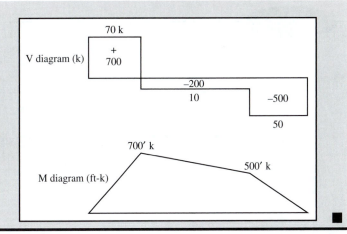

EXAMPLE 5.4

Draw shear and moment diagrams for the structure shown in Figure 5.6.

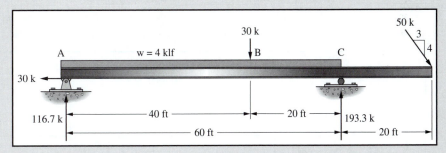

Figure 5.6

Solution.

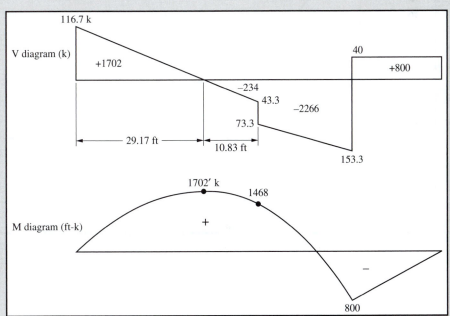

EXAMPLE 5.5

Draw shear and moment diagrams for the cantilever-type structure shown in Figure 5.7.

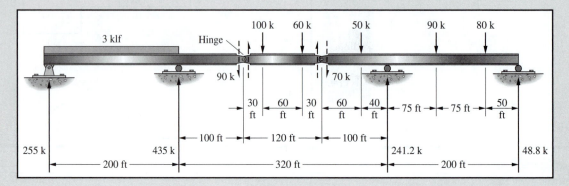

Figure 5.7

Solution.

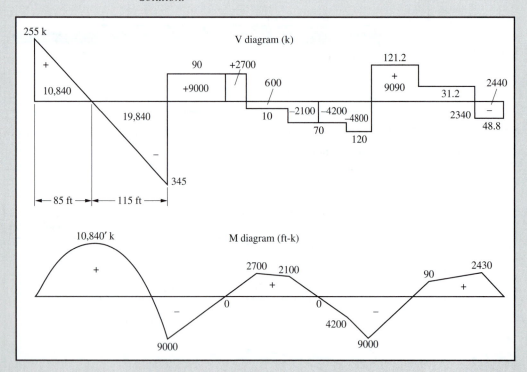

Pedestrian bridge in Pullen Park, Raleigh, North Carolina.
(Courtesy of the American Institute of Timber Construction.)

Some structures have rigid arms (such as walls) fastened to them. If horizontal or inclined loads are applied to these arms, a twist or moment will suddenly be induced in the structures at the points of attachment. The fact that moment is taken about an axis through the centroid of the section becomes important because the lever arms of the forces applied must be measured to that centroid. To draw the moment diagram at the point of attachment, it is necessary to figure the moment at an infinitesimal distance to the left of the point and then add the moment applied by the arm. The moment exactly at the point of attachment is discontinuous and cannot be figured, but the moment immediately beyond that point is available.

The usual sign convention for positive and negative moments will apply in deciding whether to add or subtract the induced moment. It can be seen in Figure 5.8 that forces to the left of a section that tend to cause clockwise moments produce tension in the bottom fibers (+ moment), whereas those forces to the left that tend to cause counterclockwise moments produce tension in the upper fibers (− moment). Similarly, a counterclockwise moment to the right of the section produces tension in the bottom fibers; a clockwise moment produces tension in the upper fibers.

Shear and moment diagrams are shown in Example 5.6 for a beam that has a moment induced at a point by a rigid arm to which a couple is applied. The moment diagram is drawn from left to right. Considering the moment of the forces to the left of a section through the beam immediately after the rigid arm is reached, it can be seen that the couple causes a clockwise or positive moment, and its value is added to the moment obtained by summation of the shear-diagram areas up to the attached arm.

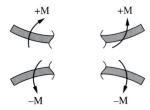

Figure 5.8

EXAMPLE 5.6

Draw shear and moment diagrams for the beam shown in Figure 5.9.

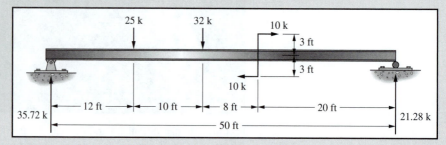

Figure 5.9

Solution.

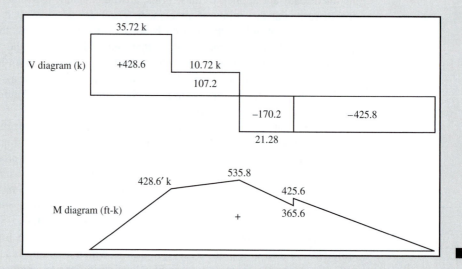

Students may at first have a little difficulty in drawing shear and moment diagrams for structures that are subjected to triangular loads. Example 5.7 is presented to demonstrate how to cope with them. The reactions are calculated for the beam shown in Figure 5.10 and the shear diagram is sketched. Note, however, that the point of zero shear is unknown and is shown in the figure as being located a distance x from the left support. The ordinate on the load diagram at this point is labeled y and can be expressed in terms of x by writing the following expression:

$$\frac{x}{30} = \frac{y}{2}$$

$$y = \frac{1}{15}x$$

At this point of zero shear the sum of the vertical forces to the left can be written as the upward reaction, 10 kips, minus the downward uniformly varying load to the left.

$$10 - \left(\frac{1}{2}\right)(x)\left(\frac{1}{15}x\right) = 0$$
$$x = 17.32 \text{ ft}$$

This value of x also can be determined by writing an expression for moment in the beam at a distance x from the support and then taking $dM/dx = 0$ of that expression and solving the result for x.

Finally, the moment at a particular point can be determined by taking (to the left or right of the point) the sum of the forces times their respective lever arms. For this example the moment at 17.32 ft is

$$M = (10)(17.32) - (10)\left(\frac{17.32}{3}\right) = 115.5' \text{ k}$$

The student may wonder why the author (once x was determined) did not just sum up the area under the shear diagram from the left support to the point of zero shear. Such a procedure is correct, but the designer must be sure that he or she determines the area properly. When partial parabolas are involved it may be necessary to determine the areas by calculus instead of with the standard parabolic formulas. As a result, it may be simpler on many occasions simply to take moments of the loads and reactions about the points where moments are desired.

In Chapter 11, Figure 11.7 presents the properties (centers of gravity and areas) of several geometric figures. These values may be quite useful to the student preparing shear and moment diagrams for complicated loading situations.

EXAMPLE 5.7

Draw the shear and moment diagrams for the beam shown in Figure 5.10.

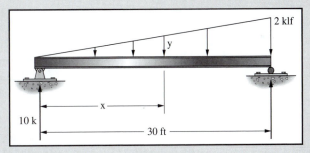

Figure 5.10

Solution.

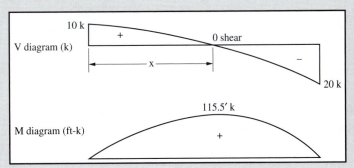

The reactions for the beam of Example 5.8 cannot be obtained using only the equations of statics. They have been computed by a method discussed in a later chapter, and the shear and moment diagrams have been drawn to show that the load, shear, and moment relationships are applicable to all structures. ∎

Draw the shear and moment diagrams for the continuous beam shown in Figure 5.11 for which the reactions are given.

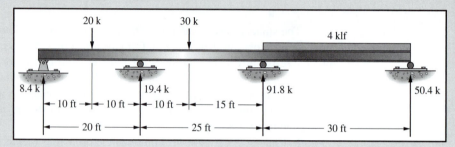

Figure 5.11

Solution.

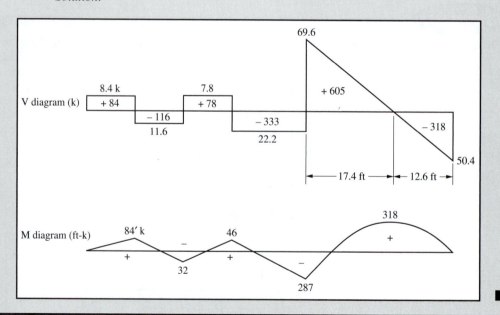

5.6 SHEAR AND MOMENT DIAGRAMS FOR STATICALLY DETERMINATE FRAMES

Shear and moment diagrams are very useful for rigid frames as well as for individual flexural members. The members of such frames cannot rotate with respect to each other at their connections. As a result, axial forces, shear forces, and bending moments are transferred between the members at the joints. These values must be accounted for in the preparation of the shear and moment diagrams.

For a first example, the frame of Example 5.9 shown in Figure 5.12 is considered. After the reactions are computed free-body diagrams are prepared for each of the members. The forces involved at the bottom of the column are obviously the reactions. Those at the top represent the effect of the rest of the frame on the top of the column and can be easily obtained by statics. Finally, the shear and moment diagrams are prepared.

EXAMPLE 5.9

Draw shear and moment diagrams for the frame shown in Figure 5.12.

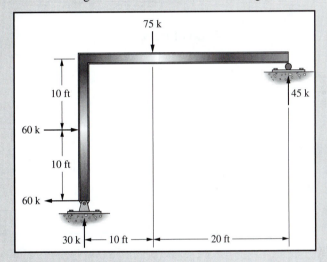

Figure 5.12

Solution.

Free Body Diagrams

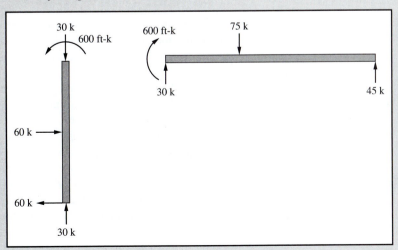

Shear Diagrams

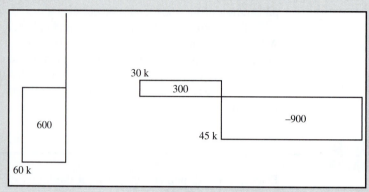

Moment Diagrams

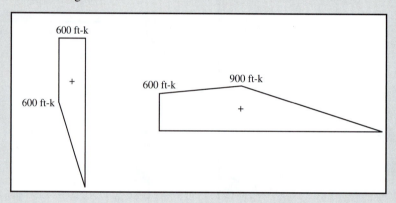

For a sign convention for this first example the author assumed that he was standing underneath the frame. Thus, the right-hand side of the column was assumed to be the bottom side. ■

Shear and moment diagrams are drawn in Example 5.10 for a frame with two columns. Here again, the author assumed that he was standing underneath the frame; therefore, the insides of the columns are the bottom sides so far as sign convention goes.

Should a frame have more than two columns, the sign convention we have been using will cause us to be confused in deciding which is the bottom side of the columns. For such a situation we can simply assume one side of all the columns (say the right-hand side) is the bottom side. This, however, will cause us to have sign convention problems at the tops of all the columns, except for the far left-hand one. That is, we will have to change sign convention as we move from the beam to the top of all the other columns. Perhaps a better procedure for multiple-column frames is simply to leave the signs off of the moment diagrams and to draw the moment diagrams on the compression sides of all the members: The results for horizontal members will be the same as previously obtained.

YKK USA Building, Dublin, Georgia (Courtesy of Britt, Peters and Associates)

Free body diagrams were not shown for the various members of this frame, as the author felt that the reader would now be capable of drawing the shear and moment diagrams directly from the external loads and reactions applied to the frame.

EXAMPLE 5.10

Draw shear and moment diagrams for the frame shown in Figure 5.13.

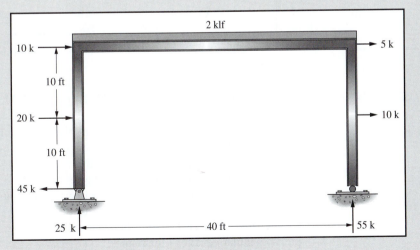

Figure 5.13

Solution. (Assuming insides of columns are bottom sides)

Shear Diagrams

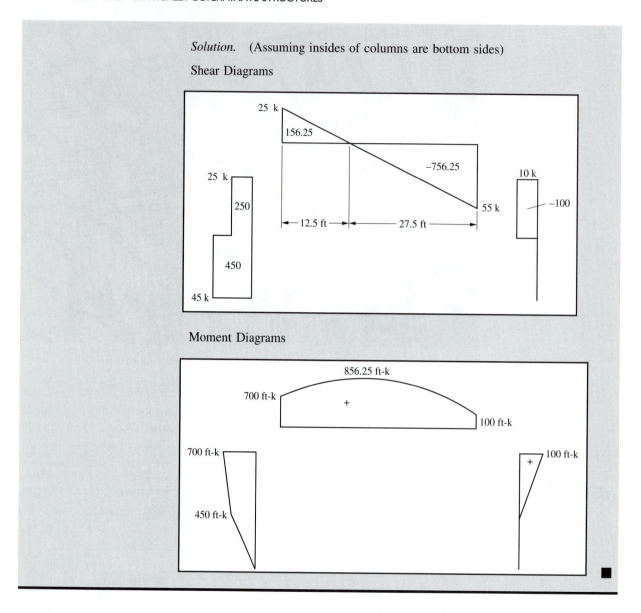

Moment Diagrams

5.7 SHEARING FORCE AND BENDING MOMENT EQUATIONS

An important part of structural engineering is the writing or derivation of equations for the shears and bending moments in different parts of a structure. The ability to write these equations is extremely important for the understanding of much of the information discussed later in this text. For instance, several of the procedures used for the determination of slopes and deflections of various points in structures require us to be able to write these types of equations. In this section, shear and moment expressions are prepared for beams. Using the same procedure, the student will easily see how to prepare similar equations for frames.

In Example 5.11, equations for shear and moment are prepared for all portions of the beam shown in Figure 5.14. Positions in the beam are located by the distance x,

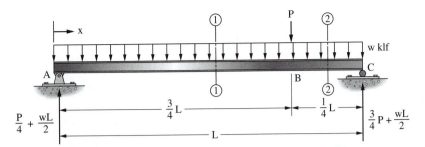

Figure 5.14

which is measured in this example from the left end of the beam. Actually, the origin of x can be moved as desired to points other than the left end during the solution. The reactions for the beam end supports have already been determined.

For the first part of the solution, section 1-1 has been passed through the beam and a free body drawn for the part of the beam to the left of the section. This free body is shown in Figure 5.15.

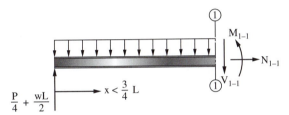

Figure 5.15   Free body to left of section 1-1 for the beam of Figure 5.14

As we move along the beam from left to right, we see that the equations we have prepared are good until a different load situation is encountered. This occurs when the concentrated load P is reached. Another section, 2-2, is passed through the member between the concentrated load and the right end and equations are prepared for that region of the beam.

Every time the load situation along a beam changes, it is necessary to write new equations. For some beams it will be necessary to write equations for several portions of the structure but only two sets are required in this example.

EXAMPLE 5.11

Prepare equations for shear and moment over the entire span of the beam shown in Figure 5.14. Use the left end of the beam as the reference point or origin for x for all of the equations.

Solution.

Considering the forces to the left of section 1-1 as shown in Figure 5.15. For $x = 0$ to $\frac{3}{4}L$ (that is from A to B).

$$V = +\frac{P}{4} + \frac{wL}{2} - wx$$

$$M = \left(\frac{P}{4} + \frac{wL}{2}\right)(x) - (wx)\left(\frac{x}{2}\right)$$

Considering the forces to the left of section 2-2.

For $x = \frac{3}{4}L$ to L (or from B to C)

$$V = \frac{P}{4} + \frac{wL}{2} - wx - P$$

$$M = \left(\frac{P}{4} + \frac{wL}{2}\right)(x) - (wx)\left(\frac{x}{2}\right) - P\left(x - \frac{3}{4}L\right) \quad \blacksquare$$

Other beams are dealt with in exactly the same manner as the beam in the example. We cut a free-body in each region after a change in load, write the equations of equilibrium for that free-body, and solve the equations of equilibrium for the shearing force and bending moment.

5.8 PROBLEMS FOR SOLUTION

Draw shearing force and bending moment diagrams for the structures in Problems 5.1 to 5.35.

5.1   (*Ans.* max V = 64 k, max M = 1120 ft-k)

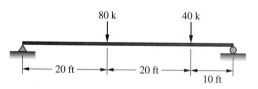

5.2   Repeat Problem 5.1 if the 80 k load is changed to 120 k.

5.3   (*Ans.* max V = 80 k, max M = −600 ft-k)

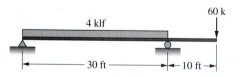

5.4   Repeat Problem 5.3 if the 4-klf load is changed to 2.5 klf.

5.5   (*Ans.* max V = 65 k, max M = −337.5 ft-k)

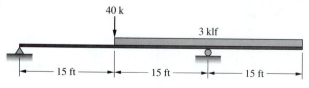

5.6

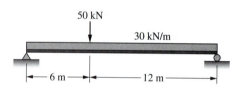

5.7   (*Ans.* max V = 64 k, max M = −1706.9 ft-k)

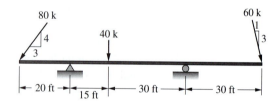

5.8

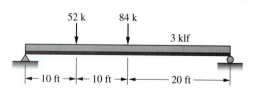

5.9   (*Ans.* max V = 120 k, max M = −1200 ft-k)

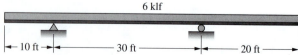

5.10

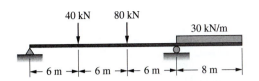

5.11   (*Ans.* max V = 145.6 ft-k, max M = −1458.5 ft-k)

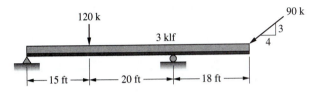

5.12

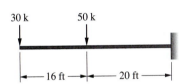

5.13   (*Ans.* max V = 204 k, max M = −4032 ft-k)

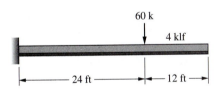

5.14

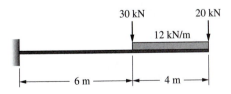

5.15   (*Ans.* max V = 90 k, max M = −900 ft-k)

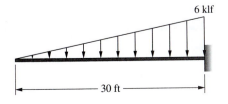

5.16

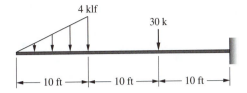

5.17   (*Ans.* max V = 210 k, max M = −3762.5 ft-k)

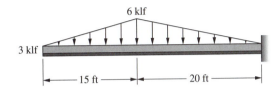

5.18

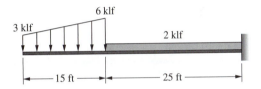

5.19   (*Ans.* max V = 53.93 k, max M = 482.1 ft-k)

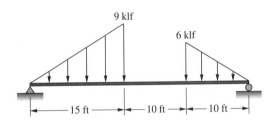

5.20

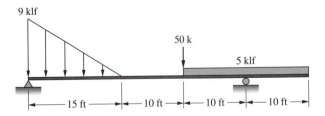

5.21   (*Ans.* max V = 70 k, max M = 453.1 ft-k)

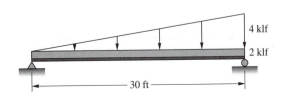

5.22

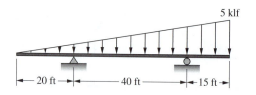

5.23 (*Ans.* max V = 317.86 kN, max M = 2526 kN·m)

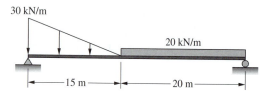

5.24

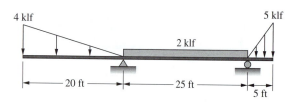

5.25 (*Ans.* max V = 118.75 k, max M = −1510.4 ft-k)

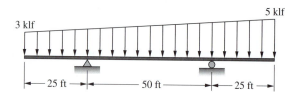

5.26

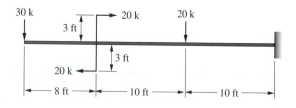

5.27 (*Ans.* max V = 29.6 k, max M = 109.5 ft-k)

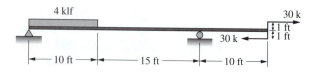

5.28

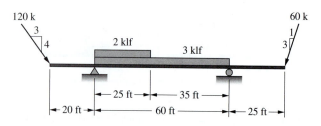

5.29 (*Ans.* max V = 140 k, max M = −2867.9 ft-k)

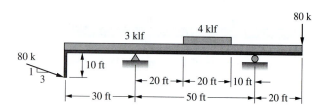

5.30

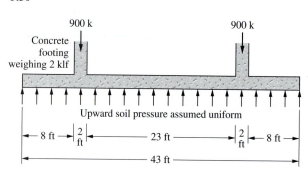

5.31 (*Ans.* max V = 96.67 k, max M = ±2933 ft-k)

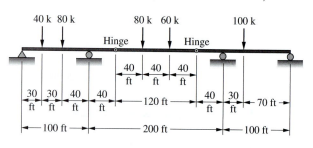

5.32

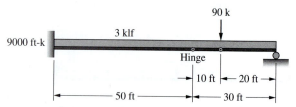

5.33   (*Ans.* max V = 105 k, max M = −3400 ft-k)

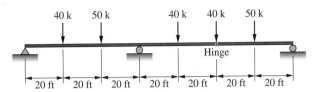

5.34   Given: Moment at interior support is −1252 ft-k.

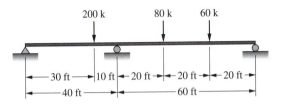

5.35   Given: Moment at fixed end is −147.3 ft-k; other reactions are shown in figure. (*Ans.* max V = 25.49 k, max M = −164.5 ft-k) (*Ans.* max V = 23.75 k, max M = −164.5 ft-k)

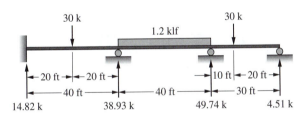

For Problems 5.36 to 5.38 draw the shear diagrams and load diagrams for the moment diagrams and dimensions given. Assume that upward forces are reactions.

5.36

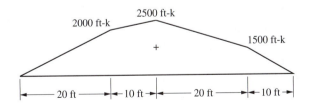

5.37   (*Ans.* Reactions and loads from left to right = 108 k, 93.33 k, 56.67 k, etc.)

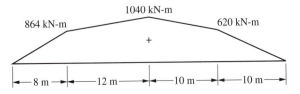

5.38

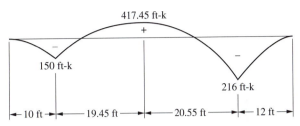

For Problems 5.39 to 5.47 draw shear and moment diagrams for the frames.

5.39   (*Ans.* max V = 103.33 k, max M = 1779.8 ft-k)

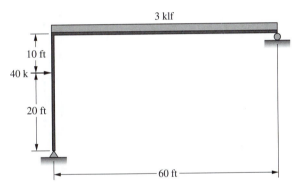

5.40

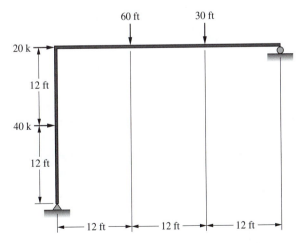

5.41   Repeat Problem 5.40 if the roller and hinge supports are swapped. (*Ans.* max V = 63.33 k, max M = 480 ft-k)

5.42

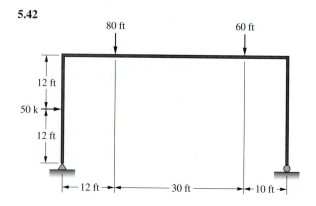

5.43 (*Ans.* max V = 106.67 k, max M = ±900 ft-k)

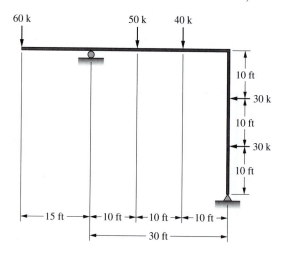

5.44

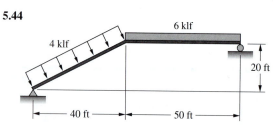

5.45 (*Ans.* max V = 112 k, max M = 1768 ft-k)

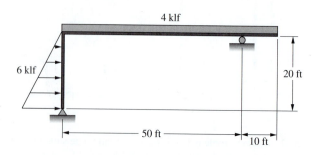

5.46

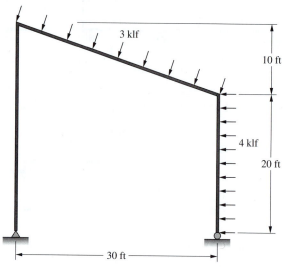

5.47 (*Ans.* max V = 67.5 ft-k, max M = −675 ft-k)

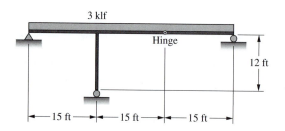

For Problems 5.48 to 5.52 write equations for shear and bending moment throughout the structures shown.

5.48 Prob. 5.1.

5.49 Prob. 5.8 (*Ans.* For x = 0 to 10 ft @ left end of beam: V = −3x + 141, M = −1.5x² + 141x)

5.50 Prob. 5.12.

5.51 Prob. 5.16 (*Ans.* For x = 0 to 10 ft @ left end: V = −0.2x², M = −0.0667x³)

5.52 Prob. 5.39.

Chapter 6

Introduction to Plane Trusses

6.1 INTRODUCTION

The Italian architect Andrea Palladio (1508–1580) is thought to have first used modern trusses, although his design basis is not known. He may have revived some old Roman designs and probably sized the truss components by some rules of thumb (perhaps old Roman rules). Palladio's extensive writing in architecture included detailed descriptions and drawings of wooden trusses quite similar to those used today. After his time, trusses were forgotten for 200 years, until they were reintroduced by Swiss designer Ulric Grubermann.

A truss is a structure formed by a group of members arranged in the shape of one or more triangles. Because the members are assumed to be connected with frictionless pins, the triangle is the only stable shape. Study of the truss of Figure 6.1(a) shows that it is impossible for the triangle to change shape under load—except through deformation of the members—unless one or more of the sides is bent or broken. Figures of four or more sides are not stable and may collapse under load, as seen in Figures 6.1(b) and 6.1(c). These structures may be deformed even if none of the members change length. We will see, however, that many stable trusses can include one or more shapes that are not triangles. Careful study reveals that trusses consist of separate groups of triangles that are connected together according to definite rules, forming non-triangular but stable figures in between.

Design engineers often are concerned with selecting either a truss or a beam to span a given opening. Should no other factors be present, the decision probably would be based on consideration of economy. The smallest amount of material will nearly always be used if a truss is selected for spanning a certain opening; however, the cost of fabrication and erection of trusses probably will be appreciably higher than that required for beams. For shorter spans, the overall cost of beams (material cost plus fabrication and erection cost)

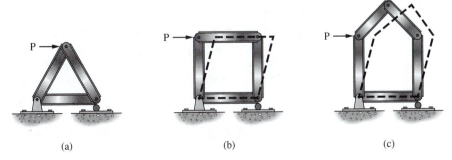

(a)	(b)	(c)

Figure 6.1 Stable truss versus unstable figures

will definitely be less, but as the spans become greater, the higher fabrication and erection costs of trusses will be more than offset by the reduction in the total weight of material used. A further advantage of a truss is that for the same amount of material it can have greater stiffness than a beam with the same span.

A lower limit for the economical span of trusses is impossible to give. Trusses may be used for spans as small as 30 to 40 ft and as large as 300 to 400 ft. Beams may be economical for some spans much greater than the lower limits mentioned for trusses.

6.2 ASSUMPTIONS FOR TRUSS ANALYSIS

The following assumptions are made to simplify the analysis of trusses:

1. *Truss members are connected with frictionless pins.* In reality, pin connections are used for very few trusses erected today, and no pins are frictionless. A heavy bolted or welded joint is very different from a frictionless pin.
2. *Truss members are straight.* If they were not straight, the axial forces would cause them to have bending moments.
3. *The displacement of the truss is small.* The applied loads cause the members to change length, which then causes the truss to deform. The deformations of a truss are not of sufficient magnitude to cause appreciable changes in the overall shape and dimensions of the truss. Special consideration may have to be given to some very long and flexible trusses.
4. *Loads are applied only at the joints.* Members are arranged so that the loads and reactions are only applied at the truss joints.

Examination of roof and bridge trusses will prove this last statement to be generally true. Beams, columns, and bracing members frame directly into the truss joints of buildings with roof trusses. Roof loads are transferred to trusses by horizontal beams, called *purlins*, which span between the trusses. The roof is supported directly by the purlins. The roof may also be supported by rafters, or sub-purlins, which run parallel to trusses and are supported by the purlins. The purlins are placed at the truss joints unless the top-chord panel lengths become exceptionally long, in which case it is sometimes economical to place purlins between the joints, although some bending will develop in the top chords. Some types of roofing, such as corrugated steel and gypsum slabs, may be laid directly on the purlins. In this case, the purlins then have to be spaced at intermediate points along the top chord to provide a proper span for the supported roof. Similarly, the loads supported by a highway bridge are transferred to the trusses at the joints by floor beams and girders underneath the roadway.

The effect of the foregoing assumptions is to produce an ideal truss, whose members have only axial forces. As illustrated in Figure 6.2(a) and (b), a member with only axial force is subjected to axial tension or compression; there is no bending moment present as shown in Figure 6.2(c). Be aware, however, that even if all the assumptions about trusses were completely true, there still would be some bending in a member because of its own weight. The weight of the member is distributed along its length rather than being concentrated at the ends. Compared to the forces caused by the applied loads, the forces caused by self-weight are small and generally can be neglected when calculating the forces in the components.

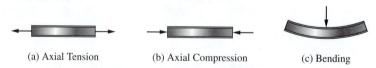

(a) Axial Tension   (b) Axial Compression   (c) Bending

Figure 6.2   Forces acting on member

Component forces obtained using some or all of these simplifying assumptions are very satisfactory in most cases, and are referred to as *primary forces*. Forces caused by conditions not considered in the primary force analysis are said to be *secondary forces*.

Truss for a conveyor at the West Virginia quarry of Pennsylvania Glass Sand Corporation (Courtesy Bethlehem Steel Corporation)

6.3 TRUSS NOTATION

A common system of denoting the members of a truss is shown in Figure 6.3. The joints are numbered from left to right. Those on the bottom are labeled L, for lower. The joints on the top are labeled U, for upper. Should there be, in more complicated trusses, joints between the lower and upper joints, they may be labeled M, for middle.

The various members of a truss often are referred to by the following names, as labeled in Figure 6.3.

1. *Chords* are those members forming the outline of the truss, such as members U_1U_2 and L_4L_5.
2. *Verticals* are named on the basis of their direction in the truss, such as members U_1L_1 and U_3L_3.
3. *Diagonals* also are named on the basis of their direction in the truss, such as members U_1L_2 and L_4U_5.
4. *End posts* are the members at the ends of the truss, such as members L_0U_1 and U_5L_6.

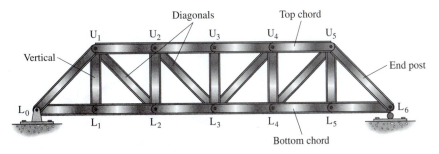

Figure 6.3   Common labeling joints of a truss

5. *Web members* include the verticals and diagonals of a truss. Most engineers consider the web members to include the end posts.

There are other notation systems frequently used for trusses. For instance, for computer-programming purposes, it is convenient to assign a number to each joint and each member of a truss. Such a system is illustrated in Section 6.11 of this chapter.

6.4 ROOF TRUSSES

The purposes of roof trusses are to support the roofs that keep out the elements (rain, snow, wind), the loads connected underneath (ducts, piping, ceiling), and their own weight.

Roof trusses can be flat or peaked. In the past, peaked roof trusses probably have been used more for short-span buildings and flatter trusses for longer spans. The trend today for both long and short spans, however, seems to be away from the peaked trusses and toward the flatter trusses. The change is predominantly due to the desired appearance of the building, and perhaps a more economical construction of roof decks. Figure 6.4 illustrates several types of roof trusses that have been used in the past.

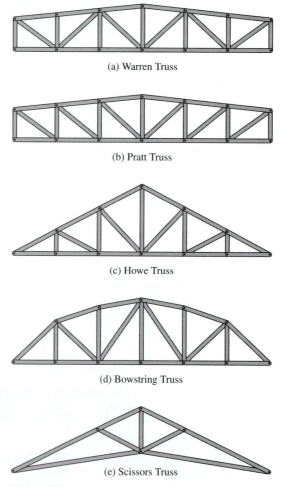

(a) Warren Truss

(b) Pratt Truss

(c) Howe Truss

(d) Bowstring Truss

(e) Scissors Truss

Figure 6.4 Several typical roof trusses

6.5 BRIDGE TRUSSES

As bridge spans become longer and loads heavier, trusses begin to be economically competitive with beams. Early bridge trusses were constructed with wood, and they had several disadvantages. First, they were subject to deterioration from wind and water. As a result covered bridges were introduced, and such structures often would last for quite a few decades. Nevertheless, wooden truss bridges, particularly railroad bridges, were subject to destruction by fire. In addition the movement of traffic loads back and forth across the bridge spans could cause a gradual loosening of the fasteners.

Willard Bridge over the Kansas River north of
Willard, Kansas (Courtesy of the American Institute
of Steel Construction, Inc.)

Because of the several preceding disadvantages, wooden truss bridges faded from use toward the end of the 19th century. Although there were some earlier iron truss bridges, structural steel bridges became predominant. Steel bridges did not require extensive protection from the elements and their joints had higher fatigue resistance.

Today existing steel truss bridges are being steadily replaced with steel, precast-concrete, or prestressed-concrete beam bridges. The age of steel truss bridges appears to be over, except for bridges with spans of more than several hundred feet, which represent a very small percentage of the total. Even for these longer spans, there is much competition from other types of structures such as cable-stayed bridges and prestressed-concrete box girder bridges.

Some highway bridges have trusses on their sides and overhead lateral bracing between the trusses. This type of bridge is called a *through bridge.* The floor system is supported by floor beams that run under the roadway and between the bottom-chord joints of the trusses. A through railroad bridge is shown in Figure 6.5(a).

(a)   (b)

Figure 6.5   A through bridge (a) and deck bridge (b)

In a *deck bridge*, the roadway is placed on top of the trusses or girders. Deck bridges have many advantages over through bridges: there is unlimited overhead horizontal and vertical clearance, future expansion is more feasible, and supporting trusses or girders can be placed close together, which reduces lateral moments in the floor system. Other advantages of the deck truss are simplified floor systems and possible reduction in the sizes of piers and abutments due to reductions in their heights. Finally, the very pleasing appearance of deck structures is another reason for their popularity. The only real disadvantage of a deck bridge is clearance beneath the bridge. The bridge may need to be set high to allow adequate clearance for ships and vehicles to pass underneath. A deck railroad bridge is shown in Figure 6.5(b). Deck bridges eliminate a sense of confinement exhibited by other types of bridges.

Sometimes short-span through-bridge trusses were so shallow that adequate depth was not available to provide overhead bracing and at the same time leave sufficient vertical clearance above the roadway for the traffic. As a result, the bracing was placed underneath the roadway. Through bridges without overhead bracing are referred to as *half-through* or *pony bridges*. One major problem with pony truss bridges is the difficulty of providing adequate lateral bracing for the top-chord compression members of the trusses. Pony trusses are not likely to be economical today because beams have captured the short-span bridge market.

Shown in Figure 6.6 are several types of bridge trusses. Parallel-chord trusses are shown in parts (a) through (d) of the figure. The Baltimore truss is said to be a subdivided truss because the unsupported lengths of some of the members have been reduced by short members called sub-diagonals and sub-verticals. The deeper a truss is made for the same-size chord members, the greater will be its resisting moment. If the depth of a truss is varied across its span in proportion to its bending moments, the result will be a lighter truss. The fabrication cost-per-pound of steel will be higher, though, than for a parallel-chord truss. As spans become longer, the reduced weight achieved by varying truss depths with bending moments will outweigh the extra fabrication costs, thus making economical the so-called curved-chord trusses. The Parker truss shown in Figure 6.6(e) is an example.

6.6 ARRANGEMENT OF TRUSS MEMBERS

The triangle has been shown to be the basic shape from which trusses are developed because it is the only stable shape. For the following discussion, remember that the members of trusses are assumed to be connected at their joints with frictionless pins.

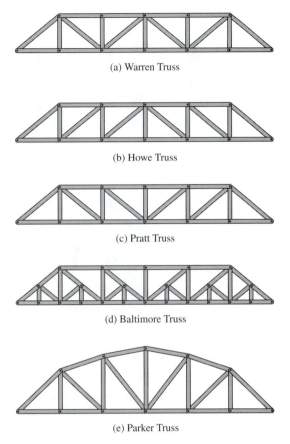

(a) Warren Truss

(b) Howe Truss

(c) Pratt Truss

(d) Baltimore Truss

(e) Parker Truss

Figure 6.6   Some typical bridge trusses

Other shapes, such as those shown in Figure 6.7(a) and (b) are obviously unstable, and may possibly collapse under load. Structures such as these, however, can be made stable by one of the following methods:

- Add members so that the unstable shapes are altered to consist of triangles. The structures of Figure 6.7(a) and (b) are stabilized in this manner in (c) and (d), respectively.
- Use a member to tie the unstable structure to a stable support. Member AB performs this function in Figure 6.7(e).
- Make rigid some or all of the joints of an unstable structure, so they become moment resisting. A figure with moment-resisting joints, however, does not coincide with the definition of a truss; the members are no longer connected with frictionless pins.

6.7 STATICAL DETERMINACY OF TRUSSES

The simplest form of truss, a single triangle, is illustrated in Figure 6.8(a). To determine the unknown forces and reaction components for this truss, it is possible to isolate the joints and write two equations of equilibrium for each. These equations of equilibrium,

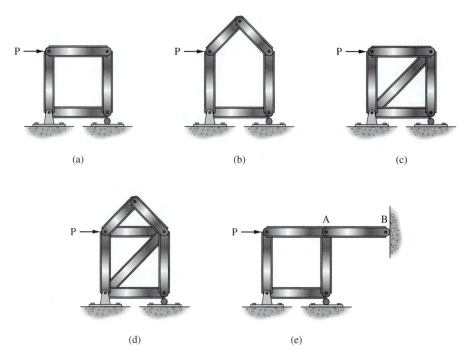

(a)   (b)   (c)

(d)   (e)

Figure 6.7   Stable and unstable arrangement of truss members

which involve the summation of vertical and horizontal forces, are:

$$\sum F_H = 0$$
$$\sum F_V = 0$$

The single-triangle truss may be expanded into a two-triangle truss by the addition of two new members and one new joint. In Figure 6.8(b), triangle ABD is added by installing new members AD and BD, and the new joint D. A further expansion with a third triangle is made in Figure 6.8(c) by the addition of members BE, DE and joint E. For each of the new joints, D and E, a new pair of equations of equilibrium is available for calculating the two new-member forces. As long as this procedure of expanding the truss is followed, the truss will be statically determinate internally. Should new members be installed without adding new joints, such as member CE in Figure 6.8(d), the truss will become statically indeterminate because no new equations of equilibrium are available to find the new-member forces.

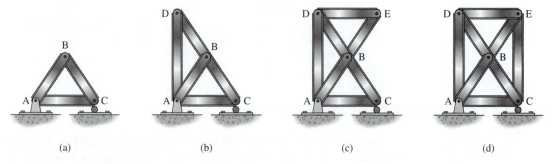

(a)   (b)   (c)   (d)

Figure 6.8   Expanding a simple truss

Brown's Bridge, Forsyth and Hall Counties, Gainesville, Georgia
(Courtesy of the American Institute of Steel Construction)

Using this information, an expression can be written for the relationship that must exist between the number of joints, the number of members, and the number of reactions for a particular truss if it is to be statically determinate internally. The identification of externally determinate structures has previously been discussed in Chapter 4. In the following discussion, m is the number of members, j is the number of joints, and r is the number of reaction components.

If the number of available equations of static equilibrium, which is 2j, is sufficient to compute the unknown forces, the structure is statically determinate. As such, the following relation may be written:

$$2j = m + r$$

This equation is more commonly written as

$$m = 2j - r$$

For this equation to be applicable, the structure must be stable externally; otherwise the results are meaningless. Therefore, r is the least number of reaction components required for external stability. Should the structure have more external reaction components than necessary for stability, and thus be statically indeterminate externally, the value of r remains the least number of reaction components required to make it stable externally. This statement means that r is equal to 3 for the usual equations of static equilibrium, plus the number of any additional equations of condition that may be available.

It is possible to build trusses that have more members than can be analyzed using equations of static equilibrium. Such trusses are statically indeterminate internally, and m will exceed 2j − r because there are more members present than are necessary for stability. The extra components are said to be *redundant members*. If m is 3 greater than 2j − r, there are three redundant members, and the truss is internally statically indeterminate to the third degree. Should m be less than 2j − r, there are not enough members present for stability.

A brief glance at a truss usually will show if it is statically indeterminate. Trusses having members that cross over each other or members that serve as the sides for more than two triangles may quite possibly be indeterminate. The 2j − r expression should be used, however, if there is any doubt about the determinacy of a truss because it is not difficult to be mistaken. Figure 6.9 shows several trusses and the application of the expression to each. The small circles on the trusses indicate the joints.

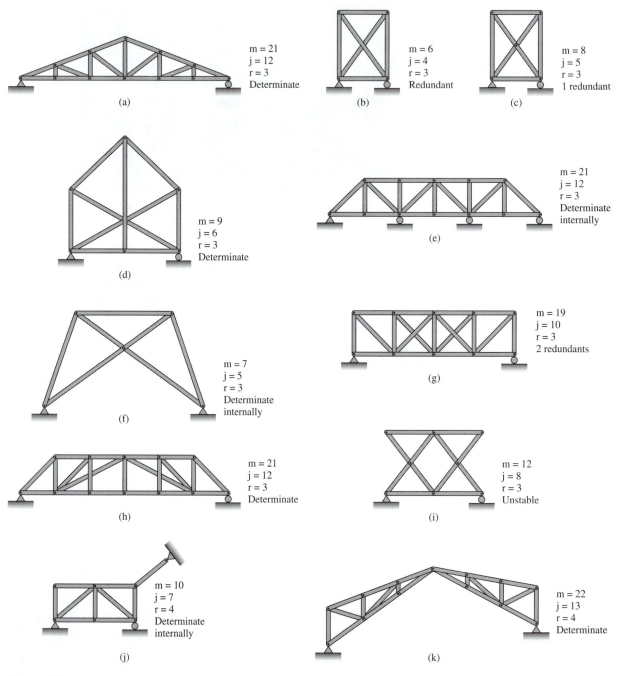

Figure 6.9

Little explanation is necessary for most of the structures shown, but some remarks may be helpful for a few. The truss of Figure 6.9(e) has five reaction components and is statically indeterminate externally to the second degree; however, two of the reaction components could be removed and leave a structure with sufficient reactions for stability.

The least number of reaction components required for stability is 3, m is 21, and j = 12. Applying the equation we obtain

$$m = 2j - r$$
$$21 = (2)(12) - 3$$
$$21 = 21 \qquad \therefore \text{ It is statically determinate internally.}$$

The truss of Fig. 6.9(j) is externally indeterminate because there are five reaction components and only four equations available. With r = 4 the structure is shown to be statically determinate internally. The three-hinged arch of Fig. 6.9(k) has four reaction components, which is the least number of reaction components required for stability; so r = 4. Application of the equation shows the arch to be statically determinate internally.

In Chap. 15 pertaining to the analysis of statically indeterminate structures, it will be seen that the values of the redundants may be obtained by applying certain simultaneous equations. The number of simultaneous equations equals the total number of redundants, whether internal, external, or both. It therefore may seem a little foolish to distinguish between internal and external determinacy. The separation is particularly questionable for some types of internally and externally redundant trusses where no solution of the reactions is possible independently of the member forces, and vice versa.

Nevertheless, if a truss is externally determinate and internally indeterminate, the reactions may be obtained using equations of static equilibrium. If the truss is externally indeterminate and internally determinate, the reactions are dependent on the internal-member forces and may not be determined by a method independent of those forces. If the truss is both externally and internally indeterminate, the solution of the member forces and reactions must be performed simultaneously. For any of these situations, it may be possible to obtain a few forces here and there by statics without going through the indeterminate procedure necessary for complete analysis. This subject is discussed in detail in later chapters.

6.8 METHODS OF ANALYSIS AND CONVENTIONS

An indispensable and essential tool in truss analysis is the ability to divide the truss into pieces, construct a free-body diagram for each piece, and then from these free-body diagrams determine the forces in the components. The free-body diagrams can be single joints in the truss or a large part of the truss. As an example, the truss of Figure 6.10 is considered. First a section is drawn completely around joint U_4. Then a vertical section 1-1 is passed through the second panel of the truss and the free-body to its left is considered. These sections are shown in Figure 6.10(a) and the free-bodies developed are shown in parts (b) and (c) of the same figure. When working with free-body diagrams such as the one shown in Figure 6.10(b), we are said to be evaluating the member forces using the method of joints. On the other hand, if we evaluate the member forces using free-body diagrams such as that which is shown in Figure 6.10(c), we are said to be using the method of sections. The methods of joints will be considered in this chapter and the method of sections will be discussed in the next chapter.

Applying the equations of static equilibrium to isolated free-bodies enables us to determine the forces in the cut members. The free-bodies must be carefully selected so that the sections do not pass through too many members whose forces are unknown. When using the method of joints, there are only two relevant equations of equilibrium for each free-body: summation of forces vertically and summation of forces horizontally. There are three applicable equations of equilibrium for each free-body when using the

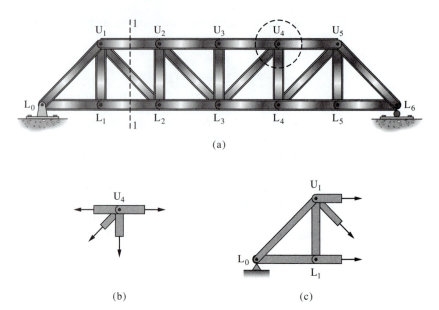

Figure 6.10 Free-body diagrams of a truss joint and a truss section

method of sections: summation of forces vertically, summation of forces horizontally, and summation of moments. On any free-body we use, there cannot be more unknown forces than there are relevant equations of static equilibrium.

After you have analyzed a few trusses, you will have little difficulty in most cases in selecting satisfactory locations for the sections. You are not encouraged to remember specific sections for specific trusses, although you will probably fall into such a habit unconsciously as time passes. At this stage, you need to consider each case individually without reference to other similar trusses.

Good practice dictates that a sign convention be adopted when analyzing trusses, and that this convention be used consistently for all trusses. By doing so, many errors inherent with changing sign conventions are eliminated. The author uses the sign convention that all unknown forces in a truss are assumed to be tensile forces. Further, tensile and compressive forces are indicated with a plus or minus sign (+ and −), respectively. This convention will be demonstrated through the example problems.

Arrows are often used to represent the character of forces. The arrows indicate what members are doing to resist the axial forces applied to them by the remainder of the truss. For example, if a particular member in a truss is acting in compression, it is pushing against the joints to which it is connected. Conversely, if a member is acting in tension, it is pulling on the joints to which it is connected. Arrows superimposed on the member indicate the action of the member on the joints as indicated in Figure 6.11.

After some practice analyzing trusses, the character of the forces in many of the members can be determined by examination. Try to picture whether a member is in tension or compression before making actual calculations: by doing so you will achieve a better understanding of the action of trusses under load. The next section demonstrates that it is possible to determine entirely by mathematical means the character as well as the numerical value of the forces.

This member is acting in tension.
(At the joints the member is being stretched and is pulling back.)

This member is acting in compression.
(At the joints the member is being compressed and is pushing back.)

Figure 6.11   Arrows indicating a member's action on its joints

6.9 METHOD OF JOINTS

An imaginary section may be passed around a joint in a truss, regardless of its location, completely isolating it from the remainder of the truss. The joint has become a free-body in equilibrium under the forces applied to it. The applicable equations of equilibrium, $\Sigma F_H = 0$ and $\Sigma F_V = 0$, may be applied to the joint to determine the unknown forces in the members meeting there. It should be evident that no more than two unknowns can be determined at a joint with these two equations.

A student learning the method of joints may initially find it necessary to draw a free-body sketch for every joint in a truss he or she is analyzing. After you have computed the forces in two or three trusses, it will be necessary to draw the diagrams for only a few joints, because you will be able to easily visualize the free-bodies involved. Drawing large sketches helps visualization. The most important thing for you to remember is that you are interested in only one joint at a time. Keep your attention away from the loads and forces at other joints. Your concern is only with the forces at the one joint on which you are working. The method of joints is illustrated through the examples that follow.

EXAMPLE 6.1

Using the method of joints, find all forces in the truss shown in Figure 6.12.

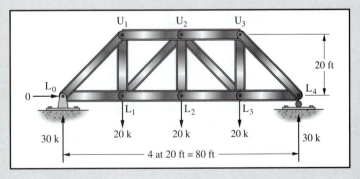

Figure 6.12

Solution.   When applying the methods of joints, there can be only two unknown bar forces at each joint since there are only two equations of equilibrium for each joint—the summation of forces horizontally and summation of forces vertically. The free-body diagram used with each joint is shown along with the equations of equilibrium. Beginning with joint L_0:

$$\sum V = 0$$
$$30 - V_{U_1 L_1} = 0$$
$$V_{L_0 U_1} = 30 \text{ k compression}$$

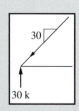

An examination of the joint shows a vertical reaction of 30 k acting upward. The equation $\sum V = 0$ indicates the members meeting there must supply 30 k downward. A member that is horizontal, such as $L_0 L_1$, can have no vertical component of force; therefore, $L_0 U_1$ must supply the entire amount and 30 k will be its vertical component. The arrow convention shows that $L_0 U_1$ is in compression. From its slope (20 : 20, or 1 : 1) the horizontal component can be seen to be 30 k also.

$$\sum H = 0$$
$$-30 + F_{L_0 L_1} = 0$$
$$F_{L_0 L_1} = 30 \text{ k tension}$$

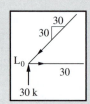

The application of the $\sum H = 0$ equation shows $L_0 U_1$ to be pushing horizontally to the left against the joint with a force of 30 k. For equilibrium, $L_0 L_1$ must pull to the right away from the joint with the same force. The arrow convention shows the force is tensile.

Considering joint U_1,

$$\sum V = 0$$
$$30 - F_{U_1 L_1} = 0$$
$$F_{U_1 L_1} = 30 \text{ k tension}$$

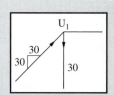

The force in $L_0 U_1$ has previously been found to be compressive with vertical and horizontal components of 30 k each. Since it is pushing upward at joint U_1 with a force of 30 k, $U_1 L_1$ (the only other member at the joint that has a vertical component) must pull down with a force of 30 k in order to satisfy the $\sum V = 0$ equation.

$$\sum H = 0$$
$$30 - F_{U_1 U_2} = 0$$
$$F_{U_1 U_2} = 30 \text{ k compression}$$

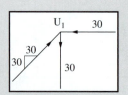

Member $L_0 U_1$ is pushing to the right horizontally with a force of 30 k. For equilibrium $U_1 U_2$ is pushing back to the left with 30 k.

Considering joint L_1,

$$\sum V = 0$$
$$30 - 20 - V_{L_1U_2} = 0$$
$$V_{L_1U_2} = 10 \text{ k compression}$$

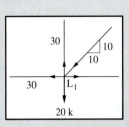

From the 1:1 slope $H_{L_1U_2} = 10 \text{ k compression}$

$$\sum H = 0$$
$$-30 - 10 - + F_{L_1L_2} = 0$$
$$F_{L_1L_2} = 40 \text{ k tension}$$

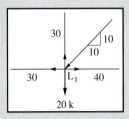

The forces, or their components, for all of the truss members may be calculated in the manner just described with the following results.

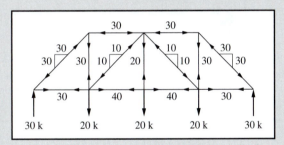

The resultant forces for inclined members may be determined from the square root of the sum of the squares of the vertical and horizontal components of force. An easier method is to write ratios comparing the resultant axial force of a member and its horizontal or vertical component with the true length of the member and its horizontal or vertical component. By letting F, H, and V represent the force and its components and L, h, and v the length and its components, the ratios of Figure 6.13 are developed.

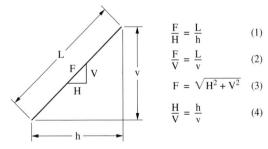

$$\frac{F}{H} = \frac{L}{h} \quad (1)$$

$$\frac{F}{V} = \frac{L}{v} \quad (2)$$

$$F = \sqrt{H^2 + V^2} \quad (3)$$

$$\frac{H}{V} = \frac{h}{v} \quad (4)$$

Figure 6.13

The method of joints may be used to compute the forces in all of the members of many trusses. The trusses of Examples 6.2 to 6.4 and all of the problems at the end of this chapter fall into this category. There are, however, a large number of trusses that need to be analyzed by a combination of the method of joints and the method of sections discussed in Chapter 7. The author likes to calculate as many forces as possible in a truss by using the method of joints. At joints where he has a little difficulty he take moments to obtain one or two forces, as described in Chapter 7. He then continues the calculations as far as possible by joints until he reaches another point of difficulty where he again take moments, and so on.

EXAMPLE 6.2

Find all the forces in the truss of Figure 6.14. Only the results are shown.

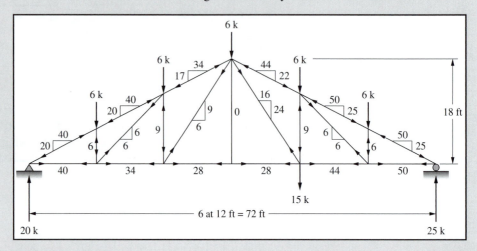

Figure 6.14

EXAMPLE 6.3

Find all forces, or their horizontal and vertical components, for the members of the truss of Figure 6.15. Use the method of joints for the solution. Only the answers are given.

Solution.

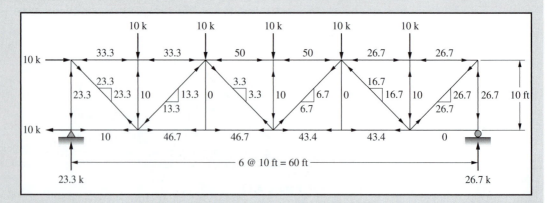

Figure 6.15

Equations may easily be written for all of the joints of a truss and those equations solved simultaneously for the unknown reaction components and member forces. Although such a procedure is followed in Example 6.4 for a three-member truss, it is extremely cumbersome for larger trusses with their numerous joints and members. An alternative procedure is to write the same joint equations in matrix form and solve those equations by a matrix procedure as described in Chapters 22 to 25 of this text.

EXAMPLE 6.4

Find all the forces in the truss shown in Figure 6.16.

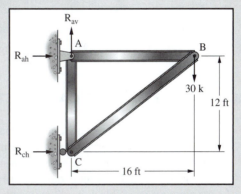

Figure 6.16

Solution.

For Joint A
$$\sum F_H = R_{ah} + F_{ab} = 0$$
$$\sum F_V = R_{av} - F_{ac} = 0$$

For Joint B
$$\sum F_H = -F_{ab} - 0.8F_{cb} = 0$$
$$\sum F_V = -0.6F_{cb} - 30 = 0$$

For joint C
$$\sum F_H = R_{ch} + 0.8F_{cb} = 0$$
$$\sum F_V = F_{ac} + 0.6F_{cb} = 0$$

Solving the equations

$$R_{ah} = -40.0 \text{ k} \leftarrow$$
$$R_{av} = +30.0 \text{ k} \uparrow$$
$$R_{ch} = +40.0 \text{ k} \rightarrow$$
$$F_{ab} = +40.0 \text{ k}$$
$$F_{cb} = -50.0 \text{ k}$$
$$F_{ac} = +30.0 \text{ k} \quad \blacksquare$$

Fairview-St. Mary's skyway system, Minneapolis, Minnesota
(Courtesy of the American Institute of Steel Contruction)

6.10 COMPUTER ANALYSIS OF STATICALLY DETERMINATE TRUSSES

Most structural analysis software uses matrix methods of analysis, which are discussed in the last few chapters of this book. These methods, which are based on displacements of the structure, require knowledge of the cross-sectional properties and material characteristics of the members. At this point in our study of structural analysis, we will be content with using the software as a "black-box" that can be used to obtain answers. Through study later in this text, we will develop an understanding of the theoretical basis of the software.

Coordinate System

When performing computer structural analysis, we must be concerned with two coordinate systems. The first of these is the global coordinate system, which is a right-hand Cartesian coordinate system in which we specify the geometry of the structure. We also use it to look at structural displacements and reaction forces. The origin of the global coordinate system is specified at some convenient location of the structure, often the lower left-hand corner. The global coordinate axes are normally aligned with the principle axes of the structure.

The second system is the local coordinate system. It relates to the forces and deformations acting on, and in, individual members of the structure. The local coordinate system used depends on the type of member being used, and is established when the software is developed. The senses of the member forces obtained from the analysis are interpreted in the local coordinate system.

Joints and Member Connectivity

When beginning an analysis, we must define the geometry of the structure. This is accomplished by first specifying the location of the joints and then information as to how the members are connected to the joints. The element used for the members must be appropriate for the analysis being conducted. When analyzing trusses we should select an element that can only carry axial forces since that is the only force theoretically resisted by a member in a truss. A graphical representation of the geometry, which is available in most software, is very useful in making sure that you have specified the geometry correctly.

Joint Restraints

A degree of freedom at a joint is a possible displacement of the joint in a particular direction. The structural supports are defined by restraining the appropriate degrees of freedom. Restraints specify that the displacement in the direction of that degree of freedom is zero. As such, there is a structural support in that direction. Also, degrees of freedom that are not considered in the analysis are restrained to zero displacement in order to remove them from the analysis. When analyzing trusses, all rotational degrees of freedom are restrained since all of the members in a truss are pin connected: one member cannot transfer moment to another member or to a joint.

Member Properties

The analyst may not feel all of these data are needed for a statically determinate truss. The computer programs, however, are general and applicable to both statically determinate and statically indeterminate structures. When analyzing trusses we must at least provide the cross-sectional area, the moment of inertia, and modulus of elasticity for each member.

6.11 EXAMPLE COMPUTER PROBLEM

Example 6.5 presents the analysis of a three-member statically determinate truss using SABLE32. The procedure used for plane trusses is the same whether statically determinate or statically indeterminate trusses are involved. This program is limited to trusses with no more than 70 joints and 70 members.

    If the student will carefully study the HELP material provided with SABLE32, he or she should have little trouble in analyzing plane trusses whether statically determinate or indeterminate. Nevertheless, the author will now walk the student through the simple steps involved in applying the program to the truss presented in Example 6.5.

EXAMPLE 6.5

Using SABLE32, determine the forces in all of the members of the truss of Figure 6.17 assuming the units used are inches and kips. For each member, the following properties are assumed: $A = 10$ sq. inches, $I = 100$ in.[4] and $E = 29 \times 10^3$ ksi.

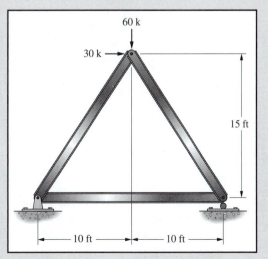

Figure 6.17

Solution.

The joints and members of the truss are numbered by a convenient system as shown in Figure 6.18.

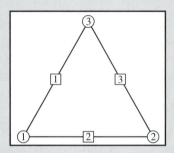

Figure 6.18

After SABLE32 is opened, the user enters the coordinates and support conditions for each truss joint. The origin selected by the author for this truss is at the far left joint and, as a result, the joint coordinates will all be positive. The first step is to click on the EDIT and JOINT data buttons on the screen. For each joint, the user clicks the NEW button and enters that joint's coordinates. After this is done, he or she needs to click on the UPDATE, SELECT, and OK buttons in that order.

As the coordinates are supplied for each joint, it is also necessary to provide information concerning the restraint provided for each joint. For instance, a hinge support provides restraint in both the x and y directions. In addition, for a plane truss no degree of freedom is assumed in the z direction for any of the joints. Thus it is necessary click on the z restraint for every joint in the truss. After all the joint data is input, click on the CLOSE button.

To see if the joint data has been inserted correctly, the user can click on the DISPLAY button followed by the buttons GEOMETRY and JOINT DATA (A similar procedure can be used to display all other data that is input.)

Next the user enters the joint numbers at the end of each member, here referring to the member numbers previously selected in Figure 6.18. When supplying the member properties, the user must be sure to click on the ACTIVE button on the left side of the screen. In addition to the joint numbers, the end conditions of those members are specified. For plane trusses, the end conditions will always be PIN PIN. These are shown on the right side of the screen. Once again the user proceeds with NEW, UPDATE, SELECT and after all the member data is specified, the user clicks the CLOSE button.

The loads applied to the truss are now entered. The sign convention used is plus for the x direction if the loads are going to the right and plus for the y direction if the loads are acting upward.

Finally the user clicks on the ANALYSIS and then the STATIC ANALYSIS buttons for this truss. The screen will show the analysis is completed, and it is necessary to click on OK. To see the results, click on DISPLAY, RESULTS, JOINT FORCES, or BEAM FORCES as desired. A printout of the results is shown below.

INPUT

Joint Location and Restraint Data

Jnt	Coordinates			Restraints		
	X	Y	Z	X	Y	Rot
1	0.000E+00	0.000E+00	0.000E+00	Y	Y	Y
2	2.000E+01	0.000E+00	0.000E+00	N	Y	Y
3	1.000E+01	1.500E+01	0.000E+00	N	N	Y

Beam location and property data

Beam	i	j	Type	Stat	Beam Properties		
					Area	lzz	E
1	1	3	P-P	A	1.000E+01	1.000E+02	2.900E+04
2	1	2	P-P	A	1.000E+01	1.000E+02	2.900E+04
3	2	3	P-P	A	1.000E+01	1.000E+02	2.900E+04

Applied joint loads

Jnt	Case	Force-X	Force-Y	Moment-Z
1	1	0.000E+00	0.000E+00	0.000E+00
2	1	0.000E+00	0.000E+00	0.000E+00
3	1	3.000E+01	−6.000E+01	0.000E+00

OUTPUT

Beam	Case	End	Axial	Shear-Y	Moment-Z
1	1	i	9.014E+00	0.000E+00	0.000E+00
		j	−9.014E+00	0.000E+00	0.000E+00
2	1	i	−3.500E+01	0.000E+00	0.000E+00
		j	3.500E+01	0.000E+00	0.000E+00
3	1	i	6.310E+01	0.000E+00	0.000E+00
		j	−6.310E+01	0.000E+00	0.000E+00

Observe that the shearing force and bending moment at each end of every member are equal to zero. This is as expected and is consistent with our assumption about behavior. Also observe from the calculated results that the axial force at one end of each member is positive whereas the axial force at the other end is negative. To determine whether a member is in axial tension or axial compression we must refer to the local coordinate system specified for the member. From the HELP information in SABLE32, we can see that the local coordinate system is as shown in Figure 6.19. These are the directions of positive forces at each end of a member.

Figure 6.19

From this sketch of the local coordinate system, we can see that a positive axial force at the left end of the member indicates axial compression whereas a positive axial force at the right end indicates axial tension. We will use the indicated results at the right end of the member, the j end, to interpret the magnitude and sense (tension or compression) of the axial forces in the members.

SABLE32 does not include any weights of the truss members in the calculations. However, you may sometimes have a truss member that you are sure has a 0 force if member weight is neglected and yet the computer shows a very, very small force. An extremely small number is printed for the member. It should actually be zero but the rounding of the numbers in the matrix will frequently leave this tiny value and it should be neglected. ∎

6.12 PROBLEMS FOR SOLUTION

For Problems 6.1 to 6.22, classify the structures as to their internal and external stability and determinacy. For statically indeterminate structures, include the degree of redundancy internally or externally. (The small circles on the trusses indicate the joints.)

6.1   (*Ans.* Statically determinate internally and externally)

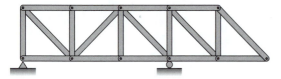

6.2

No joint

6.3   (*Ans.* Statically indeterminate internally to first degree)

No joint

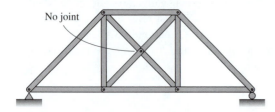

6.4

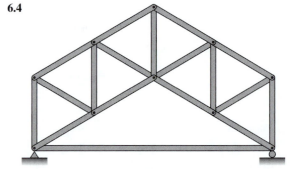

6.5   (*Ans.* Statically indeterminate externally to second degree)

6.6

No joints

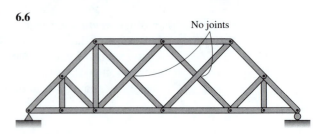

6.7   (*Ans.* Statically indeterminate internally to first degree)

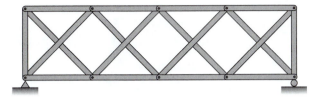

6.8

6.9   (*Ans.* Unstable)

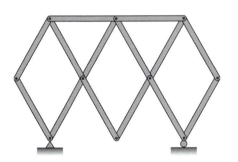

6.10

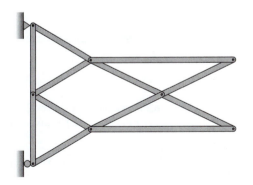

6.11   (*Ans.* Statically determinate internally and externally)

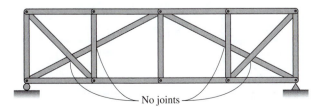

No joints

6.12

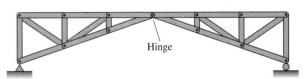

Hinge

6.13   (*Ans.* Statically indeterminate externally to second degree)

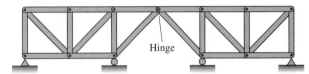

Hinge

6.14

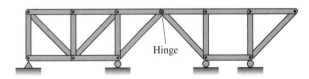

Hinge

6.15   (*Ans.* Statically determinate internally and externally)

Hinges

6.16

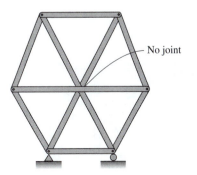

No joint

6.17 (*Ans.* Statically indeterminate internally to first degree)

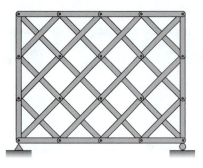

6.18

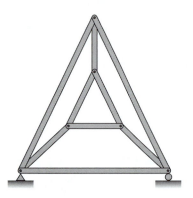

6.19 (*Ans.* Statically indeterminate internally to 4th degree)

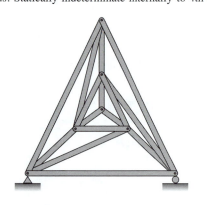

6.20

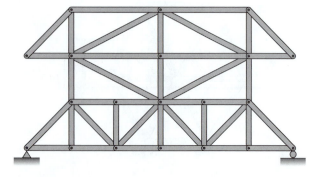

6.21 (*Ans.* Statically indeterminate internally to first degree)

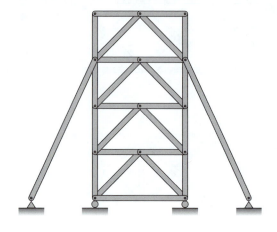

6.22

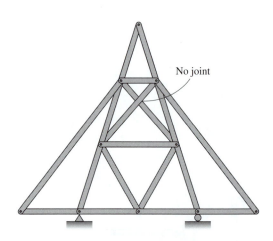

No joint

For Problems 6.23 to 6.39, compute the forces in all the members of the trusses using the method of joints.

6.23 (*Ans.* $F_{L_1L_2} = +36.67$ k, $F_{U_2U_3} = -33.33$ k, $F_{U_1L_2} = -4.71$ k)

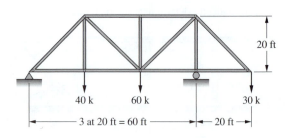

40 k   60 k   30 k

20 ft

3 at 20 ft = 60 ft   20 ft

6.24 Rework Problem 6.23 if the truss depth is reduced to 15 ft and the loads doubled.

6.25 (*Ans.* $F_{U_1U_2} = -103.66$ k, $F_{U_2U_3} = -148.73$ k, $F_{L_2L_3} = +63.75$ k)

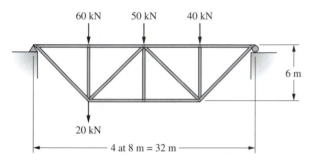

6.26

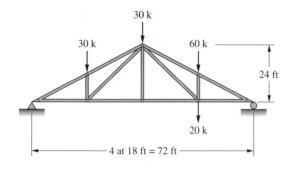

6.27 (*Ans.* $F_{CD} = +37.94$ k, $F_{CA} = +141.42$ k, $F_{BD} = -84.85$ k)

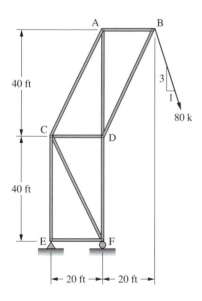

6.28

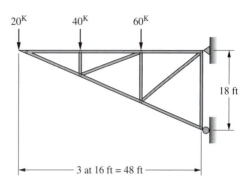

6.29 (*Ans.* $F_{L_0L_1} = +95$ k, $F_{U_1U_2} = -165$ k, $F_{U_1L_3} = +84.13$ k)

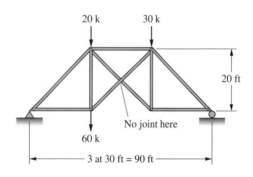

6.30

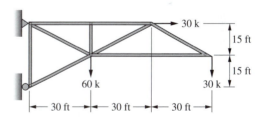

6.31 (*Ans.* $F_{L_1L_2} = -80$ k, $F_{U_2U_3} = +82.46$ k, $F_{U_3L_3} = +53.33$ k)

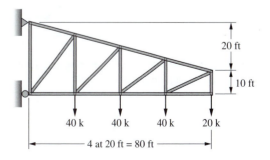

6.32

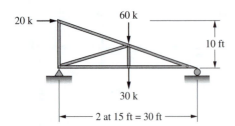

6.33 Repeat Problem 6.32 if the roller support (due to friction, corrosion, etc.) is assumed to supply one-third of the total horizontal force resistance needed, with the other two-thirds supplied by the pin support. (*Ans.* $F_{L_0L_1} = +148.33\,\text{k}$, $F_{U_1L_2} = -163.38\,\text{k}$)

6.34

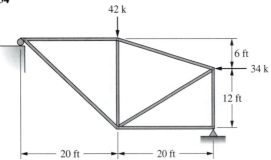

6.35 (*Ans.* $F_{U_1U_2} = -60\,\text{k}$, $F_{L_0L_1} = +55.71\,\text{k}$, $F_{U_3L_4} = -43.52\,\text{k}$)

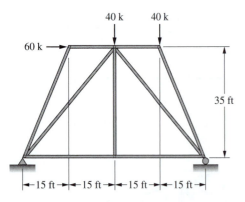

6.36 Rework Problem 6.35 if the supporting surface beneath the roller is changed as follows

6.37 (*Ans.* $F_{EF} = +50\,\text{k}$, $F_{AC} = -26\,\text{k}$, $F_{AD} = -24.41\,\text{k}$)

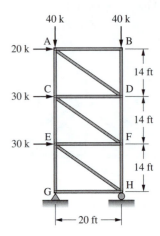

6.38

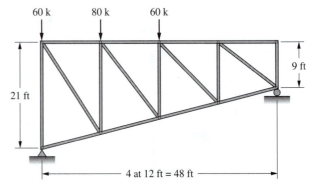

6.39 (*Ans.* $F_{U_0U_1} = -60\,\text{k}$, $F_{U_1L_1} = -105\,\text{k}$, $F_{L_2L_3} = +82.46\,\text{k}$)

For Problems 6.40 and 6.41, analyze the trusses using SABLE32 if $A = 0.05\,\text{ft}$, $I = 1\,\text{ft}^4$, and $E = 29{,}000\,\text{ksi}$ for all members.

6.40 Problem 6.32

6.41 Problem 6.35 (*Ans.* $F_{L_0L_1} = +55.71\,\text{k}$, $F_{U_2L_2} = 0$, $F_{U_2L_4} = -59.27\,\text{k}$)

Chapter 7

Plane Trusses, Continued

7.1 ANALYSIS BY THE METHOD OF SECTIONS

Applying the equations of equilibrium to free-body diagrams of sections of a truss is the basis of force computation by the method of sections, just as it was when using the method of joints in the previous chapter. When using the method of sections to determine the force in a particular member, an imaginary plane is passed completely through the truss, which cuts it into two sections as shown in Figure 7.1(a). The resulting free-body diagrams are shown in Figure 7.1(b) and (c). The location at which the sections are cut is selected so there are at least as many equations of equilibrium available as there are unknown forces.

The equations of static equilibrium may be applied to either of the free bodies to determine the magnitude of the unknown forces. Particular attention should be paid to the point about which moments are summed when applying the equations. Often moments of the forces can be taken about a point so that only one unknown force appears in the equation. As such, the value of that force can be obtained. This objective usually can be

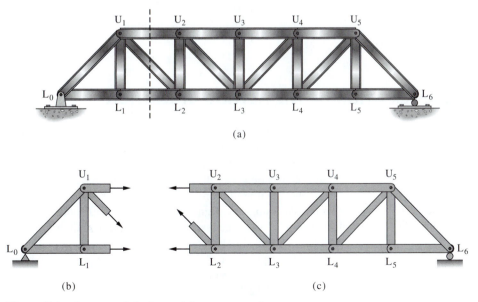

Figure 7.1 A truss and the internal forces at a section

attained by selecting a point along the line of action of one or more of the other members' forces and making it the point about which moments are summed. This point does not necessarily have to be on the cut section. Some familiar trusses have special locations for placing sections that greatly simplify the work involved. Some of these will be discussed in the pages to follow.

An advantage of the method of sections is that the force in one member of a truss can be computed in most cases without having to compute the forces in other members of the truss. If the method of joints were used, calculating the forces in other members, joint by joint from the end of the truss up to the member in question, would be necessary. Nevertheless, both methods are used in the analysis of trusses. In fact, both methods are often used at the same time. Depending upon the geometry of the truss, some member forces are more easily calculated with the method of sections than others that are more easily calculated using the method of joints.

Final truss slipped into place for Newport Bridge linking Jamestown and Newport, Rhode Island (Courtesy Bethlehem Steel Corporation)

7.2 APPLICATION OF THE METHOD OF SECTIONS

When using the method of sections, we need to establish a sign convention for the sense of the forces in the cut members just as we did when using the method of joints. As with the method of joints, assume that all unknown member forces are acting in tension. This sign convention was illustrated in Figure 7.1. Upon analysis, positive results indicate members that are in tension and negative results indicate members that are in compression. By using this sign convention, we will obtain consistent results with all of the methods of truss analysis. The errors that result from confusing the sense of member forces will be greatly reduced.

Examples 7.1 and 7.2 illustrate the computation of member forces using the method of sections. To demonstrate the principles, only the forces in selected members are computed. The forces in the other members could be computed by cutting additional sections or with the method of joints.

EXAMPLE 7.1

Find the forces in members L_1L_2 and U_2U_3 of the truss shown in Figure 7.2 using the method of sections.

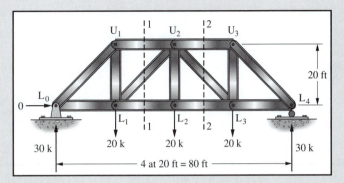

Figure 7.2

Solution.   Member L_1L_2. Section 1-1 is passed through the truss cutting L_1L_2 and as few other members as possible. The part of the truss to the left of the section is considered to be the free-body and is shown in Figure 7.3.

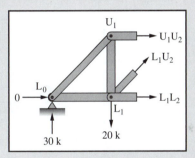

Figure 7.3

The forces acting on the free-body are the 30-kip reaction, the 20-kip load at L_1, and the axial forces in the members cut by the section (U_1U_2, L_1U_2, and L_1L_2). Moments of these forces are taken about U_2, which is the point of intersection of L_1U_2 and U_1U_2. The moment equation contains one unknown force, L_1L_2, and its value may be found by solving the equation.

$$\sum M_{U_2} = 0$$
$$(30)(40) - (20)(20) - 20F_{L_1L_2} = 0$$
$$F_{L_1L_2} = +40 \text{ k tension}$$

Member U_2U_3. Section 2-2 is passed through the truss, and the portion of the truss to the right of the section is considered to be the free-body as shown in Figure 7.4. (We could have used the part of the truss to the left of the section but it seemed a little simpler to work to the right as fewer forces were involved on that side.) Members U_2U_3, U_2L_3 and L_2L_3 are cut by the section. Taking moments at the intersection of L_2L_3 and U_2L_3 at L_3 eliminates those two members from the equation because the lines of action of their forces pass through the center of moments. The force in U_2U_3 is the only unknown appearing in the equation and its value may be determined.

$$\sum M_{L_3} = 0$$

$$-(30)(20) - 20F_{U_2U_3} = 0$$

$$F_{U_2U_3} = -30 \text{ k compression}$$

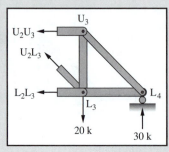

Figure 7.4 ■

EXAMPLE 7.2

Find the forces (or their horizontal and vertical components) for all of the members of the truss shown in Figure 7.5. Use both the methods of joints and sections as seems the most convenient.

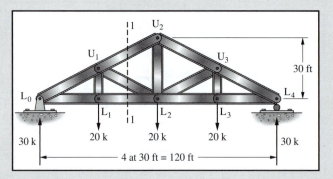

Figure 7.5

Solution. The forces of members meeting at L_0 and L_1 are quickly determined by the method of joints. To calculate the force in U_1U_2, section 1-1 is passed and moments are taken at L_2. Because U_1U_2 is an inclined member, the force is resolved into its vertical and horizontal components. The components of a force may be assumed to act anywhere along its line of action. It is convenient in this case to break the force down into its components at joint U_2, because the vertical component will pass through the center of moments and the moment equation may be solved for the horizontal component of force.

$$\sum M_{L_2} = 0$$

$$(30)(60) - (20)(30) + 30H_{U_1U_2} = 0$$

$$H_{U_1U_2} = -40 \text{ k}$$

$$V_{U_1U_2} = \left(\frac{15}{30}\right)(-40) = -20 \text{ k}$$

The forces, or their components, may now be easily obtained with the method of joints for all of the other truss members. The results are shown in Figure 7.6.

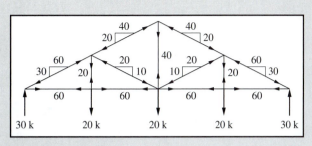

Figure 7.6 ■

Examples 7.3, 7.4, and 7.5 present the analysis of several other trusses using the joint and section methods.

EXAMPLE 7.3

Determine the forces in all of the members of the truss shown in Figure 7.7.

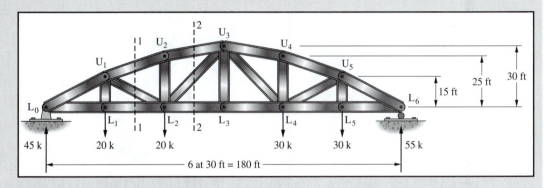

Figure 7.7

Solution. The forces meeting at joints L_0 and L_1 are determined by joints. Then section 1-1 is passed through the truss, the free-body to the left is considered, and moments are taken at L_2 to determine $H_{U_1 U_2}$.

$$\sum M_{L_2} = 0; \text{ free-body to left of section 1-1}$$
$$(45)(60) - (20)(30) + 25H_{U_1 U_2} = 0$$
$$H_{U_1 U_2} = -84 \text{ k compression}$$

The remaining forces at joints U_1 and U_2 are determined by joints. Section 2-2 is passed through the truss and with respect to the left free-body, moments are taken at U_3 to determine $L_2 L_3$.

$$\sum M_{U_3} = 0; \text{ free body to left of section 2-2}$$
$$(45)(90) - (20)(30) - (20)(60) - 30F_{L_2 L_3} = 0$$
$$F_{L_2 L_3} = +75 \text{ k tension}$$

The same procedure is continued for the rest of the truss and the results are shown in Figure 7.8.

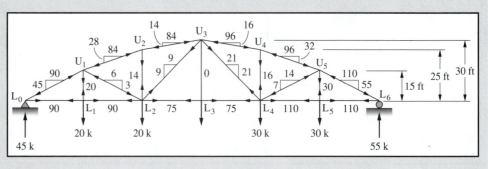

Figure 7.8

EXAMPLE 7.4

Calculate the force in member cg of the truss of Figure 7.9.

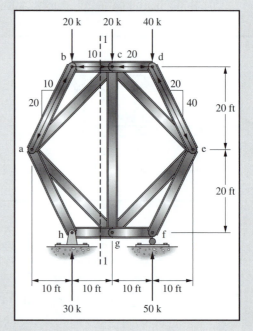

Figure 7.9

Solution.   The force in the member in question cannot be determined immediately by joints or moments. It is necessary to know the force values for several other members before the value for cg can be found. The forces in members ba, bc, dc, and de may be found by joints as shown, and the force in member ac can be found by moments. Considering section 1-1 and the free body to the left, moments may be taken about g. By noting that the force in bc is in compression and pushes against the free body from the outside, and by assuming member ac to be in tension, the following equation may be written. The unknown force is broken into its vertical and horizontal components at c.

$$\sum M_g = 0$$
$$(30)(10) - (20)(10) - (10)(40) + (H_{ac})(40) = 0$$
$$H_{ac} = +7.5 \text{ k tension}$$

Having the force in ac, the forces in ce and cg can be determined by joints as shown in Figure 7.10.

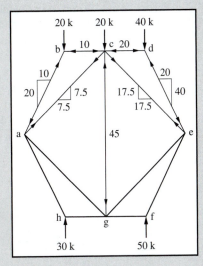

Figure 7.10

EXAMPLE 7.5

Determine the forces in all of the members of the Fink truss shown in Figure 7.11.

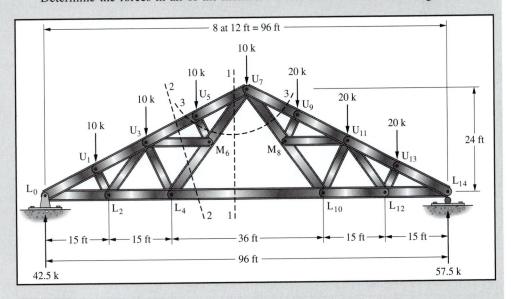

Figure 7.11

Solution.   The forces in members meeting at joints L_0, L_2, and U_1 can be found by the methods of joints and sections without any difficulty. At each of the next two joints, U_3 and L_4, there are three unknown forces that cannot be determined directly with sections. It is convenient to compute the forces in some members further over in the truss and then work back to these joints. The sections numbered 1-1, 2-2, and 3-3 may be used to advantage. From the first of these sections the forces in any of the three members cut may be obtained by moments. By using section 2-2 (see Figure 7.12) and taking moments at U_3, the force in L_4M_6 may be found. It is important to note that four members have been cut by the section and only two of them pass through the point where moments are being taken; however, the force in one of these members, L_4L_{10}, was previously found with section 1-1, and only one unknown is left in the equation. The remaining forces in the truss may be calculated by the usual methods.

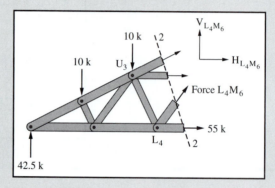

Figure 7.12

These two sections just considered (1-1 and 2-2) are sufficient for analyzing the truss, but should another approach be desired, a section such as 3-3 may be considered. From this section the force in U_5M_6 can be found by taking moments at U_7 because all of the other members cut by the section pass through U_7.

The forces in all of the members of this truss are shown in Figure 7.13.

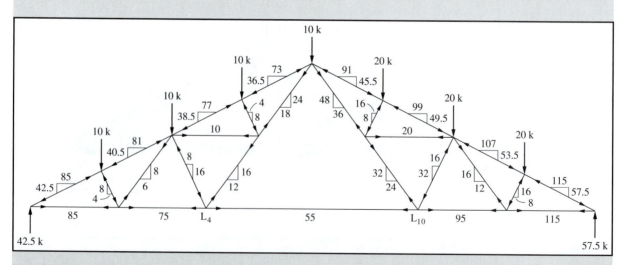

Figure 7.13

Timber trusses for tannery,
South Paris, Maine. (Courtesy of the
American Wood Preservers Institute.)

7.3 METHOD OF SHEARS

It should be obvious by this time that if a vertical section is passed through a truss and divides it into two separate free-bodies, the sum of the vertical forces to the left of the section must be equal and opposite in direction to the sum of the vertical forces to the right of the section. The summation of these forces to the left or to the right of a section has been defined as the *shear*.

The inclined members cut by a section must have vertical components of force equal and opposite to the shear along the section, because the horizontal members can have no vertical components of force. For most parallel-chord trusses there is only one inclined member in each panel, and the vertical component of force in that inclined member must be equal and opposite to the shear in the panel. The vertical components of force are computed by shears for the diagonals of the parallel-chord truss of Example 7.6.

EXAMPLE 7.6

Determine the vertical components of force in the diagonals of the truss shown in Figure 7.14. Use the method of shears.

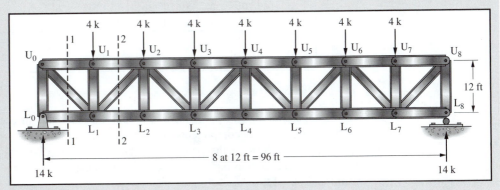

Figure 7.14

Solution. By considering section 1-1 and free-body to the left,

$$\text{shear to the left} = 14 \text{ k} \uparrow$$

$$V_{U_0L_1} = 14 \text{ k} \downarrow \text{tension (pulling away from free body)}$$

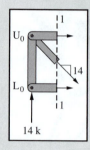

By considering section 2-2 and free-body to the left,

$$\text{shear to left} = 14 - 4 = 10 \text{ k} \uparrow$$

$$V_{L_1U_2} = 10 \text{ k} \downarrow \text{compression (pushing against free body)}$$

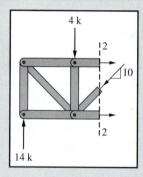

The vertical components of force in all the diagonals are as shown in Figure 7.15.

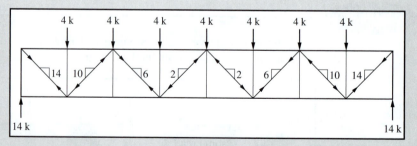

Figure 7.15

Nonparallel-chord trusses have two or more diagonals in each panel, and they all may have vertical components of force; however, their sum must be equal and opposite to the shear in the panel. If all but one of the diagonal forces in a panel are known, the remaining one may be determined by shears, as illustrated in Example 7.7. ■

EXAMPLE 7.7

Referring to sections 1-1 and 2-2 of the truss of Figure 7.7 and assuming that the forces in the chords U_1U_2 and U_2U_3 are known, find the vertical components of force in U_1L_2 and L_2U_3 by the method of shears.

Solution.   By considering section 1-1 and free-body to the left,

$$\text{shear to left} = 25 \text{ k} \uparrow$$
$$V_{U_1U_2} = 28 \text{ k} \downarrow$$
$$\text{Thus } V_{U_1L_2} = 3 \text{ k} \uparrow \text{ compression}$$

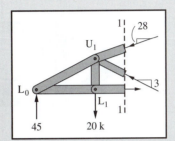

By considering section 2-2 and free-body to the right,

$$\text{shear to right} = 5 \text{ k} \downarrow$$
$$V_{U_2U_3} = 14 \text{ k} \uparrow$$
$$\text{Thus } V_{L_2U_3} = 9 \text{ k} \downarrow \text{ tension}$$

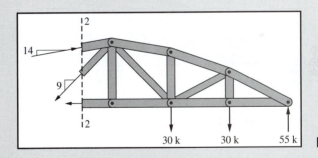

7.4 ZERO-FORCE MEMBERS

Frequently, some readily identifiable truss members have zero forces (assuming secondary forces due to member weights, load eccentricities, and so forth, are neglected). The ability to spot these members will on occasion appreciably expedite truss analysis. Zero-force members usually can be identified by making brief examinations of the truss joints. Several illustrations are presented here, with reference being made to the trusses of Figure 7.16.

1. *If only one member at a joint has a possible force in a particular direction and if there is no external load applied at the joint with a component in the direction of the member, the force in the member must be zero.* An examination of joint L_3 of

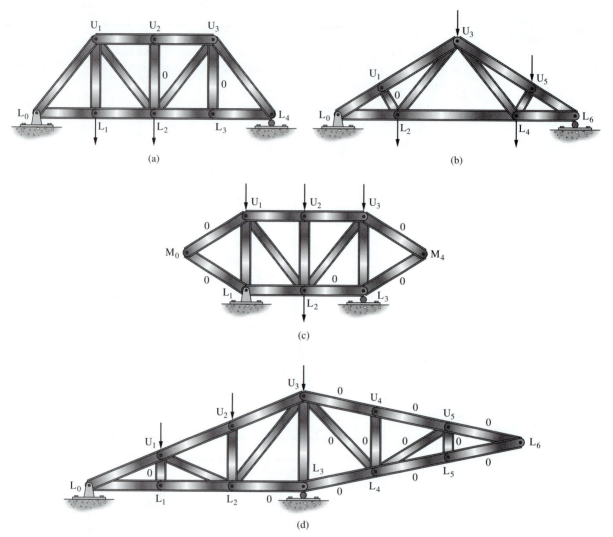

Figure 7.16   Zero-force members

the truss of part (a) of Figure 7.16 reveals that member U_3L_3 has a zero force as long as no external load with a vertical component is applied at the joint. If the member had a force, the sum of the vertical forces at L_3 could not be zero. A similar examination of joint U_2 shows member U_2L_2 has a zero force.

2. *Not only must the sum of the forces in the x and y directions at a particular joint be zero, but also the sum of the forces in any direction at the joint must be zero.* Therefore, the force in member U_1L_2 of the truss of part (b) of the figure must be zero, as there are no other forces at joint U_1 with components perpendicular to members L_0U_1 and U_1U_3.

3. *Should two members be joined (the members not being in line with each other), they both will have zero forces unless an external load is applied at the joint in the plane of the members.* In the truss of part (c) of Figure 7.16 both members M_0U_1 and M_0L_1 are zero-force members. If one of the members had a force, it

would be impossible to have a force in the other member such that both $\sum H = 0$ and $\sum V = 0$. See the following sketches.

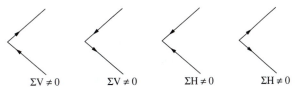

$$\sum V \neq 0 \qquad \sum V \neq 0 \qquad \sum H \neq 0 \qquad \sum H \neq 0$$

4. Based on the preceding illustrations it is easy to spot twelve zero-force members in the truss of part (d) of Figure 7.16. These members can be identified by examining the following joints in the order given: L_6, L_5, U_5, U_4, L_4, L_3, and L_1.

7.5 WHEN ASSUMPTIONS ARE NOT CORRECT

The engineer should realize that often his or her assumptions regarding the behavior of a structure (pinned joints, loads applied at joints only, frictionless rollers, or whatever) may not be entirely valid. As a result he or she should give consideration to what might happen to a structure if the assumptions made in analysis were appreciably in error. Perhaps an expansion device or roller will resist (due to friction) a large proportion of any horizontal forces present. How would this affect the member forces of a particular truss? For this very reason the truss of Problem 6.33 was included at the end of Chapter 6, where it was assumed that one-third of the horizontal load was resisted by the roller.

The types of end supports used can have an appreciable effect on the magnitude of forces in truss members caused by lateral loads.

For fairly short roof trusses generally no provisions are made for temperature expansion and contraction and both ends of the trusses are bolted down to their supports. These trusses are actually statically indeterminate, but the usual practice is to assume that the horizontal loads split equally between the supports.

For longer roof trusses, provisions for expansion and contraction are considered necessary. Usually the bolts at one end are set in a slotted hole so as to provide space for the anticipated length changes. A base plate is provided at this end on which the truss can slide.

Actually it is impossible to provide a support that has no friction. From a practical standpoint the maximum value of the horizontal reaction at the expansion end equals the vertical reaction times the coefficient of friction (one-third being a reasonable guess). Should corrosion occur, thus preventing movement (a rather likely prospect), a half-and-half split of the lateral loads may again be the best estimate.

The author once read of an interesting case in which the owners of a building with a roof supported by a series of Fink trusses decided that the middle lower chord member (L_4L_{10} in Figure 7.17) was in their way. They therefore removed the member from quite a few of the trusses and much to the designer's amazement the roof did not collapse. Apparently the roller or expansion device on one end of the trusses (perhaps bolts in a slot) permitted very little or no movement. As a result, the trusses apparently behaved as a three-hinged arch, as shown in Figure 7.17. In many situations in which the assumptions do not prove to be correct, the results are more unpleasant than they were for these Fink trusses, however.

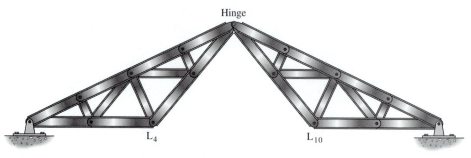

Hinge

L_4 L_{10}

Figure 7.17

7.6 SIMPLE, COMPOUND, AND COMPLEX TRUSSES

Occasionally you will hear a truss referred to as a simple truss or a compound truss. These are references to the geometric form of the truss, or the "building blocks" that form the truss. They are not references to the complexity of the analysis to determine member forces. For completeness, we will briefly discuss the categories of trusses from which these references arise. We will also discuss some issues regarding analysis of trusses in each category.

Simple Trusses

As we have seen, the first step in forming a truss is connecting three members at their ends to form a triangle. Subsequent segments are incorporated by adding two members and one joint; the new members meet at the new joint and each is pinned at its opposite ends into one of the existing joints. Trusses formed in this way are said to be *simple trusses*. The author is not suggesting, however, that all trusses formed in this manner are "simple" to analyze.

Compound Trusses

A *compound truss* is a truss made by connecting two or more simple trusses. The simple trusses may be connected by three nonparallel non-concurrent links, by one joint and one link, by a connecting truss, by two or more joints, and so on. An almost unlimited number of trusses may be formed in this way. The Fink truss shown in Figure 7.18(a) consists of the two shaded simple trusses that are connected with a joint and a link. All of the methods for checking stability and for analysis can be used on compound trusses with equal success.

Complex Trusses

There are a few trusses that are statically determinate that do not meet the requirements necessary to fall within the classification of either simple or compound trusses. Such a truss is shown in Figure 7.18(b). These are referred to as *complex trusses*.

The members of simple and compound trusses usually are arranged so that sections may be passed through three members at a time; moments are taken about the intersection of two of them and the force in the third member is found. Complex trusses may not be analyzed in this manner. Not only does the method of sections fail to simplify the analysis, the methods of joints is also of no avail. The difficulty lies in the fact that there

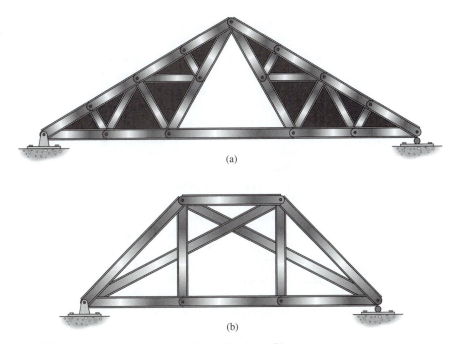

Figure 7.18 A compound truss (a) and complex truss (b)

are three or more members meeting at almost every joint. Consequently, there are too many unknowns at every location in the truss to pass a section and obtain the force in any member directly using the equations of static equilibrium.

One method of computing the forces in complex trusses is to write the equations of static equilibrium at each joint, which yields 2j equations. These equations may be solved simultaneously for the member forces and external reactions. It is often possible to calculate the external reactions initially, and their values may be used as a check against the results obtained from the solution of the simultaneous equations. This method will work for any statically determinate complex truss, but the solution of the equations is very tedious unless a digital computer is available. Note that the enclosed computer software can be used easily to analyze complex trusses, as well as other types of trusses. Should a truss be unstable, the computer will clearly indicate that fact and no analysis will be performed.

Generally, there is little need for complex trusses because it is possible to select simple or compound trusses that will serve the desired purpose equally well. Nevertheless, for a more comprehensive discussion of complex trusses you may refer to the method of substitute members described in *Theory of Structures* by S. P. Timoshenko and D. H. Young.[1]

7.7 THE ZERO-LOAD TEST

Based on the information presented up to this point, the reader will find that it is extremely difficult to determine whether complex trusses are stable or unstable without conducting a complete analysis. If such an analysis is made and all the truss joints

[1]S. P. Timoshenko and D. H. Young, *Theory of Structures*. 2nd Ed., (New York: McGraw-Hill, 1965), 92–103.

balance, the truss will be stable. If all the truss joints do not balance, the truss is unstable. The analysis procedure can be very time-consuming, however, as well as discouraging if calculations show that the structure is unstable. Indeed, we would like to know before performing an analysis. As a result, a simple method for checking stability for any type of truss is presented in this section.

A statically determinate truss has only one possible set of forces for a given loading and is thus said to have a *unique solution*. Therefore, if it is possible to show that more than one solution can be obtained for a structure for a given set of conditions, the structure is unstable.

This discussion leads to the idea of the so-called *zero-load test*. If no external loads are applied to a truss, it is logical to assume that all of the members will have zero forces. Should an assumed force (not zero) be given to one of the members of a truss that has no external loads and the forces be computed in the other members, the results must be incompatible if the truss is stable. If the calculated forces are compatible, the truss is unstable.

To illustrate this procedure, the top horizontal member of each of the three trusses of Figure 7.19 is assumed to have a tension force of X and the other member forces are computed working down from the top by joints. It will be seen that the lower three joints of the trusses of parts (a) and (b) of the figure cannot be balanced, whereas they can for the truss of part (c). A structure that has zero external loads should also have zero internal forces. Truss (c) must therefore be unstable or have *critical form*. For the first two

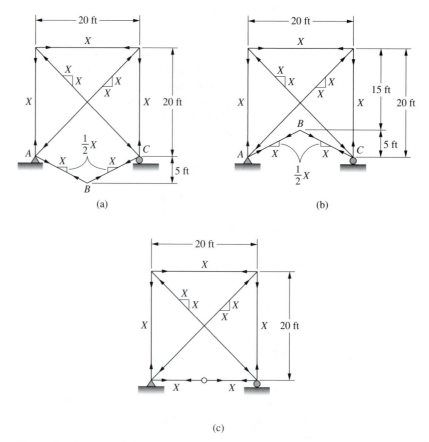

Figure 7.19 (a) Joints *A*, *B*, and *C* do not balance. Truss is stable. (b) Joints *A*, *B*, and *C* do not balance. Truss is stable. (c) All joints balance. Truss is unstable.

arrangements, (a) and (b), it is impossible to assume a set of member forces other than zero for which the joints will balance, and they must therefore be stable.[2]

7.8 STABILITY

The subject of truss stability was broached briefly in Chapter 6 and in the last section. As we develop complicated trusses such as those examined later in this chapter, the issue of stability and our ability to recognize unstable trusses becomes more important. For that reason, we will continue to discuss the stability of trusses.

The stability of a truss can always be determined through structural analysis. The members of a truss must be arranged to support the external loads. What will support the external loads satisfactorily is a rather difficult question to answer with only a glance at the truss under consideration, but an analysis of the structure will always provide the answer. If the structure is stable, the analysis will yield reasonable results and equilibrium will be satisfied at all of the joints in the truss. On the other hand, if a truss is unstable, equilibrium cannot be concurrently satisfied at all of the joints.

College fieldhouse, Largo, Maryland (Courtesy Bethlehem Steel Corporation)

Various means for quickly identifying unstable trusses are very valuable to the analyst. The careful analysis and checking of the work for a truss can be so time consuming and frustrating if it is finally discovered that the truss is unstable and all the time was wasted. Several methods for quickly identifying such structures, without the necessity of trying to analyze the structure first, are identified in the next few paragraphs. These methods are in addition to the zero-load test which was described in the last section.

Less Than 2j − r Members

A truss that has less than 2j − r members is obviously unstable internally. Nevertheless, a truss can have as many or more than 2j − r members and still be unstable. The truss of Figure 7.20(a) satisfies the 2j − r relationship. It is statically determinate and stable.

[2]G. L. Rogers and M. L. Causey, *Mechanics of Engineering Structures* (New York: John Wiley and Sons, Inc., 1962), 19–20.

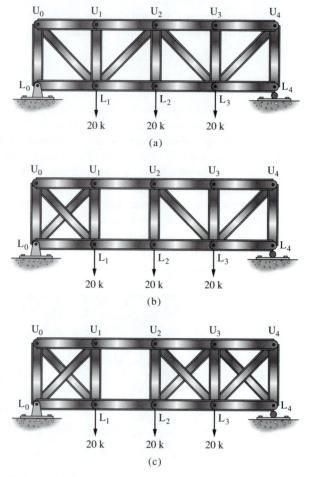

Figure 7.20 Stable (a) and unstable truss geometry (b and c)

However, if the diagonal in the second panel is removed and added to the first panel as shown in Figure 7.20(b), the truss is unstable even though the number of members remains equal to $2j - r$. The part of the truss to the left of the second panel can move with respect to the part of the truss to the right of the second panel because the second panel is a rectangle. As previously indicated, a rectangular shape is unstable unless restrained in some way.

Similarly, the addition of diagonals to the third and fourth panels, as in Figure 7.20(c), will not prevent the truss from being unstable. There are two more than $2j - r$ members, and the truss is seemingly statically indeterminate to the second degree; but it is unstable because the second panel is unstable.

Trusses Consisting of Shapes That Are Not All Triangles

As you become more familiar with trusses, in most cases you will be able to tell with a brief glance if a truss is stable or unstable. For the present, though, it may be a good idea to study trusses in detail if you think there is a possibility of instability. When a truss has some non-triangular shapes in its geometry, we should be aware that instability is indeed a possibility. The trusses shown in Figure 7.20(b) and (c) fall into this category.

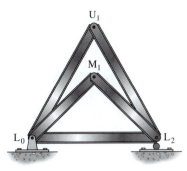

Figure 7.21 A stable truss not consisting entirely of triangles

The fallacy of this statement about triangular shapes is that there is an endless number of perfectly stable trusses that can be assembled and do not consist entirely of triangles. As an example, consider the truss shown in Figure 7.21. The basic triangle $L_0U_1L_1$ has been extended by the addition of the joint M_1 and the members L_0M_1 and M_1L_1. A stable truss is maintained, even though the shape $L_0M_1L_2U_1$ is not a triangle. Joint M_1 cannot move without changing the length of one or more members. Therefore, the truss is stable.

Compound and complex trusses often contain non-triangular shapes but are stable. The fundamental principle is that if a truss consists of shapes that are not triangular, it should be examined carefully to determine if any of the joints can displace without causing changes in length of one or more of the truss members.

Unstable Supports

A structure cannot be stable if its supports are unstable. To be stable, the truss must be supported by at least three nonparallel, non-concurrent forces. This subject was discussed in Chapter 4.

7.9 EQUATIONS OF CONDITION

On some occasions two or more separate structures are connected so that only one type of force can be transmitted through the connection. The three-hinged arch and cantilever types of structures of Chapter 4 have been shown to fall into this class because they are connected with interior hinges unable to transmit rotation.

Perhaps the simplest way to produce a hinge in a truss is by omitting a chord member in one of the panels, as shown in Figure 7.22(a). It is obvious that the moment of all the external forces on the part of the structure to the left or the right of the pin connection at joint L_3 must be zero. The truss is statically determinate because there are three statics equations and one condition equation available for calculating the four reaction components.

The omission of members in some other situations may produce equations of condition. A diagonal of the truss of Figure 7.22(b) has been omitted between the two interior supports. With no members in the panel able to have a vertical component of force, no shear can be transmitted through the panel, and an equation of condition is available. The supports on each side of the usually unstable rectangular shape prevent it from collapsing.

Practically speaking, the bars mentioned as being omitted will probably not be omitted because their omission would detract from the appearance of the structure and might frighten some users. Furthermore, the presence of these members might be useful

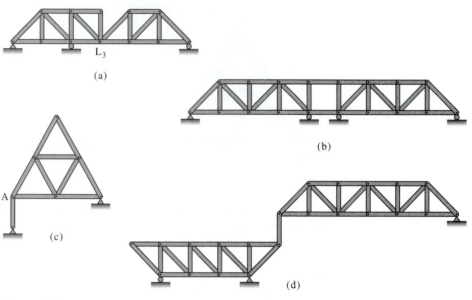

Figure 7.22

during erection. They are frequently assembled so they can be adjusted to be inactive in the completed truss.

Figure 7.22(c) and (d) present two more situations in which equations of condition are produced. In the first of these there are four reaction components, and the structure may appear to be statically indeterminate externally: however, the joint at A is pin connected and cannot transmit rotation. This equation of condition makes the structure statically determinate externally. Figure 7.22(d) shows two separate trusses that are connected by a link. The link makes available two equations of condition, because rotation may not be transmitted at either end.

7.10 PROBLEMS FOR SOLUTION

For Problems 7.1 to 7.6, determine the forces, or their vertical and horizontal components, for the indicated members using a single moment equation in each case.

The reactions are given for all of the trusses. Indicate whether the members are in compression or tension.

7.1   Members L_1L_2 and U_4U_5. (*Ans.* $F_{L_1L_2} = +200\,k$, $F_{U_4U_5} = -180\,k$)

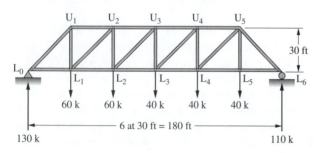

7.2   Members U_1U_2 and L_2U_3.

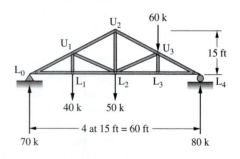

7.3   Members U_1U_2 and L_2U_3. (*Ans.* $F_{U_1U_2} = -115\,k$, $V_{L_2U_3} = -8.75\,k$)

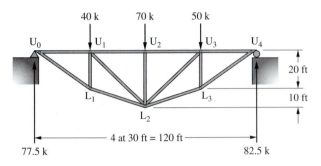

40 k   70 k   50 k

U_0 U_1 U_2 U_3 U_4

20 ft

L_1 L_3

10 ft

L_2

4 at 30 ft = 120 ft

77.5 k   82.5 k

7.4   Members AB and CD.

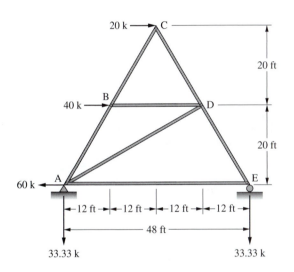

20 k   C

20 ft

B   D

40 k

20 ft

A   E

60 k

12 ft   12 ft   12 ft   12 ft

48 ft

33.33 k   33.33 k

7.5   Members U_1U_2 and L_2L_4. (*Ans.* $F_{U_1U_2} = -76.67\,k$, $H_{L_2L_4} = +43.75\,k$)

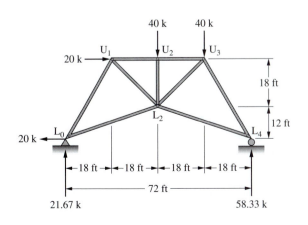

40 k   40 k

U_1 U_2 U_3

20 k

18 ft

L_2

12 ft

L_0 L_4

20 k

18 ft   18 ft   18 ft   18 ft

72 ft

21.67 k   58.33 k

7.6   Determine the force in U_1U_2 using section 1-1. Having that value, calculate the force in M_1U_2 making use of section 2-2.

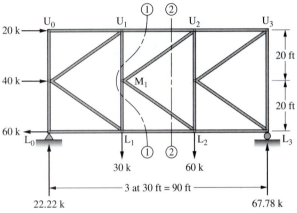

20 k U_0 U_1 ① ② U_2 U_3

20 ft

40 k M_1

20 ft

60 k L_0 L_1 ① ② L_2 L_3

20 ft

30 k   60 k

3 at 30 ft = 90 ft

22.22 k   67.78 k

For Problems 7.7 through 7.31, determine the forces in all the members of these trusses using method of sections or method of joints, as appropriate. (For Problems 7.27 and 7.28, only half of the member forces are to be determined.)

7.7   (*Ans.* $F_{U_2L_2} = +50\,k$, $V_{U_2U_3} = -45\,k$)

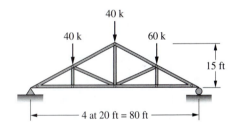

40 k

40 k   60 k

15 ft

4 at 20 ft = 80 ft

7.8

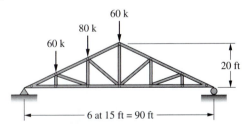

60 k

80 k

60 k

20 ft

6 at 15 ft = 90 ft

7.9   Rework Problem 7.7 if the panels are changed from 4 @ 20 ft to 4 @ 15 ft, the truss height changed from 15 to 20 ft, and the 40 k loads are doubled. (*Ans.* $F_{L_1L_2} = +172.5\,k$, $V_{U_2U_3} = -75\,k$)

7.10   Rework Problem 7.7 if a uniform load of 3 klf is applied for the entire span in addition to the loads shown. This additional load is to be applied to the bottom-chord joints.

7.11   (*Ans.* $V_{U_1 L_2} = +80.0\,k$, $H_{U_2 U_3} = -466.7\,k$)

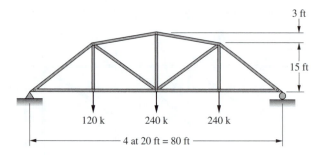

3 ft

15 ft

120 k   240 k   240 k

4 at 20 ft = 80 ft

7.12

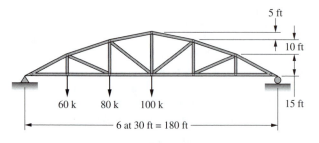

5 ft

10 ft

15 ft

60 k   80 k   100 k

6 at 30 ft = 180 ft

7.13   (*Ans.* $V_{U_2 L_3} = -33.33\,k$, $F_{U_3 U_4} = -300\,k$)

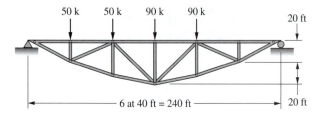

50 k   50 k   90 k   90 k

20 ft

20 ft

6 at 40 ft = 240 ft

7.14

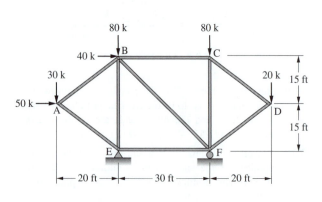

80 k   80 k

40 k   B   C

30 k

50 k   A

20 k   15 ft

D

15 ft

E   F

20 ft — 30 ft — 20 ft

7.15   (*Ans.* $F_{L_1 L_3} = -7.5\,k$, $V_{U_2 L_3} = -43.33\,k$)

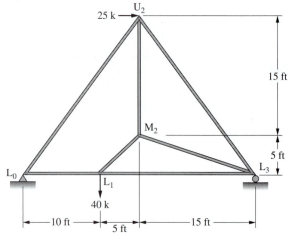

25 k → U$_2$

15 ft

M$_2$

5 ft

L$_0$ L$_1$ L$_3$

40 k

10 ft — 5 ft — 15 ft

7.16

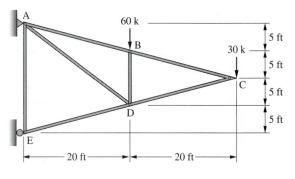

A

60 k

B   5 ft

30 k   5 ft

C

5 ft

D

5 ft

E

20 ft — 20 ft

7.17   (*Ans.* $F_{U_1 L_1} = +160\,k$, $V_{L_1 L_2} = +65\,k$)

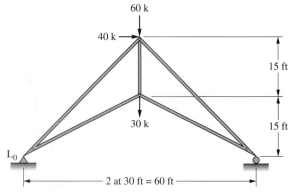

60 k

40 k →

15 ft

30 k

15 ft

L$_0$

2 at 30 ft = 60 ft

7.18

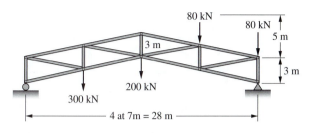

80 kN
80 kN
5 m
3 m
3 m
200 kN
300 kN
4 at 7m = 28 m

7.22

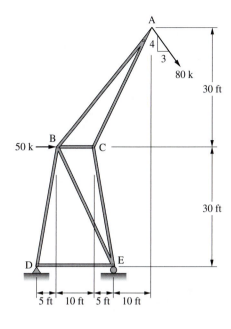

A
4
3
80 k
30 ft
B
50 k
C
30 ft
D
E
5 ft 10 ft 5 ft 10 ft

7.19 (*Ans.* $H_{BE} = +20\,k$, $F_{DE} = -20\,k$)

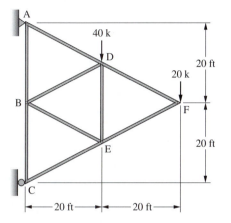

A
40 k
D
20 ft
20 k
B
F
20 ft
E
C
20 ft 20 ft

7.23 (*Ans.* $V_{U_1L_2} = -15.56\,k$, $H_{U_3U_4} = +22.22\,k$)

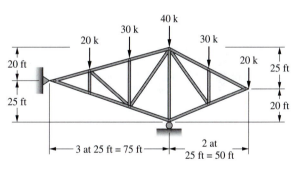

40 k
30 k
20 k
30 k
20 ft
20 k
25 ft
25 ft
20 ft
3 at 25 ft = 75 ft
2 at 25 ft = 50 ft

7.20

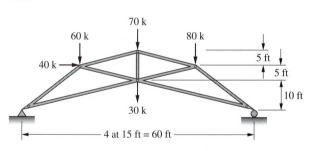

70 k
60 k
80 k
40 k
5 ft
5 ft
10 ft
30 k
4 at 15 ft = 60 ft

7.24

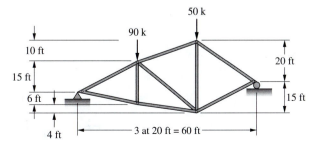

50 k
90 k
10 ft
20 ft
15 ft
6 ft
15 ft
4 ft
3 at 20 ft = 60 ft

7.21 (*Ans.* $H_{U_0L_1} = -60\,k$, $F_{U_2L_2} = -70\,k$)

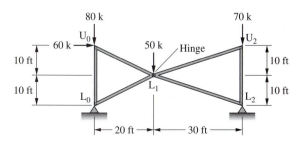

80 k
70 k
U_0
U_2
60 k
50 k
Hinge
10 ft
10 ft
10 ft
10 ft
L_0
L_1
L_2
20 ft 30 ft

7.25   (*Ans.* $F_{U_1U_4} = -28$ k, $V_{M_2L_3} = +66.67$ k)

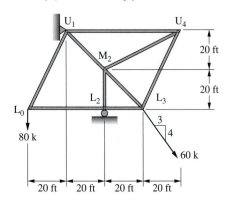

7.26

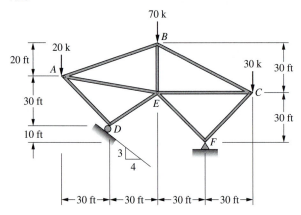

7.27   Determine forces in left half of truss including mid vertical. (*Ans.* $F_{U_2U_3} = -187.5$ k, $F_{L_3L_4} = +236.25$ k)

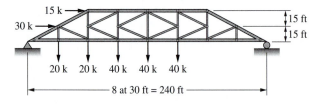

7.28   Determine forces in left half of truss.

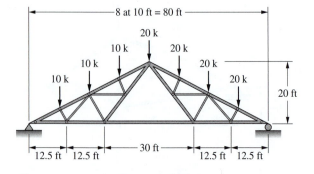

7.29   (*Ans.* $F_{AB} = -15$ k, $F_{EF} = +131.25$ k, $F_{FG} = -47.73$ k)

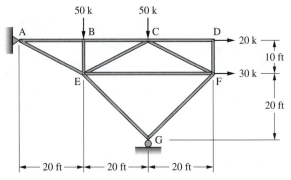

7.30   (Hint: you might draw a section around triangle DEF, take moments at the intersection of two of the cut diagonals to determine the force in the third diagonal.)

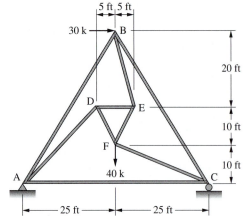

7.31   (*Ans.* $F_{U_1U_2} = +225$ k, $F_{L_2L_3} = -120$ k)

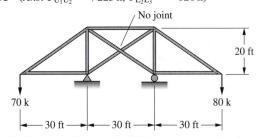

For Problems 7.32 through 7.39 use the zero-load test to determine if the members have critical form.

7.32

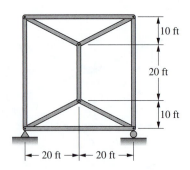

7.33   (*Ans.* Unstable)

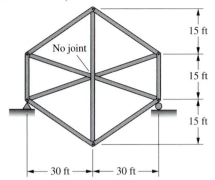

15 ft
No joint
15 ft
15 ft
30 ft
30 ft

7.34

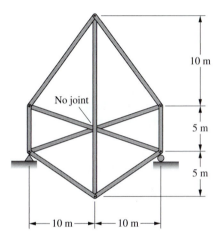

10 m
No joint
5 m
5 m
10 m
10 m

7.35   (*Ans.* Stable)

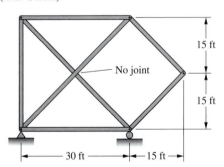

15 ft
No joint
15 ft
30 ft
15 ft

7.36

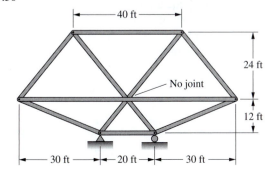

40 ft
24 ft
No joint
12 ft
30 ft
20 ft
30 ft

7.37   (*Ans.* Unstable)

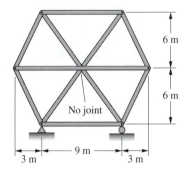

6 m
6 m
No joint
3 m
9 m
3 m

7.38

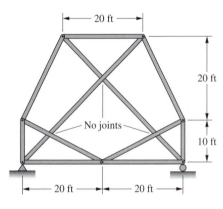

20 ft
20 ft
No joints
10 ft
20 ft
20 ft

7.39   (*Ans.* Stable)

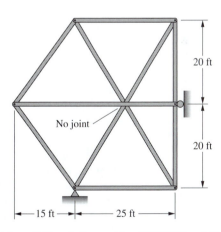

20 ft
No joint
20 ft
15 ft
25 ft

For Problems 7.40 to 7.42 use either SABLE32 or SAP2000 to repeat the analysis of the given trusses.

7.40   Problem 7.4

7.41   Problem 7.24 (*Ans.* $F_{L_0U_1} = -91.27$ k, $F_{U_1L_2} = -58.95$ k)

7.42   Problem 7.25

Chapter 8

Three-Dimensional or Space Trusses

8.1 INTRODUCTION

A very brief and elementary discussion of small space trusses is presented in this chapter. This material hopefully will help the reader develop some understanding of the behavior of space structures and to recognize that the equations of statics are as applicable in three dimensions as they are in two dimensions.

For all but the very smallest space structures, analysis by the methods of joints and moments described herein is completely unwieldy. Consequently, a large percentage of colleges do not introduce the topic until students take a course in matrix analysis.

Nearly all structures are three-dimensional in nature. However, they usually can be broken down into separate systems, each lying in a single plane at right angles to the other. Because they are at right angles, the forces in one system have no affect on the forces in the other systems. The members joining two systems together serve as members of both systems, and their total force is obtained by combining the forces developed as a part of each of those systems.

Many towers, domes, and derricks are three-dimensional structures made up of members arranged so that it is impossible to divide them into different systems, each lying in a single plane, which then could be handled individually. These systems lie in planes that are not at right angles to one another. The forces in one truss framed into another at an angle other than 90° affect the forces in that second truss. For trusses of this type we must analyze the entire structure as a unit, rather than considering the systems in various planes individually. This chapter is devoted to these types of trusses.

Structural engineers are so accustomed to visualizing structures in one plane that when they encounter space trusses they may make frequent mistakes because their minds still are operating on a single-plane basis. If the layout of a space truss is not completely clear, the construction of a small model will probably clarify the situation. Even the simplest models of paper, cardboard, or wire are helpful.

8.2 BASIC PRINCIPLES

Prior to introducing a method of analyzing space trusses, a few of the basic principles pertaining to such structures need to be considered. Like two-dimensional trusses, three-dimensional trusses are assumed to be composed of members subject to axial force only. In other words, the trusses are assumed to

- Have members that are straight between the joints;
- Have loads applied only at joints; and
- Have members whose ends are free to rotate. (Note that for this situation to be true the members would have to be connected with universal joints, or at least with several frictionless pins).

Analyses based on these assumptions usually are quite satisfactory despite the welded and bolted connections used in actual practice.

A system of forces coming together at a single point can be combined into one resultant force. The forces so resolved do not necessarily have to be in the same plane. Similarly, a single force can be resolved into component forces in each of the three directions. The coordinate directions referred to here are the X, Y, and Z coordinates that compose the Cartesian coordinate system. Recall that the Cartesian system is a right-hand orthogonal system. That means that the coordinate axes are at right angles to each other. When a force is resolved into its coordinate components, as shown in Figure 8.1, the magnitude of the components is proportional to their length projections on the axes.

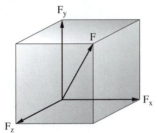

Figure 8.1   Resolution of a force into components

The values of the component forces can be computed algebraically from the following relationship:

> The force in a member is to the length of the member as the X, Y, or Z component of force is to the corresponding X, Y, or Z component of length.

This relationship can be expressed mathematically as:

$$\frac{F}{L} = \frac{F_x}{L_x} = \frac{F_y}{L_y} = \frac{F_z}{L_z}$$

$$L^2 = L_x^2 + L_y^2 + L_z^2$$

$$F^2 = F_x^2 + F_y^2 + F_z^2$$

Space trusses may be either statically determinate or statically indeterminate; consideration is given here only to those that are statically determinate. The methods developed in later chapters for statically indeterminate structures apply equally to three-dimensional and two-dimensional trusses.

8.3 EQUATIONS OF STATIC EQUILIBRIUM

There are more equations of static equilibrium available for determining the reactions of three-dimensional structures because there are two more axes to take moments about and one new axis along which to sum forces. For equilibrium, the sum of the forces along each of the three reference axes must be equal to zero, as does the sum of the moments of all the forces about each of the axes. Six equations of equilibrium are available. These equations are

$$\sum F_x = 0 \qquad \sum M_x = 0$$
$$\sum F_y = 0 \qquad \sum M_y = 0$$
$$\sum F_z = 0 \qquad \sum M_z = 0$$

The six reaction components may be determined directly from these equations.

Denver, Colorado, Convention Center (Courtesy Bethlehem Steel Corporation)

Should a structure have more than six reaction components, it is statically indeterminate externally. If there are fewer than six components of reaction, the truss is unstable. If the number of components of reaction is equal to six, the truss is statically determinate externally. Many space trusses, however, have more than six reaction components and yet are statically determinate overall. Example 8.2 shows that the reactions for this type of structure may be determined by solving them concurrently with the member forces, using only equations of static equilibrium.

As with plane trusses, the basic geometric shape of the space truss is the triangle. A triangle can be extended into a space truss by adding three members and one joint. Each of the new members frames into one of the joints of the basic triangle; the other ends come together to form a new joint. The elementary space truss formed in this manner has six members and four joints. It is called a tetrahedron, a figure with four triangular surfaces. This fundamental space truss may be enlarged by the addition of three members and one joint. For each of the joints of a space frame three equations of static equilibrium ($\sum F_x = 0$, $\sum F_y = 0$, and $\sum F_z = 0$) are available for calculating the unknown forces. If we let j equal the number of joints in the truss, m equal the number of members, and r equal the number of reaction components, we see that for a space truss to be statically determinate the following relation must be satisfied:

$$3j = m - r$$

Should there be joints in the truss where the members are all in one plane, only two equations are available at each, and it is necessary to subtract one from the left-hand side of the equation for each such joint. The omission of one member for each reaction component in excess of six will cause this equation to be satisfied, and the structure will be statically determinate internally. When this situation occurs, it is possible to compute the forces and reactions for the truss from three-dimensional static equilibrium, even though it is statically indeterminate externally.

8.4 STABILITY OF SPACE TRUSSES

The general rule for stability in regard to external forces is that the projection of the structure onto any of the three coordinate planes must itself be a stable truss. Therefore, as with two-dimensional structures, there must be at least three nonconcurrent components of reaction in any one plane. The results of reaction computations will be inconsistent for any other case.

In the preceding paragraphs, both external and internal stability and determinacy were treated as though they were completely independent subjects. The two have been separated here for clarity for the reader who has not previously encountered space trusses. Note however, from the example problems in the pages that follow that it is impossible in a majority of cases to consider the two separately. For instance, many trusses are statically indeterminate externally and statically determinate internally and can be completely analyzed using only the equations of static equilibrium. Few two-dimensional structures fall into this class.

In Section 8.3, it was stated that for a space truss to be statically determinate the $3j = m - r$ equation had to be satisfied.

This equation is not sufficient to show whether a particular space truss is stable, however. Externally the reactions must be arranged to prevent movement of the structure: internally the members must be placed to prevent the joints from moving with respect to each other.

For external stability, the reactions must be placed so that they can resist translation along and rotation about each of the three coordinate axes. To achieve this goal there must be at least six reactions and they must not intersect a common axis.

Internal stability can be achieved if the geometry of the truss is developed tetrahedron by tetrahedron, that is, by successively adding one joint and three members. In large space trusses it may be quite difficult to see if this condition has been met. An analysis of the truss, however, will provide a clear statement of stability. If we can obtain a *unique solution*, the truss is stable. If not, the truss is unstable. The zero-load test that was discussed in Section 7.7 also can be used to check stability.

8.5 SPECIAL THEOREMS APPLYING TO SPACE TRUSSES

From the principles of elementary statics, two theorems that are useful when analyzing space trusses may be developed. These are discussed in the following paragraphs.

1. The component of force in a member 90° to the direction of the member is equal to zero, because no matter how large the force may be it is equal to zero when multiplied by the cosine of 90°. A force in one plane cannot have components in a plane normal to the original plane. Furthermore, a force in one plane cannot cause moment about any axis in its plane, because it will either intersect the axis or be parallel to it.

From the foregoing principle, if several members of a truss come together at a joint, and all but one lie in the same plane, the component of force in the member normal to the plane of the other members must be equal to the sum of the components of the external forces at the joint normal to the same plane. If no external forces are present, the member has a force of zero.

2. If there is a joint in a truss to which no external loads are applied, and all but two members framing into the joint have zero-force, these two members must have zero-force unless they form a straight line.

8.6 TYPES OF SUPPORT

In Chapters 6 and 7, plane trusses were supported with rollers and hinges that provided one or two reaction forces. With three-dimensional trusses, the same types of supports are used, but the number of reaction forces can vary from one to three. These supports are described in the following paragraphs and are illustrated in Figure 8.2.

1. The *plane roller, steel ball*, and *flat plate* types of support provide resistance to movement perpendicular to the supporting surface. Thus, it has one component of reaction, which can be toward or away from the supporting surface.

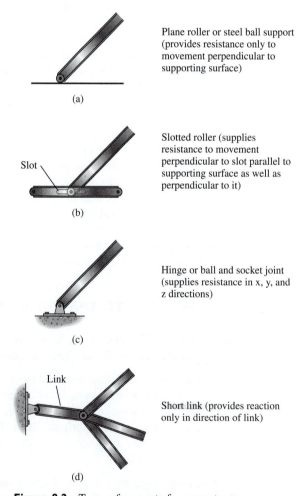

(a)

Plane roller or steel ball support (provides resistance only to movement perpendicular to supporting surface)

(b)

Slot

Slotted roller (supplies resistance to movement perpendicular to slot parallel to supporting surface as well as perpendicular to it)

(c)

Hinge or ball and socket joint (supplies resistance in x, y, and z directions)

(d)

Link

Short link (provides reaction only in direction of link)

Figure 8.2   Types of supports for space trusses

2. The *slotted roller* is free to move in one direction parallel to the supporting surface. Movement is prevented in the other direction parallel to the supporting surface as well as perpendicular to it, giving a total of two reaction components.

3. The *hinge* or *ball and socket joint* types of support provide resistance to movement in all three coordinate directions. Thus, they provide a total of three reaction components.

4. The *short link* provides resistance only in the direction of the link. As such, it provides only one reaction component and that force is parallel to the link.

This discussion indicates that it is possible to select a type of support having three reaction components or one that may have only one or two components. A little thought on the subject shows that the possibility of limiting the number of reaction components of a space truss is very advantageous. A truss that is statically indeterminate externally may have its total reaction components limited to six, making it statically determinate. The advantages of statically determinate and statically indeterminate structures are discussed in Chapter 14. For some structures, good design necessitates elimination of reaction forces in certain directions. The most obvious example occurs when a space truss is supported on walls where a reaction or thrust perpendicular to the wall is undesirable.

Broome County Veterans Memorial Arena, Binghamton, NY
(Courtesy Bethlehem Steel Corporation)

8.7 ILLUSTRATIVE EXAMPLES

Examples 8.1 and 8.2 illustrate the application of the foregoing principles to elementary space trusses. Example 8.1 considers a structure supported at three points with six reaction components, which can be computed directly. The second example presents a space truss supported at four points with seven reaction components, which cannot be solved directly. The directions in which reaction components are possible are indicated herein by dark heavy lines at the support points, as shown in the diagrams of the frames analyzed in Examples 8.1 to 8.3.

EXAMPLE 8.1

Determine the reactions and member forces in the space truss shown in Figure 8.3.

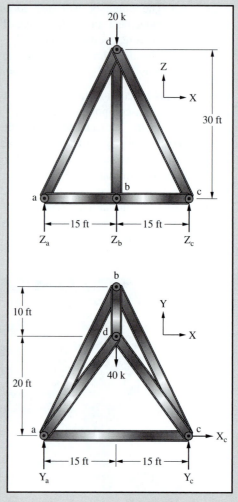

Figure 8.3

Solution.   The truss is statically determinate and stable externally because there is a total of six reaction components with three nonconcurrent reaction forces in each plane. Internally it is statically determinate, as proved with the joint equation.

$$3j = m + r$$
$$(3)(4) = 6 + 6$$
$$12 = 12$$

For a truss with three vertical reaction components, moments may be taken about an axis through any two of the components to find the third.

$$\sum M_x = 0 \quad \text{about line ac}$$
$$0 = 40(30) - 20(20) + 30Z_b$$
$$\therefore Z_b = -26.7 \text{ k} \downarrow$$

$$\sum M_y = 0 \quad \text{about line of action of } Y_a$$
$$0 = 20(15) + 26.7(15) - 30Z_c$$
$$\therefore Z_c = 23.3 \text{ k } \uparrow$$
$$\sum F_z = 0$$
$$0 = -20 - 26.7 + 23.3 + Z_a$$
$$\therefore Z_a = 23.4 \text{ k } \uparrow$$

Similarly, where there are three unknown horizontal reaction components moments may be taken about a vertical axis passing through the point of intersection of two of the components.

$$\sum M_z = 0 \quad \text{about line of action of } Z_c$$
$$0 = -40(15) + 30Y_a$$
$$\therefore Y_a = 20.0 \text{ k } \uparrow$$
$$\sum F_y = 0$$
$$0 = 20 - 40.0 + Y_c$$
$$\therefore Y_c = 20.0 \text{ k } \uparrow$$
$$\sum F_x = 0$$
$$0 = 0 + X_c$$
$$\therefore X_c = 0$$

When the reactions have been found, the member forces can readily be computed using the method of joints. At joint a, member ad is the only member having a Z component of length; therefore, its Z component must be equal and opposite to Z_a, or 23.4-kip compression. The X and Y components of ad are proportional to its components of length in those directions. Setting up a table similar to the one shown simplifies the computation of components and resultant forces.

Considering joint a, the Y component of force in member ab can be determined by joints now that the Y component of ad is known:

$$\sum F_y = 0 \quad \text{at joint a}$$
$$0 = 20 - 15.6 - Y_{ab}$$
$$\therefore Y_{ab} = -4.4 \text{ k } \downarrow$$

The other member forces are computed similarly using the method of joints. The results are shown in the table to follow:

Bar	Projection (ft)			Length (ft)	Component Forces (kips)			Force (kips)
	X	Y	Z		X	Y	Z	
ab	15	30	0	33.5	−2.2	−4.4	0	−4.92
ac	30	0	0	30.0	13.9	0	0	13.9
ad	15	20	30	39.1	−11.7	−15.6	−23.4	−30.5
bc	15	30	0	33.5	−2.2	−4.5	0	−5.02
bd	0	10	30	31.6	0	8.9	26.7	28.1
cd	15	20	30	39.1	−11.7	−15.5	−23.3	−30.3

EXAMPLE 8.2

Find all reactions and member forces of the space truss shown in Figure 8.4.

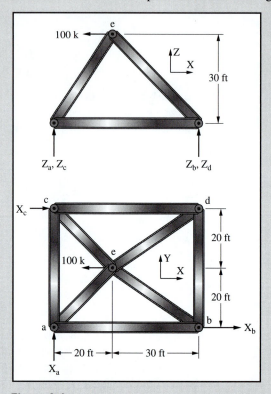

Figure 8.4

Solution.   This truss appears to be statically indeterminate externally because there are seven definite forces of reaction and only six equations of static equilibrium. Internally, however, the truss is statically determinate as shown, and the analysis may be conducted using only equations of static equilibrium.

$$3j = m + r$$
$$3(5) = 8 + 7$$
$$15 = 15$$

Although the truss is statically indeterminate externally, there are only three unknown reaction components in the XY plane, and these forces can be determined immediately. The other four components will be computed in conjunction with the member forces.

$$\sum M_z = 0 \quad \text{about line of action of } Z_a$$
$$0 = -100(20) + 40X_c$$
$$\therefore X_c = 50.0 \text{ k}$$
$$\sum F_x = 0$$
$$0 = 50 - 100.0 + X_b$$
$$\therefore X_b = 50.0 \text{ k}$$

$$\sum F_y = 0$$
$$0 = 0 + Y_a$$
$$\therefore Y_a = 0.0$$

If the value of one of the Z reaction components should be known, the values of the other three could be determined from static equilibrium. Assume Z_d has a value of S directed downward, and then compute the components of reaction in terms of S:

$$\sum M_y = 0 \quad \text{about line ac}$$
$$0 = -100(30) + 50S + 50Z_b$$
$$\therefore Z_b = 60 - S$$
$$\sum M_x = 0 \quad \text{about line ab}$$
$$0 = 40S - 40Z_c$$
$$\therefore Z_c = S$$
$$\sum F_z = 0$$
$$0 = S - S - (60 - S) + Z_a$$
$$\therefore Z_a = 60 - S$$

Checking by $\sum M_x = 0$ about line cd
$$0 = (60 - S)(40) - 40Z_a$$
$$\therefore Z_a = 60 - S$$

The calculation of member forces may now be started using the reaction forces in terms of S. These computations are continued until the forces at both ends of one bar are determined in terms of S. The two values must be equal, and they are equated to calculate the correct value of S.

The Z component of force in member de is equal to S and is in tension, whereas the Z component of force in member be is equal to 60-S and is also tensile. The Y component of force in member de is equal to

$$Y_{be} = \frac{20}{30}(60 - S) = 40 - \frac{2}{3}S$$

By summing forces in the Y direction at joint d, member bd is found to be acting in compression and has a force of 2S/3. Similarly, by summing forces in the Y direction at joint b, member bd is found to have a compressive force of $40 - 2S/3$. Because these two values are the force in member bd, they can be set equal to each other and the value of S can be computed.

$$\frac{2}{3}S = 40 - \frac{2}{3}S$$
$$\therefore S = 30 \text{ k}$$

The numerical values of the Z-reaction components can now be found from S, and the forces in the truss can be determined using the method of joints. The use of a table to work with length and force components is again convenient. The results are shown in the following table:

	Projection (ft)				Component Forces (kips)			
Bar	X	Y	Z	Length (ft)	X	Y	Z	Force (kips)
ab	50	0	0	50.0	+20	0	0	+20.0
ae	20	20	30	41.2	−20	−20	−30	−41.2
ac	0	40	0	40.0	0	+20	0	+20.0
be	30	20	30	46.9	+30	+20	+30	+46.9
bd	0	40	0	40.0	0	−20	0	−20.0
de	30	20	30	46.9	+30	+20	+30	+46.9
cd	50	0	0	50.0	−30	0	0	−30.0
ce	20	20	30	41.2	−20	−20	−30	−41.2

■

Transmission towers for the country's first 345,000-volt
transmission line, Chief Joseph-Snohomish Dam,
Washington State (Courtesy Bethlehem Steel Corporation)

8.8 SOLUTION USING SIMULTANEOUS EQUATIONS

Three simultaneous equations:

$$\sum F_x = 0$$
$$\sum F_y = 0$$
$$\sum F_z = 0$$

may be written for the forces meeting at each joint of a space truss. This results in 3j
simultaneous equations. If the truss is statically determinate, the equations may be solved

for both the member forces and the components of the reactions which are the unknowns in the simultaneous equations. Despite the large number of equations that can result, this method of solution may be rather quick for small space trusses, because of the small number of unknowns that appear in each equation.

The preparation and solution of simultaneous equations for space trusses can be simplified by making use of tension coefficients.[1] The tension coefficient for a member is equal to the force in the member divided by its length. In each of the following expressions for force components, the value F/L is replaced by T, the tension coefficient.

$$F_x = \frac{L_x}{L}F = \frac{F}{L}L_x = TL_x$$

$$F_y = \frac{L_y}{L}F = \frac{F}{L}L_y = TL_y$$

$$F_z = \frac{L_z}{L}F = \frac{F}{L}L_z = TL_z$$

When tension coefficients are used, the resulting simultaneous equations are then in terms of the tension coefficient for each member. The equations are solved for the tension coefficients, which are then multiplied by the appropriate member length to obtain the final member forces. The analysis of a space truss through use of tension coefficients and simultaneous equations is demonstrated in Example 8.3.

EXAMPLE 8.3

Using simultaneous equations, determine the forces of reaction and the member forces in the truss shown in Figure 8.5. This is the same truss analyzed in Example 8.1.

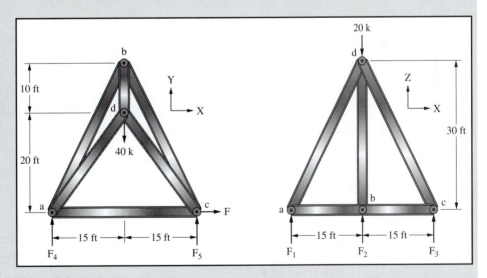

Figure 8.5

Solution.   We begin the analysis by writing the equations of equilibrium for each joint in the truss. Unknown member forces are assumed to be in tension and positive reactions are in the same direction as the coordinate axes.

---

[1]R. V. Southwell, "Primary Stress Determination in Space Frames," *Engineering 109* (1920): 165.

$$\text{Joint a} \begin{cases} 30T_{ac} + 15T_{ab} + 15T_{ad} = 0 \\ F_4 + 30T_{ab} + 20T_{ad} = 0 \\ F_1 + 30T_{ad} = 0 \end{cases}$$

$$\text{Joint b} \begin{cases} -15T_{ab} + 15T_{bc} = 0 \\ -30T_{ab} - 30T_{bc} - 10T_{bd} = 0 \\ F_2 + 30T_{bd} = 0 \end{cases}$$

$$\text{Joint c} \begin{cases} F_6 - 30T_{ac} - 15T_{bc} - 15T_{cd} = 0 \\ F_5 + 30T_{bc} + 20T_{cd} = 0 \\ F_3 + 30T_{cd} = 0 \end{cases}$$

$$\text{Joint d} \begin{cases} -15T_{ad} + 15T_{cd} = 0 \\ -20T_{ad} + 10T_{bd} - 20T_{cd} = 40 \\ -30T_{ad} - 30T_{bd} - 30T_{cd} = 20 \end{cases}$$

By solving these 12 equations simultaneously, we can determine first the values of the tension coefficients and then the reactions and the forces in each member. The results are as follows:

Member	Tension Coefficient (T)	Member Length (L)	Final Force (T·L)
ab	−0.148	33.54 ft	−4.97
ac	0.463	30.00 ft	13.89
ad	−0.778	39.05 ft	−30.37
bc	−0.148	33.54 ft	−4.97
bd	0.889	31.62 ft	28.11
cd	−0.778	39.05 ft	−30.37

8.9 EXAMPLE PROBLEM WITH SABLE32

From the example problems previously presented in this chapter, it is quite obvious that the analysis of a medium to large space truss with a pocket calculator would be a very lengthy and tedious job. As a result, all but the very simplest space trusses analyses will be carried out with computers.

The procedure for analyzing space trusses with SABLE32 is very similar to the procedure used for plane trusses in Chapter 6. After you click on the three-dimensional part of the program, only a few explanatory remarks are deemed necessary to enable you to carry out the analysis. These follow:

1. The coordinates of the truss joints are input as they were for plane trusses, noting that each joint has x, y, and z values.

2. The members of the truss are assumed to be bars fitting into ball and socket joints at their ends. As a result, there is resistance to rotation in the x, y, and z directions at those points. Unless these restraints are shown, SABLE32 will recognize the structure as unstable. The reasons for showing this restraint will become clear when we begin studying matrix analysis methods in Chapters 22–25.

3. Once again it is necessary to click on the active status when the joint load data is supplied.

EXAMPLE 8.4

Using SABLE32 repeat the analysis of the space truss of Example 8.1.

Solution.

A sketch of the truss is repeated in Figure 8.6 and the truss joints and members are numbered. In this figure, the circled numbers are joints and the member numbers are in squares.

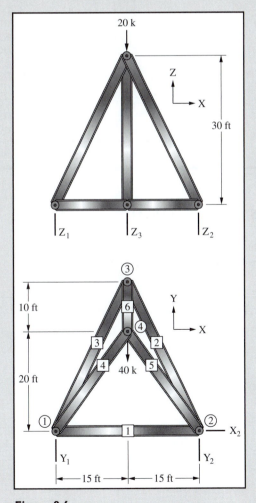

Figure 8.6

Calculated Beam End Forces

Beam	Case	End	Axial	Shear-Y	Moment-Z
		Torsion	Shear-Z	Moment-Y	
1	1	i	$-1.389E+01$	$0.000E+00$	$0.000E+00$
			$0.000E+00$	$0.000E+00$	$0.000E+00$
		j	$1.389E+01$	$0.000E+00$	$0.000E+00$
			$0.000E+00$	$0.000E+00$	$0.000E+00$
2	1	i	$4.969E+00$	$0.000E+00$	$0.000E+00$
			$0.000E+00$	$0.000E+00$	$0.000E+00$
		j	$-4.969E+00$	$0.000E+00$	$0.000E+00$
			$0.000E+00$	$0.000E+00$	$0.000E+00$
3	1	i	$4.969E+00$	$0.000E+00$	$0.000E+00$
			$0.000E+00$	$0.000E+00$	$0.000E+00$
		j	$-4.969E+00$	$0.000E+00$	$0.000E+00$
			$0.000E+00$	$0.000E+00$	$0.000E+00$
4	1	i	$3.037E+01$	$0.000E+00$	$0.000E+00$
			$0.000E+00$	$0.000E+00$	$0.000E+00$
		j	$-3.037E+01$	$0.000E+00$	$0.000E+00$
			$0.000E+00$	$0.000E+00$	$0.000E+00$
5	1	i	$-2.811E+01$	$0.000E+00$	$0.000E+00$
			$0.000E+00$	$0.000E+00$	$0.000E+00$
		j	$2.811E+01$	$0.000E+00$	$0.000E+00$
			$0.000E+00$	$0.000E+00$	$0.000E+00$
6	1	i	$3.037E+01$	$0.000E+00$	$0.000E+00$
			$0.000E+00$	$0.000E+00$	$0.000E+00$
		j	$-3.037E+01$	$0.000E+00$	$0.000E+00$
			$0.000E+00$	$0.000E+00$	$0.000E+00$

8.10 PROBLEMS FOR SOLUTION

For Problems 8.1 through 8.7, compute the reaction components and the member forces for the space trusses.

8.1 (*Ans.* $F_{ab} = +9\,k$, $F_{ac} = -11.62\,k$, $F_{bd} = -41.62\,k$)

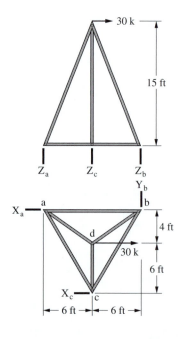

8.2

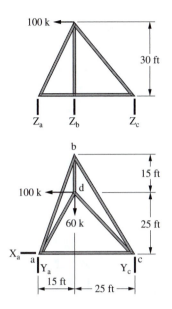

8.4

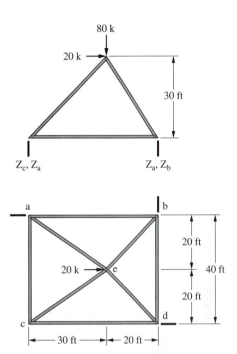

8.3   (*Ans.* $F_{ac} = +29.98$ kN, $F_{bc} = -34.96$ kN, cd $= 0$)

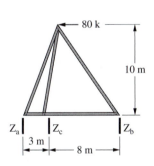

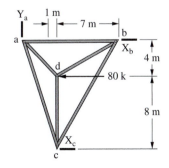

8.5   (*Ans.* $F_{ae} = +71.8$ k, $F_{ac} = -25$ k, ce $= 0$)

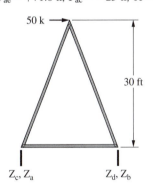

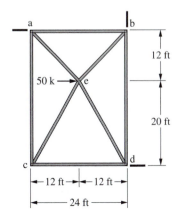

8.6

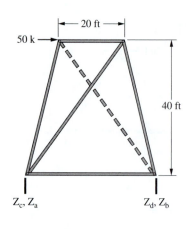

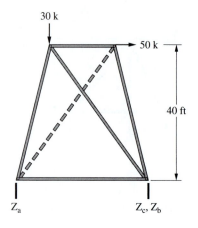

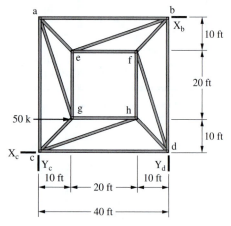

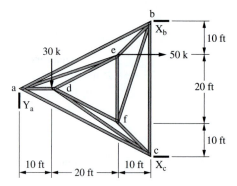

Solve Problems 8.8 to 8.9 using SABLE32 or SAP2000.

8.8 Problem 8.2

8.9 Problem 8.4 (*Ans.* $F_{ae} = -15.63$ k, $F_{cd} = +10$ k, $F_{be} = -41.23$ k)

8.7 (*Ans.* bc $= +9.82$ k, bf $= +31.8$ k, de $= +29.3$ k)

Chapter 9

Influence Lines For Beams

9.1 INTRODUCTION

Structures supporting groups of loads fixed in one position have been discussed in the previous chapters. Regardless of whether beams, frames, or trusses were being considered and whether the functions sought were shears, reactions, or member forces, the loads were stationary. In practice, however, the engineer rarely deals with structures supporting only fixed loads. Nearly all structures are subject to loads moving back and forth across their spans. Perhaps bridges with their vehicular traffic are the most noticeable examples, but industrial buildings with traveling cranes, office buildings with furniture and human loads, and frames supporting conveyor belts fall in the same category.

Each member of a structure must be designed for the most severe conditions that can possibly develop in that member. The designer places the live loads at the positions where they will produce these conditions. The critical positions for placing live loads will not be the same for every member. For example, the maximum force in one member of a bridge truss may occur when a line of trucks extends from end to end of the bridge. The maximum force in some other member, however, may occur when the trucks extend only from that member to one end of the bridge. The maximum forces in certain beams and columns of a building will occur when the live loads are concentrated in certain portions of the building. The maximum forces in other beams and columns will occur when the live loads are placed elsewhere.

On some occasions, you can determine where to place the loads to give the most critical forces by inspection. On many other occasions, however, you will need to use certain criteria or diagrams to find the locations. The most useful of these devices is the influence line.

9.2 THE INFLUENCE LINE DEFINED

The influence line, which was first used by Professor E. Winkler of Berlin in 1867, shows graphically how the movement of a unit load across a structure influences some function of the structure.[1] The functions that may be represented include reactions, shears, moments, and deflections.

---

[1]J. S. Kinney, *Indeterminate Structural Analysis* (Reading, Mass.: Addison-Wesley, 1957), Chapter 1.

Tennessee-Tombigbee Waterway Bridge in Mississippi (Courtesy of the Mississippi State Highway Department)

An influence line may be defined as a diagram whose ordinates show the magnitude and character of some function of a structure as a unit load moves across it. Each ordinate of the diagram gives the value of the function when the load is at that ordinate.

Influence lines are primarily used to determine where to place live loads to cause maximum forces. They may also be used to compute those forces. The procedure for drawing the diagrams is simply the plotting of values of the function under study as ordinates for various positions of the unit load along the span and then connecting those ordinates. You should mentally picture the load moving across the span and try to imagine what is happening to the function in question during the movement. The study of influence lines can immeasurably increase your knowledge of what happens to a structure under different loading conditions.

Study of the following sections should fix clearly in your mind what an influence line is. The actual mechanics of developing the diagrams are elementary, once the definition is completely understood. No new fundamentals are introduced here; rather, a method of recording information in a convenient and useful form is given.

9.3 INFLUENCE LINES FOR SIMPLE BEAM REACTIONS

Influence lines for the reactions of a simple beam are given in Figure 9.1. First consider the variation of the left-hand reaction, V_L, as a unit load moves from left to right across the beam. When the load is directly over the left support, $V_L = 1$. When the load is 2 ft to the right of the left support, $V_L = 18/20$, or 0.9. When the load is 4 ft to the right, $V_L = 16/20$, or 0.8, and so on.

Values of V_L are shown at 2-ft intervals as the unit load moves across the span. These values lie in a straight line because they change uniformly for equal intervals of the load. For every 2-ft interval, the ordinate changes 0.1. The values of V_R, the right-hand reaction, are plotted in a similar fashion in the same figure. For each position of the unit load, the sum of the ordinates of the two diagrams at any point equals (and for equilibrium certainly must equal) the unit load.

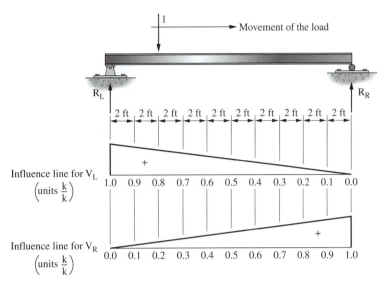

Figure 9.1 Influence line for reactions in a simple beam

9.4 INFLUENCE LINES FOR SIMPLE BEAM SHEARING FORCES

Influence lines are plotted in Figure 9.2 for the shearing force at two sections in a simple beam. The following sign convention for shearing force is used:

> Positive shearing force occurs when the sum of the transverse forces to the left of a section is up or when the sum of the forces to the right of the section is down.

This same sign convention was used for shearing force in Chapter 5. It is often referred to as the beam sign convention.

Precast-concrete Kalihiwai Bridge near Kilauea, Kauai, Hawaii
(Courtesy of the Hawaii Department of Transportation)

Placing the unit load over the left support causes no shearing force at either of the two sections. Moving the unit load 2 ft to the right of the left support results in a left-hand

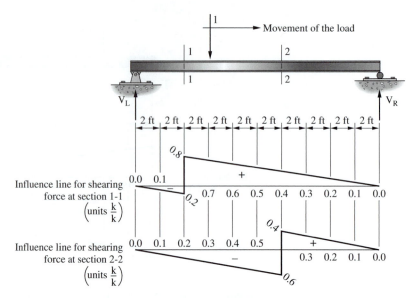

Figure 9.2 Two shearing force influence lines for a simple beam

reaction of 0.9. The sum of the forces to the left of section $1\text{-}1 = 0.9 - 1.0 = -0.1$ downward; the shearing force is -0.1. When the load is 4 ft to the right of the left support and an infinitesimal distance to the left of section 1-1, the shearing force to the left is $+0.8 - 1.0 = -0.2$. If the load is moved a very slight distance to the right of section 1-1, the sum of the forces to the left of the section becomes 0.8 upward; the shearing force is $+0.8$. Continuing to move the load across the span toward the right support results in decreasing values of the shear at section 1-1. These values are plotted for 2-ft intervals of the unit load. The influence line for shear at section 2-2 is developed in the same manner.

Observe that the slope of the shearing force influence line to the left of the section in question is equal to the slope of the influence line to the right of the section. In Figure 9.2, for instance, for the influence line at section 1-1, the slope to the left is $0.2/4 = 0.05$, while the slope to the right is $0.8/16 = 0.05$. This information is very useful in drawing other shear influence lines.

9.5 INFLUENCE LINES FOR SIMPLE BEAM MOMENTS

Influence lines for bending moment are plotted in Figure 9.3 at the same sections of the beam used in Figure 9.2 for the shearing force influence lines. To review, a positive moment causes tension in the bottom fibers of a beam. It occurs at a particular section when the sum of the external moments of all the forces to the left is clockwise, or when the sum to the right is counterclockwise. Moments are taken at each of the sections for 2-ft intervals of the unit load.

To illustrate how the calculations were made for the moment influence lines of this figure, the following illustrations are presented. If the unit load is 2 ft from the left end of the beam, the left reaction will be $+0.9$ and the right reaction $+0.1$. The moment at section 1-1 may be determined by taking moments to the left or right of the section as follows:

$$M_{\text{to the left}} = +(0.9)(4) - (1.0)(2) = +1.6$$
$$M_{\text{to the right}} = +(0.1)(16) = +1.6$$

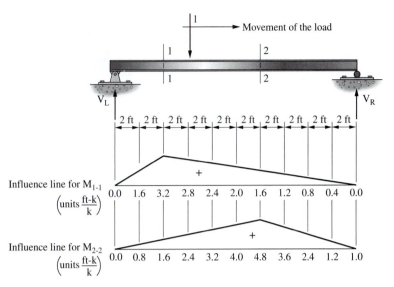

Figure 9.3   Two influence lines for bending moment in a simple beam

If the load is at section 1-1, the left reaction will be +0.8 and the right one will be +0.2. Then the moments will be

$$M_{\text{to the left}} = +(0.8)(4) = +3.2$$
$$M_{\text{to the right}} = +(0.2)(16) = +3.2$$

Several more influence lines are presented in Figure 9.4 for an overhanging beam. Included are illustrations for reactions, shears, and moments.

The major difference between shearing force and bending moment diagrams as compared with influence lines should now be clear. A shearing force or bending moment diagram shows the variation of shearing force or bending moment across an entire structure for loads fixed in one position. An influence line for shearing force or bending moment, on the other hand, shows the variation of that function at one section in the structure caused by the movement of a unit load from one end of the structure to the other.

Influence lines for functions of statically determinate structures consist of a set of straight lines. An experienced analyst will be able to compute values of the function under study at a few critical positions and connect the plotted values with straight lines. A student beginning his or her study, however, must be very careful to compute the value of the function for enough positions of the unit load. The shapes of influence lines for forces in truss members as described in Chapter 10 often are deceptive in their seeming simplicity. Plotting ordinates for several extra positions of the load is obviously better than failing to plot one essential value.

9.6 QUALITATIVE INFLUENCE LINES

Many structural analysis students initially have a great deal of trouble in preparing influence lines despite the simplicity of the mathematical calculations. The reason for this trouble is often rather perplexing to their professors. Perhaps the trouble occurs due to a lack of fully understanding the definition of influence lines. There is available a very

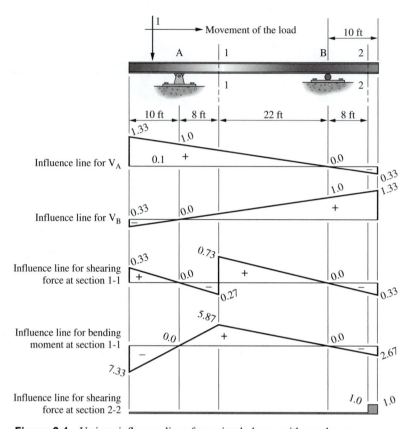

Figure 9.4   Various influence lines for a simple beam with overhangs

simple procedure that may help the students very much in their understanding and preparation of the diagrams. It involves the use of figures called qualitative influence lines. These diagrams enable the student to instantly draw the correct shape of influence lines without the need of any computations whatsoever.

The influence lines drawn in the previous sections for which numerical values were computed are referred to as *quantitative influence lines*. It is possible, however, to make rough sketches of these diagrams with sufficient accuracy for many practical purposes without computing any numerical values. These latter diagrams are referred to as *qualitative influence lines*.

A detailed discussion of the principle on which these sketches are made is given in Chapter 17 together with a consideration of their usefulness. Such a discussion is delayed until the student has had some exposure to the subject of deflections. Qualitative influence lines are based on a principle introduced by the German Professor Heinrich Müller-Breslau. This principle follows: **The deflected shape of a structure represents to some scale the influence line for a function such as reaction, shear, or moment if the function in question is allowed to act through a unit displacement.** In other words, the structure draws its own influence line when the proper displacement is applied. This principle is derived in section 17.2.

As a first example, the qualitative influence line for the left reaction of the beam of Figure 9.5(a) is considered. The constraint at the left support is removed and a displacement is introduced there in the direction of the reaction as shown in part (b) of

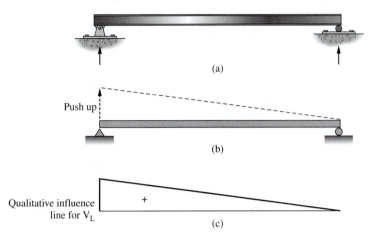

(a)

(b)

Qualitative influence
line for V_L

(c)

Figure 9.5   Qualitative influence line for reaction at left support

the figure. When the left end of the beam is pushed up, the area between the original and final position of the beam is the influence line for V_L to some scale.

In a similar manner the influence lines for the left and right reactions of the beam of Figure 9.6 are sketched.

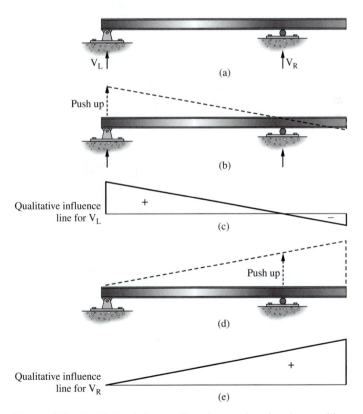

Figure 9.6   Qualitative influence lines for reactions in a beam with an overhang

As a third example, the influence line for the bending moment at section 1-1 in the beam of Figure 9.7 is considered. This diagram can be obtained by cutting the beam at the point in question and applying moments just to the left and just to the right of the cut section, as shown. In the figure the moment on each side of the section is positive with respect to the segment of the beam on that side of the section. The resulting deflected shape of the beam is the qualitative influence line for the bending moment at section 1-1.

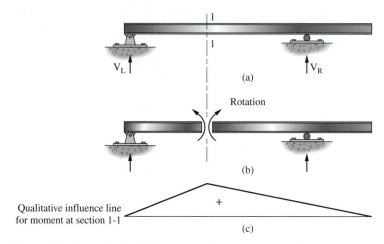

Figure 9.7 A qualitative influence line for moment in a simple beam

To draw a qualitative influence line for shear, the beam is cut at the point in question. A vertical force is applied to each side of the cut to provide positive shearing force, as shown in Figure 9.8(b). To understand the direction used for these forces, note that they are applied to the left and to the right of the cut section to

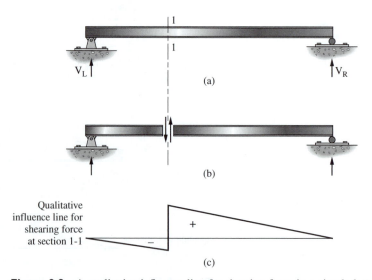

Figure 9.8 A qualitative influence line for shearing force in a simple beam

produce a positive shear for each segment. In other words, the force on the left segment is in the direction of a positive shear force applied down from the right side and vice versa.

Additional examples for qualitative influence lines are presented in Figure 9.9.

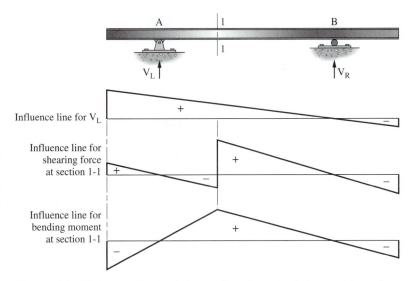

Figure 9.9 Various qualitative influence lines for a simple beam with overhangs

Overpass, Boise, Idaho (Courtesy of the American
Concrete Institute)

Müller-Breslau's principle is useful for sketching influence lines for statically determinate structures, but its greatest value is for statically indeterminate structures. Though the diagrams are drawn exactly as before, note that in Figure 9.10 they consist of curved lines instead of straight lines, as was the case for statically determinate structures.

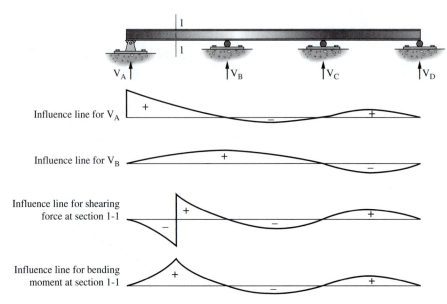

Figure 9.10   Various qualitative influence lines for a statically indeterminate 3 span beam

9.7 USES OF INFLUENCE LINES; CONCENTRATED LOADS

Influence lines are the plotted functions of structural responses for various positions of a unit load. Having an influence line for a particular function of a structure makes immediately available the value of the function for a concentrated load at any position on the structure. The beam of Figure 9.1 and the influence line for the left reaction are used to illustrate this statement. A concentrated 1-kip load placed 4 ft to the right of the left support would cause V_L to equal 0.8 kip. Should a concentrated load of 175 kips be placed in the same position, V_L would be 175 times as great, or 140 kips.

The value of a function due to a series of concentrated loads is quickly obtained by multiplying each concentrated load by the corresponding ordinate of the influence line for that function. A 150-kip load placed 6 ft from the left support in Figure 9.1 and also a 200-kip placed 16 ft from the left support would cause V_L to equal $(150)(0.7) + (200)(0.2)$, or 145 kips.

EXAMPLE 9.1

The influence lines for the left reaction and the centerline moment are shown for the simple beam of Figure 9.11. Determine the values of these functions for the several loads supported by the beam.

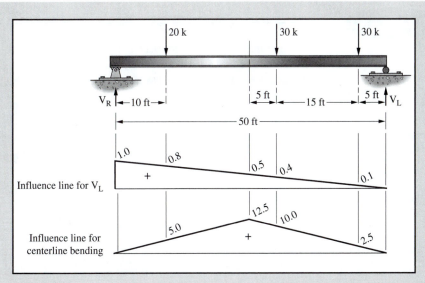

Figure 9.11

Solution.   Compute the magnitude of the left reaction. That value is equal to the summation of each load times the ordinate of the influence diagram at the location of the load.

$$V_L = 20(0.8) + 30(0.4) + 30(0.1) = 31 \text{ kips}$$

Next, compute the magnitude of the moment at the centerline. It is computed in the same manner as the left reaction.

$$M_{\mathcal{C}} = 20(5.0) + 30(10.0) + 30(2.5) = 475 \text{ k-ft} \quad \blacksquare$$

9.8 USES OF INFLUENCE LINES; UNIFORM LOADS

The value of a certain function of a structure may be obtained from the influence line when the structure is loaded with a uniform load, by multiplying the area of that part of the influence line that is opposite to the loaded part of the member by the intensity of the uniform load. The following discussion proves this statement is correct.

A uniform load of intensity w lb/ft is equivalent to a continuous series of smaller loads of (w)(0.1) lb on each 0.1 ft of the beam, or w dx lb on each length dx of the beam. Consider each length dx to be loaded with a concentrated load of magnitude w dx. The value of the function for one of these small loads is (w dx)(y) where y is the ordinate of the influence line at that point. The effect of all of these concentrated loads is equal to ∫wy dx. This expression shows that the effect of a uniform load is equal to the intensity of the uniform load, w, times the area of the influence line, ∫y dx, along the section of the structure covered by the uniform load.

EXAMPLE 9.2

Assume the beam in Figure 9.11 is loaded with a uniform load of 3 klf. Determine the magnitude of the left reaction and the centerline moment if the distributed load acts on the entire beam, and if it acts on only the left half of the beam.

Solution. Compute the values first for the load acting on the entire span of the beam.

$$V_L = (3\,\text{klf})\frac{(1\frac{k}{k})(50\,\text{ft})}{2} = 75\ \text{kips}$$

$$M_{\text{C}} = (3\,\text{klf})\frac{(12.5\frac{\text{ft-k}}{k})(50\,\text{ft})}{2} = 937.5\ \text{k-ft}$$

Now compute the same values for the uniform load when it acts only on the left half of the beam.

$$V_L = (3\,\text{klf})\frac{(1\frac{k}{k}+0.5\frac{k}{k})}{2}(25\,\text{ft}) = 56.25\ \text{kips}$$

$$M_{\text{C}} = (3\,\text{klf})\frac{(12.5\frac{\text{ft-k}}{k})(25\,\text{ft})}{2} = 468.75\ \text{k-ft} \quad\blacksquare$$

Should a structure support uniform and concentrated loads, the value of the function under study can be found by multiplying each concentrated load by its respective ordinate on the influence line and the uniform load by the area of the influence line opposite the section covered by the uniform load.

9.9 COMMON SIMPLE BEAM FORMULAS FROM INFLUENCE LINES

Several useful expressions for moments in beams can be determined with influence lines. Formulas may be developed for moment at the centerline of a simple beam in Figure 9.12(a), the beam being loaded first with a uniform load and second with a concentrated load at the centerline. From Figure 9.12(b), formulas may be developed for moment at any point in a simple beam loaded with a uniform load and for moment at any point where a concentrated load is located. These formulas follow:

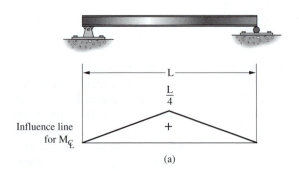

Influence line for M_{C}

(a)

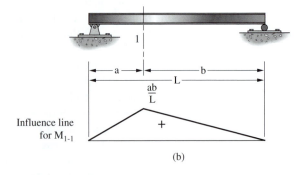

Influence line for $M_{1\text{-}1}$

(b)

Figure 9.12

(a) Loaded with a uniform load	(b) Loaded with a uniform load
$M_{\mathbb{C}} = (w)\left(\dfrac{1}{2} \times L \times \dfrac{L}{4}\right) = \dfrac{wL^2}{8}$	$M_{1\text{-}1} = (w)\left(\dfrac{1}{2} \times \dfrac{ab}{L} \times L\right) = \dfrac{wab}{2}$
Loaded with a concentrated load P at centerline:	Loaded with a concentrated load P at section 1-1:
$M_{\mathbb{C}} = \dfrac{PL}{4}$	$M_{1\text{-}1} = \dfrac{Pab}{L}$

9.10 DETERMINING MAXIMUM LOADING EFFECTS USING INFLUENCE LINES

Beams must be designed to satisfactorily support the largest shearing forces and bending moments that can be caused by the loads to which they are subjected. As an example, consider the beam shown in Figure 9.13(a) and the influence line for bending moment at section 1-1 shown in Figure 9.13(b). This is the same beam shown in Figure 9.9. We now want to determine the maximum possible bending moments at section 1-1 for a uniform dead load and a uniform live load.

We will first determine the loading to cause maximum positive bending moment. The uniform dead load, which is the weight of the structure, will be applied from one end

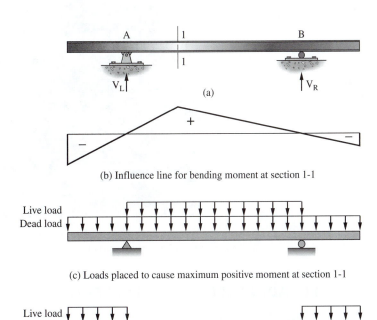

(b) Influence line for bending moment at section 1-1

(c) Loads placed to cause maximum positive moment at section 1-1

(d) Loads placed try to cause maximum negative moment at section 1-1

Figure 9.13 Placing loads to cause maximum forces

of the beam to the other as shown in Figure 9.13(c). The dead load is always acting on the entire structure. From the influence line we see that the unit load caused positive moment at section 1-1 only when it was located between the supports A and B. As such, the uniform live load is placed from A to B to determine the maximum positive bending moment, as shown in Figure 9.13(c). If there had been a concentrated live load acting with the uniform live load, it would have been placed at section 1-1 since the unit load caused the greatest positive moment when it was located there. The bending moment caused by these loads can be calculated using the ordinates on the influence line or by using the equations of static equilibrium.

The maximum negative bending moment occurring at section 1-1 can be determined in a similar fashion. For this case, the loads would be placed as shown in Figure 9.13(d). When the unit load was placed on the cantilever portions of the beam, negative bending occurred at section 1-1. As such the distributed live load was placed at those locations. If there had been a live concentrated load, it would have been placed at the left or right end of the beam—whichever had the largest negative ordinate on the influence line.

Massachusetts Eye and Ear Infirmary in Boston. (Courtesy Bethlehem Steel Corporation.)

9.11 MAXIMUM LOADING EFFECTS USING BEAM CURVATURE

In the preceding section, an influence line was used to determine the critical positions for placing live loads to cause maximum bending moments. The same results can be obtained, and perhaps obtained more easily in many situations, by considering the deflected shape or curvature of a member under load. If the live loads are placed so they cause the greatest curvature at a particular point, they will have bent the member to the greatest amount at that point. As such, the greatest bending moment will have been caused at that point.

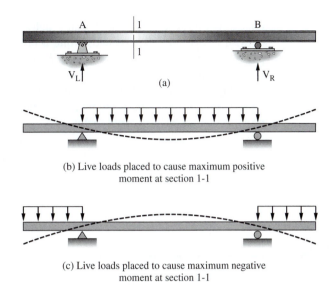

(a)

(b) Live loads placed to cause maximum positive
moment at section 1-1

(c) Live loads placed to cause maximum negative
moment at section 1-1

Figure 9.14   Obtaining maximum forces using curvature

For an illustration of this concept, let's determine the greatest positive bending moment at section 1-1 in the beam shown in Figure 9.14(a) due to the same loads considered in the last section. In Figure 9.14(b), the deflected shape of the beam is sketched, as it would be when a positive moment occurs at section 1-1. This deflected shape is shown by the dashed line. The dead load is placed all across the beam, while the live load is again placed from A to B; this location of the live load will magnify the deflected shape at section 1-1.

A similar situation is shown in Figure 9.14(c) to determine the maximum negative bending moment at section 1-1. The deflected shape of the beam, shown by the dashed line, is sketched consistent with a negative bending moment occurring at section 1-1. To magnify the negative or upward bending at section 1-1, the live load is placed on the cantilever portions of the beam, the parts outside the supports.

9.12 IMPACT LOADING

The truck and train loads applied to highway and railroad bridges are not applied gently and gradually; they are applied quickly, which causes forces to increase. As such, additional loads, called *impact loads*, must be considered. Impact loads are taken into account by increasing the live loads by some percentage, the percentage being obtained from purely empirical expressions. Numerous formulas have been presented for estimating impact. One example is the following AASHTO formula for highway bridges, in which I is the percent of impact and L is the length of the span, in feet, over which live load is placed to obtain a maximum stress. The AASHTO says that it is unnecessary to use an impact load greater than 30%, regardless of the value given by the formula. Note that as the span length increases the impact factor decreases.

$$I = \frac{50}{L + 125}$$

Impact factors for railroad bridges are larger than those for highway bridges because of the much greater vibrations caused by the wheels of a train as compared to the relatively soft rubber-tired vehicles on a highway bridge. A person need only stand near a railroad bridge for a few seconds while a fast-moving and heavily loaded freight train passes over to see the difference. Tests have shown the impact on railroad bridges will often be as high as 100% or more. Not only does a train have a direct vertical impact, or bouncing up and down, but it also has a lurching or swaying or nosing back-and-forth type of motion. For vertical impact on beams, girders, and floor beams, the AREA provides the following impact factor:

$$I = 60 - \frac{L^2}{500} \qquad L < 100 \text{ ft}$$

$$I = \frac{1,800}{L - 40} + 10 \qquad L \geq 100 \text{ ft}$$

EXAMPLE 9.3

Draw the influence line for positive moment at section 1-1 in the beam of Figure 9.15. Assuming a dead uniform load of 2 klf, a live uniform load of 3 klf, a live concentrated load (sometimes referred to as a floating load) of 40 k, and an impact factor of 30% determine the maximum positive moment at section 1-1 in the beam.

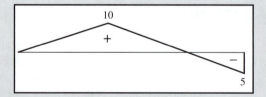

Figure 9.15

Solution.   Drawing the influence line for positive moment at section 1-1

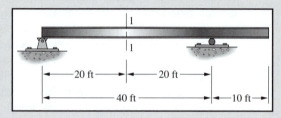

Loading the beam for maximum plus moment increasing the live loads by 30%.

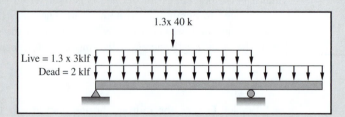

Maximum + moment at section 1-1

$$DL = 2\left[\left(\frac{1}{2}\right)(10)(40) + \left(\frac{1}{2}\right)(10)(-5)\right] = +350 \text{ ft-k}$$

$$+ \text{Uniform LL} = (1.3 \times 3)\left[\left(\frac{1}{2}\right)(10)(40)\right] = +780$$

$$+ \text{Floating LL} = (1.3 \times 40)[10] = +520$$

Ans. $= +1650 \text{ ft-k}$ ∎

9.13 PROBLEMS FOR SOLUTION

For Problems 9.1 through 9.6, draw qualitative influence lines for all of the reactions and for shear and moment at section 1-1 for each of the beams.

9.1

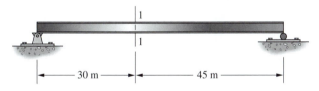

9.2

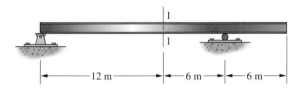

9.3

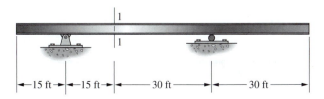

9.4

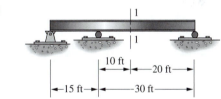

9.5

9.6

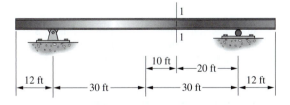

For Problems 9.7 through 9.18, draw quantitative influence lines for the situations listed.

9.7 Both reactions and the shearing force and bending moment at section 1-1 (*Ans.* Load @ 1-1: $V_L = 0.667$, $M_{1-1} = 10$)

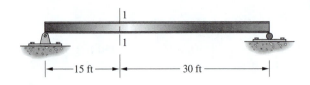

9.8 Both reactions and the shearing force and bending moment at section 1-1

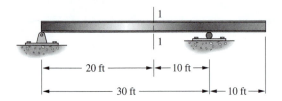

9.9 Both reactions and the shearing force and bending moment at section 1-1 (*Ans.* Load @ left free end: $V_A = 1.50$, $V_B = -0.50$, V_{1-1} @ 1-1 $= \pm 0.50$, $M_{1-1} = +3.0$)

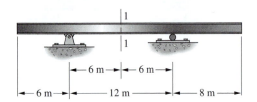

9.10 Both reactions and shear at sections 1-1 and 2-2 (just to left and right of left support), and 3-3

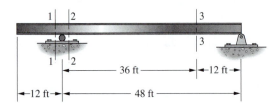

9.11 Both reactions, shear at sections 1-1 and 2-2, and moment at section 2-2 (*Ans.* Load @ right free end: $V_B = 1.50$, shear @ 2-2 $= -0.50$, $M_{2-2} = -6.00$)

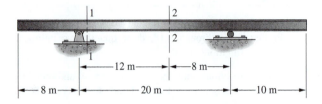

9.12 Vertical reaction and moment reaction at fixed-end, shear and moment at section 1-1

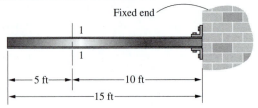

9.13 Shearing force and bending moment at sections 1-1 and 2-2 (*Ans.* Load @ free end: $M_A = -20.0$, V @ 1-1 $= 1.00$, $M_{2-2} = -5.00$)

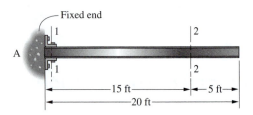

9.14 Both reactions as load moves from A to D. Also M_B and M_C.

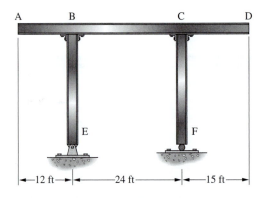

9.15 All reactions (*Ans.* Load @ hinge: $V_A = 0$, $V_B = 2.33$, $V_C = -1.33$)

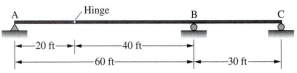

9.16 Vertical reactions at supports A and B and moment @ B

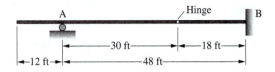

9.17 All vertical reactions and bending moments and shearing forces at section 1-1 (*Ans.* Load @ 1-1: $V_R = 0.667$, $M_{1-1} = 13.3$, $V_{1-1(left)} = -0.67$)

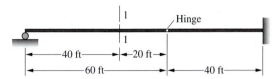

9.18 Reactions at supports A and B

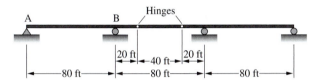

9.19 Draw influence lines for both reactions and for shear just to the left of the 18-kip load, and for moment at the 18-kip load. Determine the magnitude of each of these functions using the influence lines for the loads fixed in the positions shown in the accompanying illustration. (*Ans.* $V_A = 51.5$ k, shear just to left of 18 k load = -12.5 k, M @ 18 k load = 610 ft-k)

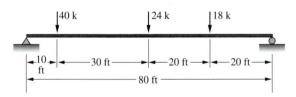

For Problems 9.20 through 9.26 using influence lines determine the quantities requested for a uniform dead load of 2 klf, a moving uniform live load of 3 klf, and a moving or floating concentrated live load of 20 kips. Assume impact = 25% in each live load case.

9.20 Maximum left reaction and maximum plus shear and moment at section 1-1

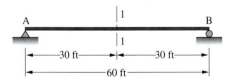

9.21 Maximum positive values of left reaction and shear and moment at section 1-1 (*Ans.* $V_A = 149.37$ k, $+V$ @ 1-1 = -2.37 k, $M_{1-1} = 1393.75$ ft-k)

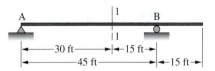

9.22 Maximum negative values of shear and moment at section 1-1, maximum negative shearing force just to left of the right hand support, and the bending moment at the right support for the beam shown.

9.23 Maximum negative shear and moment at section 1-1 (*Ans.* Maximum negative shear = -24.664 k, maximum negative moment = -180 ft-k)

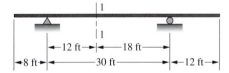

9.24 Maximum upward value of reactions at A and B and maximum negative shear and moment at section 1-1

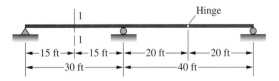

9.25 Maximum positive and negative values of shear and moment at section 3-3 in the beam of Problem 9.10 (*Ans.* Maximum negative shear = -90.375 k, maximum positive moment = $+1431$ ft-k)

9.26 Maximum positive shear at unsupported hinge, maximum negative moment at support B, and maximum downward value of reaction at support A for the beam of Problem 9.24.

Chapter 10

Truss Influence Lines and Moving Loads

10.1 INFLUENCE LINES FOR TRUSSES

Beam-type bridges of structural steel, precast reinforced concrete, or prestressed concrete have almost completely taken over the short-span bridge market. Nevertheless, the drawing of influence lines for trusses is presented in this chapter for two reasons. First, the author feels that the understanding of such diagrams gives the student a better understanding of the analysis of trusses and of the action of moving loads. Secondly, the information presented here serves as a background for the construction of influence lines for the longer span statically indeterminate bridge trusses that are still economically competitive. These latter influence lines are discussed in Chapter 17 of this text.

Truss influence lines may be drawn and used for making force calculations, or they may be sketched roughly without computing the values of the ordinates and used only for placing the moving loads to cause maximum or minimum forces.

The procedure used for preparing influence lines for trusses is closely related to the one used for beams. The exact manner of application of loads to a bridge truss from the bridge floor is described in the following section. A similar discussion could be made for the application of loads to roof trusses.

10.2 ARRANGEMENT OF BRIDGE FLOOR SYSTEMS

The arrangement of the members of a bridge floor system should be carefully studied so that the manner of application of loads to the truss will be fully understood. Probably the most common type of floor system consists of a concrete slab supported by steel stringers running parallel to the trusses. The stringers run the length of each panel and are supported at their ends by floor beams that run transverse to the roadway and frame into the panel points or joints of the truss (Figure 10.1).

The foregoing discussion apparently indicates that the stringers rest on the floor beams and the floor beams on the trusses. This method of explanation has been used to emphasize the manner in which loads are transferred from the pavement to the trusses, but the members usually are connected directly to each other on the same level. Stringers are conservatively assumed to be simply supported, but actually there is some continuity in their construction.

A 100-kip load is applied to the floor slab in the fifth panel of the truss of Figure 10.1. The load is transferred from the floor slab to the stringers, thence to the floor

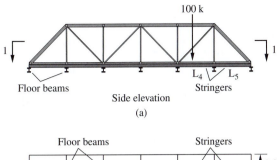

Side elevation

(a)

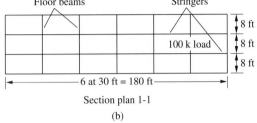

Section plan 1-1

(b)

Figure 10.1

beams, and finally to joints L_4 and L_5 of the supporting trusses. The amount of load going to each stringer depends on the position of the load between the stringers: if halfway, each stringer would carry half. Similarly, the amount of load transferred from the stringers to the floor beams depends on the longitudinal position of the load.

Figure 10.2 illustrates the calculations involved in figuring the transfer of a 100-kip load to the trusses. The final reactions shown for the floor beams represent the downward loads applied at the truss panel points. The computation of truss loads usually is a much simpler process than the one described here.

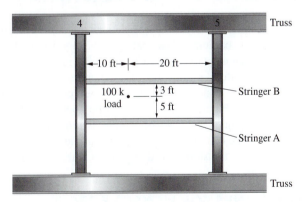

Load transferred to each stringer,

$$R_A = \frac{3}{8} \times 100 = 37.5 \, k$$

$$R_B = \frac{5}{8} \times 100 = 62.5 \, k$$

Load transferred from stringer A to floor beams,

$$R_{4\text{-}4} = \frac{20}{30} \times 37.5 = 25 \, k$$

$$R_{5\text{-}5} = \frac{10}{30} \times 37.5 = 12.5 \, k$$

Load transferred from stringer B to floor beams,

$$R_{4\text{-}4} = \frac{20}{30} \times 62.5 = 41.67\,\text{k}$$

$$R_{5\text{-}5} = \frac{10}{30} \times 62.5 = 20.83\,\text{k}$$

Floor beams loaded as follows:

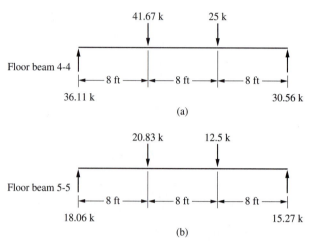

Figure 10.2

10.3 INFLUENCE LINES FOR TRUSS REACTIONS

Influence lines for reactions of simply supported trusses are used to determine the maximum loads that may be applied to the supports. Although their preparation is elementary, they offer a good introductory problem in learning the construction of influence lines for truss members.

Influence lines for the reactions at both supports of an overhanging truss are given in Figure 10.3. Loads can be applied to the truss only by the floor beams at the panel points, and floor beams are assumed to be present at each of the panel points, including the end ones. A load applied at the very end of the truss opposite the end panel point will be transferred to that panel point by the end floor beam.

10.4 INFLUENCE LINES FOR MEMBER FORCES OF PARALLEL-CHORD TRUSSES

Influence lines for forces in truss members may be constructed in the same manner as those for various beam functions. The unit load moves across the truss, and the ordinates for the force in the member under consideration may be computed for the load at each panel point. In most cases it is unnecessary to place the load at every panel point and calculate the resulting value of the force, because certain portions of influence lines can readily be seen to consist of straight lines for several panels.

One method used for calculating the forces in a chord member of a truss consists of passing an imaginary section through the truss, cutting the member in question and taking

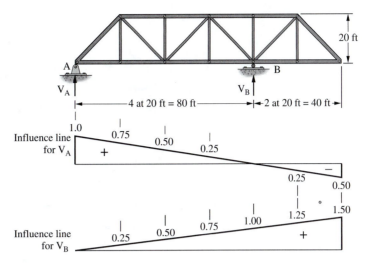

Figure 10.3

moments at the intersection of the other members cut by the section. The resulting force in the member is equal to the moment divided by the lever arm; therefore, the influence line for a chord member is the same shape as the influence line for moment at the point where moments are taken.

The truss of Figure 10.4 is used to illustrate this point. The force in member L_1L_2 is determined by passing section 1-1 and taking moments at U_1. An influence line is shown

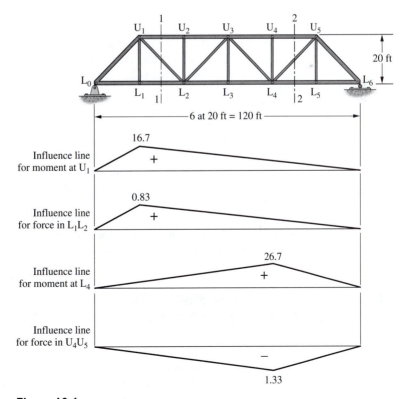

Figure 10.4

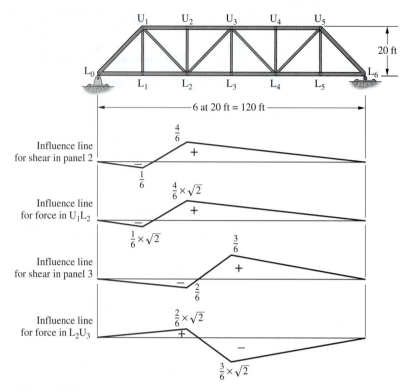

Figure 10.5

for the moment at U_1 and for the force in L_1L_2, the ordinates of the latter figure being those of the former divided by the lever arm. Similarly, section 2-2 is passed to compute the force in U_4U_5, and influence lines are shown for the moment at L_4 and for the force in U_4U_5.

The forces in the diagonals of parallel-chord trusses may be calculated from the shear in each panel. The influence line for the shear in a panel is of the same shape as the influence line for the force in the diagonal, because the vertical component of force in the diagonal is equal numerically to the shear in the panel. Figure 10.5 illustrates this fact for two of the diagonals of the same truss considered in Figure 10.4. For some positions of the unit loads the diagonals are in compression and for others they are in tension.

The vertical components of force in the diagonals can be easily converted into the actual resultant forces from their slopes as shown in Figure 10.5. The sign convention for positive and negative shears is the same as the one used previously.

10.5 INFLUENCE LINES FOR MEMBER FORCES OF NONPARALLEL-CHORD TRUSSES

Influence-line ordinates for the force in a chord member of a "curved-chord" truss may be determined by passing a vertical section through the panel and taking moments at the intersection of the diagonal and the other chord. Several such influence lines are drawn for the chords of a so-called Parker truss in Figure 10.6.

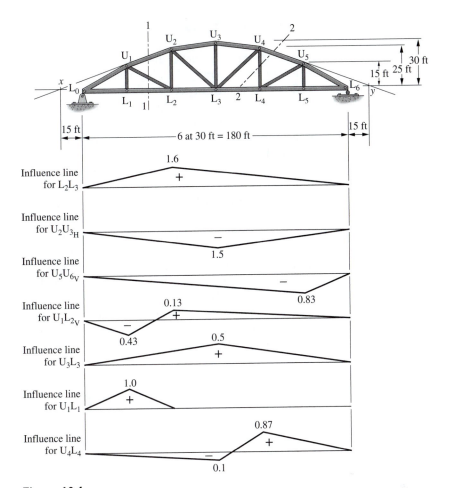

Figure 10.6

The ordinates for the influence line for force in a diagonal may be obtained by passing a vertical section through the panel and taking moments at the intersection of the two chord members, as illustrated in Figure 10.6 where the force in U_1L_2 is obtained by passing section 1-1 and taking moment at the intersection of chords U_1U_2 and L_1L_2 at point x. The influence line is drawn for the vertical component of force in the inclined member. In the following pages several more influence lines are drawn for either the vertical or horizontal components of force for inclined members.

The influence line for the midvertical U_3L_3 is obtained indirectly by computing the vertical components of force in U_2U_3 and U_3U_4. The ordinates for U_3L_3 are found by summing up these components. The influence lines for the other verticals are more easily drawn. Member U_1L_1 can have a force only when the unit load lies between L_0 and L_2. It has no force if the load is at either of these joints, but a tension of unity occurs when the load is at L_1. Influence lines for verticals such as U_4L_4 can be drawn by two methods. A section such as 2-2 may be passed through the truss and moments taken at the intersection of the chords at point y, or if the influence diagram for L_4U_5 is available, its vertical components may be used to calculate the ordinates for U_4L_4.

10.6 INFLUENCE LINES FOR K TRUSS

Figure 10.7 shows influence lines for several members of a K truss. The calculations necessary for preparing the diagrams for the chord members are equivalent to those used for the chords of trusses previously considered. The ordinates for the force in member U_3U_4 may be obtained by passing section 1-1 through the member in question and three other members (U_3M_3, M_3L_3, and L_3L_4). As the lines of action of these three forces intersect at L_3, moments may be taken there to obtain the desired force in U_3U_4.

The forces in the two diagonals of each panel may be obtained from the shear in the panel. By knowing that the horizontal components are equal and opposite, the relationship between their vertical components can be found from their slopes. If the slopes are equal, the shear to be carried divides equally between the two. The influence lines for the verticals, such as U_5M_5, may be determined from the influence lines for the adjoining diagonals, if available. On the other hand, the ordinates may be computed independently for various positions of the unit load. The reader should make a careful comparison of the influence lines for the upper and lower verticals, such as those given for M_3L_3 and U_3M_3 in the figure.

The influence line for the midvertical U_4L_4 can be developed by computing the vertical components of force in M_3L_4 and L_4M_5, or in M_3U_4 and U_4M_5, for each position of the unit load. The vertical components of force in each of these pairs of members will

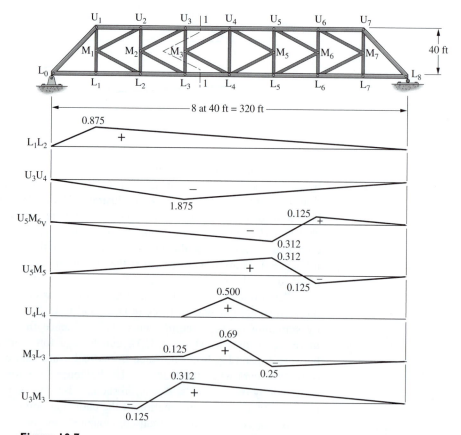

Figure 10.7

cancel each other unless the shear in panel 4 is unequal to the shear in panel 5, which is possible only when the unit load is at L_4.

10.7 DETERMINATION OF MAXIMUM FORCES

Truss members are designed to resist the maximum forces that may be caused by any combination of the dead, live, and impact loads to which the truss may be subjected. The live load probably consists of a series of moving concentrated loads representing the wheel loads of the vehicles using the structure, but for convenience in analysis an approximately equivalent uniform live load with only one or two concentrated loads often is used in their place. Live loads for which highway and railroad bridges are designed and common impact expressions are discussed in detail in Sections 10.9 and 10.10.

The dead load, representing the weight of the structure and permanent attachments, extends for the entire length of the truss, but the uniform and concentrated live loads are placed at the points on the influence line that cause maximum force of the character being studied. If tension is being studied, the live uniform load is placed along the section of the truss corresponding to the positive or tensile section of the influence line, and the live concentrated loads are placed at the maximum positive tensile ordinates on the diagram.

Members whose influence lines have both positive and negative ordinates may possibly be in tension for one combination of loads and in compression for another. A member subject to *force reversal* must be designed to resist both the maximum compressive and maximum tensile forces.

In the next few paragraphs the maximum possible forces in several members of the truss of Figure 10.8 are determined due to the following loads:

1. Dead uniform load of 1.5 klf
2. Live uniform load of 2 klf
3. Moving concentrated load of 20 kips
4. Impact of 24.4%

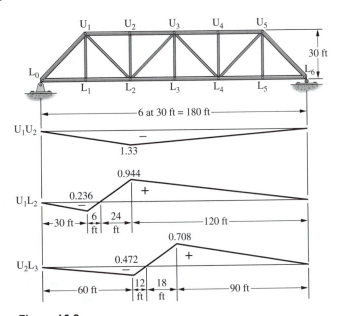

Figure 10.8

The influence lines are drawn, and the forces are computed by the exact method as described in the following paragraphs.

U_1U_2

The member is in compression for every position of the unit load; therefore, the dead uniform load and the live uniform load are placed over the entire span. The moving concentrated load of 20 kips is placed at the maximum compression ordinate on the influence line. The impact factor is multiplied by the live load forces and added to the total.

$$\begin{aligned}
\text{DL} &= (1.5)(180)(-1.33)\left(\tfrac{1}{2}\right) &&= -180.0 \\
\text{LL} &= (2)(180)(-1.33)\left(\tfrac{1}{2}\right) &&= -240.0 \\
&\quad + (20)(-1.33) &&= -26.7 \\
\text{I} &= (0.244)(-240.0 - 26.7) &&= \underline{-65.1} \\
&\quad \text{total force} &&= -511.8 \text{ k compression}
\end{aligned}$$

U_1L_2

Examination of the influence line for U_1L_2 shows that for some positions of the unit load the member is in compression, whereas for others it is in tension. The live loads should be placed over the positive portion of the diagram and the dead loads across the entire structure to obtain the largest possible tensile force. Similarly, the live loads should be placed over the negative portion of the diagram and the dead loads over the entire structure to obtain the largest possible compressive force.

Maximum tension:

$$\begin{aligned}
\text{DL} &= (1.5)(144)(+0.944)\left(\tfrac{1}{2}\right) &&= +102.0 \\
&\quad + (1.5)(36)(-0.236)\left(\tfrac{1}{2}\right) &&= -6.4 \\
\text{LL} &= (2)(144)(+0.944)\left(\tfrac{1}{2}\right) &&= +136.0 \\
&\quad + (20)(+0.944) &&= +18.9 \\
\text{I} &= (0.244)(+136.0 + 18.9) &&= \underline{+37.8} \\
&\quad \text{total force} &&= +288.3 \text{ k tension}
\end{aligned}$$

Maximum compression:

$$\begin{aligned}
\text{DL} &= (1.5)(144)(+0.944)\left(\tfrac{1}{2}\right) &&= +102.0 \\
&\quad + (1.5)(36)(-0.236)\left(\tfrac{1}{2}\right) &&= -6.4 \\
\text{LL} &= (2)(36)(-0.236)\left(\tfrac{1}{2}\right) &&= -8.5 \\
&\quad + (20)(-0.236) &&= -4.7 \\
\text{I} &= (0.244)(-8.5 - 4.7) &&= \underline{-3.2} \\
&\quad \text{total force} &&= +79.2 \text{ k tension} (\therefore \text{ no compression} \\
&&&\qquad\text{possible})
\end{aligned}$$

U_2L_3

The calculations for U_1L_2 proved it could have only tensile forces, regardless of the positioning of the live loads given. The following calculations show force reversal may occur in member U_2L_3.

Maximum tension:

$$
\begin{aligned}
\text{DL} &= (1.5)(108)(+0.708)\left(\tfrac{1}{2}\right) &&= +57.3 \\
&+ (1.5)(72)(-0.472)\left(\tfrac{1}{2}\right) &&= -25.5 \\
\text{LL} &= (2)(108)(+0.708)\left(\tfrac{1}{2}\right) &&= +76.4 \\
&+ (20)(+0.708) &&= +14.2 \\
\text{I} &= (0.244)(+76.4 + 14.2) &&= \underline{+22.1} \\
&\text{total force} &&= +144.5 \text{ k tension}
\end{aligned}
$$

Maximum compression:

$$
\begin{aligned}
\text{DL} &= (1.5)(108)(+0.708)\left(\tfrac{1}{2}\right) &&= +57.3 \\
&+ (1.5)(72)(-0.472)\left(\tfrac{1}{2}\right) &&= -25.5 \\
\text{LL} &= (2.0)(72)(-0.472)\left(\tfrac{1}{2}\right) &&= -34.0 \\
&+ (20)(-0.472) &&= -9.4 \\
\text{I} &= (0.244)(-34.0 - 9.4) &&= \underline{-10.6} \\
&\text{total force} &&= -22.2 \text{ k compression}
\end{aligned}
$$

10.8 COUNTERS IN BRIDGE TRUSSES

The fact that a member in compression is in danger of bending or buckling reduces its strength and makes its design something of a problem. The design of a 20-ft member for a tensile force of 100 kips usually will result in a smaller section than is required for a member of the same length subject to a compressive force of the same magnitude. The ability of a member to resist compressive loads depends on its stiffness, which is measured by the *slenderness ratio*. The slenderness ratio is the ratio of the length of a member to its least radius of gyration. As a section becomes longer, or as its slenderness ratio increases, the danger of buckling increases, and a larger section is required to withstand the same load.

This discussion shows there is a considerable advantage in keeping the diagonals of a truss in tension if possible. If a truss supported only dead load, it would be a simple matter to arrange the diagonals so that they were all in tension. All of the diagonals of the Pratt truss of Figure 10.9(a) would be in tension for a uniform dead load extending over the entire span. The calculations in Section 10.8, however, have shown that live loads may cause the forces in some of the diagonals of a bridge truss to alternate between tension and compression. The constant passage of trains or trucks back and forth across a bridge probably will cause the forces in some of the diagonals to change continually from tension to compression and back to tension.

The possibilities of force reversal are much greater in the diagonals near the center of a truss. The reason for this situation can be seen by referring to the truss of Figure 10.9 where a positive shear obviously causes tension in members U_1L_2 and U_2L_3. The positive dead-load shear is much smaller in panel 3 than in panel 2, and it is more likely for the live load to be in a position to cause a negative shear large enough to overcome the positive shear and produce compression in the diagonal.

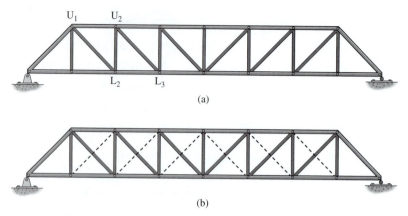

Figure 10.9

A few decades ago, when it was common for truss members to be pin-connected the diagonals were actually eyebars that were capable of resisting little compression. The same force condition exists in trusses erected today, which have diagonals consisting of a pair of small steel angles or other shapes of little stiffness. It was formerly common to add another tension-resisting diagonal to the panels where force reversal could occur, the new diagonal running across the first one and into the previously unconnected corners of the panel. These members, called *counters* or *counter diagonals*, can be seen in many older bridges across the country, but rarely in new ones.

Figure 10.9(b) shows a Pratt truss to which counters have been added in the middle six panels, the counters being represented by dotted lines. When counters have been added in a panel, both diagonals may consist of relatively slender and light members, neither being able to resist appreciable compression. With light and slender diagonals the entire shear in the panel is assumed to be resisted by the diagonal that would be in tension for that type of shear, whereas the other diagonal is relaxed or without stress. The two diagonals in a panel may be thought of as cables that can resist no compression whatsoever. If compression were applied to one of the cables, it would become limp, whereas the other one would be stretched. A truss with counters is actually statically indeterminate unless the counter is adjusted to have zero force under dead load.

If the columns and beams of a steel frame building are connected with standard or "simple beam" connections, the frames will have little resistance to lateral loads. The simplest method of providing such resistance is with "counters" or diagonal bracing such as that shown in Figure 10.10.

The reader can obviously see that in the typical building of today the use of full diagonal wind bracing will often be in the way of doors, windows, and other wall openings. Furthermore, so many modern buildings are planned with movable interior partitions that the presence of cross bracing would greatly reduce this flexibility. As a result, the diagonal bracing for such buildings is placed as much as possible around elevator shafts, stairwells, and in other walls where few openings are planned.

Today's truss bridges are designed with diagonals capable of resisting force reversals. In fact, all bridge truss members, whether subject to force reversal or not, must be capable of withstanding the large force changes that occur when vehicles move back and forth across those structures. A member that is subject to frequent force changes even though the character of the force does not change (as $+50$ to $+10$ to $+50$ kips, etc.) is in danger of a fatigue failure unless it is specifically designed for

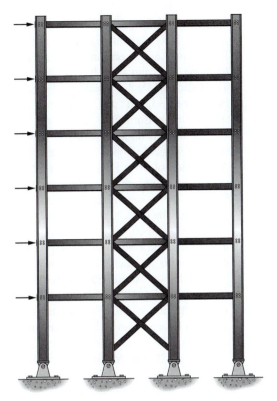

Figure 10.10

that situation. For structural steel members, tension must be involved for fatigue to be a problem.

Modern steel specifications provide a maximum permissible stress range (from high to low) for each truss member. The stress range is defined as the algebraic difference between the maximum and minimum stresses. For this calculation tensile stress is given an algebraic sign that is opposite to that of compression stress. The AASHTO and AISC specifications provide a permissible stress range that is dependent on the estimated number of cycles of stress, on a particular member, and on its type of connection. Obviously, the more critical the situation, the smaller is the permissible stress range.

10.9 LIVE LOADS FOR HIGHWAY BRIDGES

Until about 1900, bridges in the United States were "proof loaded" before they were considered acceptable for use. Highway bridges were loaded with carts filled with stone or pig iron, and railway bridges were loaded with two locomotives in tandem. Such procedures were probably very useful in identifying poor designs and/or poor workmanship, but were no guarantee against overloads and fatigue stress situations.[1]

As discussed in *America's Highways 1776–1976*, during much of the 19th century highway bridges were designed to support live loads of approximately 80 to 100 psf

---

[1]U.S. Department of Transportation, Federal Highway Administration, *America's Highways 1776–1976* (Superintendent of Documents, U.S. Government Printing Office, 1976), 429–432.

applied to the bridge decks. These loads supposedly represented large, closely spaced crowds of people walking across the bridges. In 1875, the American Society of Civil Engineers (ASCE) recommended that highway bridges should be designed to support live loads varying from 40 to 100 psf—the smaller values were to be used for very long spans. The Office of Public Roads published a circular in 1913 recommending that highway bridges be designed for a live loading of: (a) a series of electric cars, or (b) a 15-ton road roller and a uniform live load on the rest of the bridge deck.

Although highway bridges must support several different types of vehicles, the heaviest possible loads are caused by a series of trucks. In 1931, the American Association of State Highway and Transportation Officials (AASHTO) Bridge Committee issued its first printed edition of the AASHTO Standard Specification for Highway Bridges. A very important part of these specifications was the use of the truck system of live loads. The truck loads were designated as H20, H15, and H10, representing two-axle design trucks of 20, 15, and 10 tons, respectively. Each lane of a bridge was to have an H-truck placed in it and was to be preceded and followed by a series of trucks weighing three quarters as much as the basic truck.[2]

Today the AASHTO specifies that highway bridges be designed for lines of motor trucks occupying 10-ft-wide lanes. Only one truck is placed in each span for each lane. The truckloads specified are designated with an H prefix (or M if SI units are used) followed by a number indicating the total weight of the truck in tons (10^4 Newtons). The weight may be followed by another number indicating the year of the specifications. For example, an H20-44 loading indicates a 20-ton truck and the 1944 specification. A sketch of the truck and the pertinent dimensions is shown in Figure 10.11.

The selection of the particular truck loading to be used in design depends on the bridge location, anticipated traffic, and so on. These loadings may be broken down into three groups as follows.

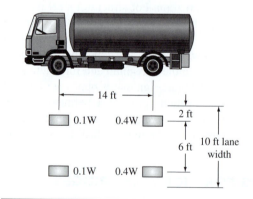

Truck	Weight (tons)	Front Axle (kips)	Front Wheel (kips)	Rear Axle (kips)	Rear Wheel (kips)
H25-44	25	10,000	5,000	40,000	20,000
H20-44	20	8,000	4,000	32,000	16,000
H15-44	15	6,000	3,000	24,000	12,000
H10-44	10	4,000	2,000	16,000	8,000

Figure 10.11   AASHTO H-Truck Loads

[2]*Ibid.*

Two-Axle Trucks: H10, H15, H20, and H25

The weight of an H truck is assumed to be distributed two-tenths to the front axle (for example, 4 tons, or 8 kips, for an H20 loading) and eight-tenths to the rear axle. The axles are spaced 14 ft on center, and the center-to-center lateral spacing of the wheels is 6 ft. Should a truck loading varying in weight from these be desired, one that has axle loads in direct proportion to the standard ones listed here may be used. A loading as small as the H10 may be used only for bridges supporting the lightest traffic.

Two-Axle Trucks Plus One-Axle Semi-Trailer: HS15-44, HS20-44, and HS25-44

For today's highway bridges carrying a great amount of truck traffic, the two-axle truck loading with a one-axle semi-trailer weighing 80% of the main truckload is commonly specified for design (Figure 10.12). The DOTs (Departments of Transportation) for many states today require that their bridges be designed for the HS25-44 trucks. This truck has 5 tons on the front axle, 20 tons on the rear axle, and 20 tons on the trailer axle. The distance from the rear truck axle to the semi-trailer axle is varied from 14 to 30 ft, depending on which spacing will cause the most critical conditions.

Uniform Lane Loading (Controls for Longer Bridges)

Computation of forces caused by a series of concentrated loads is a tedious job with a handheld calculator. This is true whether they represent two-axle trucks or two-axle trucks with semi-trailers. Therefore, a lane loading that will produce approximately the same forces frequently is used. The lane loading consists of a uniformly distributed load acting in combination with a single moving concentrated load. This load system represents a line of

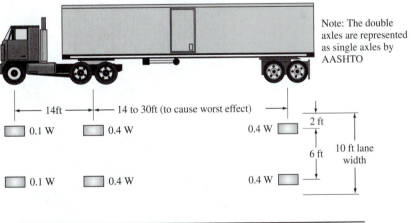

Truck	Weight (tons)	Front		Rear		Trailer	
		Axle (kips)	Wheel (kips)	Axle (kips)	Wheel (kips)	Axle (kips)	Wheel (kips)
HS25-44	25	10,000	5,000	40,000	20,000	40,000	20,000
HS20-44	20	8,000	4,000	32,000	16,000	32,000	16,000
HS15-44	15	6,000	3,000	24,000	12,000	24,000	12,000
HS10-44	10	4,000	2,000	16,000	8,000	16,000	8,000

Figure 10.12 AASHTO HS Truck Loads

medium-weight traffic with a heavy truck somewhere in the line. The uniformly distributed load, per foot of traffic lane, is equal to 0.016 times the total weight of the truck to which the load is to be roughly equivalent. The concentrated moving load is equal to 0.45 times the truck weight for bending moment calculations and 0.65 times the truck weight for shearing force calculations. These values for an H20 loading would be as follows:

Lane Load	0.016(20 tons) = 640 lb/ft of lane
Concentrated Load for Moment	0.45(20 tons) = 18 kips
Concentrated Load for Shear	0.65(20 tons) = 26 kips

For continuous spans, another concentrated load of equal weight is to be placed in one of the other spans at a location to cause maximum negative bending moment to occur. For positive moment, only one concentrated load is to be used per lane, with the uniform load placed on as many spans as necessary to produce the maximum positive bending moment.

The lane loading is more convenient, but it should not be used unless it produces bending moments and shearing forces that are equal to or greater than those produced by the corresponding H loading. Using information presented later in this chapter, in simple spans the equivalent lane loading for the HS20-44 truck will produce greater bending moments when the span length exceeds 145 ft and greater shearing forces when the span length exceeds 128 ft. Appendix A of the AASHTO specifications contains tables that give the maximum shears and moments in simple spans for the various H truck loads or for their equivalent lane loads, whichever controls.

John F. Fitzgerald Expressway, Mystic River Bridge, Boston, Massachusetts (Courtesy of the American Institute of Steel Construction, Inc.)

The possibility of having a continuous series of heavily loaded trucks in every lane of a bridge that has more than two lanes does not seem as great for a bridge that has only two lanes. The AASHTO therefore permits the values caused by full loads in every lane to be reduced by a certain factor if the bridge has more than two lanes.

Interstate Highway System Loading

Another loading system can be used instead of the HS20-44 in the design of structures for the Interstate Highway System. This alternate system, which consists of a pair of 24-kip axle loads spaced 4 ft on center, is critical for short spans only. It is possible to show that this loading will produce maximum bending moments in simple spans ranging from 11.5 to 37 ft in length and maximum shearing forces in simple spans ranging from 6 to 22 ft in length. For other spans, the HS20-44 loading or its equivalent lane loading will be the more critical load.

10.10 LIVE LOADS FOR RAILWAY BRIDGES

Railway bridges are commonly analyzed for a series of loads devised by Theodore Cooper in 1894. His loads, which are referred to as E loads, represent two locomotives with their tenders followed by a line of freight cars as shown in Figure 10.13(a). A series of concentrated loads is used for the locomotives, and a uniform load represents the freight cars, as pictured in Figure 10.13(c). The E-40 train is assumed to have a 40-kip load on the driving axle of the engine. Today bridges are designed based on an E-80 loading or larger.[3]

If information is available for one E loading, the information for any other E loading can be obtained by direct proportion. The axle loads of an E-75 are 75/40 of those for an E-40; those for an E-60 are 60/72 of those for an E-72; and so on. As such, the axle loads for an E-80 load are twice those shown in Figure 10.13.

The American Railway Engineering Association also specifies an alternate loading. That loading is shown in Figure 10.14. This load or the E-80 load—whichever causes the greatest stress in the components—is to be used.

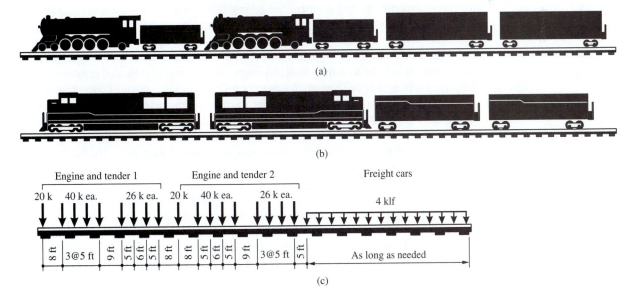

Figure 10.13 Cooper E-40 Railway Load

[3]American Railway Engineering Association, *AREA Manual*, 1996.

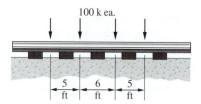

Figure 10.14   Alternate railroad loading

As we can see from the modern locomotives in Figure 10.13(b), Cooper's loads do not accurately picture today's trains. Nevertheless, they are still in general use despite the availability of several more modern and more realistic loads, such as Dr. D. B. Steinman's M-60 loading.[4]

10.11 MAXIMUM VALUES FOR MOVING LOADS

In the pages of this chapter we have repeatedly indicated that to design a structure supporting moving loads, the engineer must determine where to place the loads to cause maximum forces at various points in the structure. If one can place the loads at the positions causing maximum forces to occur, one need not worry about any other positions the loads might take on the structure.

If a structure is loaded with a uniform live load and not more than one or two moving concentrated loads, the critical positions for the loads will be obvious from the influence lines. If, however, the structure is to support a series of concentrated loads of varying magnitudes, such as groups of truck or train wheels, the problem is not as simple. The influence line provides an indication of the approximate positions for placing the loads, because it is reasonable to assume that the heaviest loads should be grouped near the largest ordinates of the diagram.

Space is not taken herein to consider all of the possible situations that might be faced in structural analysis. The author feels, however, that the determination of the absolute maximum bending moment caused in a beam by a series of concentrated loads is so frequently encountered by the engineer that discussion of this topic is warranted.

The absolute maximum bending moment in a simple beam usually is thought of as occurring at the beam centerline. Maximum bending moment does occur at the centerline if the beam is loaded with a uniform load or a single concentrated load located at the centerline. A beam, however, may be required to support a series of varying moving concentrated loads such as the wheels of a train, and the absolute maximum moment in all probability will occur at some position other than the centerline. To calculate the moment, it is necessary to find the point where it occurs and the position of the loads causing it. To assume that the largest moment is developed at the centerline of long-span beams is reasonable, but for short-span beams, this assumption may be considerably in error. It is therefore necessary to have a definite procedure for determining absolute maximum moment.

The bending moment diagram for a simple beam loaded with concentrated loads will consist of straight lines regardless of the position of the loads. Therefore, the absolute maximum moment occurring during the movement of these loads across the span will occur at one of the loads, usually at the load nearest the center-of-gravity of the group of loads. The beam shown in Figure 10.15 with the series of loads P_1, P_2, P_3, and so on is studied in the following paragraphs.

[4]"Locomotive Loadings for Railway Bridges," *Transactions of the American Society of Civil Engineers* *86* (1923): 606–636.

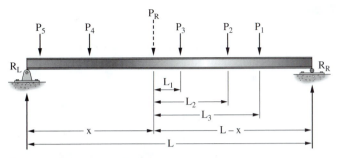

Figure 10.15 A simple beam with numerous applied loads

The load P_3 is assumed to be the one nearest the center of gravity of the loads on the span. It is located a distance L_1 from P_R, which is the resultant of all the loads on the span. The left reaction, R_L, is located a distance x from P_R. In the following paragraphs, maximum bending moment is assumed to occur at P_3, and a definite method is developed for placing this load to cause the maximum.

The bending moment at P_3 may be written as follows:

$$M = V_R(L - x - L_1) - P_2(L_2 - L_1) - P_1(L_3 - L_1)$$

After substituting the value of V_R, $P_R x/L$, we obtain the following equation:

$$M = \left(\frac{P_R x}{L}\right)(L - x - L_1) - P_2(L_2 - L_1) - P_1(L_3 - L_1)$$

We want to find the value of x for which the moment at P_3 will be a maximum. Maximum bending moment at P_3, which occurs when the shearing force is zero, may be found by differentiating the bending moment equation with respect to x, setting the result equal to zero, and solving for x.

$$\frac{dM}{dx} = L - 2x - L_1 = 0$$

$$\therefore x = \frac{L}{2} - \frac{L_1}{2} = \frac{1}{2}(L - L_1)$$

From the preceding derivation a general rule for absolute maximum moment may be stated:

> *Maximum moment in a beam loaded with a moving series of concentrated loads usually will occur at the load nearest the center of gravity of the loads on the beam when the center of gravity of the loads on the beam is the same distance on one side of the centerline of the beam as the load nearest the center of gravity of the loads is on the other side.*

Should the load nearest the center-of-gravity of the loads be relatively small, the absolute maximum moment may occur at some other load nearby. Occasionally two or three loads have to be considered to find the greatest value. Nevertheless, the problem is not a difficult one because another moment criteria not described herein—the average load to the left must be equal to the average load to the right—must be satisfied. There will be little trouble determining which of the nearby loads will govern. Actually, it can be shown that the absolute maximum moment occurs under the load that would be placed at the centerline of the beam to cause maximum moment there, when that wheel is placed as far on one side of the beam centerline as the center of gravity of all the loads is on the other.[5]

[5]A. Jakkula, and H. K. Stephenson, *Fundamentals of Structural Analysis* (New York: Van Nostrand, 1953), 241–242.

EXAMPLE 10.1

Determine the absolute maximum moment that can occur in the 50-ft simple beam shown in Figure 10.16 as the series of concentrated loads shown moves across the span.

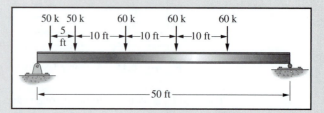

Figure 10.16

Solution.   The center of gravity of the loads is determined from:

$$\frac{50(5) + 60(15) + 60(25) + 60(35)}{50 + 50 + 60 + 60 + 60} = 16.96 \text{ ft from left load}$$

Then the loads are placed as follows and the shearing force and bending moment diagrams are drawn. The absolutely largest moment that can occur for these loads is 1980 ft-k.

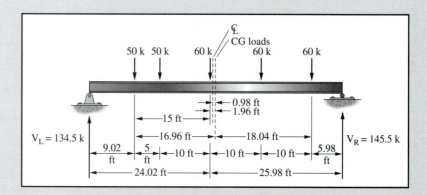

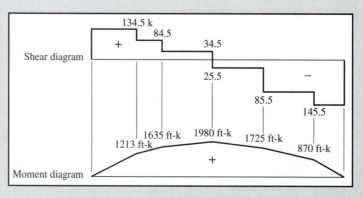

10.12 PROBLEMS FOR SOLUTION

For Problems 10.1 through 10.18, draw influence lines for the members indicated. Vertical or horizontal components are satisfactory.

10.1 L_0U_1, L_0L_1, U_1L_1, U_1U_2 (*Ans.* L_0L_1 $+1.125$ at L_1, U_1L_1 $+1.00$ at L_1, U_1U_2 -1.50 at L_2)

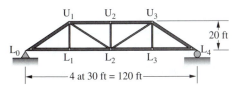

10.2 L_0L_1, U_2U_3, U_2L_3

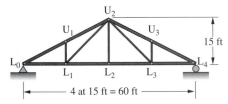

10.3 U_2U_3, L_2L_3, L_2U_3 and L_4U_5 (*Ans.* U_2U_3 -2.00 at L_2, L_2U_{3V} $+0.333$ at L_2, L_4U_5 $+0.667$ at L_4)

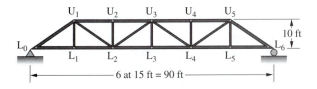

10.4 U_0L_1, L_1U_2, and U_2L_2 as a unit load moves across top of truss

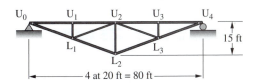

10.5 U_2U_3, L_3U_4, U_3L_3, and L_4L_5 (*Ans.* U_2U_{3H} -0.938 at L_3, L_3U_{4V} $+0.3125$ at L_3, L_4L_5 $+0.952$ at L_4)

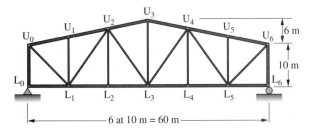

10.6 U_1L_2, U_2U_3, and L_2L_3

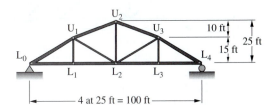

10.7 Members L_3L_4, U_1U_2 and U_2L_2 of the truss of Problem 10.6 (*Ans.* L_3L_4 $+1.25$ at L_3, U_1U_{2H} -1.00 at L_2, U_2L_2 $+0.80$ at L_2)

10.8 U_1U_3, L_2L_4, L_2U_3, L_4U_5

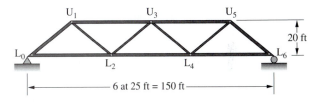

10.9 U_0L_1, U_1U_2, U_2L_3, L_3U_4 as unit load moves across top of truss (*Ans.* U_0L_{1H} $+1.042$ at U_1, U_1U_2 -1.67 at U_2, L_3U_{4V} -0.333 at U_4)

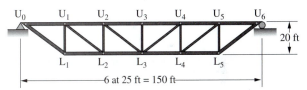

10.10 U_1U_2, U_1L_1, U_1L_2, U_3L_4 as unit load moves across top of truss

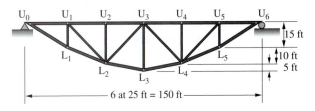

10.11 L_0U_1, U_1L_2, U_2U_3 (*Ans.* L_0U_{1V} -0.50 at L_1, U_1L_{2V} -1.00 at L_4, U_2U_3 $+3.33$ at L_4)

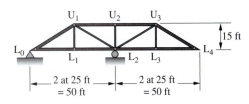

10.12 L_0U_1, U_1L_2, U_3L_4, U_5U_6

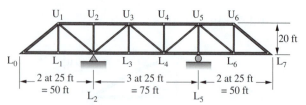

10.13 L_0U_1, U_1L_2, U_3L_4, U_5U_6 (*Ans.* $U_1L_2 - 0.5$ at L_1, U_3L_{4V} $- 0.375$ at L_3, $U_5U_{6H} + 3.33$ at L_8)

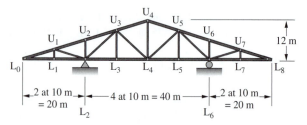

10.14 U_1L_2, L_2L_3, U_3L_4

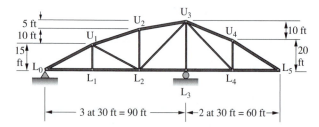

10.15 U_1U_2, U_2L_3, U_3L_3 (*Ans.* $U_1U_{2H} - 1.60$ at L_2, U_2L_{3V} $+ 0.25$ at L_3, $U_3L_3 + 0.500$ at L_3)

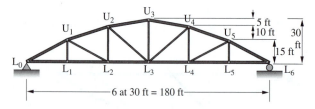

10.16 U_2U_4, L_3L_5, U_4L_5, L_5U_6 as unit load moves across top of truss

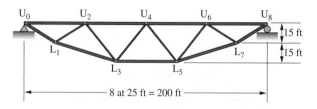

10.17 U_1U_2, U_1L_1, L_2U_3 Assume unit load moves across top of truss (*Ans.* $U_1U_2 - 2.00$ at U_2, $U_1L_1 - 0.875$ at U_1, $L_2L_{3V} + 0.583$ at U_2)

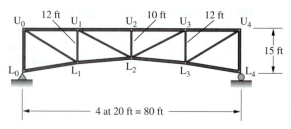

10.18 U_2U_3, M_1L_2, M_2L_2, U_3L_3

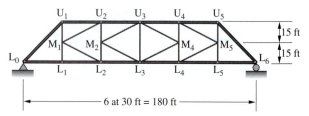

10.19 Determine the maximum and minimum resultant forces in member L_2U_3 of the truss of Problem 10.3 for a uniform dead load of 1 klf, a moving uniform load of 2 klf, a moving concentrated load of 20 k and an impact factor of 30% (*Ans.* $+30.16$ k, -100.25 k)

10.20 Is force reversal possible in member U_1L_2 of the truss of Problem 10.15 for the loads and impact factor used in Problem 10.19? Show resultant forces.

10.21 Determine the absolute maximum shear and moment possible in a 30-ft simple beam due to the load system shown. It is anticipated that the reader will determine the maximum shear by a trial and error process and the maximum moment by using the general rule developed in Section 10.11 of this chapter (*Ans.* 75.33 k, 501.5 ft-k)

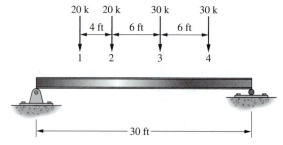

10.22 A simple beam of 20-m span supports a pair of 60-kN moving concentrated loads 4 m apart. Compute the maximum possible moment at the centerline of the beam and the absolute maximum moment in the beam.

10.23 What is the maximum possible moment that can occur in an 80-ft simple beam as the load system shown moves across the span? (*Ans.* 2792.6 ft-k)

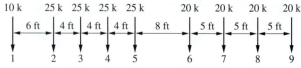

Chapter 11

Deflections and Angle Changes Using Geometric Methods

11.1 INTRODUCTION

This chapter and the next two are concerned with the elastic deformations of structures. Both the linear displacements of points (deflections) and the rotational displacements of lines (slopes) are considered. The word *elastic* is used to mean that:

1. Stresses are proportional to strains;
2. There is a linear variation of stress from the neutral axis of a beam to its extreme fibers; and
3. The members will return to their original geometry after loads are removed.

The deformations of structures are caused by bending moments, by axial forces, and by shearing forces. For beams and frames, the largest values are caused by bending moments, whereas for trusses the largest values are caused by axial forces. Deflections caused by shearing forces are neglected in this text, as they are quite small in almost all beam-like structures. Deflections caused by shearing forces, as a percentage of beam deflections, increase as the ratio of beam depth to span increases. For the usual depth/span ratio of 1/12 to 1/6, the percentages of shear deflections to bending deflections vary from about 1% to 8%. For a depth/span ratio of one-quarter, the percentages can be as high as 15% to 18%.[1]

In this chapter, displacements are computed using the moment-area method. This method is referred to as geometric method because the deformations are obtained directly from the strains in the structure. Chapter 12 provides a continuation of geometric methods by presenting the elastic weight and conjugate beam methods. In Chapter 13 displacements are determined with *energy methods*, which are based on the conservation of energy principle. Both the geometric and energy procedures will provide identical results.

11.2 SKETCHING DEFORMED SHAPES OF STRUCTURES

Before learning methods for calculating structural displacements, there is considerable benefit in learning to qualitatively sketch the expected deformed shapes of structures under load. Understanding the displacement behavior of structural systems is a very

[1]C. K. Wang, *Intermediate Structural Analysis* (New York: McGraw-Hill Book Company, 1983), 750.

important part of understanding how structures perform. A structural analyst should sketch the anticipated deformed shape of structures under load before making actual calculations. Such a practice provides an appreciation of the behavior of the structure and provides a qualitative check of the magnitudes and directions of the computed displacements.

To sketch the anticipated deformed shape of a structural system, there are only a few general rules to follow. Some of these rules apply to beams and columns and others apply to the joints between the components. By applying these simple rules, we can obtain reasonable qualitative indications of the deflection response of beams and frames. Only by applying the quantitative methods discussed in this and later chapters can we obtain the actual deflections.

Rules for Members

1. A member deforms in the direction of the load applied to it.
2. Deflections of loaded members are sketched first. Deflections of unloaded members are sketched after the deflections of the joints are sketched.

(a) Point of maximum deflection is some where to the left of this off-center load.

(b) Tangent at the fixed end is horizontal and the right end deflects upward.

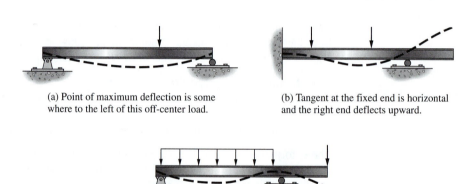

(c) Without calculations we do not know whether the deflection at the right end is up or down. The concentrated load tends to push the right end down while the uniform load tends to push it up.

(d) Note the upward deflection in the third span.

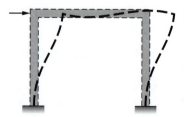

(e) The frame sways to the right.

Figure 11.1   Qualitative deformed shapes of some structures under load

3. Unless there is a hinge between a member and a joint, the end of the member and the joint displace in the same manner.
4. Members with smaller stiffnesses (EI/L) tend to deform more than do members with higher stiffnesses. That is to say, long slender members deform more than short stocky ones.
5. When sketching the qualitative deformed shape of a structure, the beams and columns are assumed to retain their original lengths.

Rules for Joints

1. A joint in a structure is assumed to be rigid. A rigid joint can displace but it cannot deform—the joint does not change size or shape as it displaces. The relative orientation of the ends of the members connected to a joint is the same before and after displacement of the joint.
2. A joint can only displace in accordance with the external supports acting on it. A joint at a fixed support can neither translate nor rotate. A joint at a pin support can rotate but it cannot translate. A joint at a roller can rotate, cannot translate perpendicular to the surface on which the rollers bears, and can translate parallel to the surface on which the roller bears.

In Figure 11.1, the approximate deflected shapes of several loaded structures are sketched by applying these rules. In each case, the member weight is neglected. You should note that the corners of the frame of part (e) of the figure are free to rotate, but the angles between the members meeting there are assumed to be constant. If the moment diagrams have previously been prepared, they can be helpful in making the sketches where we have both positive and negative moments.

Several examples for sketching the qualitative deformed shapes of structures follow. In these examples, the thought process for preparing each sketch is presented.

EXAMPLE 11.1

Consider the three-span continuous beam of Figure 11.2, which is subjected to a concentrated force on one span and a distributed force on another span. Sketch the qualitative deflected shape for this beam.

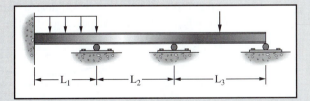

Figure 11.2

Solution.   Sketch the left span first. The left side cannot rotate because it is connected to a fixed joint. The load is acting downward so the span will tend to deflect downward. The right end can rotate, but neither the left nor right ends can translate vertically. After the member deformation is sketched, sketch the displacement of the joints. Then sketch the deformation of the right span. Both ends of the right span can rotate but neither can translate. The result to this point is shown in the following figure.

Now sketch the deformed shape of the middle span. Because the member has no external load acting on it, it deforms only in response to the displacement of the joints to which it is connected. The slope of the members connected at a particular joint must be the same.

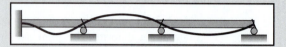

This is the qualitative deflected shape of the beam.   ■

EXAMPLE 11.2

Sketch the qualitative deflected shape of the overhanging beam of Figure 11.3, which is subjected to a uniformly distributed load.

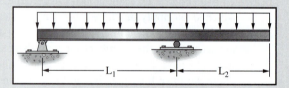

Figure 11.3

Solution.   Sketching the deflected shape of this beam is a little more involved. The load on the left span tends to cause the joint at the right support to rotate counter-clockwise while the load on the right span tends to cause that same joint to rotate clockwise. As the longer span will tend to dominate the rotation; we can sketch the deformation accordingly.

Sketch the left span first and show the resulting rotation of the joints. The load is acting downward so the span will tend to deflect downward. Both ends can rotate, but neither end can translate vertically. As the member deformation is sketched, sketch the displacement of the joints.

Next, sketch the deformation of the right span, the cantilever span. There is not a support at the right side of the span so that end will displace in response to the applied load. Recall that the geometry of the joints does not change so the tangents of the deflected shapes of the two spans at the right support must be the same.

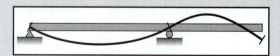

This is the qualitative deflected shape of the beam. Whether the right end moves upward or downward will depend on the magnitude of the loads and relative lengths of the two spans. ■

EXAMPLE 11.3

Sketch the qualitative deformed shape of the braced frame shown in Figure 11.4. Because this is a braced frame, the joints will not translate relative to one another.

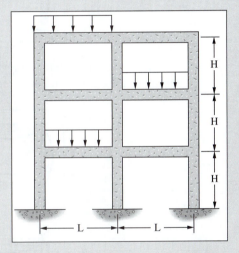

Figure 11.4

Solution. Sketch the deformed shape of the loaded members first. At the same time indicate the rotation of the joints. The loaded members will tend to deform in the direction of the applied loads.

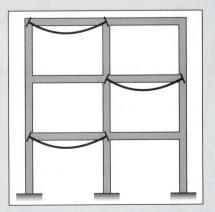

Now sketch the deformed shape of the other beams taking into account the displacements of the joints to which they are connected. Because the beams are fully connected at the joints, the tangents of the deformed shape of the beams connected at a joint must be the same.

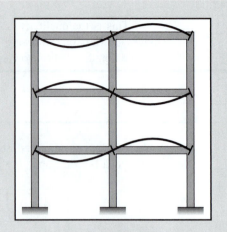

Now sketch the deformations of the columns. Recall that the joints do not deform so the right angles between the beams and columns must be maintained.

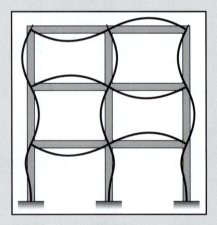

The result is the qualitative deformed shape of the braced frame.   ■

11.3 REASONS FOR COMPUTING DEFLECTIONS

The members of all structures are made up of materials that deform when loaded. If deflections exceed allowable values, they may detract from the appearance of the structures and the materials attached to the members may be damaged. For example, a floor beam that deflects too much may cause cracks in the ceiling below, or if it supports concrete or tile floors, it may cause cracks in the floors. In addition, the use of a floor supported by beams that "give" appreciably does not inspire confidence, although the beams may be perfectly safe. Excessive vibration may occur in a floor of this type, particularly if it supports machinery.

Standard American practice is to limit deflections caused by live load to 1/360 of the span lengths. This figure probably originated for beams supporting plaster ceilings and was thought to be sufficient to prevent plaster cracks. The reader should realize that a large part of the deflections in a building are due to dead load, most of which will have occurred before plaster is applied.

The 1/360 deflection is only one of many maximum deflection values in use because of different loading situations, different designers, and different specifications. For situations in which precise and delicate machinery is supported, maximum deflections may be limited to 1/1500 or 1/2000 of the span lengths. The 2002 AASHTO specifications limit deflections in steel beams and girders due to live load and impact to 1/800 of the span length. The value, which is applicable to both simple and continuous spans, is preferably reduced to 1/1000 for bridges in urban areas that are used in part by pedestrians. Corresponding AASHTO values for cantilevered arms are 1/300 and 1/375.

Members subject to large downward deflections often are unsightly and may even cause users of the structure to be frightened. Such members may be *cambered* so their displacements do not appear to be so large. The members are constructed of such a shape that they will become straight under some loading condition (usually dead load). A simple beam would be constructed with a slight convex bend so that under gravity loads it would become straight as assumed in the calculations. Some designers take into account both dead and live loads when figuring the amount of camber to be installed.

Houston Ship Channel Bridge, Houston, Texas (Courtesy of the Texas State Department of Highways and Public Transportation)

Despite the importance of deflection calculations, rarely will structural deformations be computed—even for statically indeterminate structures—for the purpose of modifying the original dimensions on which computations are based. The deformations of the materials used in ordinary work are quite small when compared to the overall dimensions of the structure. For example, the strain (ε) that occurs in a steel section that has a modulus of elasticity (E) of 29×10^6 psi when the stress (σ) is 20,000 psi is only

$$\varepsilon = \frac{\sigma}{E} = \frac{20 \times 10^3}{29 \times 10^6} = 0.000690$$

This value is only 0.0690 percent of the member length.

Historically, several methods are available for determining deflections. A structural engineer should be familiar with several of these. For some structures one method may be easier to apply; for others another method is more satisfactory. In addition, the ability to evaluate a structural system by more than one method is important for checking results.

In this chapter and the next two, the following methods for computing slopes and deflections are presented: (a) moment-area theorems, (b) elastic weights, (c) conjugate beam, (d) virtual work and (e) Castigliano's second theorem. As shown in later chapters, these methods may be used for computing the reactions for statically indeterminate beams, frames, and trusses.

11.4 THE MOMENT-AREA THEOREMS

The moment-area method for calculating deflections was presented by Charles E. Greene of the University of Michigan in about 1873. Under changing loads the neutral axis of a member changes in shape according to the positions and magnitudes of the loads. The *elastic curve* of a member is the shape the neutral axis takes under temporary loads. Professor Greene's theorems are based on the shape of the elastic curve of a member and the relationship between bending moment and the rate of change of slope at a point on the curve.

To develop the theorems, the simple beam of Figure 11.5 is considered. Under the loads P_1 to P_4 it deflects downward as indicated in the figure.

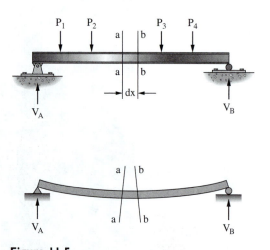

Figure 11.5

The segment of length dx, bounded on its ends by line a-a and line b-b, is shown in Figure 11.6. The size, degree of curvature, and distortion of the segment are tremendously exaggerated so that the slopes and deflections to be discussed can be easily seen. In Figure 11.6(a), line ac lies along the neutral axis of the beam and is unchanged in length. Line ce is drawn parallel to ab. Therefore, be is equal to ac and de represents the increase in length of be, the bottom fiber of the segment dx. Shown in Figure 11.6(b) is an enlarged view of triangle cde and the angle dθ. The angle dθ is the change in slope of the tangent to the elastic curve between the left and right ends of the section. Sufficient information is now available to determine dθ. In the derivation to follow, remember that the angle dθ

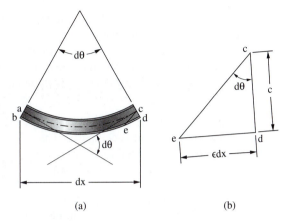

(a)   (b)

Figure 11.6

being considered is very small. For a very small angle, the sine of the angle, the tangent of the angle, and the angle in radians are nearly identical, which permits their values to be used interchangeably. It is worthwhile to check a set of natural trigonometry tables to see the range of angles for which the functions are almost identical.

The bending moments developed by the external loads are positive and cause shortening of the upper beam fibers and lengthening of the lower fibers. The changes in fiber dimensions have caused the change in slope $d\theta$. The modulus of elasticity of the beam material is known, and the stress at any point can be determined by the elastic flexure formula. Therefore, the strain in any fiber can be found because it is equal to the stress divided by the modulus of elasticity. The value of $d\theta$ may be expressed as follows:

$$\tan(d\theta) = \frac{\text{strain}}{c} = \frac{ed}{cd} = d\theta \text{ in radians (rad)}$$

$$d\theta = \frac{\varepsilon \, dx}{c}$$

By substituting σ/E for ε, we find that:

$$d\theta = \frac{(\sigma/E) \, dx}{c}$$

However, stress can be calculated using the elastic flexure equation, so this equation can be expressed as:

$$d\theta = \frac{\left(\dfrac{Mc}{EI}\right) dx}{c} = \frac{M \, dx}{EI}$$

This equation represents the change in slope of the elastic curve between the two ends of a segment that has a length dx. The total change in slope from one point A in the beam to another point B can be expressed as the integral of $d\theta$ over the length AB, namely:

$$\theta_{AB} = \int_A^B \frac{M \, dx}{EI}$$

This equation represents the area of the M/EI diagram between the points A and B. The M/EI diagram is simply the bending moment diagram divided by EI. From this discussion the first moment-area theorem may be expressed as follows:

   The change in slope between the tangents to the elastic curve at two points on a member is equal to the area of the M/EI diagram between the two points.

Once we have a method by which changes in slopes between tangents to the elastic curve at various points may be determined, only a small extension is needed to develop a method for computing deflections between the tangents. In a segment of length dx, the neutral axis changes direction by an amount $d\theta$. The deflection $d\delta$, of one point on the beam with respect to the tangent at another point due to this angle change, then, is equal to:

$$d\Delta = x \, d\theta$$

In this equation, x is the distance from the point at which deflection is desired to the point at which the tangent is computed. The value of $d\theta$ from Eq. 10.4 can be substituted into this expression to obtain:

$$d\Delta = x \frac{M \, dx}{EI} = \frac{Mx \, dx}{EI}$$

To determine the total deflection from the tangent at one point, A, to the tangent at another point on the beam, B, the preceding equation can be integrated over the distance AB, namely:

$$\Delta_{AB} = \int_A^B \frac{Mx \, dx}{EI}$$

The preceding equation is a mathematical statement of the second moment-area theorem which is:

The deflection of a tangent to the elastic curve of a beam at one point with respect to a tangent at another point is equal to the moment of the M/EI diagram between the two points taken about the point at which deflection is desired.

Stick-welding decking on a shopping center mall in Charlotte, North Carolina (Courtesy Lincoln Electric Company)

11.5 APPLICATION OF THE MOMENT-AREA THEOREMS

In the paragraphs that follow, you will see that the moment area method is most conveniently used for determining slopes and deflections for beams when the slope of the elastic curve at one or more points is known. Such beams include cantilevered beams, where the slope at the fixed end does not change. The method is applied quite easily to beams loaded with concentrated loads, because the moment diagrams consist of straight lines. These diagrams can be broken down into simple triangles and rectangles, which facilitates the mathematics. Beams supporting uniform loads or uniformly varying loads may be handled, but the mathematics is slightly more difficult.

The properties of the several figures shown in Figure 11.7 are quite useful for solving the problems in this chapter and the next.

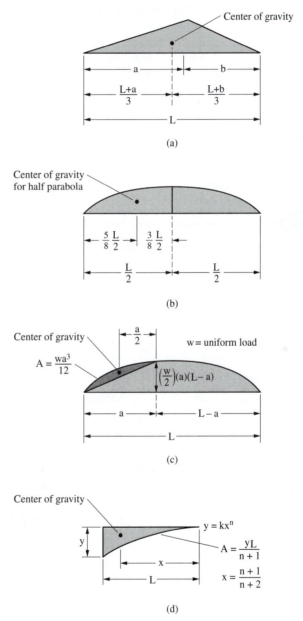

Figure 11.7 Frequently used area properties.

Examples 11.4 to 11.9 illustrate the application of the moment-area theorems. On occasion, the mathematics may be simplified by drawing the moment diagram and making the calculations in terms of symbols. Such symbols would include P for a concentrated load, w for a uniform load, and L for span length. This concept is illustrated in Examples 11.4 and 11.6. The numerical values of each of the symbols are substituted in the final step to obtain the slope or deflection desired.

When solving problems, care must be taken to use consistent units in the calculations. Further, to prevent mistakes when applying the moment-area theorems, *always remember that the slopes and deflections obtained are with respect to tangents to the elastic curve at the points being considered. The theorems do not directly give the slope or deflection at a point in the beam as compared to the horizontal (except in one or two special cases). Rather, the theorems give the change in slope of the elastic curve from one point to another, or the deflection of the tangent at one point with respect to the tangent at another point.*

If a beam or frame has several loads applied, the M/EI diagram may be inconvenient to handle. This is especially true for beams and frames supporting both concentrated and uniform loads. When distributed loads and concentrated loads are applied concurrently, the M/EI diagram becomes rather complex, which causes difficulty in computing the needed properties of the areas involved. The calculations may be simplified by drawing a separate M/EI diagram for each of the loads and determining the slopes and deflections caused by each load separately. The final result for a particular point can be found by adding the values obtained for the individual loads. The principle of superposition, which was presented in Section 4.12 of Chapter 4 of this book, applies to the moment-area method, and to any of the other methods for computing deflections as long as we are in the elastic range.

EXAMPLE 11.4

Determine the slope and deflection of the right end of the cantilevered beam shown in Figure 11.8.

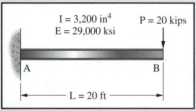

Figure 11.8

Solution.   The tangent to the elastic curve at the fixed end is horizontal. Therefore, the changes in slope and deflection of the tangent at the free end, with respect to a tangent at the fixed end, are the slope and deflection of the free end. The bending moment diagram, and the associated M/EI diagram, for the beam are as follows

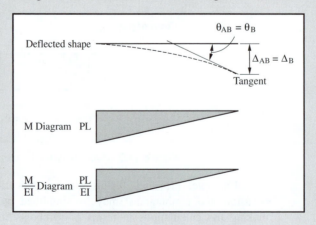

The slope at B is equal the area under the M/EI diagram from A to B

$$\theta_B = \left(\frac{1}{2}\right) L \left(\frac{PL}{EI}\right) = \frac{PL^2}{2EI} = \frac{(20)(1000)[20(12)]^2}{2(3,200)(29 \times 10^6)}$$

$$\theta_B = 0.00621 \text{ rad} = 0.36° \ \seardiag$$

The deflection at B is equal to the moment of the M/EI diagram between A and B taken at B where deflection is desired.

$$\Delta_B = \left(\frac{1}{2}\right) L \left(\frac{PL}{EI}\right) \left(\frac{2L}{3}\right) = \frac{PL^3}{3EI} = \frac{(20)(1000)[20(12)]^3}{3(3,200)(29 \times 10^6)}$$

$$\Delta_B = 0.99 \text{ inches} \downarrow \quad \blacksquare$$

EXAMPLE 11.5

Determine the slope and deflection of the beam at point B, which is 10 ft from the left end of the beam of Figure 11.9.

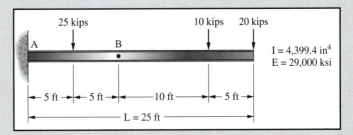

Figure 11.9

Solution.   The left end of the beam is again fixed. As such the slope at B is equal to the area of the M/EI diagram from A to B. The M/EI diagram is broken down into convenient triangles, as shown, for making the calculations.

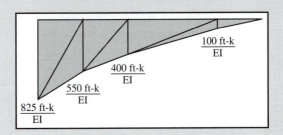

$$\theta_B = \frac{\frac{1}{2}(825)(5) + \frac{1}{2}(550)(5) + \frac{1}{2}(550)(5) + \frac{1}{2}(400)(5)}{EI} = \frac{5812.5 \text{ k-ft}^2}{EI}$$

$$\theta_B = \frac{5812.5(12)^2(1000)}{(29 \times 10^6)(4399.4)} = 0.00656 \text{ rad} = 0.38° \ \seardiag$$

The deflection at B is equal to the first moment of the M/EI diagram between A and B taken about B.

$$\Delta_B = \frac{\frac{1}{2}(825)(5)(8.33) + \frac{1}{2}(550)(5)(6.67) + \frac{1}{2}(550)(5)(3.33) + \frac{1}{2}(400)(5)(1.67)}{EI}$$

$$= \frac{32{,}600\,\text{k-ft}^3}{EI}$$

$$\Delta_B = \frac{32{,}600(12)^3(1000)}{(29 \times 10^6)(4399.4)} = 0.442\,\text{inches} \downarrow \quad \blacksquare$$

EXAMPLE 11.6

Determine the slope and deflection at the free end of the cantilevered beam shown in Figure 11.10.

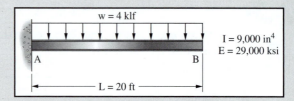

Figure 11.10

Solution.   The M/EI diagram used for the beam is shown here.

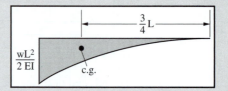

Carefully note the units used to solve the slope and deflection equations developed. Inches and pounds are used throughout the equations. The value of w, then, is 4000/12 lb/in. and not just 4000 lb/ft. Be careful with the units for the distributed load, since the slope or deflection can easily be miscalculated by a multiple of 12.

As before, the slope at the left side is horizontal because that end is fixed. The slope of the elastic curve at the free end B, then, is equal to the area of the M/EI diagram between the left end and B.

$$\theta_B = \left(\frac{1}{3}\right)L\left(\frac{wL^2}{2EI}\right) = \frac{wL^3}{6EI}$$

$$\theta_B = \frac{\left(\dfrac{4000}{12}\right)[20(12)]^3}{(6)(29 \times 10^6)(9000)} = 0.00294\,\text{rad} = 0.17°$$

The deflection at A is equal to the moment of the M/EI diagram between the left end and A taken about A.

$$\Delta_B = \left(\frac{1}{3}\right) L \left(\frac{wL^2}{2EI}\right)\left(\frac{3L}{4}\right) = \frac{wL^4}{8EI}$$

$$\Delta_B = \frac{\left(\dfrac{4000}{12}\right)[20(12)]^4}{(8)(29 \times 10^6)(9000)} = 0.530 \text{ inches} \downarrow \quad \blacksquare$$

EXAMPLE 11.7

Compute the slope and deflection at the free end of the cantilevered beam shown in Figure 11.11. The size of the beam and thus its moment of inertia have been increased near the support where the bending moment is so much larger than it is near the free end.

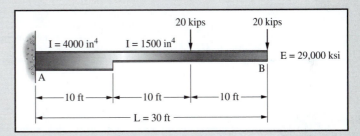

Figure 11.11

Solution.   The moment diagram and the corresponding M/EI diagram are shown below. The M/EI diagram is drawn by keeping the constant E as a symbol and dividing the ordinates by the proper moments of inertia. The resulting figure is conveniently divided into triangles and the computations made as before.

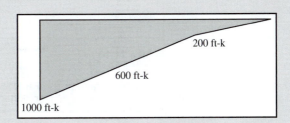

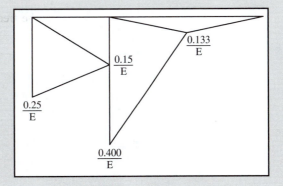

The slope at the left end of the beam is horizontal because that end is a fixed support. The slope of the elastic curve at B then, is equal to the area of the M/EI diagram between the left end and B.

$$\theta_B = \frac{\frac{1}{2}(10)(0.25) + \frac{1}{2}(10)(0.15) + \frac{1}{2}(10)(0.40) + \frac{1}{2}(20)(0.133)}{E} = \frac{5.333\,\text{k-ft}^2}{E}$$

$$\theta_B = \frac{5.333(12)^2(1000)}{(29 \times 10^6)} = 0.0265\,\text{rad} = 1.52°$$

Again the deflection at B is equal to the first moment of the M/EI diagram taken at B.

$$\Delta_B = \frac{\frac{1}{2}(10)(0.25)(26.67) + \frac{1}{2}(10)(0.15)(23.33) + \frac{1}{2}(10)(0.40)(16.67) + \frac{1}{2}(20)(0.133)(10)}{E}$$

$$= \frac{97.47\,\text{k-ft}^3}{E} = \frac{97.47(12)^3(1000)}{(29 \times 10^6)} = 5.81\,\text{inches} \downarrow \quad \blacksquare$$

EXAMPLE 11.8

Compute the deflection at the centerline of the uniformly loaded simply supported beam shown in Figure 11.12.

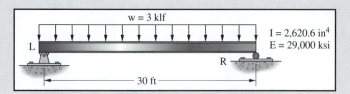

Figure 11.12

Solution.   The M/EI diagram used in the solution is shown here.

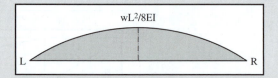

The tangents to the elastic curve at each end of the beam are inclined. Computing the deflection between a tangent at the centerline and one of the end tangents is a simple matter, but the result is not the actual deflection at the centerline of the beam. To obtain the correct deflection, a somewhat roundabout procedure is used. This indirect procedure is described and demonstrated in the following paragraphs. It is not as practical as the other methods that are presented for such situations in the next two chapters.

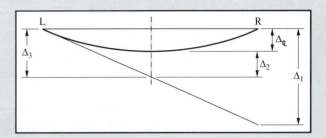

First, the deflection of the tangent at the right end from the tangent at the left end, Δ_1, is found:

$$\Delta_1 = \left(\frac{2}{3}\right)L\left(\frac{wL^2}{8EI}\right)\left(\frac{L}{2}\right) = \frac{wL^4}{24EI}$$

Next, the deflection of a tangent at the centerline from a tangent at L, Δ_2, is found:

$$\Delta_2 = \left(\frac{2}{3}\right)\left(\frac{L}{2}\right)\left(\frac{wL^2}{8EI}\right)\left(\frac{3}{8}\right)\left(\frac{L}{2}\right) = \frac{wL^4}{128EI}$$

Then, by proportions, the distance from the original chord between L and R and the tangent at $\frac{L}{2}$, Δ_3, can be computed:

$$\Delta_3 = \frac{1}{2}\delta_1 = \left(\frac{1}{2}\right)\left(\frac{wL^4}{24EI}\right) = \frac{wL^4}{48EI}$$

The difference between Δ_3 and Δ_2 is the centerline deflection.

$$\Delta_\mathbb{C} = \Delta_3 - \Delta_2 = \frac{wL^4}{48EI} - \frac{wL^4}{128EI} = \frac{5wL^4}{384EI}$$

$$\Delta_\mathbb{C} = \frac{5\left(\frac{3000}{12}\right)[30(12)]^4}{384(2620.6)(29 \times 10^6)} = 0.719\,\text{inches} \downarrow \ \blacksquare$$

11.6 ANALYSIS OF FIXED-END BEAMS

Example 11.9 shows that the moment area theorems may be used to determine the moments at the ends of a beam fixed at both ends. Such a member, an example of which is shown in Figure 11.13, is statically indeterminate to the third degree. This type of calculation is useful to us when we are using the moment distribution method of Chapters 20 and 21 of this text to analyze statically indeterminate beams and frames. It is particularly useful for handling situations when the moments of inertia vary along the members, when loads are uniformly varying and for other similar situations.

EXAMPLE 11.9

Determine the moments at the ends of the fixed-ended beam shown in Figure 11.13 for which E and I are constant.

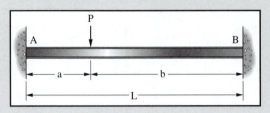

Figure 11.13

Solution. Examination of the beam reveals no change of slope and no deflection of the tangent at A from the tangent at B; therefore, the total area of the M/EI diagram from A to B is zero, and the moment of the M/EI diagram about either end is zero.

The M/EI diagram may be drawn in two parts: the simple beam moment diagram, the ordinates of which are known, and the moment diagram due to the unknown end moments M_A and M_B. For each of the latter moments, a triangular-shaped diagram may be drawn and the two combined into one trapezoid. The moment-area theorems are written to express the change in slope and deflection from B to A. Each of the two equations contains the two unknowns M_A and M_B, and the equations are solved simultaneously.

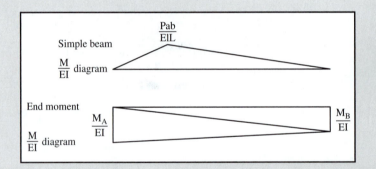

Theorem 1:

$$\left(\frac{1}{2}\right)\left(\frac{Pab}{EIL}\right)(L) + \left(\frac{1}{2}\right)\left(\frac{M_A}{EI}\right)(L) + \left(\frac{1}{2}\right)\left(\frac{M_b}{EI}\right)(L) = 0$$

$$\frac{Pab}{2EI} + \frac{M_AL}{2EI} + \frac{M_BL}{2EI} = 0 \tag{1}$$

Theorem 2 (taking moments about left end A):

$$\left(\frac{Pab}{2EI}\right)\left(\frac{L+a}{3}\right) + \left(\frac{M_AL}{2EI}\right)\left(\frac{1}{3}L\right) + \left(\frac{M_BL}{2EI}\right)\left(\frac{2}{3}L\right) = 0$$

$$\frac{Pabl + Pa^2b}{6EI} + \frac{M_AL^2}{6EI} + \frac{M_BL^2}{3EI} = 0 \tag{2}$$

By solving equations (1) and (2) simultaneously for M_A and M_B,

$$M_A = -\frac{Pab^2}{L^2} \qquad M_B = -\frac{Pa^2b}{L^2} \quad \blacksquare$$

Westinghouse Rapid Transit, Pittsburgh, Pennsylvania. (Courtesy Bethlehem Steel Corporation.)

11.7 MAXWELL'S LAW OF RECIPROCAL DEFLECTIONS

The deflections of two points in a beam have a surprising relationship to each other. This relationship was first published by James Clerk Maxwell in 1864. Maxwell's law may be stated as follows:

> *The deflection at one point A in a structure due to a load applied at another point B is exactly the same as the deflection at B if the same load is applied at A.*

The rule is perfectly general and applies to any type of structure, whether it is a truss, beam, or frame, which is made up of elastic materials following Hooke's law. The displacements may be caused by flexure, by shear, or by torsion. When preparing influence lines for continuous structures, when analyzing statically indeterminate structures and with model-analysis problems, this useful tool is frequently applied.

The law is not only applicable to the deflections in all of these types of structures but is also applicable to rotations. For instance, a unit couple at A will produce a rotation at B equal to the rotation caused at A if the same couple is applied at B.

Example 11.10 demonstrates that the law is correct for a simple cantilevered beam in which the deflections at two points are determined using the moment-area method.

EXAMPLE 11.10

Demonstrate Maxwell's law of reciprocal deflections by comparing the deflection at point A caused by a 10-kip load applied first at B with the deflection at B caused by a 10-kip load at A. The cantilever beam used in this example and the first loading condition are shown in Figure 11.14.

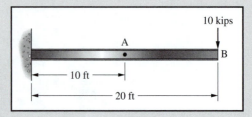

Figure 11.14

Solution. The M/EI diagram for this beam with a 10-kip load applied at B is shown next. The M/EI diagram has been divided into simple triangles to facilitate the computations.

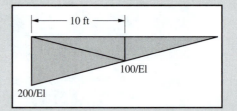

Because the left side is a fixed support, the deflection at A is equal to the first moment of the area under the M/EI diagram between the left support and A taken about A.

$$\Delta_A = \frac{200(10)(6.667)}{2EI} + \frac{100(10)(3.333)}{2EI} = \frac{8{,}332 \text{ ft}^3\text{-k}}{EI}$$

When the 10-kip load is applied at A, the loaded beam and the corresponding M/EI diagram are as shown in Figure 11.15.

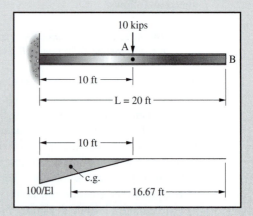

Figure 11.15

The deflection caused at B by this load is the first moment of the M/EI diagram about B.

$$\Delta_B = \frac{100(10)(16.667)}{2EI} = \frac{8,335 \text{ ft}^3\text{-k}}{EI}$$

These two computed deflection are the same, which demonstrates Maxwell's law of reciprocal deflections: *The deflection at A caused by a load at B is equal to the deflection at B caused by the same load at A.* ■

11.8 PROBLEMS FOR SOLUTION

For problems 11.1 through 11.7 qualitatively sketch the deformed shape of the structures for the given loads.

11.1

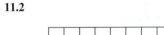

11.2

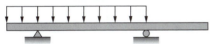

11.3

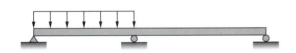

11.4

11.5

11.6

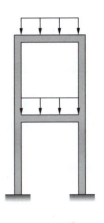

11.7

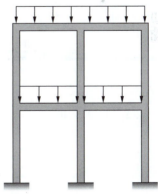

11.8 to 11.22 using the moment-area method, determine the quantities asked.

11.8 θ_A, θ_B, Δ_A, Δ_B. $E = 29 \times 10^6$ psi. $I = 1200$ in.^{4}.

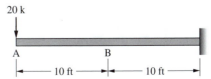

20 k

A   B

|← 10 ft →|← 10 ft →|

11.9 θ_A, θ_B, Δ_A, Δ_B. E $= 29 \times 10^6$ psi. I $= 2500$ in.^{4}. (*Ans.* $\theta_A = 0.0109$ rads /, $\Delta_A = 1.87$ in. ↓, $\Delta_B = 0.636$ in. ↓)

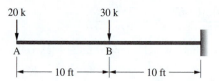

11.10 θ_A, θ_B, Δ_A, Δ_B. E $= 29 \times 10^6$ psi. I $= 1500$ in.^{4}.

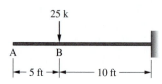

11.11 θ_A, Δ_A. E $= 29 \times 10^6$ psi. I $= 1140$ in.^{4}. (*Ans.* $\theta_A = 0.002509$ rads /, $\Delta_A = 0.271$ in. ↓)

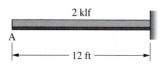

11.12 θ_A, Δ_A. E $= 29 \times 10^6$. I $= 1000$ in.^{4}.

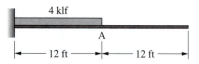

11.13 θ_A, θ_B, Δ_A, Δ_B. E $= 29 \times 10^6$ psi. I $= 2100$ in.^{4}. (*Ans.* $\theta_B = 0.01430$ rads \, $\Delta_A = 1.03$ in. ↓, $\Delta_B = 3.02$ in. ↓)

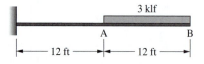

11.14 θ_A, θ_B, Δ_A, Δ_B. E $= 29 \times 10^6$ psi. I $= 5200$ in.^{4}.

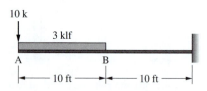

11.15 θ_A, Δ_A. E $= 29 \times 10^6$ psi. I $= 3000$ in.^{4}. (*Ans.* $\theta_A = 0.00393$ rads \, $\Delta_A = 0.310$ in. ↓)

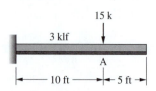

11.16 θ_A, θ_B, θ_C, Δ_C. E $= 29 \times 10^6$ psi. I $= 1320$ in.^{4}.

11.17 θ_A, Δ_A. E $= 29 \times 10^6$ psi. I $= 600$ in.^{4}. (*Ans.* $\theta_A = 0.00233$ rads \, $\Delta_A = 0.335$ in. ↓)

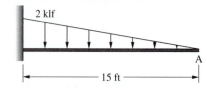

11.18 θ_A, θ_B, Δ_A, Δ_B. E $= 29 \times 10^6$ psi.

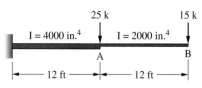

11.19 θ_A, Δ_A. E $= 29 \times 10^6$. (*Ans.* $\theta_A = 0.00432$ rads /, $\Delta_A = 0.720$ in. ↓)

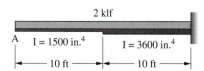

11.20 θ_A, Δ_B. E $= 200\,000$ MPa. I $= 3 \times 10^8$ mm^4.

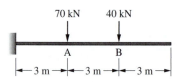

11.21 θ_A, Δ_A. E $= 200\,000$ MPa. (*Ans.* $\theta_A = 0.0322$ rads /, $\Delta_A = 273$ mm ↓)

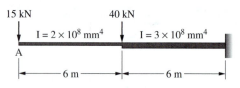

11.22 Δ_A. E $= 29 \times 10^6$ psi. I $= 1500$ in.^{4}.

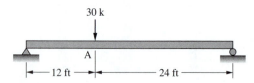

For Problems 11.23 to 11.27, compute fixed end moments for the beams given. E and I constant except as shown.

11.23 (*Ans.* $M_A = -135$ ft-k, $M_B = -45$ ft-k)

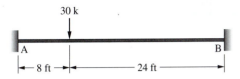

11.24

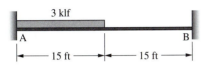

11.25 (*Ans.* $M_A = M_B = -366.8$ ft-k)

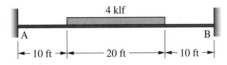

11.26

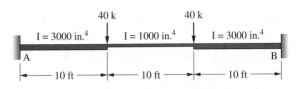

11.27 (*Ans.* $M_A = -323.5$ ft-k, $M_B = -105.6$ ft-k)

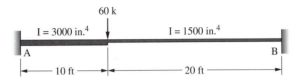

Chapter 12

Deflections and Angle Changes Using Geometric Methods Continued

12.1 THE METHOD OF ELASTIC WEIGHTS

A careful study of the procedure used in applying the moment-area theorems will reveal a simpler and more practical method of computing slopes and deflections for most beams. In reviewing this procedure the beam and M/EI diagram of Figure 12.1 are considered.

By letting A equal the area of the M/EI diagram, the deflection of the tangent at R from the tangent at L equals Ay, and the change in slope between the two tangents is A.

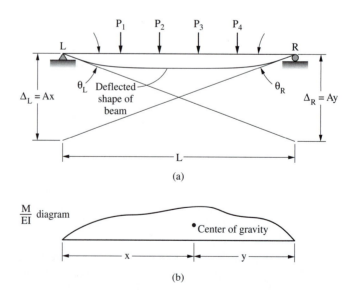

Figure 12.1 Beam for discussing method of elastic weights

An imaginary beam is loaded with the M/EI diagram, as shown in Figure 12.2 and the reactions R_L and R_R are determined. They equal Ay/L and Ax/L, respectively.

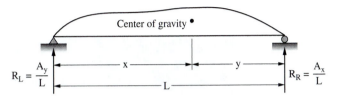

Figure 12.2   Reactions for beam loaded with M/EI diagram

In Figure 12.2 the slopes of the tangents to the elastic curve at each end of the beam (θ_L and θ_R) are equal to the deflections between the tangents at each end divided by the span length, as follows:

$$\theta_L = \frac{\Delta_R}{L} \quad \theta_R = \frac{\Delta_L}{L}$$

The values of Δ_L and Δ_R have previously been found to equal A_y and A_x, respectively, and may be substituted in these expressions:

$$\theta_L = \frac{A_y}{L} \quad \theta_R = \frac{A_x}{L}$$

The end slopes are exactly the same as the reactions for the beam in Figure 12.2. At either end of the fictitious beam the shear equals the reaction and thus the slope in the actual beam. Further experiments will show that the shear at any point in the beam loaded with the M/EI diagram equals the slope at that point in the actual beam.

A similar argument can be made concerning the computation of deflections, and it will be found that the deflection at any point in the actual beam equals the moment at that point in the fictitious beam. In detail, the two theorems of elastic weights may be stated as follows:

1. *The slope of the elastic curve of a simple beam at a point, measured with respect to a chord between the supports, equals the shear at that point if the beam is loaded with the M/EI diagram.*
2. *The deflection of the elastic curve of a simple beam at a point, measured with respect to a chord between the supports, equals the moment at that point if the beam is loaded with the M/EI diagram.*

12.2 APPLICATION OF THE METHOD OF ELASTIC WEIGHTS

The method of elastic weights in its present form is applicable only to beams simply supported at each end. It will be found in using the method that maximum deflections in the actual beam occur at points of zero shear in the imaginary beam. The reasoning is the same as that presented for shear and moment diagrams in Chapter 5, where maximum moments were found to occur at points of zero shear.

Consideration has not been given to the subject of sign conventions for either of the elastic-weight theorems. Study of the shears and moments on the fictitious beam reveals the directions of slopes and deflections. A positive shear in the fictitious beam shows the left side is being pushed up with respect to the right side, or the beam is sloping downward from left to right. Similarly, a positive moment indicates downward deflection.

Precast concrete, Kitchens of Sara Lee, Inc., Deerfield, Illinois.
(Courtesy of the Portland Cement Association.)

Examples 12.1 to 12.3 illustrate the application of elastic weights.

EXAMPLE 12.1

Determine the slope and deflection at the 30 k load in the beam of Figure 12.3 if $E = 29 \times 10^6$ psi and $I = 1500$ in.^{4}.

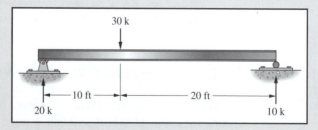

Figure 12.3

Solution.   Drawing M/EI diagram and loading it onto the actual beam and calculating beam reactions for that loading.

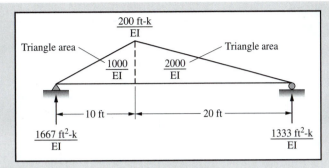

$$\theta_{30} = \text{``shear'' at } 30\,\text{k load} = \frac{1667 - 1000}{EI}$$

$$= \frac{667\,\text{ft}^2\text{k}}{EI} = \frac{(667)(144)(1000)}{(29 \times 10^6)(1500)} = 0.00221\,\text{radians} \,\backslash$$

$$\Delta_{30} = \text{``moment'' at } 30\,\text{k load} = \frac{(1667)(10) - (1000)(3.33)}{EI}$$

$$= \frac{13.333\,\text{ft}^3\,\text{k}}{EI} = \frac{(13.333)(1728)(1000)}{(29 \times 10^6)(1500)} = 0.477\,\text{in.} \downarrow \ \blacksquare$$

EXAMPLE 12.2

Determine the deflection at the centerline of the beam shown in Figure 12.4.

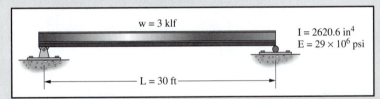

Figure 12.4

Solution.

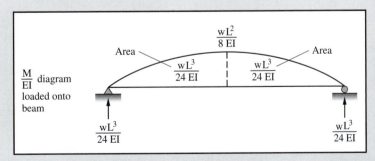

Deflection at centerline:

$$= \text{moment}_{\text{\footnotesize ₵}} = \left(\frac{wL^3}{24EI}\right)\left(\frac{L}{2}\right) - \left(\frac{wL^3}{24EI}\right)\left(\frac{3}{8}\right)\left(\frac{L}{2}\right)$$

$$\Delta_{\text{\footnotesize ₵}} = \frac{5wL^4}{384EI} = \frac{(5)(3000/12)(30 \times 12)}{(384)(29 \times 10^6)(2620.6)} = 0.719\,\text{in.} \downarrow \ \blacksquare$$

EXAMPLE 12.3

Determine the maximum deflection in the beam of Figure 12.3 which is repeated in Figure 12.5 and is loaded with its M/EI diagram.

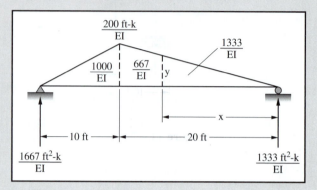

Figure 12.5

Solution. Maximum deflection occurs when the "shear" in the actual beam loaded with the M/EI diagram equals zero. Obviously this point falls somewhere in the right hand 20 ft of the beam a distance x from the right support. The value of x can be determined as follows noting that the ordinate y, shown in Figure 12.5, can be determined by proportions in terms of x.

$$\frac{y}{200} = \frac{x}{20}$$
$$y = 10x$$

The area of the M/EI diagram to the right

$$= \left(\frac{1}{2}y\right)(x) = \left(\frac{1}{2}\right)(10x)(x) = 1333$$
$$5x^2 = 1333$$
$$x = 16.33\,\text{ft}$$

Taking moments to the right of y (we could work to left but it is slightly more complicated).

$$= \frac{(1333)(16.33) - (1333)\left(\dfrac{16.33}{3}\right)}{EI}$$
$$= \frac{14.512\,\text{ft}^3\text{-k}}{EI} = \frac{(14.512)(1728)(1000)}{(29 \times 10^6)(1500)}$$
$$\Delta_{\text{max}} = 0.576\,\text{in.}\ \downarrow\ \blacksquare$$

EXAMPLE 12.4

Compute the centerline deflection for the simple beam shown in Figure 12.6.

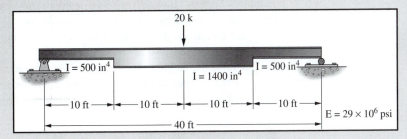

Figure 12.6

Solution. The moment of inertia is larger in the middle of this beam, where bending moments are larger, than they are nearer the beam ends. Consequently the M/EI diagram must reflect this fact. Therefore in the solution, the numerical value of I of 1400 in.[4] is divided into the moments along the center 20 ft part of the beam and the I of 500 in.[4] is divided into the moments along the end 10 ft sections. The rest of the calculations are handled exactly as they were for the members previously considered.

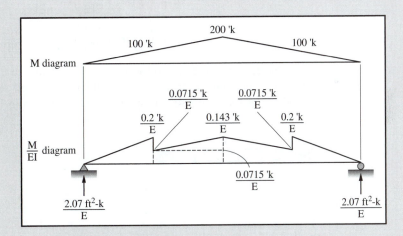

$$\Delta_{\mathbb{C}} = \left(\frac{2.07}{E}\right)(20) - \left(\frac{1}{2}\right)(10)\left(\frac{0.2}{E}\right)(13.33) - (10)\left(\frac{0.0715}{E}\right)(5)$$

$$- \left(\frac{1}{2}\right)(10)\left(\frac{0.0715}{E}\right)(3.33)$$

$$= \frac{23.3}{E} = \frac{(23.3)(1728)(1000)}{29 \times 10^6} = 1.39 \text{ in.} \downarrow \quad \blacksquare$$

12.3 LIMITATIONS OF THE ELASTIC-WEIGHT METHOD

The method of elastic weights was developed for simple beams, and in its present form will not work for cantilevered beams, overhanging beams, fixed-ended beams, and continuous beams. However, we will learn in the next few pages that by making very simple changes to the member supports, not including the ones for simple end supported beams, we can make the method work for these other beams just as well as it does for simple end supported beams.

To show the support changes that must be made for a cantilever beam to make this revised method work, the moment-area theorems are used to determine the correct slopes and deflections in the uniformly loaded cantilever beam of Figure 12.7.

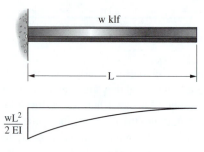

Figure 12.7 Cantilever beam loaded with M/EI diagram for a uniform load

$$\theta_B = \left(\frac{1}{3}\right)(L)\left(\frac{wL^2}{2EI}\right) = \frac{wL^3}{6EI}$$

$$\Delta_B = \left(\frac{1}{3}\right)(L)\left(\frac{wL^2}{2EI}\right)\left(\frac{3}{4}L\right) = \frac{wL^4}{8EI}$$

If the elastic-weight method was used in an attempt to find the slope and deflection at the ends of the same beam, the result would be slopes and deflections of zero at the free end and $wl^3/6EI$ and $wl^4/24EI$ at the fixed end, as shown in Figure 12.8.

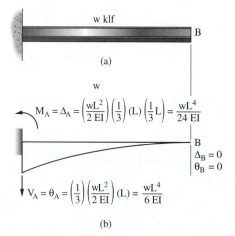

Figure 12.8 Slope and deflection found using method of elastic weights

The slope and deflection at the fixed end A must be zero; however, application of elastic weights to the beam results in both shear and moment, falsely indicating slope and deflection.

If the fixed end of the beam was moved to the free end and the resulting beam loaded with the M/EI diagram, the shears and moments would correspond exactly to the slopes and deflections on the actual beam as found by the moment-area method.

12.4 CONJUGATE-BEAM METHOD

This revised method, which is called the conjugate-beam method, makes use of an analogous or "conjugate" beam to be handled by elastic weights in place of the actual beam, to which it cannot be correctly applied. The shear and moment in the imaginary beam, loaded with the M/EI diagram, must correspond exactly with the slope and deflection of the actual beam.

The correct mathematical relationship is obtained for a beam simply supported if it is loaded "as is" with the M/EI diagram. If the elastic-weight method is applied to other types of beams, the largest moments due to the M/EI loading occur at the supports, incorrectly indicating that the largest deflections occur at those points. For elastic weights to be applied correctly, use must be made of substitute beams, or conjugate beams, that have the supports changed so the correct relationships are obtained.

The loads and properties of the true beam have no effect on the manner in which the conjugate beam is supported. The only factors affecting the supports of the imaginary beam are the supports of the actual beam. The lengths of the two beams are equal. The following paragraphs present a discussion of what the various types of beam supports must become in the conjugate beam so that the elastic-weight method will apply. The mathematical proof of these relationships is explained in detail in books on strength of materials.

Free End

The free end of a beam slopes and deflects when the beam is loaded. The conjugate beam must have both shear and moment at that end when it is loaded with the M/EI diagram. The only type of end support having both shear and moment is the fixed end. *A free end in the actual beam becomes a fixed end in the conjugate beam.*

Fixed End

A similar discussion in reverse order can be made for a fixed end. No slope or deflection can occur at a fixed end, and there must not be any shear or moment in the conjugate beam at that point. *A fixed end in the actual beam becomes a free end in the conjugate beam.*

Simple End Support

A simple end slopes but does not deflect when the beam is loaded. The imaginary beam will have shear but no moment at that point, a situation that can occur only at a simple support. *A simple end support in the actual beam remains a simple end support in the conjugate beam.*

Simple Interior Support

There is no deflection at either a simple interior or a simple end support. Both types may slope when the beam is loaded, but the situations are somewhat different. The slope at a simple interior support is continuous across the support; that is, no sudden change of

Figure 12.9   Qualitative deflections for a beam continuous over one support

slope occurs. This condition is not present at a simple end support where the slope suddenly begins, as shown in the deflection curve for the beam of Figure 12.9. If there is no change of slope at a simple interior support, there can be no change of shear at the

Actual beam   Conjugate beam

Figure 12.10   Typical beams and their conjugates

corresponding support in the conjugate beam. Any type of external support at this point would cause a change in the shear; therefore, an internal pin (or unsupported hinge) is required. *A simple interior support in the actual beam becomes an unsupported internal hinge in the conjugate beam.*

Internal Hinge

At an unsupported internal hinge there is both slope and deflection, which means that the corresponding support in the conjugate beam must have shear and moment. *An internal hinge in the actual beam becomes a simple support in the conjugate beam.*

12.5 SUMMARY OF CONJUGATE BEAMS

Figure 12.10 shows several types of common beams and their corresponding conjugate beams.

12.6 EQUILIBRIUM

The reactions, moments, and shears of the conjugate beam are easily computed by statics because the conjugate beam is always statically determinate, even though the real beam may be statically indeterminate. Sometimes the conjugate beam may appear to be completely unstable. The most conspicuous example is the conjugate beam for the fixed-end beam as shown in Figure 12.11 which has no supports whatsoever. On second glance, the areas of the M/EI diagram are seen to be so precisely balanced between downward and upward loads (positive and negative areas of the diagram, respectively) as to require no supports. Any supports seemingly required would have zero reactions, and the proper shears and moments are supplied to coincide with the true slopes and deflections. Even a real beam continuous over several simple supports has a conjugate beam that is simply end-supported.

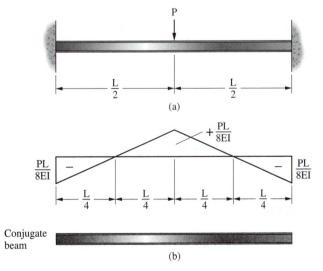

Figure 12.11   A fixed end beam whose conjugate appears

12.7 SUMMARY OF BEAM RELATIONS

A brief summary of the relations that exist between loads, shears, moments, slope changes, slopes, and deflections is presented in Figure 12.12. The relations are shown for a uniformly loaded beam but are applicable to any type of loading. For the two sets of curves shown the ordinate on one curve equals the slope at that point on the following curve. It is obvious from these figures that the same mathematical relations that exist between load, shear, and moment hold for M/EI loading, slope, and deflection.

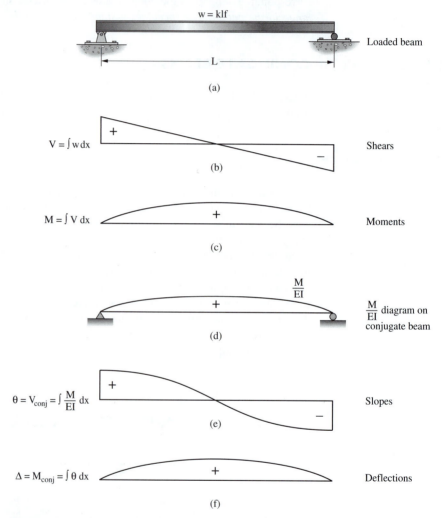

Figure 12.12   Summary of relations for conjugate beam method

12.8 APPLICATION OF THE CONJUGATE METHOD TO BEAMS

Examples 12.5 and 12.6 illustrate the conjugate method of calculating slopes and deflections for beams. The symbols and units used in applying the method are in general the same as those used for the moment-area and elastic-weight methods. Maximum deflections occur at points of zero shear on the conjugate structure. For example, the point of zero shear in the beam of Figure 12.11 is the centerline. The deflection is as follows:

$$\Delta_{\mathbb{C}} = \text{moment}_{\mathbb{C}}$$

$$= \left(\frac{1}{2}\right)\left(\frac{L}{4}\right)\left(\frac{PL}{8EI}\right)\left(\frac{5}{12}L\right) - \left(\frac{1}{2}\right)\left(\frac{L}{4}\right)\left(\frac{PL}{8EI}\right)\left(\frac{L}{12}\right)$$

$$= \frac{PL^3}{192EI} \downarrow$$

EXAMPLE 12.5

Determine the slope and deflection of point A in the beam of Figure 12.13.

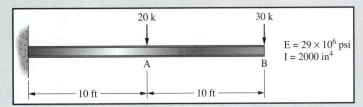

Figure 12.13

Solution.

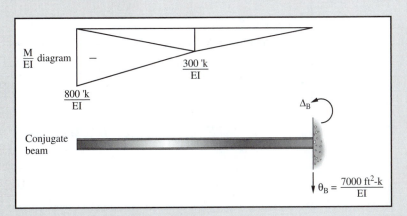

Slope

$$\theta_A = \frac{\left(\frac{1}{2}\right)(300)(10) + \left(\frac{1}{2}\right)(800)(10)}{EI} = \frac{5500\,\text{ft}^2\text{-k}}{EI}$$

$$= \frac{(5500)(144)(1000)}{(29 \times 10^6)(2000)} = 0.0136\,\text{rad} = 0.78°\, \backslash$$

Deflection

$$\Delta_A = \frac{\left(\frac{1}{2}\right)(300)(10)(3.33) + \left(\frac{1}{2}\right)(800)(10)(6.67)}{EI} = \frac{31{,}667\,\text{ft}^2\text{-k}}{EI}$$

$$= \frac{(31{,}667)(1728)(1000)}{(29 \times 10^6)(2000)} = 0.943\,\text{in.}\downarrow \quad\blacksquare$$

EXAMPLE 12.6

Determine deflections at points A and B in the overhanging beam shown in Figure 12.14.

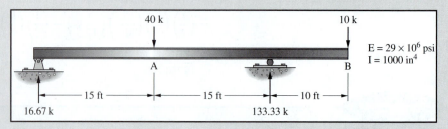

Figure 12.14

Solution.   The M/EI diagram is drawn and placed on the conjugate beam, which has an interior hinge. The reactions are determined as they were for the cantilever-type structures of Chapter 4. The portion of the beam to the left of the hinge is considered a simple beam, and its reactions are determined. The reaction at the hinge is applied as a concentrated load acting at the end of the cantilever to the right of the hinge in the opposite direction, and the reactions at the fixed end are determined. To simplify the mathematics, a separate moment diagram is drawn for each of the concentrated loads.

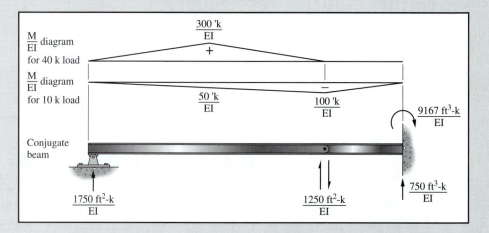

$$\Delta_A = \frac{(1750)(15) - \left(\frac{1}{2}\right)(15)(300)(5) + \left(\frac{1}{2}\right)(15)(50)(5)}{EI} = \frac{16{,}875 \text{ ft}^3\text{-k}}{EI} = 1.01 \text{ in.} \downarrow$$

$$\Delta_B = \frac{-(1250)(10) + \left(\frac{1}{2}\right)(10)(100)(6.67)}{EI} = -\frac{9167 \text{ ft}^3\text{-k}}{EI} = 0.546 \text{ in.} \uparrow \quad ■$$

12.9 LONG-TERM DEFLECTIONS

Under sustained loads concrete will continue to deform for long periods of time. This additional deformation is called *creep*, or *plastic flow*. If a compressive load is applied to a concrete member, an immediate or elastic shortening occurs. If the load is left in place for a long time, the member will continue to shorten over a period of several years and the final deformation may be as much as two or three (or more) times the initial deformation. Creep is dependent on such items as humidity, temperature, curing conditions, age of concrete at time of loading, ratio of stress to strength, and other items.

When sustained loads are applied to reinforced-concrete beams, their compression sides will become shorter and shorter over time and the result will be larger deflections. The American Concrete Institute states that the total long-term deflection in a particular member should be estimated by: (1) calculating the instantaneous deflection caused by all the loads; (2) computing the part of the instantaneous deflection that is caused by the sustained loads; (3) multiplying this value by an empirical factor from the ACI Code, which is dependent on the time elapsed; and (4) adding this value to the instantaneous deflection.[1]

The sustained loads for a building include the dead load plus some percentage of the live load. For an apartment house or for an office building perhaps only 20 to 25% of the live load should be considered as being sustained, while as much as 70% to 80% of the live load of a warehouse might fall into this category.

A similar discussion is possible for timber structures. Seasoned timber members subjected to long-term loads develop a permanent deformation or sag approximately equal to twice the deflection computed for short-term loads of the same magnitude.

12.10 APPLICATION OF THE CONJUGATE METHOD TO FRAMES

The slopes and deflections of frames can be computed with the conjugate method much as they were for beams. The procedure, however, is rather confusing and is not recommended for the usual frame. The virtual work method, described in the next chapter, seems to the author to be much more logical and simpler to apply.

Should the conjugate method be applied to frames, there are several factors to be accounted for which do not arise in beam calculations. These include such items as: the effects of conjugate supports at frame corners, possible swaying of the frame, and the effect of rotations of the frame joints on slopes and deflections at other points. As a result of these complications no, further reference is made herein to the subject.

12.11 PROBLEMS FOR SOLUTION

For Problems 12.1 to 12.21 use the conjugate beam method to determine the information requested for each beam. $E = 29 \times 10^6$ psi unless otherwise noted as it is for Problem 12.12.

12.1 θ_A and Δ_A. $I = 1500$ in.[4]. (*Ans.* $\theta_A = 0.00318$ rads \, $\Delta_A = 0.915$ in. ↓)

12.2 Maximum deflection in the beam of Problem 12.1.

12.3 θ_A, θ_B, Δ_A and Δ_B. $I = 1870$ in.[4]. (*Ans.* $\theta_B = 0.0139$ rads /, $\Delta_A = 3.00$ in. ↓)

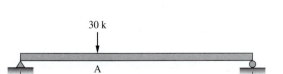

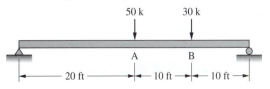

[1]Building Code Requirements for Reinforced Concrete, ACI 318-05 (Farmington Hills, Michigan: American Concrete Institute), Sect. 9.5, pp 112–113.

12.4 Maximum deflection in the beam of Problem 12.3.

12.5 θ_A and Δ_A. $I = 1870$ in.^{4}. (*Ans.* $\theta_A = 0.0128$ rads \,
$\Delta_A = 2.89$ in. ↓)

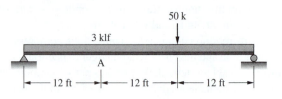

12.6 θ_A and Δ_A. $I = 2170$ in.^{4}.

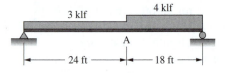

12.7 θ_A and Δ_A. $I = 1750$ in.^{4}. (*Ans.* $\theta_A = 0.000399$ rads /,
$\Delta_A = 0.934$ in. ↓)

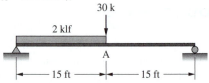

12.8 Δ_A and Δ_B. $I = 1600$ in.^{4}.

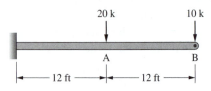

12.9 θ_A and Δ_A. $I = 1250$ in.^{4}. (*Ans.* $\theta_A = 0.00814$ rads /,
$\Delta_A = 1.17$ in. ↓)

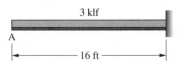

12.10 θ_A, θ_B, Δ_A and Δ_B. $I = 1100$ in.^{4}.

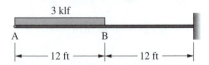

12.11 θ_A and Δ_A. $I = 513$ in.^{4}. (*Ans.* $\theta_A = 0.0141$ rads /,
$\Delta_A = 2.44$ in.)

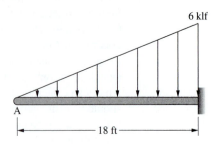

12.12 θ_A and Δ_A. $E = 200\,000$ MPa. I values shown on figure.

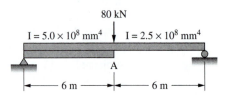

12.13 θ_A and Δ_A. I values shown on figure. (*Ans.* $\theta_A = 0.00517$ rads \, $\Delta_A = 3.56$ in. ↓)

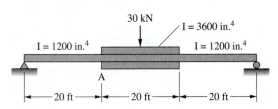

12.14 θ_A and Δ_B. $I = 1650$ in.^{4}.

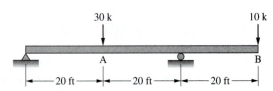

12.15 θ_A, θ_B, Δ_A, and Δ_B. $I = 2250$ in.^{4}. (*Ans.* $\theta_A = 0.00294$ rads \, $\Delta_B = 0.677$ in. ↓)

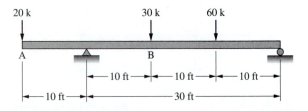

12.16 θ_A and Δ_A. $I = 1800$ in.^{4}.

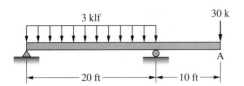

12.17 θ_A, θ_B, Δ_A, and Δ_B. $I = 1650$ in.^{4}. (*Ans.* $\theta_B = -0.00875$ rads /, $\Delta_A = 0.763$ in. ↓)

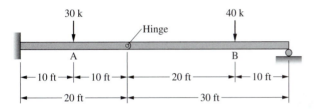

12.18 θ_A and Δ_A. Reactions given. $I = 2690$ in.^{4}.

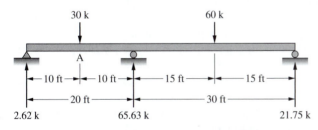

12.19 θ_A, θ_B, Δ_A, and Δ_B. (*Ans.* $\theta_A = -0.00400$ rads /, $\Delta_A = 0.381$ in. ↓)

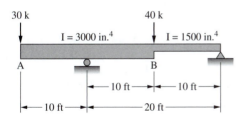

12.20 θ_A and Δ_A. $I = 2650$ in.^{4}.

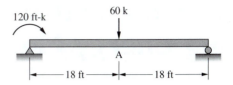

12.21 θ_A and Δ_A. (*Ans.* $\theta_A = 0.0248$ rad \, $\Delta_A = 2.57$ in. ↓) (Suggestion in preparing M/EI diagram: Take the part of the beam for which the depth varies and divide it into 2 ft sections. Then compute the I at the center of those sections and use the calculated values to draw the M/EI diagram. The width of the member is 12 in.)

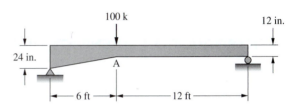

Chapter 13

Deflection and Angle Changes Using Energy Methods

13.1 INTRODUCTION TO ENERGY METHODS

In Chapters 11 and 12 deflections and angle changes were computed using geometric methods (moment area and the conjugate beam method). These procedures are satisfactory for many structures, including some rather complicated ones.

In this chapter it will be shown that the same deflections and angle changes may be determined using the principle of conservation of energy. Two energy methods, virtual work and Castigliano's theorem, will be introduced. For some complicated structures, these methods may be more desirable than the geometric methods because of the simplicity with which the expressions may be set up for making the solutions. In addition, the energy methods are applicable to more types of structures.

13.2 CONSERVATION OF ENERGY PRINCIPLE

Before the conservation of energy principle is introduced, a few comments are presented concerning the concept of work. For this discussion *work* is defined as the product of a force and its displacement in the direction in which the force is acting. Should the force be constant during the displacement, the work will equal the force times the total displacement as follows:

$$W = F\Delta$$

On many practical occasions the force changes in magnitude as does the deformation. For such a situation it is necessary to sum up the small increments of work for which the force can be assumed to be constant, that is

$$W = \int F \, d\Delta$$

The *conservation of energy principle* is the basis of all energy methods. When a set of external loads is applied to a deformable structure the points of load application move, with the result that the members or elements making up the structure become deformed. According to the conservation of energy principle, the work done by the external loads, W_e, equals the work done by the internal forces acting on the elements of the structure, W_i. Thus

$$W_e = W_i$$

As the deformation of a structure takes place, the internal work—commonly referred to as strain energy—is stored within the structure as potential energy. If the elastic limit of the material is not exceeded, the elastic strain energy will be sufficient to return the structure to its original undeformed state when the loads are removed. Should a structure be subjected to more than one load the total energy stored in the body will equal the sum of the energies stored in the structure by each of the loads.

It is to be clearly noted that the conservation of energy principle will be applicable only when static loads are applied to elastic systems. If the loads are not applied gradually, acceleration may occur and some of the external work will be transferred into kinetic energy. If inelastic strains are present, some of the energy will be lost in the form of heat.

13.3 VIRTUAL WORK OR COMPLEMENTARY VIRTUAL WORK METHOD

With the virtual work principle, a system of forces in equilibrium is related to a compatible system of displacements in a structure. The word *virtual* means not in fact, but equivalent. Thus, the virtual quantities discussed in this chapter do not exist in a real sense. When we speak of a virtual displacement we are speaking of a fictitious displacement imposed on a structure. *The work performed by a set of real forces during a virtual displacement is called virtual work.*

We will see in the discussion to follow that for each virtual work theorem there is a corresponding theorem that is based on complementary virtual work. As a result the method is sometimes referred to as complementary virtual work.

Virtual work is based on the principle of virtual velocities, which was introduced by Johann Bernoulli of Switzerland in 1717. The virtual work theorem can be stated as follows:

If a displacement is applied to a deformable body that is in equilibrium under a known load or loads, the external work performed by the existing load or loads due to this new displacement will equal the internal work performed by the stresses existing in the body that were caused by the original load or loads.

The virtual work or complementary work method also is often referred to as the method of work or the dummy unit-load method. The name *dummy unit-load method* is used because a fictitious or dummy unit load is used in the solution.

Virtual work is based on the law of conservation of energy. To make use of this law in the derivation to follow, it is necessary to make the following assumptions:

1. The external and internal forces are in equilibrium.
2. The elastic limit of the material is not exceeded.
3. There is no movement of the supports.

Should a load F_1 be gradually applied to the bar shown in Figure 13.1 so that the load and the bar's deformation (or increase in length) increase together from zero to their total values (F_1 and Δ_1), the external work done will equal the average load $F_1/2$ times the total displacement Δ_1.

$$W_e = \frac{F_1}{2}\Delta_1$$

If a load F_1 is gradually applied to the bar, it will elongate an amount Δ_1. Therefore, the forces that undergo displacement during this elongation will perform external work equal to $W_e = 1/2F_1\Delta_1$, as shown by the triangular area with horizontal cross-hatched lines in Figure 13.2.

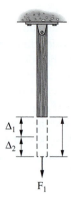

F_1 **Figure 13.1**   Axially loaded straight bar

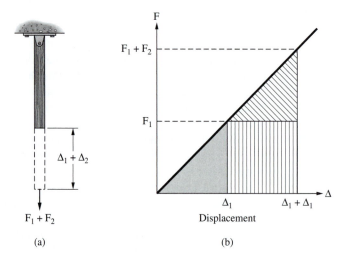

(a)   (b)

Figure 13.2   Force-Displacement relationship for an elastic bar with multiple loads

Should another gradually applied force F_2 be added to the bar causing an additional displacement Δ_2, additional external work will be performed. The force F_1 will be present for all of the Δ_2 displacement and the work will be $F_1\Delta_2$. This work is represented in Figure 13.2 by the rectangle with vertical lines. The gradually applied force F_2 will perform an additional amount of work equal to $1/2F_2\Delta_2$. Thus, the total external work is

$$W_e = \frac{1}{2}F_1\Delta_1 + F_1\Delta_2 + \frac{1}{2}F_2\Delta_2$$

Another type of work is complementary work W^*. It is represented by the area above the curve. Complementary work does not really have a physical meaning as does the external work W_e, but we can see that for a gradually applied load

$$W_e + W^* = F\Delta$$

Thus, W^* is the complement of the work W_e because it completes the rectangle, and as long as Hooke's law applies, $W_e = W^*$. Should the bar material's behavior not be linear, however, the load deformation diagram will be a curve and W_e and W^* will be unequal, as seen in Figure 13.3.

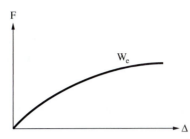

Figure 13.3 Load-deformation curve for a nonlinear member

13.4 TRUSS DEFLECTIONS BY VIRTUAL WORK

In this section the virtual work method is used to determine truss deflections, making use of the dummy unit-load procedure. To determine displacements by this procedure two loading systems are considered. The first system consists of the structure subjected to the force or forces for which the deflections are to be calculated. The second system consists of the structure subjected only to a fictitious unit or dummy load acting at the point and in the direction in which displacement is desired.

The dummy structure with the unit load in place is now subjected to the deformations of the real truss when it is subjected to the real forces, after which the resulting internal and external work are equated. For this calculation the only external load that performs work is the unit load, and as it remains constant, the external work equals $1 \times \Delta$ at location and in direction of unit load.

Next we need to write an expression for the internal work. The truss of Figure 13.4 will be considered for this discussion. Loads P_1 to P_3 are applied to the truss as shown and cause forces in the truss members. Each member of the truss shortens or lengthens depending on the character of its force. These internal deformations cause external deflections, and each of the external loads moves through a short distance. The law of conservation of energy as it applies to the truss may now be stated in detail. The external work performed by the loads P_1 to P_3 as they move through their respective truss deflections equals the internal work performed by the member forces as they move through their respective changes in length.

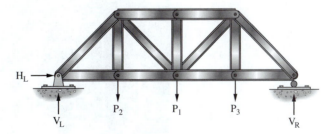

Figure 13.4 A typical truss for which deflections are computed

To write an expression for the internal work performed by a truss member, it is necessary to develop an expression for the deformation of the member. For this purpose the bar of Figure 13.5 is considered.

The force applied to the bar causes it to elongate by an amount ΔL. The elongation may be computed from the properties of the bar. The strain or elongation

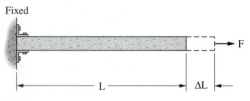

Fixed

L   ΔL   F

Figure 13.5

per unit length, ϵ, is equal to the total elongation divided by the length of the bar and also is equal to the stress intensity divided by the modulus of elasticity. An expression for ΔL may be developed as follows:

$$E = \frac{\sigma}{\epsilon} = \frac{F/A}{\Delta L/L}$$

$$\Delta L = \frac{FL}{AE}$$

In accordance with previous assumptions, the members of a truss have only axial force. These forces are referred to as F forces, and each member will change in length by an amount equal to its FL/AE value.

It is desired that an expression for the deflection at a joint in the truss of Figure 13.4 be developed. A convenient means of developing such an expression is to remove the external loads from the truss, place a unit load in the desired direction of deflection at the joint where deflection is desired, replace the external loads, and write an expression for the internal and external work performed by the unit load and the forces it causes when the external loads are replaced.

Aluminum trusses, Reynolds Hangar, Byrd Field, Richmond, Virginia. (Courtesy of Reynolds Metals Company.)

The forces caused in the truss members by the unit load are called μ forces. They cause small deformations of the members and small external deformations of the truss. When the external loads are returned to the truss, the force in each of the members changes by the appropriate F force, and the deformation of each member changes by its FL/AE value. The truss deflects and the unit load is carried through a distance Δ. The external work performed by the unit load when the external loads are returned to the structure may be expressed as follows:

$$W_e = 1 \times \Delta$$

Internally the μ force in each member is carried through a distance $\Delta L = FL/AE$. The internal work performed by all of the μ forces as they move through these distances is

$$W_i = \sum \mu \frac{FL}{AE}$$

By equating the internal and external work, the deflection at a joint in the truss may be expressed as follows:

$$\Delta = \sum \frac{F\mu L}{AE}$$

The μ values are actually dimensionless since the unit load is assumed to be dimensionless.

13.5 APPLICATION OF VIRTUAL WORK TO TRUSSES

Examples 13.1 and 13.2 illustrate the application of virtual work to trusses. In each case the forces due to the external loads are computed initially. Second, the external loads are removed and a unit load is placed at the point and in the direction in which deflection is desired (not necessarily horizontal or vertical). The forces due to the unit load are determined, and, finally, the value of $F\mu L/AE$ for each of the members is found. To simplify the numerous multiplications, a table is used. The modulus of elasticity is carried through as a constant until the summation is made for all of the members, at which time its numerical value is used. Should there be members of different Es, it is necessary that their actual or their relative values be used for the individual multiplications. A positive value of $\sum(F\mu L/AE)$ indicates a deflection in the direction of the unit load.

EXAMPLE 13.1

Determine the horizontal and vertical components of deflection at joint L_2 in the truss shown in Figure 13.6. Circled figures are areas, in square inches. $E = 29 \times 10^3$ ksi.

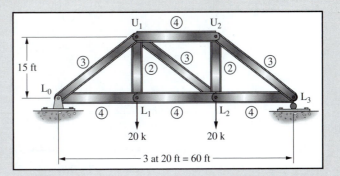

Figure 13.6

Solution.

Forces due to external loads (F forces)

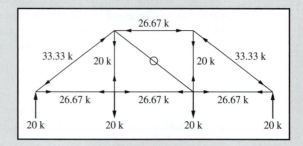

Forces due to a unit vertical load placed at joint L_2 (μ_v forces)

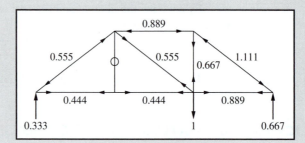

Forces due to a unit horizontal load placed at joint L_2 (μ_H forces)

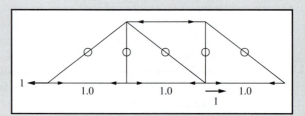

TABLE 13.1 COMPUTING $\frac{F\mu L}{A}$ VALUES (NOTICE E HAS BEEN LEFT OUT UNTIL THE LAST STEP BECAUSE IT IS CONSTANT FOR ALL MEMBERS.)

Member	L (in.)	A (in.²)	$\frac{L}{A}$	F (kips.)	μ_v	$\frac{F\mu_v L}{A}$ (kips/in.)	μ_H	$\frac{F\mu_H L}{A}$ (kips/in.)
L_0L_1	240	4	60	+26.67	+0.444	+710	+1.0	+1600
L_1L_2	240	4	60	+26.67	+0.494	+710	+1.0	+1600
L_2L_3	240	4	60	+26.67	+0.889	+1423	+1.0	+1600
L_0U_1	300	3	100	−33.33	−0.555	+1850	0	0
U_1U_2	240	4	60	−26.67	−0.889	+1423	0	0
U_2L_3	300	3	100	−33.33	−1.111	+3703	0	0
U_1L_1	180	2	90	+20.00	0	0	0	0
U_1L_2	300	3	100	0	+0.555	0	0	0
U_2L_2	180	2	90	+20.00	+0.667	+1201	0	0

$$\Sigma = +11,020 \text{ (kips/in.)} \qquad \Sigma = +1600 \text{ (kips/in.)}$$

$$(\Delta_{L_2})_V = +\frac{11,020}{E} = +\frac{11,020}{29 \times 10^3} = +0.38 \text{ in.} \downarrow$$

$$(\Delta_{L_2})_H = +\frac{1600}{E} = +\frac{1600}{29 \times 10^3} = +0.055 \text{ in.} \rightarrow$$

Since both displacements are positive, they are in the direction of the applied unit loads. ∎

Recreation Center, Lander College, Greenwood, South Carolina (Courtesy of Britt, Peters and Associates)

EXAMPLE 13.2

Determine the vertical component of deflection of joint L_4 in the truss of Figure 13.7 with the method of virtual work. The circled number next to each bar is the area of the bar in square inches; $E = 29 \times 10^3$ ksi.

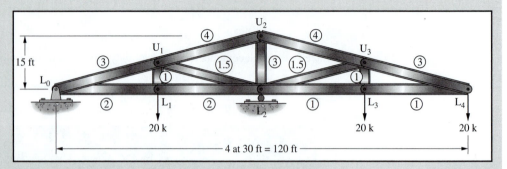

Figure 13.7

Solution. Forces due to external loads (F forces)

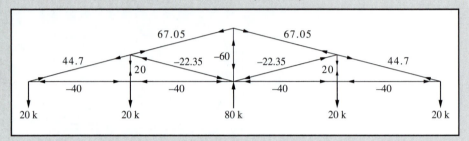

Forces due to vertical unit load placed at L_4 (μ_v forces)

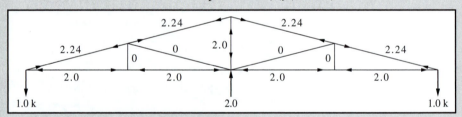

TABLE 13.2 COMPUTING $\frac{F\mu_v L}{A}$ VALUES

Member	L (in)	A (in²)	L/A	F (kips)	μ_v (kips)	$\frac{F\mu_v L}{A}$ (kips/in.)
L_0L_1	180	2.0	90.0	−40.0	−2.0	+7,200
L_1L_2	180	2.0	90.0	−40	−2.0	+7,200
L_2L_3	180	1.0	180.0	−40	−2.0	+14,400
L_3L_4	180	1.0	180.0	−40	−2.0	+14,400
L_0U_1	202	3.0	67.3	44.7	+2.24	+6,730
U_1U_2	202	4.0	50.5	67.05	+2.24	+7,575
U_2U_3	202	4.0	50.5	67.05	+2.24	+7,575
U_3U_4	202	3.0	67.3	44.7	+2.24	+6,730
U_1L_1	90	1.0	90.0	20.0	0	0
U_1L_2	202	1.5	134.5	−22.35	0	0
U_2L_2	180	3.0	60.0	−60.0	−2.0	+7,200
L_2U_3	202	1.5	134.5	−22.35	0	0
U_3L_3	90	1.0	90.0	20.0	0	0
						+79,010

$$(\Delta_{L_4})_V = +\frac{79,010}{E} = +2.72 \text{ in. } \downarrow \quad \blacksquare$$

13.6 DEFLECTIONS OF BEAMS AND FRAMES BY VIRTUAL WORK

The law of conservation of energy may be used to develop an expression for the deflection at any point in a beam or frame. In the following derivation each fiber of the structure is considered to be a "bar" or member similar to the members of the trusses considered in the preceding sections. The summation of the internal work performed by the force in each of the "bars" must equal the external work performed by the loads.

For the following discussion the beam of Figure 13.8(a) is considered. Part (b) of the figure shows the beam cross section. It is desired to know the vertical deflection Δ at point A in the beam caused by the external loads P_1 to P_3. If the loads were removed from the beam and a vertical unit load placed at A, small forces and deformations would be developed in the "bars," and a small deflection would occur at A. Replacing the external loads would cause increases in the "bar" forces and deformations, and the unit load at A would deflect an additional amount Δ. The internal work performed by the unit load forces, as they are carried through the additional "bar" deformations, equals the external work performed by the unit load as it is carried through the additional deflection Δ.

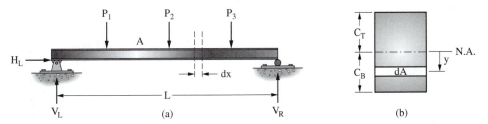

Figure 13.8 System for development of virtual work for beam

The following symbols are used in writing an expression for the internal work performed in a dx length of the beam: M is the moment at any section in the beam due to the external loads and m is the moment at any section due to the unit load. The stress in a differential area of the beam cross section due to the unit load can be found from the flexure formula as follows:

$$\text{stress in dA} = \frac{my}{I}$$

$$\text{total force in dA} = \frac{my}{I}\,dA$$

The dA area has a thickness or length of dx that deforms by $\epsilon\,dx$ when the external loads are returned to the structure. The deformation is as follows:

$$\text{stress due to external loads} = \sigma = \frac{My}{I}$$

$$\text{deformation of dx length} = \epsilon\,dx = \frac{\sigma}{E}\,dx = \frac{My}{EI}\,dx$$

The total force in dA due to the unit load (my/I) dA is carried through this deformation, and the work it performs is as follows:

$$\text{work in dA} = \left(\frac{my}{I}\,dA\right)\left(\frac{My}{EI}\,dx\right) = \frac{Mmy^2}{EI^2}\,dA\,dx$$

The total work performed on the cross section equals the summation of the work in each dA area in the cross section.

$$\text{work in a dx length} = \int_{C_b}^{C_t} \frac{Mmy^2}{EI^2} \, dA \, dx = \frac{Mm}{EI^2} \left(\int_{C_b}^{C_t} y^2 \, dA \right) dx$$

The expression $\int y^2 \, dA$ is a familiar one, being the moment of inertia of the section, and the equation becomes

$$\text{work} = \frac{Mm}{EI} \, dx$$

It is now possible to determine the internal work performed in the entire beam because it equals the integral from 0 to L of the preceding expression.

$$W_i = \int_0^L \frac{Mm}{EI} \, dx$$

The external work performed by the unit load as it is carried through the distance Δ is $1 \times \Delta$. By equating the external work and the internal work, an expression for the deflection at any point in the beam is obtained.

$$W_e = W_i$$

$$1 \times \Delta = \int_0^L \frac{Mm}{EI} \, dx$$

$$\Delta = \int_0^L \frac{Mm}{EI} \, dx$$

13.7 EXAMPLE PROBLEMS FOR BEAMS AND FRAMES

Examples 13.3 to 13.8 illustrate the application of virtual work to beams and frames. To apply the method, a unit load is placed at the point and in the direction in which deflection is desired. Expressions are written for M and m throughout the structure and the results are integrated from 0 to L. It is rarely possible to write one expression for M or one expression for m that is correct in all parts of the structure. As an illustration, consider the beam of Figure 13.9 and let the deflection under P_2 be desired. A unit load is placed at this point in the figure. The reactions V_L and V_R are due to loads P_1 and P_2, whereas v_L and v_R are due to the unit load.

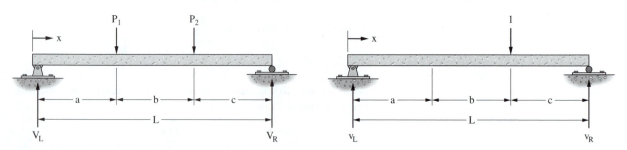

Figure 13.9

The values of M and m *are written with respect to the distance* x from the left support. From the left support to the unit load m may be represented by one expression, $v_L x$, but the expression for M is not constant for the full distance. Its value is $V_L x$ from the left support to P_1 and $V_L x - P_1(x - a)$ from P_1 to P_2. The integration will be made from 0 to a for $V_L x$ and $v_L x$, and from a to $a + b$ for $V_L x - P_1(x - a)$ and $v_L x$. Moment expressions for all parts of the beam are shown as follows. The left support is used as the origin of x.

For $x = 0$ to a:

$$M = V_L x$$

$$m = v_L x$$

For $x = a$ to $a + b$

$$M = V_L x - P_1(x - a)$$

$$m = v_L x$$

For $x = a + b$ to L

$$M = V_L x - P_1(x - a) - P_2(x - a - b)$$

$$m = v_L x - 1(x - a - b)$$

$$\Delta = \int_0^a \frac{Mm}{EI} \, dx + \int_a^{a+b} \frac{Mm}{EI} \, dx + \int_{a+b}^L \frac{Mm}{EI} \, dx$$

A positive sign is used for a moment that causes tension in the bottom fibers of a beam. If the result of integration is positive, the direction used for the unit load is the direction of the deflection.

EXAMPLE 13.3

Determine the deflection at point A for the beam shown in Figure 13.10 by virtual work. $E = 29 \times 10^6$ psi. $I = 5000$ in.^{4}.

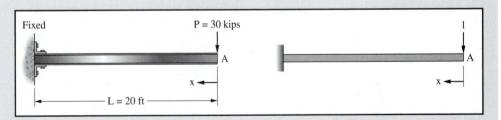

Fixed   P = 30 kips   1

A   A

x   x

L = 20 ft

Figure 13.10

Solution. An expression is written for M (the moment due to the 30-kip load) at any point a distance x from the free end. A unit load is placed at the free end and an expression is written for the moment m it causes at any point. The origin of x may be selected at any point as long as the same point is used for writing M and m for that portion of the beam.

For $x = 0$ to 20 ft

$$M = -Px$$

$$m = -1x$$

$$\Delta_A = \int_0^L \frac{Mm}{EI}\ dx = \int_0^L \frac{(-Px)(-1x)}{EI}\ dx = \int_0^L \frac{Px^2}{EI}\ dx$$

$$= \frac{P}{EI}\left[\frac{x^3}{3}\right]_0^L = \frac{PL^3}{3EI}$$

$$= \frac{(30{,}000)(20 \times 12)^3}{(3)(29 \times 10^6)(5000)} = 0.955\ \text{in.}\ \downarrow \quad \blacksquare$$

EXAMPLE 13.4

Determine the deflection at point A on the beam shown in Figure 13.11. $E = 29 \times 10^6$ psi. $I = 5000$ in.^{4}.

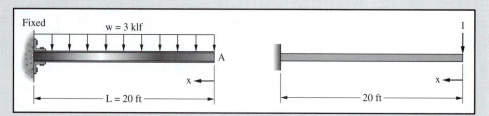

Fixed   w = 3 klf   1

A

x   x

L = 20 ft   20 ft

Figure 13.11

Solution.

For $x = 0$ to 20 ft

$$M = -(w)(x)\left(\frac{x}{2}\right) = -\frac{wx^2}{2}$$

$$m = -1x$$

$$Mm = +\frac{wx^3}{2}$$

$$\Delta_A = \int_0^L \frac{Mm}{EI}\ dx = \int_0^L \frac{wx^3}{2EI}\ dx$$

$$= \frac{w}{2EI}\left[\frac{x^4}{4}\right]_0^L = \frac{wL^4}{8EI}$$

$$= \frac{(3000/12)(20 \times 12)^4}{(8)(29 \times 10^6)(5000)} = 0.714\ \text{in.}\ \downarrow \quad \blacksquare$$

EXAMPLE 13.5

Determine the deflection at point B in the beam shown in Figure 13.12. $E = 29 \times 10^6$ psi. $I = 1000$ in.^{4}.

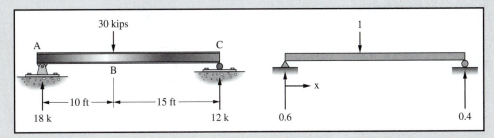

Figure 13.12

Solution.   It is necessary to write one expression for M from A to B and another from B to C. The same is true for m. Frequently it is possible to simplify the mathematics by using different origins of x for different sections of the beam. The same final results would have been obtained if for the M and m expressions from B to C the origin had been taken at C.

For $x = 0$ to 10 ft:

$$M = 18x$$
$$m = 0.6x$$
$$Mm = 10.8x^2$$

For $x = 10$ ft to 25 ft:

$$M = 18x - (30)(x - 10) = -12x + 300$$

$$m = 0.6x - 1(x - 10) = -0.4x + 10$$

$$Mm = +4.8x^2 - 240x + 3000$$

$$\Delta_B = \int_0^{10} \frac{10.8x^2}{EI}\ dx + \int_{10}^{25} \frac{(4.8x^2 - 240x + 3000)}{EI}\ dx$$

$$= \frac{1}{EI}[3.6x^3]_0^{10} + \frac{1}{EI}[1.6x^3 - 120x^2 + 3000x]_{10}^{25}$$

$$= \frac{3600}{EI} + \frac{25,000 - 19,600}{EI}$$

$$= \frac{9000\ \text{ft}^3\text{-k}}{EI} = \frac{(9000)(1728)(1000)}{(29 \times 10^6)(1000)} = 0.536\ \text{in.}\ \downarrow\quad ■$$

EXAMPLE 13.6

Determine the deflection at point B in the beam of Figure 13.13. $E = 29 \times 10^6$ psi. $I = 1000$ in.^{4}.

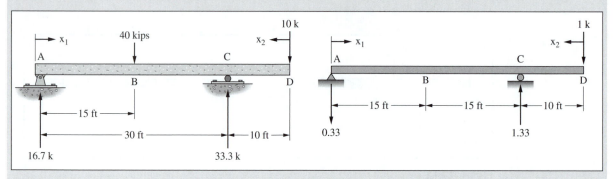

Figure 13.13

Solution. In writing the moment expressions from the left support to the right support, the left support is used for the origin of x. For the overhanging portion of the beam, the right end is used as the origin.

For $x_1 = 0$ to 15:

$$M = 16.7x$$
$$m = -0.33x$$
$$Mm = -5.56x^2$$

For $x_1 = 15$ to 30:

$$M = 16.7x - (40)(x - 15)$$
$$= -23.3x + 600$$
$$m = -0.33x$$
$$Mm = +7.77x^2 - 200x$$

For $x_2 = 0$ to 10 coming back from right end

$$M = -10x$$
$$m = -x$$
$$Mm = +10x^2$$

$$\Delta_D = \frac{1}{EI} \int_0^{15} (-5.56x^2) dx_1 + \frac{1}{EI} \int_{15}^{30} (7.77x_1^2 - 200x_1) dx_1 + \frac{1}{EI} \int_0^{10} (10x_2^2)\, dx_2$$

$$= \frac{1}{EI} [-1.85x_1^3]_0^{15} + \frac{1}{EI} [2.59x_1^3 - 100x_1^2]_{15}^{30} + \frac{1}{EI} [3.33x_2^3]_0^{10}$$

$$= \frac{-6250}{EI} + \frac{-20,000 + 13,750}{EI} + \frac{3330}{EI}$$

$$= -\frac{9170 \text{ ft}^3\text{-k}}{EI} = -0.546 \text{ in. } \uparrow \quad \blacksquare$$

EXAMPLE 13.7

Find the horizontal deflection at D in the frame shown in Figure 13.14. $E = 29 \times 10^6$ psi.

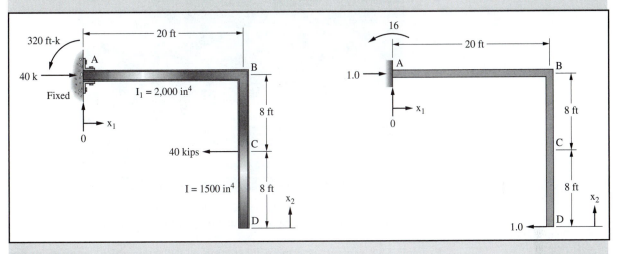

Figure 13.14

Solution. A unit load acting horizontally to the left is placed at D. With A as the origin for x_1, moment expressions are written for member AB. For the vertical member, the origin is taken at D, one pair of M and m expressions is written for the portion of the member from D to C, and another pair is written for the portion of the beam from C to B. Members AB and BD do not have the same moments of inertia, and it is necessary to carry them separately as shown in the calculations.

For $x_1 = 0$ to 20 ft:

$$M = -320$$
$$m = -16$$
$$Mm = +5120$$

For $x_2 = 0$ to 8 ft:

$$M = 0$$
$$m = -x$$
$$Mm = 0$$

For $x_2 = 8$ to 16:

$$M = -40(x_2 - 8) = -40x_2 + 320$$
$$m = -x_2$$
$$Mm = 40x_2^2 - 320x_2$$

$$\Delta_D = \int_0^{20} \left(\frac{5120}{EI_1}\right) dx_1 + \int_8^{16} \left(\frac{40x_2^2 - 320x_2}{EI_2}\right) dx_2$$

$$= \frac{1}{EI_1}[5120x_1]_0^{20} + \frac{1}{EI_2}\left(\frac{40}{3}x_2^3 - 160x_2^2\right)_8^{16}$$

$$= \frac{102,400 \text{ ft}^3\text{-k}}{EI_1} + \frac{17,066 \text{ ft}^3\text{-k}}{EI_2} = +3.73 \text{ in.} \leftarrow \quad \blacksquare$$

I-40, I-240 interchange, Oklahoma City, Oklahoma. (Courtesy of
the State of Oklahoma Department of Transportation.)

Example 13.8 presents one more illustration of deflection calculations using the
virtual work procedure. Its purpose is to provide the reader with a little additional practice
writing the M and m expressions. For convenience, the inclined 60 k load applied to the
structure is broken into its vertical and horizontal components as shown in Figure 13.15.
You should note also that the moment expressions are written separately for each of the
x_1, x_2, and x_3 directions shown in the figure.

EXAMPLE 13.8

Determine the horizontal deflection at the 60 k load for the frame shown in Figure 13.15.
Assume that each of the members has an $I = 1250$ in.[4] and an $E = 29 \times 10^6$ psi.

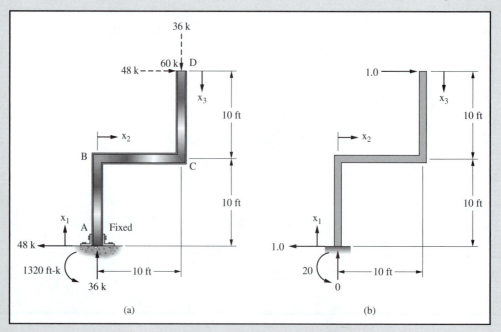

Figure 13.15

Solution.

For $x_1 = 0$ to 10 ft

$$M = 48x_1 - 1320$$
$$m = 1x_1 - 20$$
$$Mm = 48x_1^2 - 2280x_1 + 26,400$$

For $x_2 = 0$ to 10 ft

$$M = 36x_2 + 480 - 1320 = 36x_2 - 840$$
$$m = 10 - 20 = -10$$
$$Mm = -360x_2 + 8400$$

For $x_3 = 0$ to 10 ft

$$M = 48x$$
$$m = 1x$$
$$Mm = 48x_3^2$$

After integration

$$\Delta_{horiz} = (16x_1^3 - 1140x_1^2 + 26,400)_0^{10} + (-180x_2^2 + 8400x_2)_0^{10} + (16x_3^3)_0^{10}$$
$$= 248,000 \text{ ft}^3\text{-k}$$
$$= \frac{(248,000)(1728)(1000)}{(29 \times 10^6)(1250)} = 11.82 \text{ in.} \rightarrow \quad \blacksquare$$

13.8 ROTATIONS OR ANGLE CHANGES BY VIRTUAL WORK

Virtual work may be used to determine the rotations or slopes at various points in a structure. To find the slope at point A in the beam of Figure 13.16 a unit couple is applied at A, the external loads being removed from the structure. The value of moment at any point in the beam caused by the couple is m. Replacing the external loads will cause an additional moment, at any point, of M.

If the application of the loads causes the beam to rotate through an angle θ at A, the external work performed by the couple equals $1 \times \theta$. The internal work or the internal elastic energy stored is $\int (Mm/EI) \, dx$.

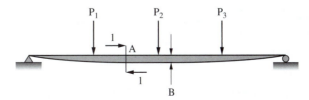

Figure 13.16

$$\theta = \int \frac{Mm}{EI}\ dx$$

If a clockwise couple is assumed at the position where slope is desired and the result of integration is positive, the rotation is clockwise. Examples 13.9 and 13.10 illustrate the determination of slopes by virtual work. The slopes obtained are in radians.

EXAMPLE 13.9

Find the slope at the free end A of the beam shown in Figure 13.17. E and I are constant.

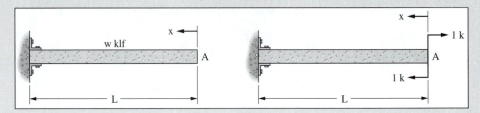

Figure 13.17

Solution.

For x = 0 to L

$$M = -\frac{wx^2}{2}$$

$$m = -1$$

$$Mm = \frac{wx^2}{2}$$

$$\theta_A = \int_0^L \frac{wx^2}{2EI}\ dx$$

$$\theta_A = \left[\frac{wx^3}{6EI}\right]_0^L = +\frac{wL^3}{6EI}\ \text{clockwise} \curvearrowright \ \blacksquare$$

EXAMPLE 13.10

Find the slope at the 30-kip load in the beam of Figure 13.18. E and I are constant.

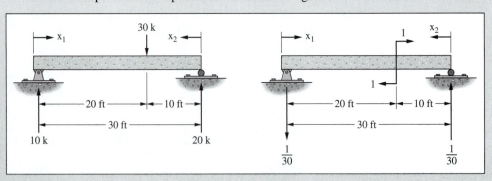

Figure 13.18

For $x_1 = 0$ to 20 ft:

$$M = 10x_1$$

$$m = -\frac{1}{30}x_1$$

$$Mm = -\frac{1}{3}x_1^2$$

For $x_2 = 0$ to 10 ft:

$$M = 20x_2$$

$$m = -\frac{1}{30}x_2$$

$$Mm = \frac{2}{3}x_2^2$$

$$\theta = \int_0^{20}\left(-\frac{1x_1^2}{3EI}\right)dx_1 + \int_0^{10}\left(\frac{2x_2^2}{3EI}\right)dx_2$$

$$= \left[-\frac{x_1^3}{9EI}\right]_0^{20} + \left[\frac{2x_2^3}{9EI}\right]_0^{10} = -\frac{888\ \text{ft}^2\text{-k}}{EI} + \frac{222\ \text{ft}^2\text{-k}}{EI}$$

$$= -\frac{666\ \text{ft}^2\text{-k}}{EI}\ \text{counterclockwise} \quad \blacksquare$$

13.9 INTRODUCTION TO CASTIGLIANO'S THEOREMS

Alberto Castigliano, an Italian railway engineer, published an original and elaborate book in 1879 on the study of statically indeterminate structures.[1] In this book were included the two theorems that are known today as Castigliano's first and second theorems. The first theorem, commonly known as the method of least work, is discussed in Chapter 16. It has played an important role historically in the development of the analysis of statically indeterminate structures.

Castigliano's second theorem, which provides an important method for computing deflections, is considered in this section and the next. Its application involves equating the deflection to the first partial derivative of the total internal work of the structure with respect to a load at the point where deflection is desired. In detail, the theorem can be stated as follows:

Theorem

For any linear elastic structure subject to a given set of loads, constant temperature, and unyielding supports, the first partial derivative of the strain energy with respect to a particular force will equal the displacement of that force in the direction of its application.

In presenting the virtual-work method for beams and frames in Section 13.6, the following equation was derived.

$$W = \int \frac{Mm}{EI}\ dx$$

---

[1]A. Castigliano, *Théorie de l' équilibre des systèmes élastique et ses applications*, Turin, 1879. (There was an English translation by E. S. Andrews entitled *Elastic Stresses in Structures*, published by Scott, Greenwood & Son in London in 1919. This translation was then reprinted in the United States by Dover Publications, Inc., of New York and entitled *The Theory of Equilibrium of Elastic Systems and Its Application*.)

to express the total internal work accomplished by the stresses caused by a unit load placed at the point and in the direction of a desired deflection when the external loads have been replaced on the beam.

The internal work of the actual stresses caused by gradually applied external loads can be determined in a similar manner. The work in each fiber equals the average stress, $(My/2I)(dA)$ (as the external loads vary from zero to full value), times the total strain in the fiber, $(My/EI)\,dx$. Integrating the product of these two expressions for the cross section of the member and throughout the structure gives the total internal work or strain energy stored for the entire structure.

$$W = \int_0^L \frac{M^2\,dx}{2EI}$$

In a similar manner the internal work in the members of a truss due to a set of gradually applied loads can be shown to equal

$$W = \sum \frac{F^2L}{2AE}$$

These expressions will be used frequently in the following section, in which a detailed discussion of Castigliano's second theorem is presented.

13.10 CASTIGLIANO'S SECOND THEOREM

As a general rule the other methods of deflection computation (such as moment–area or virtual-work) are a little easier to use and more popular than Castigliano's second theorem. For certain structures, however, this method is very useful, and from the standpoint of the student its study serves as an excellent background for the very important first theorem.

The derivation of the second theorem (presented in this section) is very similar to that given by Kinney.[2] Shown in Figure 13.19 is a beam that has been subjected to the gradually applied loads P_1 and P_2. These loads cause the deflections Δ_1 and Δ_2. We want to determine the magnitude of the deflection Δ_1 at load P_1.

The external work performed during the application of the loads must equal the average load times the deflection. It also must equal the strain energy of the beam.

$$W = \frac{P_1\Delta_1}{2} + \frac{P_2\Delta_2}{2} \tag{13.1}$$

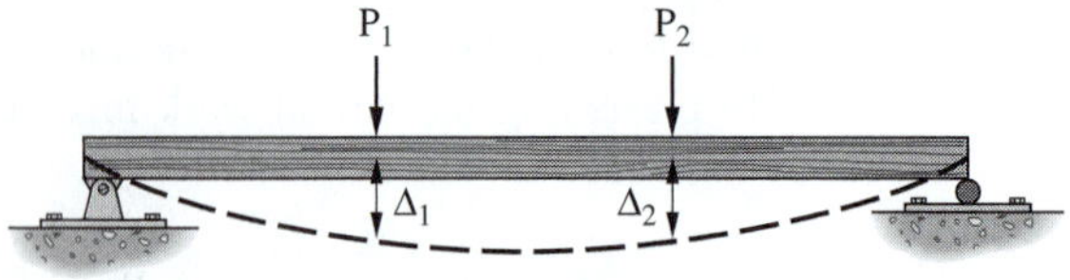

Figure 13.19   Beam for development of Castigliano's second theorem

Should the load P_1 be increased by the small amount dP_1, the beam will deflect an additional amount. Figure 13.20 illustrates the additional deflections, $d\Delta_1$ and $d\Delta_2$, at each of the loads.

[2]J. S. Kinney, *Indeterminate Structural Analysis* (Reading, Mass.: Addison-Wesley, 1957), 84–86.

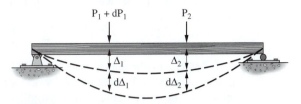

Figure 13.20 Addition deflection caused by dP_1

The additional work performed, or strain energy stored, during the application of dP_1 is as follows:

$$dW = \left(P_1 + \frac{dP_1}{2}\right)d\Delta_1 + P_2 d\Delta_2$$

This equation represents the variation in work caused by a small variation in load. Performing the indicated multiplication and neglecting the product of the differentials, which are negligibly small, we find that:

$$dW = P_1 d\Delta_1 + P_2 d\Delta_2 \qquad (13.2)$$

Now, instead of adding dP_1 after the loads P_1 and P_2, let's place all three loads on the beam at the same time. In doing so, the work done by the loads can be represented by

$$W = \left(\frac{P_1 + dP_1}{2}\right)(\Delta_1 + d\Delta_1) + \left(\frac{P_2}{2}\right)(\Delta_2 + d\Delta_2)$$

Again, if we perform the indicated multiplication and neglect the product of the differentials, we obtain the equation:

$$W = \frac{P_1\Delta_1}{2} + \frac{P_1 d\Delta_1}{2} + \frac{dP_1 \Delta_1}{2} + \frac{P_2\Delta_2}{2} + \frac{P_2 d\Delta_2}{2} \qquad (13.3)$$

The change in the work done in this case can be obtained by subtracting Equation 13.1 from Equation 13.3. In doing so we find that:

$$dW = \frac{P_1\Delta_1}{2} + \frac{P_1 d\Delta_1}{2} + \frac{dP_1 \Delta_1}{2} + \frac{P_2\Delta_2}{2} + \frac{P_2 d\Delta_2}{2} - \frac{P_1\Delta_1}{2} - \frac{P_2\Delta_2}{2}$$

$$\qquad (13.4)$$

$$dW = \frac{P_1 d\Delta_1}{2} + \frac{dP_1 \Delta_1}{2} + \frac{P_2 d\Delta_2}{2}$$

From Equation 13.2, we find that the $P_2 d\Delta_2$ is

$$P_2 d\Delta_2 = dW - Pd\Delta_1$$

By substituting this result into Equation 13.4, and solving for the deflection Δ_1, the deflection we wanted to compute, we find that

$$\Delta_1 = \frac{dW}{dP_1}$$

Since the work done by the external loads is equal to the strain energy in the structure caused by the loads, and since more than one load is usually applied to a structure, this deflection can be written as a partial derivative in terms of strain energy as follows:

$$\Delta = \frac{\partial W}{\partial P} = \frac{\partial U}{\partial P}$$

This is a mathematical statement of Castigliano's second theorem. By substituting into this expression the total internal work or strain energy $W = \int \frac{M^2}{2EI} dx$, the deflection at a point in a beam or frame using Castigliano's second theorem can be written as

$$\Delta = \frac{\partial}{\partial P} \int \frac{M^2}{2EI} dx$$

When applying the method represented by this equation, M would be squared and integrated; then the first partial derivative would be taken. If M is rather complicated, as it frequently is, the process becomes very tedious. For this reason usually it is simpler to differentiate under the integral sign with the following results:

$$\Delta = \int \frac{M}{EI} \left(\frac{\partial M}{\partial P} \right) dx$$

For a truss, the corresponding relationships are:

$$\Delta = \frac{\partial}{\partial P} \sum \frac{F^2 l}{2AE}$$

$$\Delta = \sum \frac{Fl}{AE} \left(\frac{\partial F}{\partial P} \right)$$

Examples 13.11 through 13.13 illustrate the application of Castigliano's second theorem. In applying the theorem, the load at the point where the deflection is desired is referred to as P. After the operations required in the equation are completed, the numerical value of P is replaced in the expression. Should there be no load at the point or in the direction in which deflection is desired, an imaginary force P will be placed there in the direction desired. After the operation is completed, the correct value of P (zero) will be substituted in the expression. This principle is demonstrated in Example 13.11.

If slope or rotation is desired in a structure, the partial derivative is taken with respect to an assumed moment M acting at the point where rotation is desired. A positive sign on the answer indicates that the rotation is in the assumed direction of the moment P.

Burro Creek Bridge, Wickieup, Mohave County, Arizona. (Courtesy of the American Institute of Steel Construction, Inc.)

EXAMPLE 13.11

Determine the vertical deflection at the 10-kip load in the beam shown in Figure 13.21.

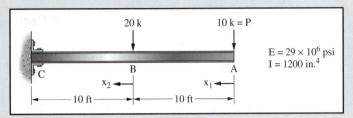

Figure 13.21

Solution.   By letting the 10-kip load be P.

TABLE 13.3

Section	M	$\dfrac{\partial M}{\partial P}$	$M\dfrac{\partial M}{\partial P}$	$\displaystyle\int M\left(\dfrac{\partial M}{\partial P}\right)\dfrac{dx}{EI}$
A to B	$-Px$	$-x$	$+Px^2$	$\displaystyle\int_0^{10} (Px^2)\dfrac{dx}{EI} = \dfrac{333P}{EI}$
B to C	$-P(x+10) - 20x$	$-x - 10$	$+Px^2 + 20Px + 20x^2$ $+100P + 200x$	$\displaystyle\int_0^{10} (Px^2 + 20Px + 20x^2$ $+100P + 200x)\dfrac{dx}{EI}$ $= \dfrac{2333P + 16{,}667}{EI}$
Σ				$\dfrac{2666P + 16{,}667}{EI}$

$$\Delta_v = \frac{2666P + 16{,}667}{EI} = \frac{43{,}333 \text{ ft}^3\text{-k}}{EI} = 2.15 \text{ in.} \downarrow \quad \blacksquare$$

EXAMPLE 13.12

Determine the vertical deflection at the free end of the cantilever beam shown in Figure 13.22.

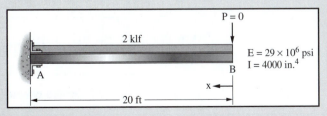

Figure 13.22

Solution.   Placing an imaginary load P at the free end.

TABLE 13.4

Section	M	$\dfrac{\partial M}{\partial P}$	$M\dfrac{\partial M}{\partial P}$	$\int M\left(\dfrac{\partial M}{\partial P}\right)\dfrac{dx}{EI}$
B to A	$-Px - x^2$	$-x$	$Px^2 + x^3$	$\displaystyle\int_0^{20}\left(Px^2 - x^3\right)\dfrac{dx}{EI} = \dfrac{40{,}000}{EI}$
Σ				$\dfrac{40{,}000}{EI}$

$$\Delta = 0.596 \text{ in. } \downarrow \quad \blacksquare$$

EXAMPLE 13.13

Determine the vertical deflection at the 30-kip load in the beam of Figure 13.23.

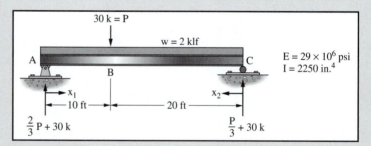

Figure 13.23

Solution.   The 30-kip load is replaced with P. Note that the beam reactions are computed with a load of P at point B and not 30 kips.

TABLE 13.5

Section	M	$\dfrac{\partial M}{\partial P}$	$M\dfrac{\partial M}{\partial P}$	$\int M\left(\dfrac{\partial M}{\partial P}\right)\dfrac{dx}{EI}$
A to B	$\dfrac{2}{3}Px_1 + 30x_1 - x_1^2$	$\dfrac{2}{3}x_1$	$\dfrac{4}{9}Px_1^2 + 20x_1^2 - \dfrac{2}{3}x_1^3$	$\dfrac{1}{EI}\displaystyle\int_0^{10}\left(\dfrac{4}{9}Px_1^2 + 20x_1^2 - \dfrac{2}{3}x_1^3\right)dx_1$
C to B	$\dfrac{P}{3}x_2 + 30x_2 - x_2^2$	$\dfrac{x_2}{3}$	$\dfrac{Px_2^2}{9} + 10x_2^2 - \dfrac{x_2^3}{3}$	$\dfrac{1}{EI}\displaystyle\int_0^{20}\left(\dfrac{Px_2^2}{9} + 10x_2^2 - \dfrac{x_2^3}{3}\right)dx_2$

Performing the indicated operations $\Delta = 0.838$ in. $\downarrow$ $\blacksquare$

EXAMPLE 13.14

Find the slope at the free end of the cantilever beam of Example 13.12. The beam is drawn again in Figure 13.24.

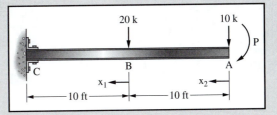

Figure 13.24

Solution. A clockwise moment of P is assumed to be acting at the free end as shown in the figure.

TABLE 13.6

Section	M	$\dfrac{\partial M}{\partial P}$	$M\dfrac{\partial M}{\partial P}$	$\displaystyle\int M\left(\dfrac{\partial M}{\partial P}\right)\dfrac{dx}{EI}$
A to B	$-10x - P$	-1	$10x + P$	$\dfrac{1}{EI}\displaystyle\int_0^{10}(10x + P)\,dx$
B to C	$-30x - P - 100$	-1	$30x + P + 100$	$\dfrac{1}{EI}\displaystyle\int_0^{10}(30x + P + 100)\,dx$

Performing the indicated operations $\theta = +0.0124$ rad$\curvearrowleft$ ■

13.11 PROBLEMS FOR SOLUTION

For Problems 13.1 through 13.6 use the virtual-work method to determine the deflection of each of the joints marked on the trusses shown in the accompanying illustrations. The circled figures are areas in square inches and $E = 29 \times 10^6$ psi unless otherwise indicated.

13.1 U_2 vertical, U_2 horizontal (*Ans.* 0.405 in.↓, 0.221 in. ←)

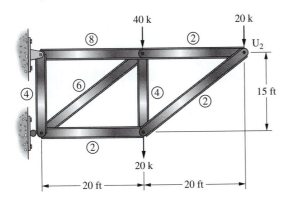

13.2 L_2 vertical, L_4 horizontal

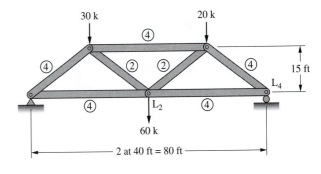

13.3 U_2 vertical, U_1 horizontal (*Ans.* 28.0 mm ↓, 9.5 mm →)

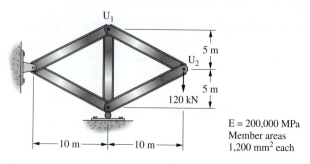

E = 200,000 MPa
Member areas
1,200 mm² each

13.4 L_3 vertical

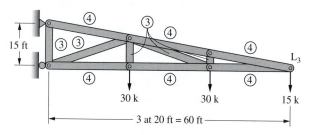

13.5 U_1 vertical, L_1 horizontal (*Ans.* 0.858 in. ↓, 0.702 in. →)

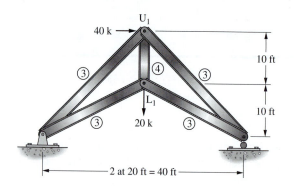

13.6 L_1 vertical, L_1 horizontal

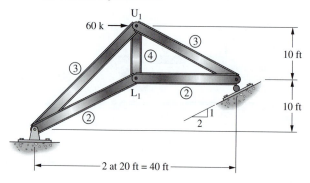

Use the virtual-work method for solving Problems 13.7 through 13.30. $E = 29 \times 10^6$ psi for all problems unless otherwise indicated.

13.7   Determine the deflection at points A and B of the structure shown in the accompanying illustration. $I = 2000$ in.⁴. (*Ans.* 3.97 in. ↓, 1.24 in. ↓,)

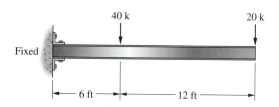

13.8   Determine the deflection underneath each of the concentrated loads shown in the accompanying illustration. $I = 3500$ in.⁴.

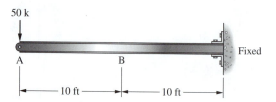

13.9   Find the slope and deflection at the free end of the beam shown in the accompanying illustration. $I = 5000$ in.⁴. (*Ans.* 0.00503 rads \, 0.628 in. ↓)

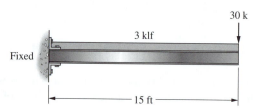

13.10   Find the deflections at point A for the beam shown in the accompanying illustration. $I = 3500$ in.⁴.

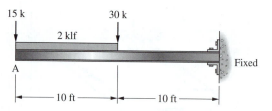

13.11   Determine the slope and deflection at point A in the beam of Problem 11.11. (*Ans.* 0.00251 rads /, 0.271 in. ↓)

13.12 Determine the slope and deflection 12 ft from the right end of Problem 11.12.

13.13 Find the slope and deflection at the 40 kN load in the beam of Problem 11.20. (*Ans.* 0.01725 rads \, 74.25 mm ↓)

13.14 Calculate the slope and deflection at a point 3 m from the left support of the beam shown in the accompanying illustration. $I = 2.5 \times 10^8 \, \text{mm}^4$, $E = 200\,000 \, \text{MPa}$.

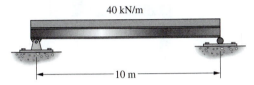

40 kN/m

10 m

13.15 Determine θ_A and Δ_C for the beam of Problem 11.16. (*Ans.* 0.0106 rads \, 1.70 in. ↓)

13.16 Compute θ_A and Δ_A for the beam of Problem 11.19.

13.17 Compute θ_A and Δ_A for the beam of Problem 12.1. (*Ans.* 0.00318 rads \, 0.915 in. ↓)

13.18 Determine the deflection at the centerline and the slope at the right end of the beam shown in the accompanying illustration. $I = 2100 \, \text{in.}^4$.

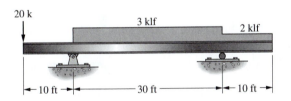

20 k

3 klf

2 klf

10 ft   30 ft   10 ft

13.19 Calculate the deflection at the 80-kip load in the structure shown in the accompanying illustration. (*Ans.* 2.12 in. ↓)

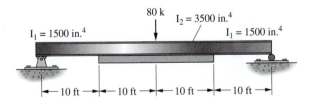

80 k $I_2 = 3500 \, \text{in.}^4$

$I_1 = 1500 \, \text{in.}^4$ $I_1 = 1500 \, \text{in.}^4$

10 ft   10 ft   10 ft   10 ft

13.20 Calculate the slope and deflection at the 60-kip load on the beam shown in the accompanying illustration.

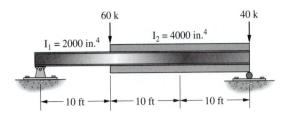

60 k   40 k $I_2 = 4000 \, \text{in.}^4$

$I_1 = 2000 \, \text{in.}^4$

10 ft   10 ft   10 ft

13.21 Calculate the slope and deflection at the 60-kN load on the structure shown in the accompanying illustration. $I = 1.46 \times 10^9 \, \text{mm}^4$. $E = 200\,000 \, \text{MPa}$ (*Ans.* 0.00628 rads /, 39.96 mm ↓)

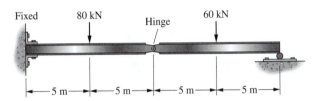

Fixed   80 kN   Hinge   60 kN

5 m   5 m   5 m   5 m

13.22 Find the deflection 24 ft from the left end of the simple beam shown in the accompanying illustration. $I = 2250 \, \text{in.}^4$.

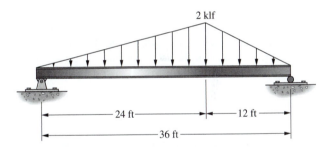

2 klf

24 ft   12 ft

36 ft

13.23 Calculate the slope and deflection at the free end of the beam shown in the accompanying illustration. $I = 4250 \, \text{in.}^4$. (*Ans.* 0.00175 rads /, 0.442 in. ↑)

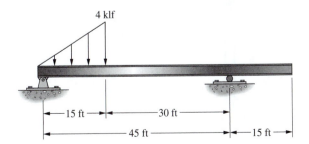

4 klf

15 ft   30 ft

45 ft   15 ft

13.24 Find the vertical deflection at point A and the horizontal deflection at point B on the frame shown in the accompanying illustration. $I = 3000$ in.^{4}.

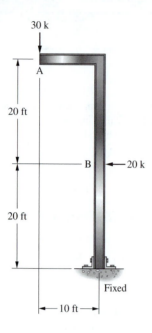

13.25 Determine the horizontal deflection at the 20 k load and the vertical deflection at the 30 k load for the frame shown in the accompanying illustration. (*Ans.* 1.212 in. →, 0.198 in. ↓)

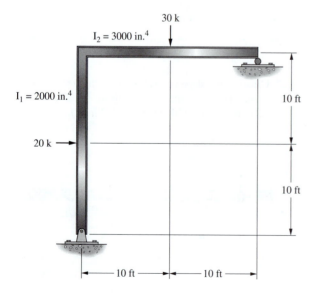

13.26 Find the vertical and horizontal components of deflection at the 50 kip load in the frame shown in the accompanying illustration. $I = 3250$ in.^{4}.

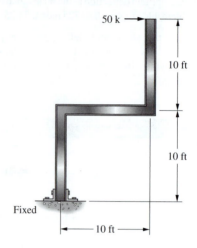

13.27 Determine the vertical and horizontal components of deflection at the free end of the frame shown in the accompanying illustration. $I = 2.60 \times 10^9$ mm^4. $E = 200\,000$ MPa (*Ans.* 149.4 mm ↓, 71.13 mm ←)

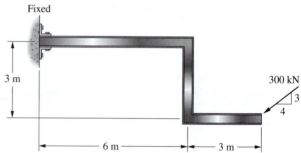

13.28 Find the horizontal deflection at point A and the vertical deflection at point B on the frame shown in the accompanying illustration. $I = 1200$ in.^{4}.

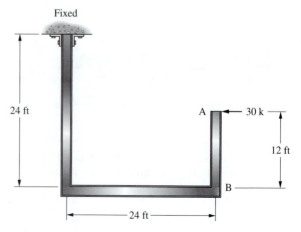

13.29 Calculate the horizontal deflection at the roller support and the vertical deflection at the beam center line for the frame shown in the accompanying illustration. $I = 1850$ in.^{4}. (*Ans.* 1.57 in. $\rightarrow$, 0.626 in. $\downarrow$)

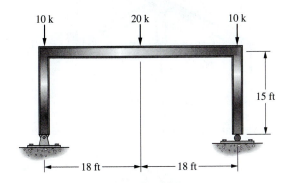

13.30 Find the horizontal deflections at points A and B on the frame shown in the accompanying illustration. $I = 2750$ in.^{4}.

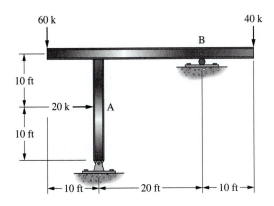

For Problems 13.31 through 13.42, use Castigliano's second theorem to determine the slopes and deflections indicated. $E = 29 \times 10^6$ psi for the problems having customary units and 200,000 MPa for those with SI units. Similarly circled values are cross-sectional areas in square inches or in square millimeters.

13.31 Deflection at each load; $I = 3250$ in.^{4}. (*Ans.* 3.76 in. $\downarrow$ @ 10 k loads, 1.29 in. $\downarrow$ @ 20 k load)

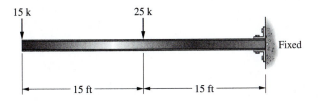

13.32 Slope and deflection at free end; $I = 1750$ in.^{4}.

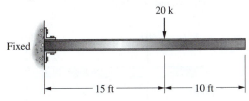

13.33 Slope and deflection at 30 k load; $I = 5500$ in.^{4}. (*Ans.* 0.00933 rads /, 1.48 in. $\downarrow$)

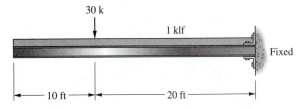

13.34 Deflection at each end of uniform load; $I = 1250$ in.^{4}.

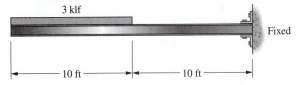

13.35 Deflection at each load; $I = 4.0 \times 10^8$ mm^4 (*Ans.* 26.4 mm $\downarrow$ @ 20 kN load, 9 mm $\downarrow$ @ 30 kN load)

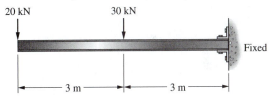

13.36 Slope and deflection at 100-kN load. $I = 2000 \times 10^8$ mm^4

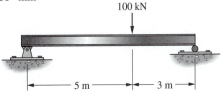

13.37 Rework Problem 13.10. (*Ans.* 1.69 in. $\downarrow$)

13.38 Deflection at each load; $I = 2250$ in.^{4}.

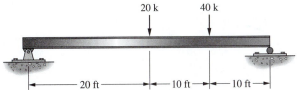

13.39 Slope and deflection at the 60 k load: $I = 3100\,\text{in.}^4$. (*Ans.* 0.00040 rads /, 0.048 in. ↓)

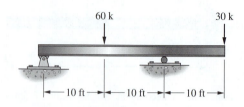

13.40 Vertical deflection at joint L_1

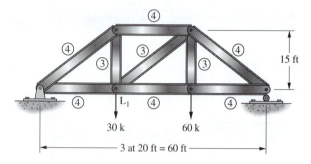

13.41 Vertical and horizontal deflection at L_0 (*Ans.* $\Delta_H = 0.0993\,\text{in.} \rightarrow$)

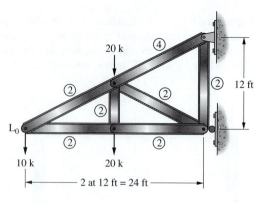

13.42 Vertical deflection at L_1 and horizontal deflection at L_2

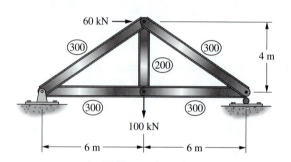

PART TWO

STATICALLY INDETERMINATE STRUCTURES

CLASSICAL METHODS

Chapter 14

Introduction to Statically Indeterminate Structures

14.1 INTRODUCTION

When a structure has too many external reactions and/or internal forces to be determined using only equations of static equilibrium (including any equations of condition), it is statically indeterminate. A load placed on one part of a statically indeterminate or continuous structure will cause shearing forces, bending moments, and deflections in other parts of the structure. In other words, loads applied to a column affect the beams, slabs, and other columns and vice versa. This is often true, but not necessarily true, with statically determinate structures.

Up to this point, the text has been so completely devoted to statically determinate structures that the reader may have been falsely led to believe that statically determinate beams and trusses are the rule in modern structures. In truth, it is difficult to find an ideal, simply supported beam. Probably the best place to look for one would be in a textbook on structures, for bolted or welded beam-to-column connections do not produce ideal simple supports with zero moments.

The same holds for statically determinate trusses. Bolted or welded joints are not frictionless pins, as previously assumed. The other assumptions made about trusses in the earlier chapters of this book are not altogether true either, and thus in a strict sense all trusses are statically indeterminate because they have some bending and secondary forces.

Almost all cast-in-place reinforced-concrete structures are statically indeterminate. The concrete for a large part of a concrete floor, including the support beams, and girders, and perhaps parts of the columns, may be placed at the same time. The reinforcing bars extend from member to member, as from one span of a beam into the next. When there are construction joints, the reinforcing bars are left protruding from the older concrete so they may be lapped or spliced to the bars in the newer concrete. In addition, the old concrete is cleaned and perhaps roughened so that the newer concrete will bond to it as well as possible. The result of all these facts is that reinforced-concrete structures are generally monolithic or continuous and thus are statically indeterminate.

About the only way a statically determinate reinforced-concrete structure could be built would be with individual sections precast at a concrete plant and assembled on the job site. Even these structures could have some continuity in their joints.

Until the early part of the 20th century, statically indeterminate structures were avoided as much as possible by most American engineers. However, three great developments completely changed the picture. These developments were monolithic reinforced-concrete structures, arc welding of steel structures, and modern methods of analysis.

14.2 CONTINUOUS STRUCTURES

As the spans of simple structures become longer, the bending moments increase rapidly. If the weight of a structure per unit of length remained constant, regardless of the span, the dead-load moment would vary in proportion to the square of the span length ($M_{max} = wL^2/8$). This proportion, however, is not correct, because the weight of structures must increase with longer spans to be strong enough to resist the increased bending moments. Therefore, the dead-load moment increases at a greater rate than does the square of the span.

For economy, it pays in long spans to introduce types of structures that have smaller moments than the tremendous ones that occur in long-span, simply supported structures. One type of structure that considerably reduces bending moments, cantilever-type construction, was introduced in Chapter 4. Two other moment-reducing structures are discussed in the following paragraphs.

In some locations, a beam with fixed ends, rather than one with simple supports, may be possible. A comparison of the moments developed in a uniformly loaded simple beam with those in a uniformly loaded, fixed-ended beam is made in Figure 14.1.

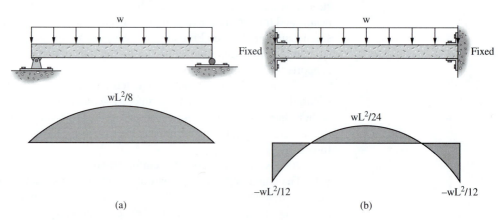

Figure 14.1   A simple beam (a) and a fixed ended beam (b)

The maximum bending moment in the fixed-ended beam is only two thirds of that in the simply supported beam. Usually it is difficult to fix the ends, particularly in the case of bridges. For this reason, *flanking spans* are often used, as illustrated in Figure 14.2. These spans will partially restrain the interior supports, thus tending to reduce the moment in the center span. This figure compares the bending moments that occur in three uniformly loaded simple beams (spans of 60 ft, 100 ft, and 60 ft) with the moments of a uniformly loaded beam continuous over the same three spans.

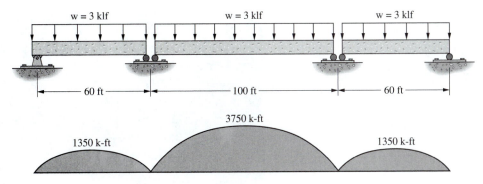

(a) Moment diagrams if three simple beams are used

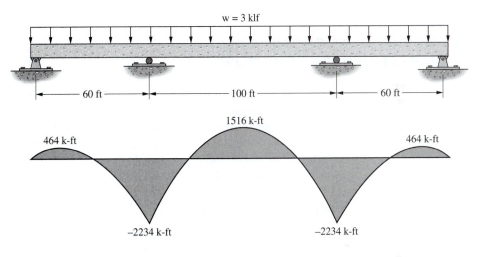

(b) Moment diagrams if one continuous beam is used

Figure 14.2   Comparison of bending moments in three simple beams versus one continuous beam

    The maximum bending moment in the continuous beam is approximately 40% less than that for the simple beams. Unfortunately, there will not be a corresponding 40% reduction in total cost. The cost-reduction probably is only a small percentage of the total cost of the structure because such items as foundations, connections, and floor systems are not reduced a great deal by the moment reduction. Varying the lengths of the flanking spans will change the magnitude of the largest moment occurring in the continuous member. For a constant uniform load over the 3 spans, the smallest moment will occur when the flanking spans are from 0.3 to 0.4 times as long as the center span.

    In the foregoing discussion, the maximum moments developed in beams were reduced appreciably by continuity. This reduction occurs when beams are rigidly connected to each other or where beams and columns are rigidly connected. There is a continuity of action in resisting a load applied to any part of a continuous structure because the load is resisted by the combined efforts of all the members of the frame.

Colorado River arch bridge, Utah Route 95 (Courtesy of the Utah Department of Transportation)

14.3 ADVANTAGES OF STATICALLY INDETERMINATE STRUCTURES

When comparing statically indeterminate structures with statically determinate structures, the first consideration for most engineers would likely be cost. However, making a general economic statement favoring one type of structure over another is impossible without reservation. Each structure presents a different and unique situation, and all factors must be considered—economic or otherwise. In general, statically indeterminate structures have the following advantages.

Savings in Materials

The smaller bending moments developed often enable the engineer to select smaller members for the structural components. The material saving could possibly be as high as 10 to 20% for highway bridges. Because of the large number of force reversals that occur in railroad bridges, the cost saving is closer to 10%.

A structural member of a given size can support more load if it is part of a continuous structure than if it is simply supported. The continuity permits the use of smaller members for the same loads and spans or increased spacing of supports for the same size members. The possibility of fewer columns in buildings, or fewer piers in bridges, may permit a reduction in overall costs.

Continuous concrete or steel structures are cheaper without the joints, pins, and so on required to make them statically determinate, as was frequently the practice in past years. Monolithic reinforced-concrete structures are erected so that they are naturally continuous and statically indeterminate. Installing the hinges and other devices necessary to make them statically determinate would be a difficult construction problem, and very expensive. Furthermore, if a building frame consisted of columns and simply supported beams, objectionable diagonal bracing between the joints would be necessary to provide a stable frame with sufficient rigidity.

Larger Safety Factors

Statically indeterminate structures often have higher safety factors than statically determinate structures. When parts of statically indeterminate steel or reinforced-concrete structures are overstressed they will often have the ability to redistribute portions of those stresses to less-stressed areas. Statically determinate structures generally do not have this ability.[1] If the bending moments in a component of a statically determinate structure reach the ultimate moment capacity of that component, the structure would fail. This is not the case for statically indeterminate structures since load may be redistributed to other parts of the structure.

As can be clearly shown, a statically indeterminate beam or frame normally will not collapse when the ultimate moment capacity is reached at just one location. Instead, there is a redistribution of the moments in the structure. This behavior is quite similar to the case in which three men are walking along with a log on their shoulders and one of the men gets tired and lowers his shoulder just a little. The result is redistribution of load to the other men with corresponding changes in the reactions, shearing forces and bending moments along the log.

Greater Rigidity and Smaller Deflections

Statically indeterminate structures are more rigid and have smaller deflections than statically determinate structures. Because of their continuity, they are stiffer and have greater stability against all types of loads (horizontal, vertical, moving, etc.).

More Attractive Structures

It is difficult to imagine statically determinate structures having the gracefulness and beauty of many statically indeterminate arches and rigid frames being erected today.

Broadway Street Bridge showing cantilever erection, Kansas City, Missouri (Courtesy of USX Corporation)

Adaptation to Cantilever Erection

The cantilever method of erecting bridges is of particular value where conditions underneath (probably marine traffic or deep water) hinder the erection of falsework. Continuous statically indeterminate bridges and cantilever-type bridges are conveniently erected by the cantilever method.

[1] J. C. McCormac and J. K. Nelson, Jr. *Structural Steel Design LRFD Method*, 3rd ed. (upper Saddle River, N. J.: Prentice Hall, 2003), 221–231.

14.4 DISADVANTAGES OF STATICALLY INDETERMINATE STRUCTURES

A comparison of statically determinate and statically indeterminate structures shows the latter have several disadvantages that make their use undesirable on many occasions. These disadvantages are discussed in the following paragraphs.

Support Settlement

Statically indeterminate structures are not desirable where foundation conditions are poor, because seemingly minor support settlements or rotations may cause major changes in the bending moments, shearing forces, reaction forces, and member forces. Where statically indeterminate bridges are used despite the presence of poor foundation conditions, it is occasionally necessary to physically measure the dead-load reactions. The supports of the bridge are jacked up or down until the calculated reaction forces are obtained. The support is then built to that elevation.

Development of Other Stresses

Support settlement is not the only condition that causes stress variations in statically indeterminate structures. Variation in the relative positions of members caused by temperature changes, poor fabrication, or internal deformation of members in the structure under load may cause significant force changes throughout the structure.

Difficulty of Analysis and Design

The forces in statically indeterminate structures depend not only on their dimensions but also on their elastic and cross-sectional properties (moduli of elasticity, moments of inertia, and areas). This situation presents a design difficulty: the forces cannot be determined until the member sizes are known, and the member sizes cannot be determined until their forces are known. The problem is handled by assuming member sizes and computing the forces, designing the members for these forces and computing the forces for the new sizes, and so on, until the final design is obtained. Design by this method—*the method of successive approximations*—takes more time than the design of a comparable statically determinate structure, but the extra cost is only a small part of the total cost of the structure. Such a design is best done by interaction between the designer and the computer. Interactive computing is now used extensively in the aircraft and automobile industries.

Stress Reversals

Generally, more force reversals occur in statically indeterminate structures than in statically determinate structures. Additional material may be required at certain sections to resist the different force conditions and to prevent fatigue failures.

14.5 METHODS OF ANALYZING STATICALLY INDETERMINATE STRUCTURES

Statically indeterminate structures contain more unknown forces than there are equations of static equilibrium. As such, they cannot be analyzed using only the equations of static equilibrium; additional equations are needed. Forces beyond those needed to maintain a stable structure are redundant forces. The redundant forces can be forces of reaction or forces in members that make up the structure. There are two general approaches used to find the magnitude of these redundant forces: *force methods* and *displacement methods.* The bases of these methods are discussed in this section.

Force Methods

With force methods, equations of condition involving displacement at each of the redundant forces in the structure are introduced to provide the additional equations necessary for solution. Equations for displacement at and in the direction of the redundant forces are written in terms of the redundant forces; one equation is written for the displacement condition at each redundant force. The resulting equations are solved for the redundant forces, which must be sufficiently large to satisfy the boundary conditions. As we will soon see, the boundary conditions do not necessarily have to be zero displacement. Force methods are also called *flexibility methods* or *compatibility methods.*

In 1864, James Clerk Maxwell published the first consistent force method for analyzing statically indeterminate structures. His method was based on a consideration of deflections, but the presentation (which included the reciprocal deflection theorem) was rather brief and attracted little attention. Ten years later Otto Mohr independently extended the theory to almost its present stage of development. Analysis of redundant structures with the use of deflection computations is often referred to as the *Maxwell–Mohr method* or the *method of consistent distortions.*[2,3]

The force methods of structural analysis are somewhat useful for analyzing beams, frames, and trusses that are statically indeterminate to the first or second degree. They also are convenient for some single-story frames with unusual dimensions. For structures that are highly statically indeterminate, such as multistory buildings and large complex trusses, other methods are more appropriate and useful. These methods, which include moment distribution and the matrix methods, are more satisfactory and are introduced in later chapters. As such, the force methods have been almost completely superseded by the methods of analysis described in Part Three of this text. Nevertheless, study of the force methods will provide an understanding of the behavior of statically indeterminate structures that might not otherwise be obtained.

Displacement or Stiffness Methods

In the displacement methods of analysis, the displacement of the joints (rotations and translations) necessary to describe fully the deformed shape of the structure are used in the equations instead of the redundant forces used in the force methods. When the simultaneous equations that result are solved, these displacements are determined and then substituted into the original equations to determine the various internal forces. The most commonly used displacement method is the matrix method discussed in Chapters 22 through 25.

---

[2]J. I. Parcel and R. B. B. Moorman, *Analysis of Statically Indeterminate Structures* (New York: Wiley, 1955), 48.

[3]J. S. Kinney, *Indeterminate Structural Analysis* (Reading, Mass.: Addison-Wesley, 1957), 12–13.

14.6 LOOKING AHEAD

In Chapters 15–18 classical methods of analyzing statically indeterminate structures are presented. These methods include consistent distortions, Castigliano's theorems, and slope deflection. The first two methods are force methods, whereas the latter method is a displacement method of analysis. These methods are primarily of historical interest and are almost never used in practice. However, they do form the basis for modern methods of analysis.

The author has placed in Part Three (Chapters 19–25) methods for analyzing statically indeterminate structures that are commonly used today in the structural engineering profession. First, several approximate methods of analysis are introduced, and then moment distribution and the matrix methods are discussed at length. Truthfully speaking computer analyses based on matrix methods are almost the "only game in town" today.

Chapter 15

Force Methods of Analyzing Statically Indeterminate Structures

15.1 BEAMS AND FRAMES WITH ONE REDUNDANT

The force method of analyzing statically indeterminate structures is often referred to as the *method of consistent distortions*. For a first illustration of consistent distortions the two-span beam of Figure 15.1(a) is considered. The beam is assumed to consist of a material which follows Hooke's law. This statically indeterminate structure supports the loads P_1 and P_2 and is in turn supported by reaction components at points A, B, and C. Removal of support B would leave a statically determinate beam, proving the structure to be statically indeterminate to the first degree. With this support removed it is a simple matter to find the deflection at B, Δ_B, in Figure 15.1(b), caused by the external loads.

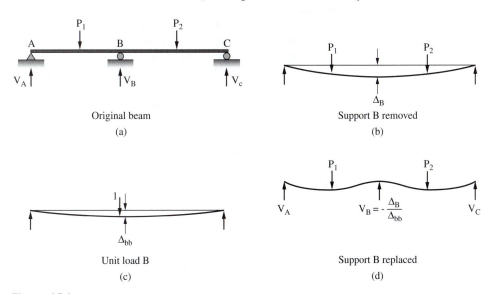

Original beam

(a)

Support B removed

(b)

Unit load B

(c)

Support B replaced

(d)

Figure 15.1

If the external loads are removed from the beam and a unit load is placed at B, a deflection at B equal to Δ_{bb} will be developed, as indicated in Figure 15.1(c). Deflections due to external loads are denoted with capital letters herein. The deflection at point C on a

I-180 viaduct, Cheyenne, Wyoming. (Courtesy of the Wyoming State Highway Department.)

beam due to external loads would be Δ_C. Deflections due to the imaginary unit load are denoted with two small letters. The first letter indicates the location of the deflection and the second letter indicates the location of the unit load. The deflection at E caused by a unit load at B would be Δ_{eb}. Displacements due to unit loads often are called *flexibility coefficients*.

Support B is unyielding, and its removal is merely a convenient assumption. An upward force is present at B and it is sufficient to prevent any deflection, or, continuing with the fictitious line of reasoning, there is a force at B that is large enough to push point B back to its original nondeflected position. The distance the support must be pushed is Δ_B.

A unit load at B causes a deflection at B equal to Δ_{bb} and a 10-kip load at B will cause a deflection of $10\Delta_{bb}$. Similarly, an upward reaction at B of V_B will push B up an amount $V_B\Delta_{bb}$. The total deflection at B due to the external loads and the reaction V_B is zero and may be expressed as follows:

$$\Delta_B + V_B\Delta_{bb} = 0$$

$$V_B = -\frac{\Delta_B}{\Delta_{bb}}$$

The minus sign in this expression indicates V_B is in the opposite direction from the downward unit load. If the solution of the expression yields a positive value, the reaction is in the same direction as the unit load. Examples 15.1 to 15.4 illustrate the force method of computing the reactions for statically indeterminate beams having one redundant reaction component. Example 15.5 shows that the method may be extended to include statically indeterminate frames as well. The necessary deflections for the first four examples are determined with the conjugate-beam procedure, while those for Example 15.5 are determined by virtual work. After the value of the redundant reaction in each problem is found, the other reactions are determined by statics, and shear and moment diagrams are drawn.

EXAMPLE 15.1

Determine the reactions and draw shear and moment diagrams for the two-span beam of Figure 15.2; assume V_B to be the redundant; E and I are constant.

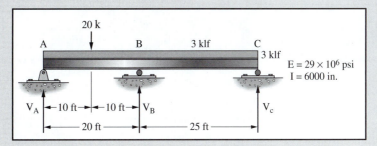

Figure 15.2

Solution.   Separating the uniform load and the concentrated load and drawing their $\frac{M}{EI}$ diagrams.

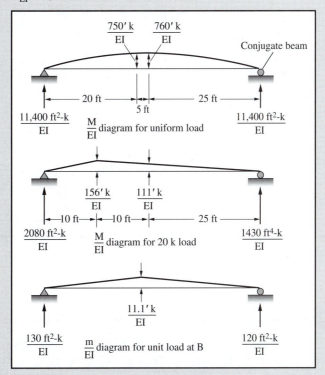

Computing Δ_B, Δ_{bb}, and V_B

$$EI\Delta_B = (20)(11,400) - \left(\tfrac{1}{2}\right)(20)(750)(6.67) - \frac{(3)(20)^3}{12}(10)$$
$$+ (1430)(25) - \left(\tfrac{1}{2}\right)(25)(111)(8.33)$$

$$EI\Delta_B = 182,100 \text{ ft}^3\text{-k}$$

$$EI\Delta_{bb} = (20)(130) - \left(\tfrac{1}{2}\right)(20)(11.1)(6.67)$$

$$EI\Delta_{bb} = 1860 \text{ ft}^3$$

$$V_B = -\frac{\Delta_B}{\Delta_{bb}} = -\frac{182,100}{1860} = 98 \text{ k } \uparrow$$

Computing reactions at A and C by statics,

$$\sum M_A = 0$$
$$(20)(10) + (45)(3)(22.5) - (20)(98) - 45V_C = 0$$
$$V_C = 28.5 \text{ k } \uparrow$$

$$\sum V = 0$$
$$20 + (3)(45) - 98 - 28.5 - V_A = 0$$
$$V_A = 28.5 \text{ k } \uparrow$$

Shear and moment diagrams:

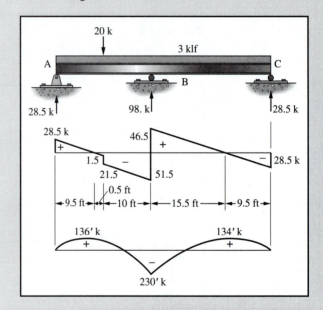

Determine the reactions and draw shear and moment diagrams for the propped beam shown in Figure 15.3. Consider V_B to be the redundant; E and I are constant.

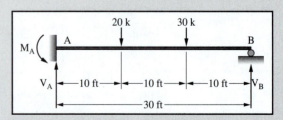

Figure 15.3

Solution.

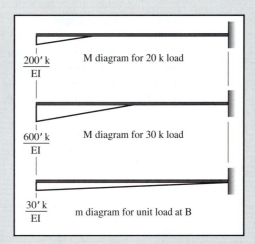

$$EI\Delta_B = \left(\tfrac{1}{2}\right)(200)(10)(26.67) + \left(\tfrac{1}{2}\right)(600)(20)(23.33) = 166,670 \text{ ft}^3\text{-k}$$
$$EI\Delta_{bb} = \left(\tfrac{1}{2}\right)(30)(30)(20) = 9000 \text{ ft}^3$$
$$V_B = -\frac{166,670}{9000} = 18.5 \text{ k} \uparrow$$

By statics

$$V_A = 31.5 \text{ k} \uparrow \text{ and } M_A = 245 \text{ ft-k} \circlearrowright$$

Shear and moment diagrams:

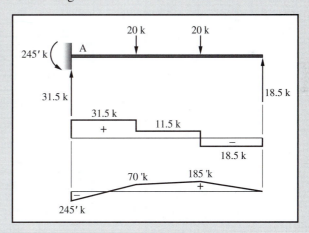

EXAMPLE 15.3

Rework Example 15.2 by using the resisting moment at the fixed end as the redundant.

Solution. Any one of the reactions may be considered to be the redundant and taken out, provided a stable structure remains. If the resisting moment at A is removed, a simple support remains and the beam loads cause the tangent to the elastic curve to rotate an amount θ_A. A brief discussion of this condition will reveal a method of determining the moment.

The value of θ_A equals the shear at A in the conjugate beam loaded with the M/EI diagram. If a unit moment is applied at A, the tangent to the elastic curve will rotate an amount θ_{aa}, which can also be found with the conjugate beam procedure. The tangent to the elastic curve at A actually does not rotate; therefore, when M_A is replaced, it must be of sufficient magnitude to rotate the tangent back to its original horizontal position. The following expression equating θ_A to zero may be written and solved for the redundant M_A.

$$\theta_A + M_A\theta_{aa} = 0$$

$$M_A = -\frac{\theta_A}{\theta_{aa}}$$

Solution.

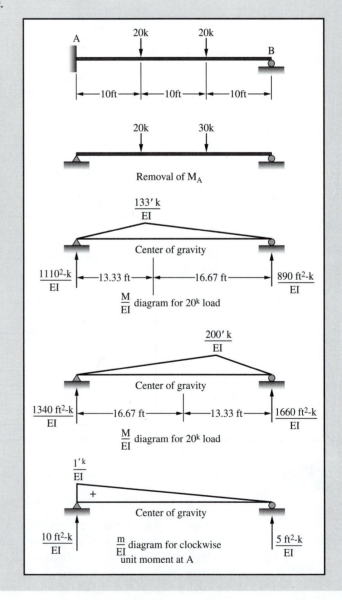

From the $\dfrac{M}{EI}$ diagrams

$$\theta_A = \frac{1110 + 1340}{EI} = \frac{2450 \ \text{ft}^2\text{-k}}{EI}$$

$$\theta_{aa} = \frac{10 \ \text{ft}}{EI}$$

$$M_A = -\frac{\theta_A}{\theta_{aa}} = -\frac{2450}{10} = 245 \ \text{ft-k} \ \circlearrowright \quad \blacksquare$$

EXAMPLE 15.4

Find the reactions and draw the shear and moment diagrams for the two-span beam shown in Figure 15.4. Assume the moment at the interior support B to be the redundant.

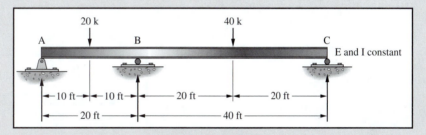

Figure 15.4

Solution. Removal of moment from the interior support changes the support into a hinge, and the beams on each side are free to rotate independently as indicated by the angles θ_{b_1} and θ_{b_2} in the deflection curve shown. The numerical values of the angles can be found by placing the M/El diagram on the conjugate structure and computing the shear on each side of the support. In the actual beam there is no change of slope of the tangent to the elastic curve from a small distance to the left of B to a small distance to the right of B.

The angle represented in the diagram by θ_B is the angle between the tangents to the elastic curve on each side of the support (that is, $\theta_{b_1} + \theta_{b_2}$). The actual moment M_B, when replaced, must be of sufficient magnitude to bring the tangents back together or reduce θ_B to zero. A unit moment applied on each side of the hinge produces a change of slope of θ_{bb}; therefore, an expression for the magnitude of M_b is:

$$M_B = -\frac{\theta_B}{\theta_{bb}}$$

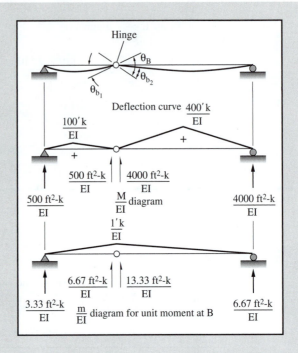

Solution.

$$\theta_{b_1} = \frac{500 \text{ ft}^2\text{-k}}{EI}$$

$$\theta_{b_2} = \frac{4000 \text{ ft}^2\text{-k}}{EI}$$

$$\theta_B = \theta_{b_1} + \theta_{b_2} = \frac{4500 \text{ ft}^2\text{-k}}{EI}$$

$$\theta_{bb} = 6.67 + 13.33 = \frac{20 \text{ ft}^2\text{-k}}{EI}$$

$$M_B = -\frac{\theta_B}{\theta_{bb}} = -\frac{4500}{20} = -225 \text{ ft-k}$$

By statics the following reactions are found:

$$V_A = -1.25 \text{ k}$$

$$V_B = -46.87 \text{ k}$$

$$V_C = -14.38 \text{ k}$$

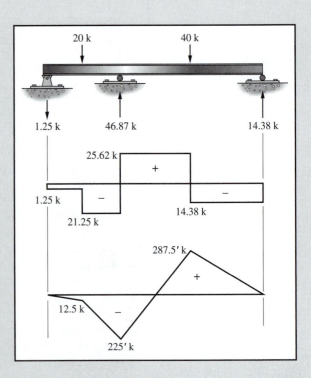

Compute the reactions and draw the moment diagram for the structure shown in Figure 15.5.

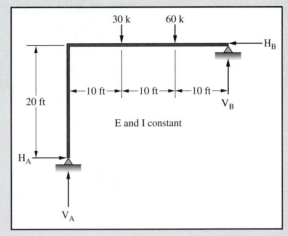

Figure 15.5

Solution.   Remove H_A as the redundant. This will change A to a roller type of support.

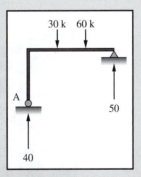

Compute the horizontal deflection at A by virtual work. The result is

$$\Delta_A = \frac{86{,}600 \text{ ft}^3\text{-k}}{EI} \leftarrow$$

Apply a unit horizontal load at A and determine the horizontal deflection Δ_{aa}.

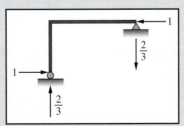

The result is

$$\Delta_{aa} = +\frac{6667 \text{ ft}^3\text{-k}}{EI} \rightarrow$$

$$\Delta_A + H_A \Delta_{aa} = 0$$

$$H_A = -\frac{\Delta_A}{\Delta_{aa}} = -\frac{-86,660}{+6667} = +13 \text{ k} \rightarrow$$

Compute the remaining reactions by statics.

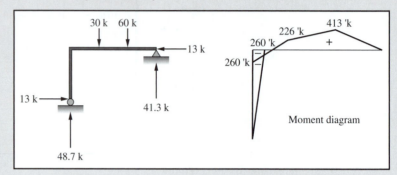

Raritan River Bridge in New Jersey. (Courtesy of Steinman, Boynton, Gronquist & Birdsall Consulting Engineers.)

15.2 BEAMS AND FRAMES WITH TWO OR MORE REDUNDANTS

The force method of analyzing beams and frames with one redundant may be extended to beams and frames having two or more redundants. The continuous beam of Figure 15.6, which has two redundant reactions, is considered here.

To make the beam statically determinate, it is necessary to remove two supports. Supports B and C are assumed to be removed, and their deflections Δ_B and Δ_C due to the external loads are computed. The external loads are theoretically removed from the beam; a unit load is placed at B; and the deflections at B and C—Δ_{bb} and Δ_{cb}—are found. The unit load is moved to C, and the deflections at the two points Δ_{bc} and Δ_{cc} are determined.

The reactions at supports B and C push these points up until they are in their original positions of zero deflection. The reaction V_B will raise B an amount $V_b\Delta_{bb}$ and C an amount $V_B\Delta_{cb}$. The reaction V_C raises C by $V_C\Delta_{cc}$ and B by $V_C\Delta_{bc}$.

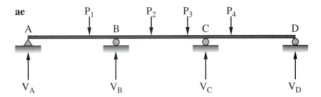

Figure 15.6

I-81 river relief route interchange, Harrisburg, Pennsylvania.
(Courtesy of Gannett Fleming.)

An equation may be written for the deflection at each of the supports. Both equations contain the two unknowns V_B and V_C, and their values may be obtained by solving the equations simultaneously.

$$\Delta_B + V_B\Delta_{bb} + V_C\Delta_{bc} = 0$$
$$\Delta_C + V_B\Delta_{cb} + V_C\Delta_{cc} = 0$$

The force method of computing redundant reactions may be extended indefinitely for structures with any number of redundants. The calculations become quite lengthy, however, if there are more than two or three redundants. Considering the beam of Figure 15.7 the following expressions may be written:

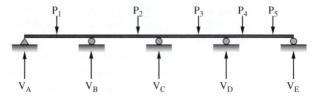

Figure 15.7

$$\Delta_B + V_B \Delta_{bb} + V_C \Delta_{bc} + V_D \Delta_{bd} = 0$$
$$\Delta_C + V_B \Delta_{cb} + V_C \Delta_{cc} + V_D \Delta_{cd} = 0$$
$$\Delta_D + V_B \Delta_{db} + V_C \Delta_{dc} + V_D \Delta_{dd} = 0$$

15.3 SUPPORT SETTLEMENT

Continuous beams with unyielding supports have been considered in the preceding sections. Should the supports settle or deflect from their theoretical positions major changes may occur in the reactions, shears, moments, and stresses. Whatever the factors causing displacement (weak foundations, temperature changes, poor erection or fabrication, and so on), analysis may be made with the deflection expressions previously developed for continuous beams.

An expression for deflection at point B in the two-span beam of Figure 15.1 was written in Section 15.1. The expression was developed on the assumption that support B was temporarily removed from the structure, allowing point B to deflect, after which the support was replaced. The reaction at B, V_B, was assumed to be of a sufficient magnitude to push B up to its original position of zero deflection. Should B actually settle 1.0 in., V_B will be smaller because it will only have to push B up an amount $\Delta_B - 1.0$ in., and the deflection expression may be written as

$$\Delta_B - 1.0 + V_B \Delta_{bb} = 0$$

If three men are walking along with a log on their shoulders (a statically indeterminate situation) and one of them lowers his shoulder slightly, he will not have to support as much of the total weight as before. He has, in effect, backed out from under the log and thrown more of its weight to the other men. The settlement of a support in a statically indeterminate continuous beam has the same effect.

The values of Δ_B and Δ_{bb} must be calculated in inches if the support movement is given in inches; they are calculated in feet if the support movement is given in feet, and so on. Example 15.6 illustrates the analysis of the two-span beam of Example 15.1 with the assumption of a $\frac{3}{4}$-in. settlement of the interior support. The moment diagram is drawn after settlement occurs and is compared with the diagram before settlement. *The seemingly small displacement has completely changed the moment picture.*

EXAMPLE 15.6

Determine the reactions and draw shear and moment diagrams for the beam of Example 15.1, which is reproduced in Figure 15.8, if support B settles $\frac{3}{4}$ in.

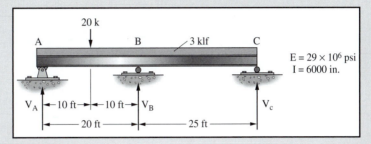

Figure 15.8

Solution.   The values of Δ_B and Δ_{bb} previously found are computed in inches, and the effect of the support settlement on V_B is determined. By statics the new values of V_A and V_C are found and the shear and moment diagrams are drawn. The moment diagram before settlement is repeated to illustrate the striking changes. The reader will now understand why structural engineers are reluctant to use statically indeterminate structures when foundation conditions are poor. Settlements may cause all sorts of changes and problems.

$$\Delta_B = \frac{182,100\ \text{ft}^3\text{-k}}{EI} = 1.81\ \text{in.}$$

$$\Delta_{bb} = \frac{1860\ \text{ft}^3\text{-k}}{EI} = 0.0185\ \text{in.}$$

$$\Delta_B - 0.750 + V_B \Delta_{bb} = 0$$

$$V_B = -\frac{1.81 - 0.750}{0.0185} = 57.3\ \text{k} \uparrow$$

$$V_A = 51.2\ \text{k} \uparrow$$

$$V_C = 46.5\ \text{k} \uparrow$$

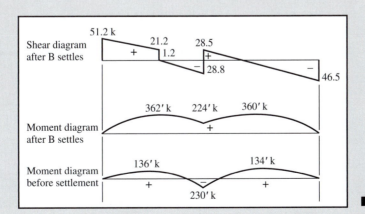

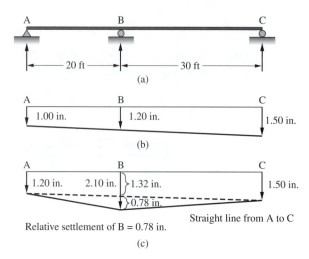

Figure 15.9

When several or all of the supports are displaced, the analysis may be conducted on the basis of relative settlement values. For example, if all of the supports of the beam of Figure 15.9(a) were to settle 1.5 in., the stress conditions would theoretically be unchanged. If the supports settle different amounts but remain in a straight line, as illustrated in Figure 15.9(b), the situation theoretically is the same as before settlement.

Where inconsistent settlements occur and the supports no longer lie on a straight line, the stress conditions change because the beam is distorted. The situation may be handled by drawing a line through the displaced positions of two of the supports, usually the end ones. The distances of the other supports from this line are determined and used in the calculations, as illustrated in Figure 15.9(c).

It is assumed that the supports of the three-span beam of Figure 15.10(a) settle as follows: A is 1.25 in., B is 2.40 in., C is 2.75 in., and D is 1.10 in. A diagram of these settlements is plotted in Figure 15.10(b) and the relative settlements of supports B and C are determined.

The solution of the two simultaneous equations that follow will yield the values of V_B and V_C. By statics the values of V_A and V_D may then be computed.

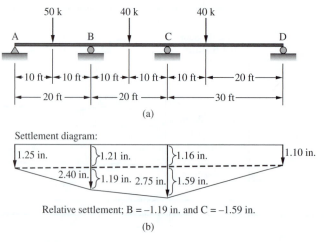

Figure 15.10

$$\Delta_B - 1.19 + V_B\Delta_{bb} + V_C\Delta_{bc} = 0$$
$$\Delta_C - 1.59 + V_B\Delta_{cb} + V_C\Delta_{cc} = 0$$

EXAMPLE 15.7

Determine the reactions for the beam of Figure 15.10 assuming the supports settle as shown in that figure. Draw resulting shear and moment diagrams.

Solution.
Computing deflections at supports B and C.

$$\Delta_B = \frac{514{,}350 \text{ ft}^3\text{-k}}{EI} = 3.76 \text{ in.}$$

$$\Delta_C = \frac{625{,}300 \text{ ft}^3\text{-k}}{EI} = 4.57 \text{ in.}$$

$$\Delta_{bb} = \frac{4765 \text{ ft}^3\text{-k}}{EI} = 0.0349 \text{ in.}$$

$$\Delta_{cc} = \frac{6820 \text{ ft}^3\text{-k}}{EI} = 0.0498 \text{ in.}$$

$$\Delta_{bc} = \Delta_{db} = \frac{5140 \text{ ft}^3\text{-k}}{EI} = 0.0374 \text{ in.}$$

Writing the equations for final deflections at the two supports

$$\Delta_B + V_B\Delta_{bb} + V_C\Delta_{bc} = \left\{ \begin{array}{l} \text{relative} \\ \text{settlement} \\ \text{of B} \end{array} \right\}$$

$$3.76 + 0.0349V_B + 0.0374V_C = 1.19 \qquad \textbf{(15.1)}$$

$$\Delta_C + V_B\Delta_{cb} + V_C\Delta_{cc} = \Delta_C$$

$$4.57 + 0.0374V_B + 0.0498V_C = 1.59 \qquad \textbf{(15.2)}$$

Solving Eqs. (15.1) and (15.2) simultaneously gives values of V_B and V_C, and by statics V_A and V_D are determined.

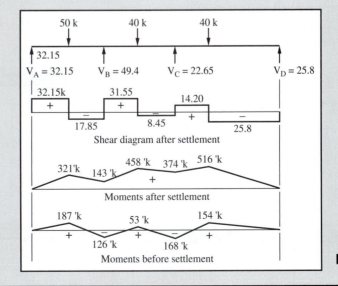

50 k   40 k   40 k

32.15

$V_A = 32.15$ $V_B = 49.4$ $V_C = 22.65$ $V_D = 25.8$

32.15k   31.55

+   14.20

+   −

17.85   8.45   +   −

25.8

Shear diagram after settlement

321'k   143 'k   458 'k   374 'k   516 'k

+

Moments after settlement

187 'k   53 'k   154 'k

+   −   +   −   +

126 'k   168 'k

Moments before settlement

15.4 PROBLEMS FOR SOLUTION

For Problems 15.1 to 15.23 compute the reactions and draw shear and moment diagrams for the continuous beams or frames; E and I are constant unless noted otherwise. The method of consistent distortions is to be used.

15.1 (*Ans.* $V_B = 96.28$ k $\downarrow$, $M_B = -262.8$ ft-k)

15.2

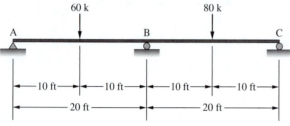

15.3 (*Ans.* $V_B = 102.37$ k $\uparrow$, $M_B = -368.5$ ft-k)

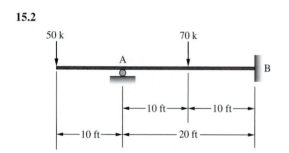

15.4

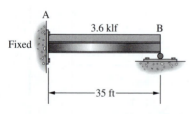

15.5 (*Ans.* $V_A = 7.5$ k $\uparrow$, $M_B = -562.5$ ft-k)

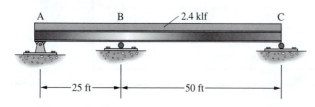

15.6

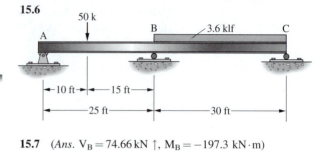

15.7 (*Ans.* $V_B = 74.66$ kN $\uparrow$, $M_B = -197.3$ kN·m)

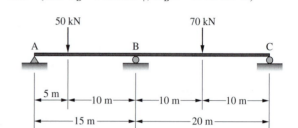

15.8

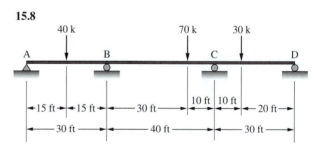

15.9 (*Ans.* $V_A = 9.42$ k $\uparrow$, $M_B = -291.6$ ft-k)

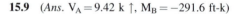
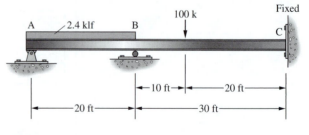

15.10

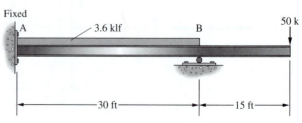

15.11 (*Ans.* $V_B = 64.59$ k $\uparrow$, $M_C = -268.9$ ft-k)

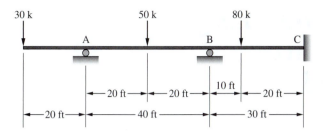

15.18

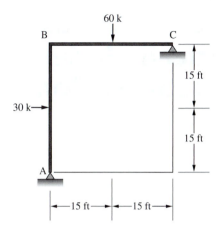

15.12 The beam of Problem 15.1 assuming support B settles 2.50 in. $E = 29 \times 10^6$ psi. $I = 1200$ in.^{4}

15.13 The beam of Problem 15.1 assuming the following support settlements: A = 4.00 in., B = 2.00 in., and C = 3.50 in. $E = 29 \times 10^6$ psi. $I = 1200$ in.^{4} (*Ans.* $V_B = 122.66$ k, $M_B = -526.6$ ft-k)

15.14 The beam of Problem 15.8 assuming the following support settlements: A = 1.00 in., B = 3.00 in., C = 1.50 in., and D = 2.00 in. $E = 29 \times 10^6$ psi. $I = 3200$ in.^{4}

15.19 (*Ans.* $V_C = 17.55$ k $\uparrow$, $H_A = 13.7$ k $\leftarrow$)

15.15 (*Ans.* $V_B = 14.00$ k $\uparrow$, $M_A = -160$ ft-k $\circlearrowleft$)

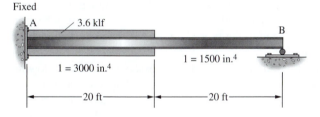

15.16

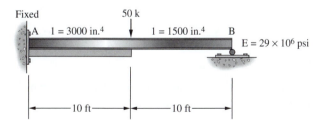

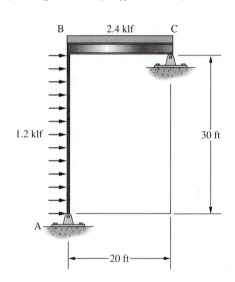

15.20

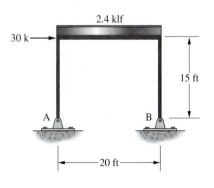

15.17 (*Ans.* $V_C = 19.54$ k, $M_B = -313.6$ ft-k)

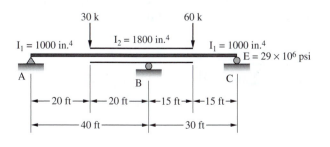

Chapter 16

Force Methods for Analyzing Statically Indeterminate Structures Continued

16.1 ANALYSIS OF EXTERNALLY REDUNDANT TRUSSES

Trusses may be statically indeterminate because of redundant reactions, redundant members, or a combination of redundant reactions and members. Externally redundant trusses will be considered initially, and they will be analyzed on the basis of deflection computations in a manner closely related to the procedure used in the preceding chapter for statically indeterminate beams.

The two-span continuous truss of Figure 16.1 is considered for the following discussion. One reaction component, for example, V_B, is removed, and the deflection at that point caused by the external loads is determined. Next the external loads are removed from the truss, and the deflection at support point B due to a unit load at that point is determined. The reaction is replaced, and it supplies the force necessary to push the support back to its original position. The familiar deflection expression is then written as follows:

$$\Delta_B + V_B \Delta_{bb} = 0$$

$$V_B = -\frac{\Delta_B}{\Delta_{bb}}$$

The forces in the truss members due to the external loads, when the redundant is removed, are not the correct final forces and are referred to herein as the F' forces. The deflection at the removed support due to the external loads can be computed by $\Sigma(F'\mu L/AE)$. The deflection caused at the support by placing a unit load there can be found by applying the same virtual work expression, except the unit load is now the external load and the forces caused are the same as the μ forces. The deflection at the support due to the unit load is $\Sigma(\mu^2 L/AE)$, and the redundant reaction may be expressed as follows:

$$V_B = -\frac{\sum (F'\mu_B L/AE)}{\sum (\mu_B^2 L/AE)}$$

Example 16.1 illustrates the complete analysis of a two-span truss by the method just described. After the redundant reaction is found, the other reactions and the final member forces may be determined by statics. However, another method is available for finding the final forces and should be used as a mathematics check. When the redundant reaction V_B is returned to the truss, it causes the force in each member to change by V_B times its μ force value. The final force in a member becomes

$$F = F' + V_B \mu$$

Rio Grande Gorge Bridge, Taos County, New Mexico. (Courtesy of the American Institute of Steel Construction, Inc.)

EXAMPLE 16.1

Compute the reactions and member forces for the two-span continuous truss shown in Figure 16.1. Circled figures are member areas, in square inches.

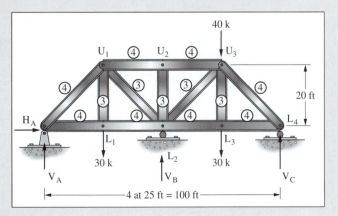

Figure 16.1

Solution.   Remove center support as the redundant and compute F′ forces.

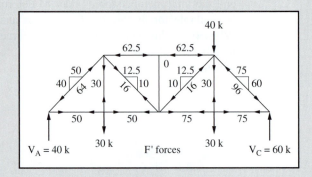

Remove the external loads and place a unit load at the center support. Then compute the μ forces.

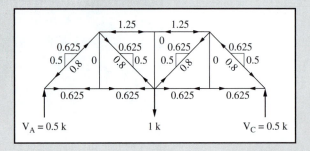

Computing the values of Δ_B and Δ_{bb} in Table 16.1.

$$V_B = -\frac{\dfrac{35{,}690}{E}}{\dfrac{637.6}{E}} = -56.0\text{k} \uparrow$$

The following reactions and member forces are found by statics in order to check the final values in the table.

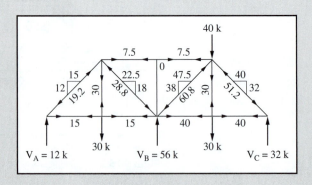

TABLE 16.1

Member	L (in.)	A (in.²)	$\dfrac{L}{A}$	F' (kips)	μ	$\Delta_B = \dfrac{F'\mu L}{AE}$	$\Delta_{bb} = \dfrac{\mu^2 L}{AE}$	$F = F' + V_B\mu$
L_0L_1	300	4	75	+50	+0.625	+2340	+29.2	+15.0
L_1L_2	300	4	75	+50	+0.625	+2340	+29.2	+15.0
L_2L_3	300	4	75	+75	+0.625	+3510	+29.2	+40.0
L_3L_4	300	4	75	+75	+0.625	+3510	+29.2	+40.0
L_0U_1	384	4	96	−64	−0.800	+4920	+61.4	−19.2
U_1U_2	300	4	75	−62.5	−1.25	+5850	+117.0	+7.5
U_2U_3	300	4	75	−62.5	−1.25	+5850	+117.0	+7.5
U_3L_4	384	4	96	−96	−0.800	+7370	+61.4	−51.2
U_1L_1	240	3	80	+30	0	0	0	+30.0
U_1L_2	384	3	128	+16	+0.800	+1640	+ 82.0	−28.8
U_2L_2	240	3	80	0	0	0	0	0
L_2U_3	384	3	128	−16	+0.800	−1640	+ 82.0	− 60.8
U_3L_3	240	3	80	+30	0	0	0	+30.0
Σ						35,690/E	637.6/E	

It should be evident that the deflection procedure may be used to analyze trusses that have two or more redundant reactions. The truss of Figure 16.2 is continuous over three spans, and the reactions at the interior supports V_B and V_C are considered to be the redundants. The following expressions, previously written for a three-span continuous beam, are applicable to the truss:

$$\Delta_B + V_B\Delta_{bb} + V_C\Delta_{bc} = 0$$
$$\Delta_C + V_B\Delta_{cb} + V_C\Delta_{cc} = 0$$

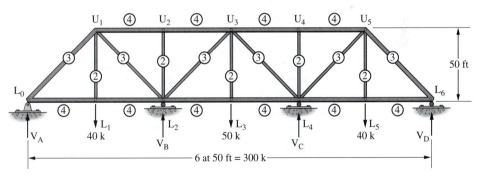

Figure 16.2

The forces due to a unit load at B are called the μ_B forces; those due to a unit load at C are called the μ_C forces. A unit load at B will cause a deflection at C equal to $\Sigma(\mu_B\mu_C L/AE)$, and a unit load at C causes the same deflection at B, $\Sigma(\mu_C\mu_B L/AE)$, which is another illustration of Maxwell's law. The deflection expressions become

$$\Sigma\frac{F'\mu_B L}{AE} + V_B\Sigma\frac{\mu_B^2 L}{AE} + V_C\Sigma\frac{\mu_B\mu_C L}{AE} = 0$$
$$\Sigma\frac{F'\mu_C L}{AE} + V_B\Sigma\frac{\mu_C\mu_B L}{AE} + V_C\Sigma\frac{\mu_C^2 L}{AE} = 0$$

A simultaneous solution of these equations will yield the values of the redundants. Should support settlement occur, the deflections would have to be worked out numerically in the same units used for the settlements.

Tennessee River Bridge, Stevenson, Alabama. (Courtesy of USX Corporation.)

16.2 ANALYSIS OF INTERNALLY REDUNDANT TRUSSES

The truss of Figure 16.3 has one more member than necessary for stability and is therefore statically indeterminate internally to the first degree, as can be proved by applying the equation

$$m = 2j - 3$$

Internally redundant trusses may be analyzed in a manner closely related to the one used for externally redundant trusses for a truss statically indeterminate to the first degree. One member is assumed to be the redundant and is theoretically cut or removed from the structure. The remaining members must form a statically determinate and stable truss. The F' forces in these members are assumed to be of a nature causing the joints at the ends of the removed member to pull apart, the distance being $\sum(F'\mu L/AE)$.

The redundant member is replaced in the truss and is assumed to have a unit tensile force. The μ force in each of the members caused by the redundant's force of $+1$ are computed and they will cause the joints to be pulled together an amount equal to $\sum(\mu^2 L/AE)$. If the redundant has an actual force of X, the joints will be pulled together an amount equal to $X\sum(\mu^2 L/AE)$.

If the member had been sawed in half, the F' forces would have opened a gap of $\sum(F'\mu L/AE)$; therefore, X must be sufficient to close the gap, and the following expressions may be written:

$$X\sum \frac{\mu^2 L}{AE} + \sum \frac{F'\mu L}{AE} = 0$$

$$X = -\frac{\sum(F'\mu L/AE)}{\sum(\mu^2 L/AE)}$$

The application of this method of analyzing internally redundant trusses is illustrated by Example 16.2. After the truss force in the redundant member is found, the force in any other member equals its F′ force plus X times its μ force. Final forces may also be calculated by statics as a check on the mathematics.

EXAMPLE 16.2

Determine the forces in the members of the internally redundant truss shown in Figure 16.3. Members U_1L_1, U_1L_2, L_1U_2, and U_2L_2 have areas of 1 in.^{2} The areas are 2 in.^{2} for each of the other members. $E = 29 \times 10^6$ psi.

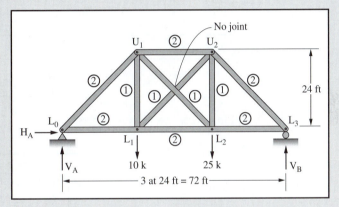

Figure 16.3

Solution. Assume L_1U_2 to be the redundant, remove it, and compute the F′ forces.

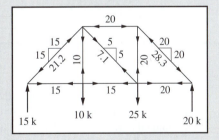

Replace L_1U_2 with a force of $+1$ and compute the μ forces.

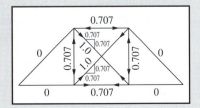

Setting up Table 16.2 and computing X.

TABLE 16.2

Member	L (in.)	A (in.²)	$\dfrac{L}{A}$	F' (kips)	μ	$\dfrac{F'\mu L}{AE}$	$\dfrac{\mu^2 L}{AE}$	$F = F'' + X\mu$ (kips)
L_0L_1	288	2	144	+15	0	0	0	+15.00
L_1L_2	288	2	144	+15	−0.707	−1530	+72	+13.47
L_2L_3	288	2	144	+20	0	0	0	+20.00
L_0U_1	408	2	204	−21.2	0	0	0	−21.20
U_1U_2	288	2	144	−20	−0.707	+2040	+72	−21.53
U_2L_3	408	2	204	−28.3	0	0	0	−28.30
U_1L_1	288	1	288	+10	−0.707	−2040	+144	+8.47
U_1L_2	408	1	408	+7.1	+1.0	+2900	+408	+9.26
L_1U_2	408	1	408	0	+1.0	0	+408	+2.16
U_2L_2	288	1	288	+20	−0.707	−4070	+144	+18.47
Σ						−2700	+1248	

$$X = -\frac{\sum (F'\mu L/AE)}{\sum (\mu^2 L/AE)} = -\frac{-2700}{+1248} = +2.16 \text{ k}$$

Final forces in truss members.

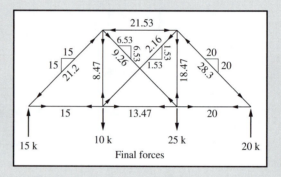

Final forces

For trusses that have more than one redundant internally, simultaneous equations are necessary in the solution. Two members, with forces of X_A and X_B, assumed to be the redundants are theoretically cut. The F' forces in the remaining truss members pull the cut places apart by $\sum(F'\mu_A L/AE)$ and $\sum(F'\mu_B L/AE)$, respectively. Replacing the first redundant member with a force of $+1$ causes μ_A forces in the truss members and causes the gaps to close by $\sum(\mu_A^2 L/AE)$ and $\sum \mu_A \mu_B L/AE$. Repeating the process with the other redundant causes μ_B forces and additional gap closings of $\sum(\mu_B \mu_A L/AE)$ and $\sum(\mu_B^2 L/AE)$. The redundant forces must be sufficient to close the gaps, permitting the writing of the following equations:

$$\sum \frac{F'\mu_A L}{AE} + X_A \sum \frac{\mu_A^2 L}{AE} + X_B \sum \frac{\mu_A \mu_B L}{AE} = 0$$

$$\sum \frac{F'\mu_B L}{AE} + X_A \sum \frac{\mu_B \mu_A L}{AE} + X_B \sum \frac{\mu_B^2 L}{AE} = 0$$

16.3 ANALYSIS OF TRUSSES REDUNDANT INTERNALLY AND EXTERNALLY

Deflection equations have been written so frequently in the past few sections that the reader probably is able to set up his or her own equations for types of statically indeterminate beams and trusses not previously encountered. Nevertheless, one more group of equations is developed here—those necessary for the analysis of a truss that is statically indeterminate internally and externally. For the following discussion the truss of Figure 16.4, which has two redundant members and one redundant reaction component, will be considered.

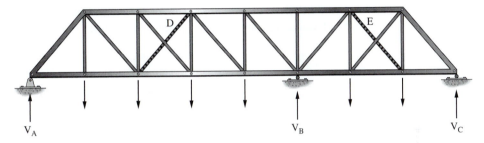

Figure 16.4

The diagonals lettered D and E and the interior reaction V_B are removed from the truss, which leaves a statically determinate structure. The openings of the gaps in the cut members and the deflections at the interior support due to the external loads may be computed from the following:

$$\Delta_B = \sum \frac{F'\mu_B L}{AE} \quad \Delta_D = \sum \frac{F'\mu_D L}{AE} \quad \Delta_E = \sum \frac{F'\mu_E L}{AE}$$

Placing a unit load at the interior support will cause deflections at the gaps in the cut members as well as at the point of application.

$$\Delta_{bb} = \sum \frac{\mu_B^2 L}{AE} \quad \Delta_{db} = \sum \frac{\mu_B \mu_D L}{AE} \quad \Delta_{eb} = \sum \frac{\mu_B \mu_E L}{AE}$$

Delaware River Turnpike Bridge. (Courtesy of USX Corporation.)

Replacing member D and assuming it to have a positive unit tensile force will cause the following deflections:

$$\Delta_{bd} = \sum \frac{\mu_B \mu_D L}{AE} \quad \Delta_{dd} = \sum \frac{\mu_D^2 L}{AE} \quad \Delta_{ed} = \sum \frac{\mu_E \mu_D L}{AE}$$

Similarly, replacement of member E with a force of +1 will cause these deflections:

$$\Delta_{be} = \sum \frac{\mu_B \mu_E L}{AE} \quad \Delta_{de} = \sum \frac{\mu_D \mu_E L}{AE} \quad \Delta_{ee} = \sum \frac{\mu_E^2 L}{AE}$$

Computation of these sets of deflections permits the calculation of the numerical values of the redundants, because the total deflection at each may be equated to zero.

$$\Delta_B + V_B \Delta_{bb} + X_D \Delta_{bd} + X_E \Delta_{be} = 0$$
$$\Delta_D + V_B \Delta_{db} + X_D \Delta_{dd} + X_E \Delta_{de} = 0$$
$$\Delta_E + V_B \Delta_{eb} + X_D \Delta_{ed} + X_E \Delta_{ee} = 0$$

Nothing new is involved in the solution of this type of problem, and space is not taken for the lengthy calculations necessary for an illustrative example.

16.4 TEMPERATURE CHANGES, SHRINKAGE, FABRICATION ERRORS, AND SO ON

Structures are subject to deformations due not only to external loads but also to temperature changes, support settlements, inaccuracies in fabrication dimensions, shrinkage in reinforced concrete members caused by drying and plastic flow, and so forth. Such deformations in statically indeterminate structures can cause the development of large additional forces in the members. For an illustration it is assumed that the top-chord members of the truss of Figure 16.5 may be exposed to the sun much more than are the other members. As a

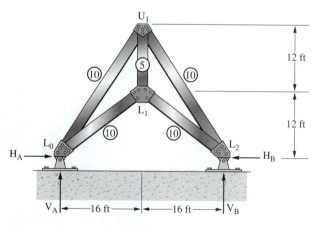

Figure 16.5 Truss used for discussion of effects of temperature changes

result, on a hot sunny day they may have much higher temperatures than the other members and the member forces may undergo some appreciable changes.

Problems such as these may be handled exactly as were the previous problems of this chapter. The changes in the length of each of the members due to temperature are computed. (These values, which correspond to the F′L/AE values, each equal the temperature change times the coefficient of expansion of the material times the member length.) The redundant is removed from the structure, a unit load is placed at the support in the direction of the redundant, and the μ forces are computed. Then the values $\sum(F'\mu L/AE)$ and $\sum(\mu^2 L/AE)$ are determined in the same units and the usual deflection expression is written. Such a problem is illustrated in Example 16.3

EXAMPLE 16.3

The top-chord members of the statically indeterminate truss of Figure 16.5 are assumed to increase in temperature by 60°F. If $E = 29 \times 10^6$ psi and the coefficient of linear temperature expansion is 0.0000065/°F, determine the forces induced in each of the truss members. The circled figures are areas in square inches.

Solution.   Assume H_B is the redundant and compute the μ forces.

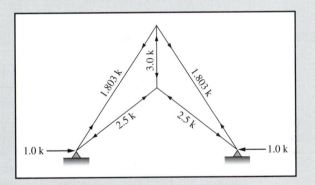

TABLE 16.3

Member	L (in.)	A (in.²)	$\frac{L}{A}$	μ	$\frac{\mu^2 L}{AE}$	$\Delta L = \Delta t \cdot \text{coeff} \cdot L$ (equivalent to $\frac{F'L}{AE}$)	$\mu(\Delta L) = \frac{F'\mu L}{AE}$
L_0L_1	240	10	24	−2.5	+150	—	—
L_1L_2	240	10	24	−2.5	+150	—	—
L_0U_1	346	10	34.6	+1.803	+112.48	(60)(0.0000065)(346) = 0.1349	0.2433
U_1L_2	346	10	34.6	+1.803	+112.48	(60)(0.0000065)(346) = 0.1349	0.2433
U_0L_1	144	5	28.8	−3.0	+259.2	—	—
					$\Sigma = +\dfrac{784.16}{E}$		$\Sigma = 0.4866$

$$\Delta_{bb} = + \frac{(784.16)(1000)}{29 \times 10^6} = +0.02704 \text{ in.}$$

$$\Delta_B = +0.4866 \text{ in.}$$

$$\Delta_B + H_B \Delta_{bb} = 0$$

$$H_B = -\frac{0.4866}{0.02704} = -18.02 \text{ k} \rightarrow$$

The final forces in the truss members due to the temperature change are as follows:

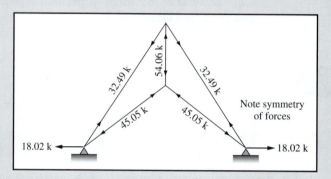

16.5 CASTIGLIANO'S FIRST THEOREM

Castigliano's first theorem, commonly known as the *method of least work*, has played an important role in the development of structural analysis through the years and is occasionally used today. It is closely related to the method of consistent distortions discussed in Chapter 15, and is rather effective for the analysis of statically indeterminate structures, particularly trusses and composite structures. (Composite structures are defined here as structures that have some members with axial stresses only and other members with axial stresses and bending stresses.) Although applicable to beams and frames, other methods such as the moment-distribution method (discussed in Chaps. 20 and 21) are usually more satisfactory. The method of least work has the disadvantage that it is not applicable in its usual form to forces caused by displacements due to temperature changes, support settlements, and fabrication errors.

In Chapter 13 it was shown that the first partial derivative of the total internal work with respect to a load P (real or imaginary) applied at a point in a structure equaled the deflection in the direction of P. For this discussion, the continuous beam of Figure 16.6 and its vertical reaction at support B, V_B, are considered.

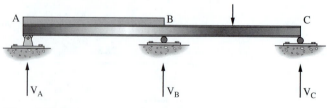

Figure 16.6

If the first partial derivative of the work in this beam is taken with respect to the reaction V_B, the deflection at B will be obtained but that deflection is zero.

$$\frac{\partial W}{\partial P} = 0$$

This is a statement of Castigliano's first theorem. Equations of this type can be written for each point of constraint of a statically indeterminate structure. A structure will deform in a manner consistent with its physical limitations or so that the internal work of deformation will be at a minimum.

The columns and girders meeting at a joint in a building will all deflect the same amount: the smallest possible value. Neglecting the effect of the other ends of these members, it can be seen that each member does no more work than necessary, and the total work performed by all of the members at the joint is the least possible.

From the foregoing discussion the theorem of least work may be stated: *The internal work accomplished by each member or each portion of a statically indeterminate structure subjected to a set of external loads is the least possible amount necessary to maintain equilibrium in supporting the loads.*

On some occasions (particularly for continuous beams and frames), the least-work method is very laborious to apply. Consequently, readers often express rather strong opinions as to what they think of the term "least work."

To analyze a statically indeterminate structure with Castigliano's theorem, certain members are assumed to be the redundants and are considered removed from the structure. The removal of the members must be sufficient to leave a statically determinate and stable base structure. The F' forces in the structure are determined for the external loads; the redundants are replaced as loads X_1, X_2, and so on; and the forces they cause are determined.

The total internal work of deformation may be set up in terms of the F' forces and the forces caused by the redundant loads. The result is differentiated successively with respect to the redundants. The derivatives are made equal to zero in order to determine the values of the redundants.

Examples 16.4 to 16.9 illustrate the analysis of statically indeterminate structures by least work. Although the least-work and consistent-distortion methods are the most general methods for analyzing various types of statically indeterminate structures, they are not frequently used today because other methods such as moment distribution and computer programs are so much simpler to apply.

EXAMPLE 16.4

Determine the reaction at support C in the beam of Figure 16.7 by least work: E and I are constant.

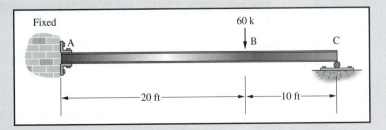

Figure 16.7

Solution.

The reaction at C is assumed to be V_C and the other reactions are determined as follows:

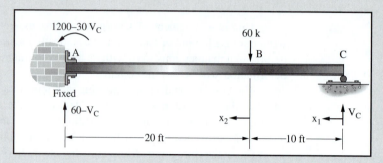

TABLE 16.4

Section	M	$\dfrac{\partial M}{\partial V_c}$	$M\dfrac{\partial M}{\partial V_c}$	$\displaystyle\int M\left(\dfrac{\partial M}{\partial V_c}\right)\dfrac{dx}{EI}=0$
C to B	$V_c x_1$	x_1	$V_c x_1^2$	$\dfrac{I}{EI}\displaystyle\int_0^{10}\left(V_c x_1^2\right)dx_1$
B to A	$V_c x_2 + 10V_c$ $-60x_2$	$x_2 + 10$	$V_c x_2^2 + 20V_c x_2$ $-60x_2^2 +$ $100V_c - 600x_2$	$\dfrac{1}{EI}\displaystyle\int_0^{20}\left(V_c x_2^2 + 20V_c x_2\right.$ $-60x_2^2 + 100V_c$ $\left.-600x_2\right)dx_2$
Σ				$9000V_c - 280.000 = 0$

$$V_c = 31.1 \text{ k} \uparrow \qquad \blacksquare$$

EXAMPLE 16.5

Determine the value of the reaction at support C, Figure 16.8 by the method of least work.

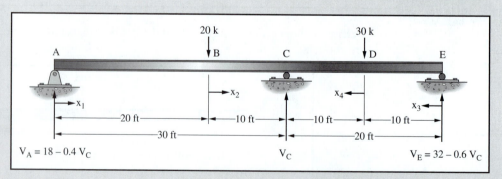

Figure 16.8

Solution.

TABLE 16.5

Section	M	$\dfrac{\partial M}{\partial V_c}$	$\int M\left(\dfrac{\partial M}{\partial V_B}\right)\dfrac{dx}{EI}=0$
A to B	$18x_1 - 0.4V_c x_1$	$-0.4x_1$	$\displaystyle\int_0^{20}(-7.2x_1^2 + 0.16V_c x_1^2)dx_1$
B to C	$-2x_2 - 0.4V_c x_2 - 8V_c + 360$	$-0.4x_2 - 8$	$\displaystyle\int_0^{10}(+0.8x_2^2 + 0.16V_c x_2^2 + 6.4V_c x_2 - 128x_2 + 64V_c - 2880)\,dx_2$
E to D	$32x_3 - 0.6V_c x_3$	$-0.6x_3$	$\displaystyle\int_0^{10}(-19.2x_3^2 + 0.36V_c x_3^2)\,dx_3$
D to C	$2x_4 - 0.6V_c x_4 - 6V_c + 320$	$-0.6x_4 - 6$	$\displaystyle\int_0^{10}(-1.2x_4^2 + 0.36V_c x_4^2 + 7.2V_c x_4 - 204x_4 + 36V_c - 1920)\,dx_4$

By integrating the $\int M(\partial M/\partial V_B)(dx/EI)$ expressions and substituting the values of the proper limits, the result for the entire beam is

$$-90,333 + 2400.3V_C = 0$$
$$V_C = +37.6 \text{ k} \uparrow \quad \blacksquare$$

EXAMPLE 16.6

Determine the force in member CD of the truss shown in Figure 16.9. Circled values are areas in square inches. E is constant.

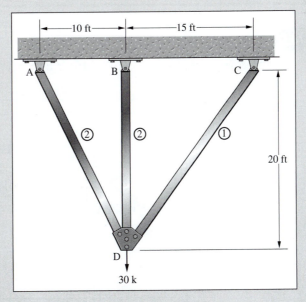

Figure 16.9

Solution.
Member CD is assumed to be the redundant and is assigned a force of T. The other forces are determined by joints with the following results.

$$\frac{\partial W}{\partial T} = \frac{\partial}{\partial T}\sum\frac{F^2 L}{2AE} = \sum F\frac{\partial F}{\partial T}\frac{L}{AE}$$

TABLE 16.6

Member	L (in.)	A (in.²)	$\frac{L}{A}$	F	$\frac{\partial F}{\partial T}$	$F\left(\frac{\partial F}{\partial T}\right)\frac{L}{AE}$
AD	268	2	134	+1.34T	+1.34	+240T
BD	240	2	120	−2T + 30	−2	+480T − 7200
CD	300	1	300	+T	+1	+300T
Σ						1020T − 7200 = 0

$$T = +7.06 \text{ k}$$

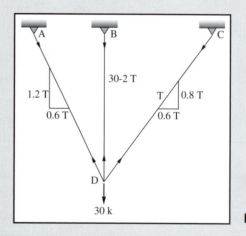

Analyze the truss of Figure 16.10 by the least-work method. Circled numbers are member areas, in square inches.

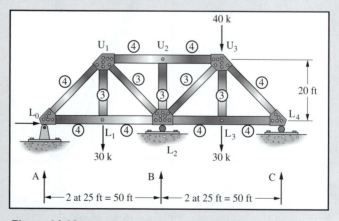

Figure 16.10

Solution.
Remove the center support and compute the F′ forces.

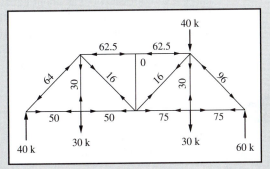

Replace the center support and determine its effect on member forces in terms of V_B.

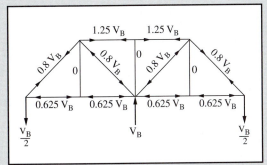

Compute the values in Table 16.7.

$$E\frac{\partial W}{\partial V_B} = \frac{FL}{A}\frac{\partial F}{\partial V_B}$$

$$-35.738 + 638.2V_B = 0$$

$$V_B = +56 \text{ k} \uparrow$$

TABLE 16.7

Member	L (in.)	A (in.)^2	$\dfrac{L}{A}$	F(kips)	$\dfrac{FL}{A}$	$\dfrac{\partial F}{\partial V_B}$	$\dfrac{FL}{A}\dfrac{\partial F}{\partial V_B}$
L_0L_1	300	4	75	$+50 \;\; -0.625V_B$	$+3750 - 46.9V_B$	-0.625	$-2345 + 29.3V_B$
L_1L_2	300	4	75	$+50 \;\; -0.625V_B$	$+3750 - 46.9V_B$	-0.625	$-2345 + 29.3V_B$
L_2L_3	300	4	75	$+75 \;\; -0.625V_B$	$+5625 - 46.9V_B$	-0.625	$-3520 + 29.3V_B$
L_3L_4	300	4	75	$+75 \;\; -0.625V_B$	$+5625 - 46.9V_B$	-0.625	$-3520 + 29.3V_B$
L_0U_1	384	4	96	$-64 \;\; +0.8V_B$	$-6144 + 76.8V_B$	$+0.8$	$-4915 + 61.4V_B$
U_1U_2	300	4	75	$-62.5 + 1.25V_B$	$-4687 + 93.8V_B$	$+1.25$	$-5860 + 117.2V_B$
U_2U_3	300	4	75	$-62.5 + 1.25V_B$	$-4687 + 93.8V_B$	$+1.25$	$-5860 + 117.2V_B$
U_3L_4	384	4	96	$-96 \;\; +0.8V_B$	$-9216 + 76.8V_B$	$+0.8$	$-7373 + 61.4V_B$
U_1L_1	240	3	80	$+30 \;\; +0$	$+2400 + 0$	0	$0 + 0$
U_1L_2	384	3	128	$+16 \;\; -0.8V_B$	$+2048 - 102.4V_B$	-0.8	$-1638 + 81.9V_B$
U_2L_2	240	3	80	$0 \;\; +0$	$0 + 0$	0	$0 + 0$
L_2U_3	384	3	128	$-16 \;\; -0.8V_B$	$-2048 - 102.4V_B$	-0.8	$+1638 + 81.9V_B$
U_3L_3	240	3	80	$+30 \;\; +0$	$+2400 + 0$	0	$0 + 0$
Σ							$-35,738$ $+638.2V_B$

It should be obvious from the preceding example problems that the amount of work involved in the analysis of statically indeterminate trusses by the methods of least work and consistent distortions is about equal.

Least work is particularly useful for analyzing composite structures, such as those to be considered in Examples 16.8 and 16.9. In these types of structures both bending and truss action take place. The reader will be convinced of the advantage of least work for the analysis of composite structures if he or she attempts to solve the following two problems by consistent distortions.

EXAMPLE 16.8

Using the least-work procedure, calculate the force in the steel rod of the composite structure shown in Figure 16.11.

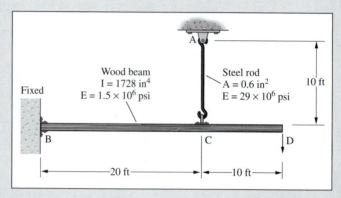

Figure 16.11

Solution.

The rod is assumed to be the redundant with a force of T and the values of $\Sigma F(\partial F/\partial T)(L/AE)$ and $\int M(\partial M/\partial T)(dx/EI)$ are computed. Relative values of E of 29 and 1.5 are used for the steel and wood, respectively.

TABLE 16.8

Member	L (in.)	A (in.²)	$\dfrac{L}{A}$	$\dfrac{L}{AE}$	F (kip)	$\dfrac{\partial F}{\partial T}$	$F\left(\dfrac{\partial F}{\partial T}\right)\dfrac{L}{AE}$
AC	120	0.6	200	6.90	+T	+1	+6.90T
Σ							+6.90T

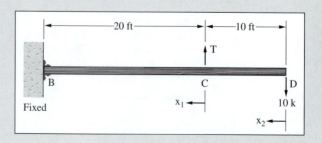

TABLE 16.9

Section	M	$\dfrac{\partial M}{\partial T}$	$M\dfrac{\partial M}{\partial T}$	$\int M\left(\dfrac{\partial M}{\partial T}\right)\dfrac{dx}{EI}$
D to C	$-10x$	0	0	0
C to B	$Tx - 10x$ $- 100$	x	$Tx^2 - 10x^2 - 100x$	$\dfrac{\displaystyle\int_0^{20} (Tx^2 - 10x^2 - 100x)\ dx}{(1.5)(1728)}$
Σ				$1.03T - 18$

To change the value of $1.03T - 18$ to inches it is necessary to multiply it by 1728×1000 and divide by 1×10^6 as 29 was used for E instead of 29×10^6. To change the value $6.90T$ to inches it is necessary to multiply it by 1000 and divide by 1×10^6. The only difference in the two conversions is the 1728; therefore the expression can be written as follows to change them to the same units.

$$\int M\left(\frac{\partial M}{\partial T}\right)\frac{dx}{EI} + \sum F\left(\frac{\partial F}{\partial T}\right)\frac{1}{AE} = (1.03T - 18)(1728) + 6.90T = 0$$

$$T = +17.4 \text{ k} \quad \blacksquare$$

EXAMPLE 16.9

Find the forces in all members of the king post truss shown in Figure 16.12.

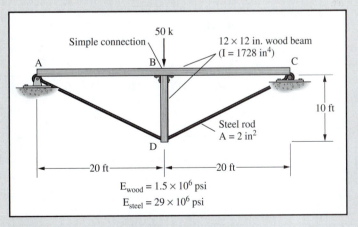

Figure 16.12

Solution.
By letting BD be the redundant with a force of F, the deflection of the beam at B is found in terms of F.

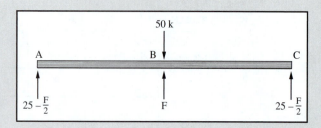

From A to B and from C to B:

$$M = 25x - \frac{F}{2}x \qquad \frac{\partial M}{\partial F} = -\frac{x}{2}$$

$$\int M\frac{\partial M}{\partial F}\frac{dx}{EI} = 2\int_0^{20}\frac{(-12.5x^2 + 0.25Fx^2)\,dx}{EI}$$

$$\frac{2(-12.5x^3/3 + 0.25Fx^3/3)_0^{20}}{EI} = \frac{-66,600 + 1333F}{EI}$$

$$\frac{-66,600 + 1333F}{1.5} = -44,400 + 889F$$

Determine the forces in the various members in terms of the unknown force F.

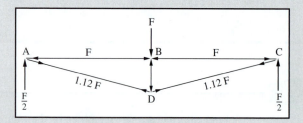

TABLE 16.10

Member	L (in.)	A (in.²)	E	$\dfrac{L}{AE}$	F′ (kips)	$\dfrac{F'L}{AE}$	$\dfrac{\partial F'}{\partial F}$	$\dfrac{\partial F'}{\partial F} \times \dfrac{F'L}{AE}$
AC	480	144	1.5×10^6	2.22	$-F$	$-2.22F$	-1	$+2.22F$
BD	120	144	1.5×10^6	0.555	$-F$	$-0.555F$	-1	$+0.555F$
AD	268	2	29×10^6	4.62	$+1.12F$	$+5.19F$	$+1.12$	$+5.80F$
DC	268	2	29×10^6	4.62	$+1.12F$	$+5.19F$	$+1.12$	$+5.80F$
Σ								14.375F

$$-44,400 + 889F + 14.375F = 0$$

By changing these values to equivalent units and solving for F, F = 49.2 k
Final forces:

$$AC = -(1)(49.2) = -49.2 \text{ k} \quad AD = +(1.12)(49.2) = +55.1 \text{ k}$$

$$BD = -(1)(49.2) = -49.2 \text{ k} \quad DC = +(1.12)(49.2) = +55.1 \text{ k} \quad \blacksquare$$

Castigliano's first theorem is not as easy to use as the second theorem, or, for that matter, as easy as other energy methods such as virtual work, because strain energy is more easily formulated in terms of load than in terms of displacement. However, for some structural responses, and for the development of system matrices when using matrix methods (which we will study beginning in Chapter 22) Castigliano's first theorem is the easiest method to use. This method is particularly useful for the evaluation of structures with nonlinear response.

16.6 ANALYSIS USING COMPUTERS

As should be quite apparent to you by now, the analysis of statically indeterminate structural systems is very tedious, especially when structures of any significant size are involved. In many cases, hand solutions are not feasible even if computational software is used. Computer programs are usually employed in engineering offices when analyzing large structures. Analysis of statically indeterminate structures using a computer is demonstrated in Example 16.10 where SABLE32 is used to analyze a statically indeterminate truss. The procedure used is identical with that used for statically determinate trusses.

EXAMPLE 16.10

Determine the forces in the truss shown in Figure 16.13 using the computer program SABLE32. This is the same truss that was analyzed in Example 16.2. The number by each member is the area of that member.

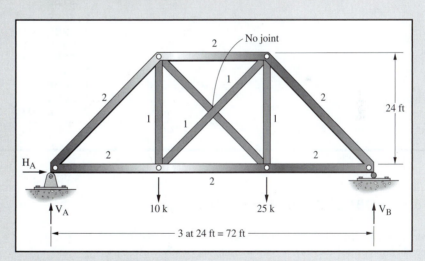

Figure 16.13

Solution.
This truss was modeled in the same manner as the trusses in the previous chapters. The flexural degrees of freedom at the ends of each member were released as were the torsional degrees of freedom at the beginning of each member. The resulting structural model is shown.

Arrows are shown on two of the members to indicate the i and j ends as used in the computer solution. The i end is located at the start of the arrow while the j end is at the far end.

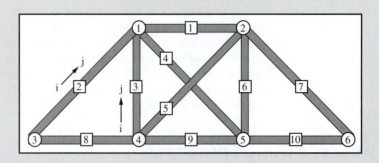

Member end forces

Beam	Case	End	Axial	Shear-Y	Moment-Z
1	1	i	2.154E+01	0.000E+00	0.000E+00
		j	−2.154E+01	0.000E+00	0.000E+00
2	1	i	2.121E+01	0.000E+00	0.000E+00
		j	−2.121E+01	0.000E+00	0.000E+00
3	1	i	−8.457E+00	0.000E+00	0.000E+00
		j	8.457E+00	0.000E+00	0.000E+00
4	1	i	−9.253E+00	0.000E+00	0.000E+00
		j	9.253E+00	0.000E+00	0.000E+00
5	1	i	−2.182E+00	0.000E+00	0.000E+00
		j	2.182E+00	0.000E+00	0.000E+00
6	1	i	−1.846E+01	0.000E+00	0.000E+00
		j	1.846E+01	0.000E+00	0.000E+00
7	1	i	2.828E+01	0.000E+00	0.000E+00
		j	−2.828E+01	0.000E+00	0.000E+00
8	1	i	−1.500E+01	0.000E+00	0.000E+00
		j	1.500E+01	0.000E+00	0.000E+00
9	1	i	−1.346E+01	0.000E+00	0.000E+00
		j	1.346E+01	0.000E+00	0.000E+00
10	1	i	−2.000E+01	0.000E+00	0.000E+00
		j	2.000E+01	0.000E+00	0.000E+00

16.7 PROBLEMS FOR SOLUTION

For Problems 16.1 to 16.13 determine the reactions and member forces for the trusses shown using the method of consistent distortions.

16.1 (*Ans.* $L_1L_2 = +25.93$ k, $U_1L_2 = −36.66$ k)

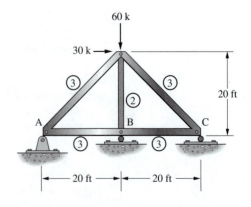

16.2

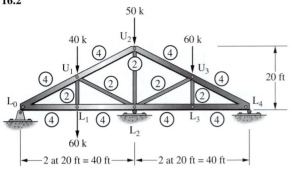

16.6 All areas are equal.

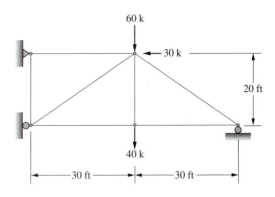

16.3 All areas are equal. (*Ans.* $V_R = 45.5\,\text{k}\uparrow$, $U_0U_1 = +8.4\,\text{k}$, $U_2L_2 = -26.8\text{k}$)

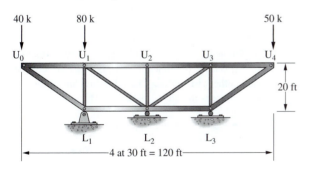

16.7 (*Ans.* $V_L = 49.0\,\text{k}\uparrow$, $U_0U_1 = -61.25\,\text{k}$, $U_2L_3 = -113.6\ \text{k}$)

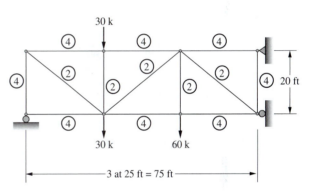

16.4 All areas are equal.

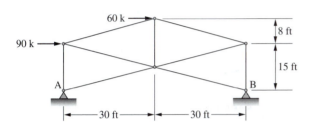

16.5 (*Ans.* $V_B = 64.89\,\text{kN}\uparrow$, $L_0L_1 = +35.05\,\text{kN}$)

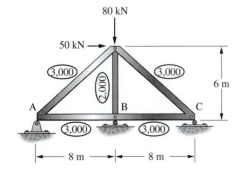

16.8

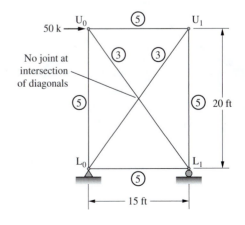

16.9   (*Ans.* $L_1U_2 = -144.22$ k, $L_0L_1 = -49.05$) All areas are equal.

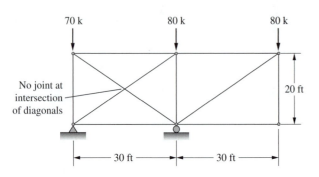

16.10

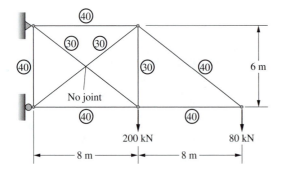

16.11   (*Ans.* $H_L = 19.5$ k $\rightarrow$, $M_0U_1 = +23.44$ k, $M_2L_2 = -13.7$ k)

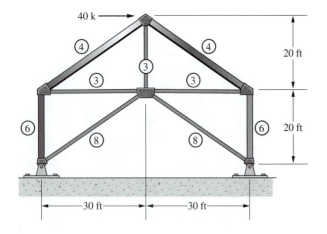

16.12

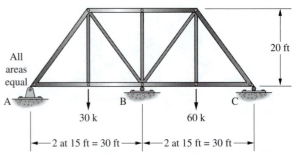

16.13   (*Ans.* $H_B = 18.19$ k $\rightarrow$, $U_0U_1 = -36.38$ k, $L_1L_2 = +21.83$ k)

Determine the forces in all the members of the truss shown in the accompanying illustration if the top-chord members, U_0U_1 and U_1U_2, have an increase in temperature of 75°F and no change in temperature in other members. Coefficient of linear expansion $\epsilon = 0.0000065$. $E = 29 \times 10^6$ psi.

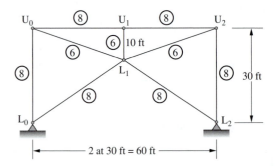

16.14   For Problems 16.14 to 16.22 analyze the structures using Castigliano's first theorem. E and I are constant unless otherwise noted. Circled values on trusses are member areas in square inches.

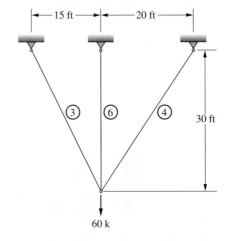

16.15 (*Ans.* $L_0L_2 = +25.2$ k, $U_1M_1 = +6.6$ k) All areas are equal.

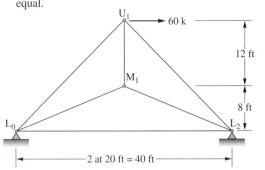

16.16

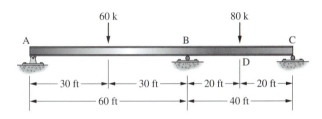

16.17 (*Ans.* $V_A = 49.0$ k ↑, $V_B = 105.6$ k ↑)

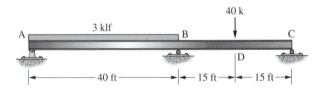

16.18

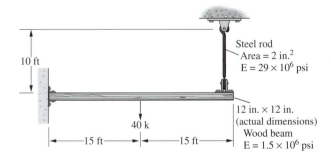

16.19 (*Ans.* $U_0U_1 = +36.1$ k, $U_2L_2 = +6.48$ k)

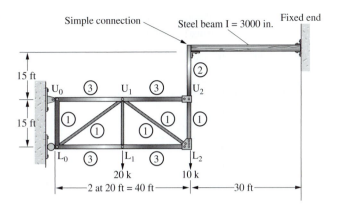

16.20

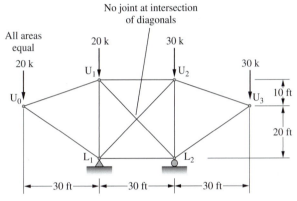

16.21 (*Ans.* $H_L = 31.27$ k, $U_0L_0 = +38.31$ k, $U_0L_1 = -47.88$ k)

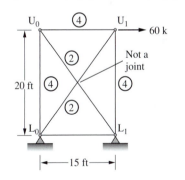

16.22

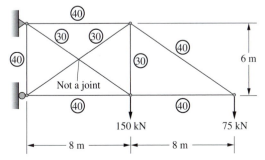

For Problems 16.23 through 16.30 use SABLE32 or SAP 2000 and repeat the following problems.

16.23 Problem 16.1 (*Ans.* $L_0U_1 = +5.68$ k, $L_1L_2 = +25.98$ k)

16.24 Problem 16.2

16.25 Problem 16.5 (*Ans.* $L_0L_1 = +35.07$ k, $U_1L_2 = -43.84$ k)

16.26 Problem 16.15

16.27 Problem 16.20 (*Ans.* $U_1L_1 = -14.761$ k, $L_1U_2 = -2.692$ k, $L_1L_2 = -18.10$ k)

16.28 Problem 16.22

16.29 (*Ans.* $L_0U_1 = -13.25$ k, $U_2L_2 = -6.983$ k, $L_2L_3 = +1.945$)

16.30

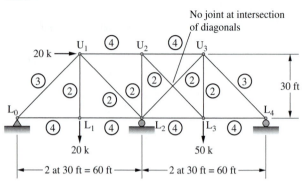

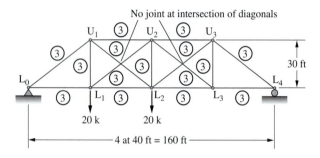

Chapter 17

Influence Lines for Statically Indeterminate Structures

17.1 INFLUENCE LINES FOR STATICALLY INDETERMINATE BEAMS

The uses of influence lines for statically indeterminate structures are the same as those for statically determinate structures. They enable the designer to locate the critical positions for placing live loads and to compute forces for various positions of the loads. Influence lines for statically indeterminate structures are not as simple to draw as they are for statically determinate structures. For the latter case it is possible to compute the ordinates for a few controlling points by statics and connect those values with a set of straight lines. Unfortunately, influence lines for continuous structures require the computation of ordinates at a large number of points because the diagrams are either curved or made up of a series of chords. The chord-shaped diagram occurs where loads can only be transferred to the structure at intervals.

The problem of preparing the diagrams is not as difficult as the preceding paragraph seems to indicate because a large percentage of the work may be eliminated by applying Maxwell's law of reciprocal deflections. The preparation of an influence line for the interior reaction of the two-span beam of Figure 17.1 is considered in the following paragraphs.

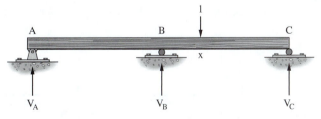

Figure 17.1

The procedure for calculating V_B has been to remove it from the beam and then compute Δ_B and Δ_{bb} and substitute their values in the usual formula. The same procedure may be used in drawing an influence line for V_B. A unit load is placed at some point x causing Δ_B to equal Δ_{bx}, from which the following expression is written:

$$V_B = -\frac{\Delta_B}{\Delta_{bb}} = -\frac{\Delta_{bx}}{\Delta_{bb}}$$

At first glance it appears that the unit load will have to be placed at numerous points on the beam and the value of Δ_{bx} laboriously computed for each. A study of the deflections caused by a unit load at point x, however, proves these computations to be unnecessary. By Maxwell's law the deflection at B due to a unit load at x(Δ_{bx}) is identical with the deflection at x due to a unit load at B(Δ_{xb}). The expression for V_B becomes

$$V_B = -\frac{\Delta_{xb}}{\Delta_{bb}}$$

It is now evident that the unit load need only be placed at B, and the deflections at various points across the beam computed. Dividing each of these values by Δ_{bb} gives the ordinates for the influence line. If a deflection curve is plotted for the beam for a unit load at B (support B being removed), an influence line for V_B may be obtained by dividing each of the deflection ordinates by Δ_{bb}. Another way of expressing this principle is as follows: If a unit deflection is caused at a support for which the influence line is desired, the beam will draw its own influence line because the deflection at any point in the beam is the ordinate of the influence line at that point for the reaction in question.

Maxwell's presentation of his theorem in 1864 was so brief that its value was not fully appreciated until 1886 when Heinrich Müller-Breslau clearly showed its true worth as described in the preceding paragraph.[1] Müller-Breslau's principle may be stated in detail as follows: **The deflected shape of a structure represents to some scale the influence line for a function such as stress, shear, moment, or reaction component if the function is allowed to act through a unit displacement.** This principle, which is applicable to statically determinate and indeterminate beams, frames, and trusses, is proved in the next section of this chapter.

The influence line for the reaction at the interior support of a two-span beam is presented in Example 17.1. Influence lines also are shown for the end reactions, the values for ordinates having been obtained by statics from those computed for the interior reaction. The conjugate-beam procedure is one excellent method of determining the beam deflections necessary for preparing the diagrams.

EXAMPLE 17.1

Draw influence lines for reactions at each support of the structure shown in Figure 17.2.

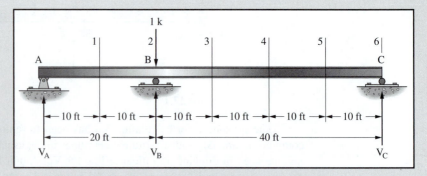

Figure 17.2

[1]J. S. Kinney, *Indeterminate Structural Analysis* (Reading, Mass.: Addison-Wesley, 1957), 14.

Solution.   Remove V_B, place a unit load at B, and compute the deflections caused at 10-ft intervals by the conjugate-beam method:

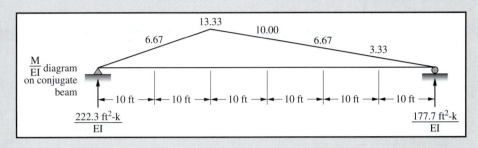

$$\Delta_1 = (222.3)(10) - \left(\tfrac{1}{2}\right)(10)(6.67)(3.33) = 2112$$

$$\Delta_2 = (222.3)(20) - \left(\tfrac{1}{2}\right)(20)(13.33)(6.67) = 3550$$

$$\Delta_3 = (177.7)(30) - \left(\tfrac{1}{2}\right)(30)(10.00)(10.00) = 3831$$

$$\Delta_4 = (177.7)(20) - \left(\tfrac{1}{2}\right)(20)(6.67)(6.67) = 3109$$

$$\Delta_5 = (177.7)(10) - \left(\tfrac{1}{2}\right)(10)(3.33)(3.33) = 1722$$

Noting $\Delta_{bb} = \Delta_2$, the value of the influence-line ordinates for V_B are found by dividing each deflection by Δ_2.

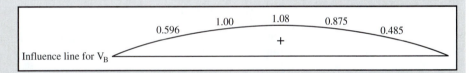

Having the values of V_B for various positions of the unit load, the values of V_A and V_C for each load position can be determined by statics with the following results:

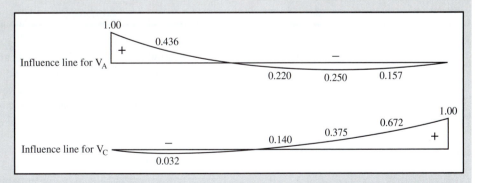

For checking results, support C is removed, a unit load is placed there, and the resulting deflections are computed at 10-ft intervals.

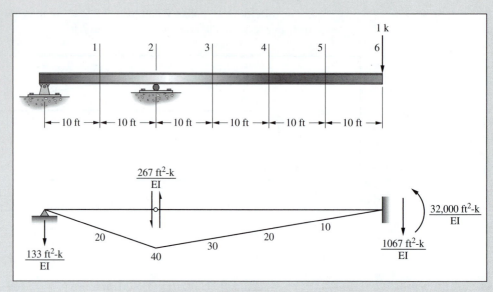

$$\Delta_1 = -(133)(10) + \left(\tfrac{1}{2}\right)(10)(20)(3.33) = -1000$$
$$\Delta_2 = -(133)(20) + \left(\tfrac{1}{2}\right)(20)(40)(6.67) = 0$$
$$\Delta_3 = +32{,}000 - (1067)(30) + \left(\tfrac{1}{2}\right)(30)(30)(10) = +4500$$
$$\Delta_4 = +32{,}000 - (1067)(20) + \left(\tfrac{1}{2}\right)(20)(20)(6.67) = +12{,}000$$
$$\Delta_5 = +32{,}000 - (1067)(10) + \left(\tfrac{1}{2}\right)(10)(10)(3.33) = +21{,}500$$
$$\Delta_6 = +32{,}000$$

Since $\Delta_{cc} = \Delta_6$, the ordinates of the influence line for V_C are determined by dividing each deflection by Δ_6.

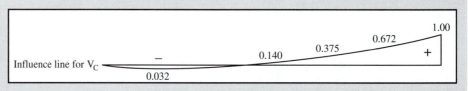

The next problem is to draw the influence lines for beams continuous over three spans, which have two redundants. For this discussion the beam of Figure 17.3 is considered, and the reactions V_B and V_C are assumed to be the redundants.

It will be necessary to remove the redundants and compute the deflections at various sections in the beam for a unit load at B and also for a unit load at C.

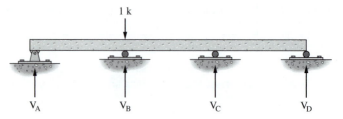

Figure 17.3

By Maxwell's law a unit load at any point x causes a deflection at $B(\Delta_{bx})$ equal to the deflection at x due to a unit load at $B(\Delta_{xb})$. Similarly, $\Delta_{cx} = \Delta_{xc}$. After computing Δ_{xb} and Δ_{xc} at the several sections, their values at each section may be substituted into the following simultaneous equations, whose solution will yield the values of V_B and V_C.

$$\Delta_{xb} + V_B\Delta_{bb} + V_C\Delta_{bc} = 0$$
$$\Delta_{xc} + V_B\Delta_{cb} + V_C\Delta_{cc} = 0$$

The simultaneous equations are solved quickly, even though a large number of ordinates are being computed, because the only variables in the equations are Δ_{xb} and Δ_{xc}. After the influence lines are prepared for the redundant reactions of a beam, the ordinates for any other function (moment, shear, and so on) can be determined by statics. Example 17.2 illustrates the calculations necessary for preparing influence lines for several functions of a three-span continuous beam.

EXAMPLE 17.2

Draw influence lines for V_B, V_C, V_D, M_7, and shear at section 6 in Figure 17.4.

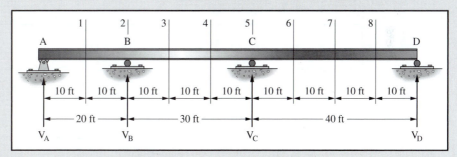

Figure 17.4

Solution.   Remove V_B and V_C, place a unit load at B, and load the conjugate beam with the M/EI diagram.

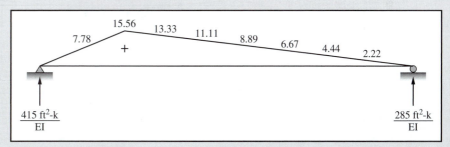

Place a unit load at C and load the conjugate beam with the M/EI diagram.

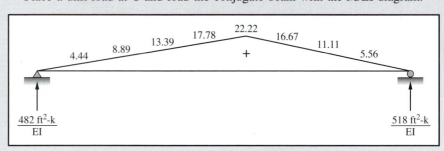

Compute the values of Δ_{xb} and Δ_{xc}, from which V_B and V_C are obtained by solving the simultaneous equations. Ordinates for M_7, V_D, and shear at section 6 are obtained by statics and shown in Table 17.1.

TABLE 17.1

Section	Δ_{xb}	Δ_{xc}	V_B	V_C	V_D	M_7	V_6
1	4020	4746	+0.646	−0.0735	+0.008	+0.160	−0.008
2	7255 = Δ_{bb}	9040 = Δ_{bc}	+1.00	0	0	0	0
3	9100	12.460	+0.855	+0.320	−0.0344	−0.688	+0.0344
4	9630	14.540	+0.432	+0.720	−0.0513	−1.026	+0.0513
5	9040 = Δ_{cb}	14.825 = Δ_{cc}	0	+1.00	0	0	0
6	7550	13.040	−0.227	+1.02	+0.1504	+3.008	−0.1504
							+0.8496
7	5404	9618	−0.257	+0.805	+0.388	+7.76	+0.612
8	2813	5088	−0.159	+0.440	+0.688	+3.68	+0.316

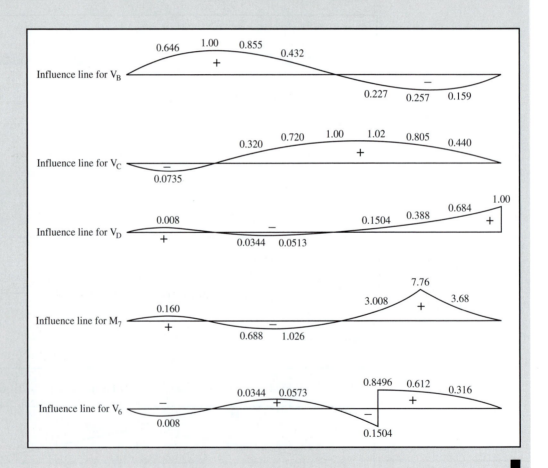

Influence lines for the reactions, shears, moments, and so on for frames can be prepared exactly as they are for the statically indeterminate beams just considered. Space is not taken to present such calculations. The next section of this chapter, which is concerned with the preparation of qualitative influence lines, will show the reader how to place live loads in frames so as to cause maximum values.

17.2 QUALITATIVE INFLUENCE LINES

Müller-Breslau's principle is based upon Castigliano's theorem of least work, which was presented in Chapter 16 of this text. This theorem can be expressed as follows: **when a displacement is induced into a structure, the total virtual work done by all the active forces equals zero.**

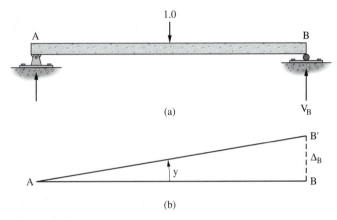

Figure 17.5

Proof of Müller-Breslau's principle can be made by considering the beam of Figure 17.5(a) when it is subjected to a moving unit load. To determine the magnitude of the reaction V_B, we can remove the support at B and allow V_B to move through a small distance Δ_B, as shown in part (b) of the figure. Notice the beam's position is now represented by the line AB′ in the figure and the unit load has been moved through the distance y.

Writing the virtual work expression for the active forces on the beam

$$(V_B)(\Delta_B) = (1.0)(y)$$
$$V_B = \frac{y}{\Delta_B}$$

Should Δ_B be given a unit value, V_B will equal y.

$$V_B = \frac{y}{1.0} = y$$

You can now see that y is the ordinate of the deflected beam at the unit load and also is the value of the right-hand reaction V_B due to that moving unit load. Therefore, the deflected beam position AB′ represents the influence line for V_B. Similar proofs can be developed for the influence lines for other functions of a structure, such as shear and moments.

Müller-Breslau's principle is of such importance that space is taken to emphasize its value. The shape of the usual influence line needed for continuous structures is so simple to obtain from his principle that in many situations it is unnecessary to perform the rather tedious labor needed to compute the numerical values of the ordinates. It is possible to sketch the diagram roughly with sufficient accuracy to locate the critical positions for live load for various functions of the structure. This possibility is of particular importance for building frames, as will be illustrated in subsequent paragraphs.

If the influence line is desired for the left reaction of the continuous beam of Figure 17.6(a), its general shape can be determined by letting the reaction act upward through a unit distance as shown in Figure 17.6(b) of the figure. If the left end of the beam were pushed up, the beam would take the shape shown. This distorted shape can be easily sketched, remembering the other supports are unyielding. Influence lines obtained by sketching are said to be *qualitative influence lines*, whereas the exact ones are said to be *quantitative influence lines*. The influence line for V_C in Figure 17.6(c) is another example of qualitative sketching for reaction components.

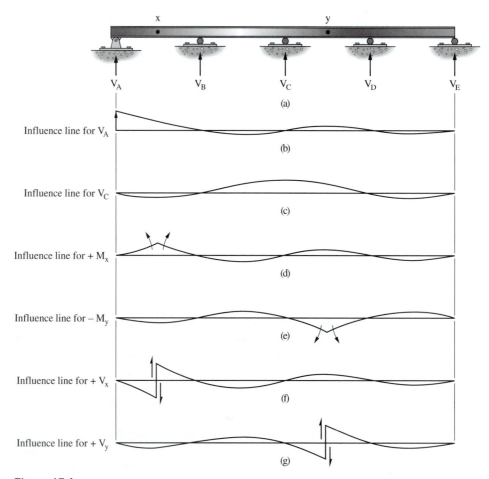

Figure 17.6

Figure 17.6(d) shows the influence line for positive moment at point x near the center of the left-hand span. The beam is assumed to have a pin or hinge inserted at x and a couple applied adjacent to each side of the pin that will cause compression on the top fibers (plus moment). Twisting the beam on each side of the pin causes the left span to take the shape indicated, and the deflected shape of the remainder of the beam may be roughly sketched. A similar procedure is used to draw the influence line for negative moment at point y in the third span, except that a moment couple is applied at the assumed pin, which will tend to cause compression on the bottom beam fibers, corresponding with negative moment.

Finally, qualitative influence lines are drawn for positive shear at points x and y. At point x the beam is assumed to be cut, and two vertical forces of the nature required to give positive shear are applied to the beam on the sides of the cut section. The beam will take the shape shown in Figure 17.6(f). The same procedure is used to draw the diagram for positive shear at point y.

From these diagrams considerable information is available concerning critical live-loading conditions. If a maximum positive value of V_A was desired for a uniform live load, the load would be placed in spans 1 and 3, where the diagram has positive ordinates; if maximum negative moment was required at point x, spans 2 and 4 would be loaded, and so on.

Qualitative influence lines are particularly valuable for determining critical load positions for buildings, as shown by the moment influence line for the building frame of Figure 17.7. In drawing the diagrams for an entire frame, the joints are assumed to be free to rotate, but the members at each joint are assumed to be rigidly connected to each other so that the angles between them do not change during rotation. The influence line for this figure is sketched for positive moment at the center of beam AB.

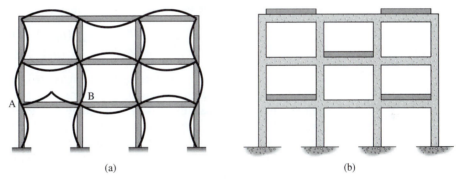

(a)   (b)

Figure 17.7 Qualitative influence line for positive moment at center of span AB. To obtain maximum positive moment at ₵ span AB, place live load as shown.

The spans that should be loaded to cause maximum positive moment are obvious from the diagram. It should be noted that loads on a beam more than approximately three spans away have little effect on the function under consideration. This fact can be seen in the influence lines of Example 17.2, where the ordinates even two spans away are quite small.

A warning should be given regarding qualitative influence lines. They should be drawn for functions near the center of spans or at the supports, but for sections near one-fourth points they may not be sketched without a good deal of study. Near the one-fourth point of a span is a point called the *fixed point*, at which the influence line changes in type. The subject of fixed points is discussed at some length in the book *Continuous Frames of Reinforced Concrete* by H. Cross and N. D. Morgan.[2]

[2]H. Cross and N. D. Morgan, *Continuous Frames of Reinforced Concrete* (New York: Wiley, 1932).

17.3 INFLUENCE LINES FOR STATICALLY INDETERMINATE TRUSSES

For analyzing statically indeterminate trusses, influence lines are necessary to determine the critical positions for live loads, as they were for statically determinate trusses.

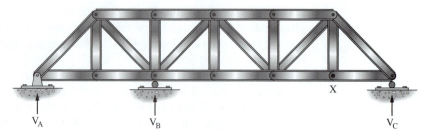

Figure 17.8

The discussion of the details of construction of these diagrams for statically indeterminate trusses is quite similar to the one presented for statically indeterminate beams in Sections 17.1 and 17.2. To prepare the influence line for a reaction of a continuous truss, the support is removed and a unit load is placed at the support point. For this position of the unit load, the deflection at each of the truss joints is determined. For example, the preparation of an influence line for the interior reaction of the truss of Figure 17.8 is considered. The value of the reaction when the unit load is at joint x may be expressed as follows:

$$V_B = -\frac{\Delta_{xb}}{\Delta_{bb}} = -\frac{\sum (\mu_x \mu_B L/AE)}{\sum (\mu_B^2 L/AE)}$$

After the influence line for V_B has been plotted, the influence line for another reaction may be prepared by repeating the process of removing it as the redundant, introducing a unit load there, and computing the necessary deflections. A simpler procedure is to compute the other reactions, or any other functions for which influence lines are desired, by statics after the diagram for V_B is prepared. This method is employed in Example 17.3 for a two-span truss for which influence lines are desired for the reactions and several member forces.

EXAMPLE 17.3

Draw influence lines for the three vertical reactions and for forces in members U_1U_2, L_0U_1, and U_1L_2 (Figure 17.9). Circled figures are member areas, in square inches.

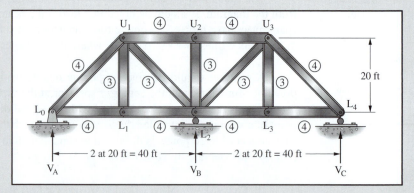

Figure 17.9

Solution.   Remove the interior support and compute the forces for a unit load at L_1 and at L_2. (Note that the deflection at L_1 caused by a unit load at L_2 is the same as the deflection caused at L_3 due to symmetry.) The calculation of these deflections is presented in Table 17.2.

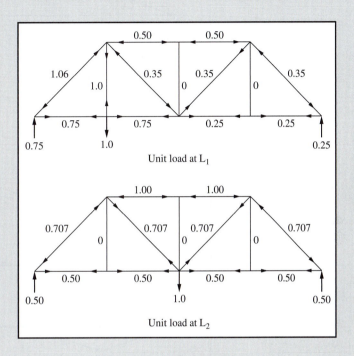

TABLE 17.2

Memeber	L(in.)	A(in.^{2})	$\dfrac{L}{A}$	μ_B	μ_A	$\dfrac{\mu_B^2 L}{AE}$	$\dfrac{\mu_B \mu_A L}{AE}$
L_0L_1	240	4	60	+0.50	+0.75	+15	+22.5
L_1L_2	240	4	60	+0.50	+0.75	+15	+22.5
L_2L_3	240	4	60	+0.50	+0.25	+15	+7.5
L_3L_4	240	4	60	+0.50	+0.25	+15	+7.5
L_0U_1	340	4	85	−0.707	−1.06	+42.5	+63.6
U_1U_2	240	4	60	−1.00	−0.50	+60	+30.0
U_2U_3	240	4	60	−1.00	−0.50	+60	+30.0
U_3L_4	340	4	85	−0.707	−0.35	+42.5	+21.0
U_1L_1	240	3	80	0	+1.00	0	0
U_1L_2	340	3	113	+0.707	−0.35	+56.5	−28.0
U_2L_2	240	3	80	0	0	0	0
L_2U_3	340	3	113	+0.707	+0.35	+56.5	+28.0
U_3L_3	240	3	80	0	0	0	0
Σ						$\dfrac{378}{E}$	$\dfrac{204.6}{E}$

Divide each of the values by Δ_{bb} to obtain the influence-line ordinates for V_B and compute the ordinates for the other influence diagrams by statics.

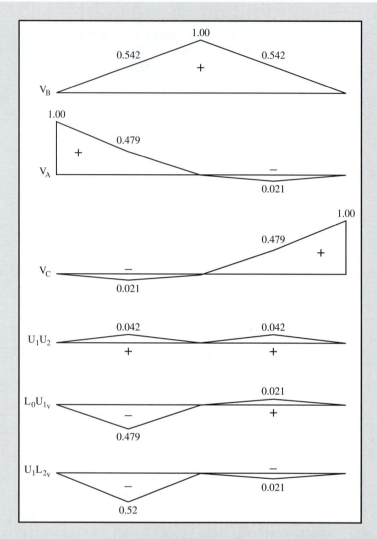

Example 17.4 shows that influence lines for members of an internally redundant truss may be prepared by an almost identical procedure. The member assumed to be the redundant is given a unit force, and deflections caused thereby at each of the joints are calculated. The ordinates of the diagram for the member are obtained by dividing each of these deflections by the deflection at the member. All other influence lines are prepared by statics. ■

EXAMPLE 17.4

Prepare influence lines for force in members L_1U_2, U_1U_2, and U_2L_2 of the truss of Figure 16.3, which is reproduced in Figure 17.10. Circled figures are member areas, in square inches.

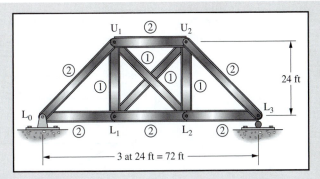

Figure 17.10

Solution. Remove L_1U_2 as the redundant and compute the forces caused by unit loads at L_1 and L_2; replace L_1U_2 with a force of $+1$ and compute the forces in the remaining truss members. The necessary deflection calculations are shown in Table 17.3.

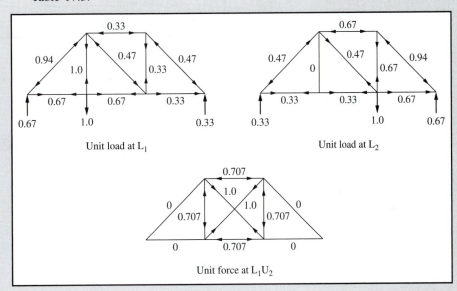

TABLE 17.3

Memeber	L(in.)	A(in.²)	$\dfrac{L}{A}$	μ_{L_1}	μ_{L_2}	μ_A	$\dfrac{\mu_{L_1}\mu_A L}{AE}$	$\dfrac{\mu_{L_2}\mu_A L}{AE}$	$\dfrac{\mu_A^2 L}{AE}$
L_0L_1	288	2	144	+0.67	+0.33	0	0	0	0
L_1L_2	288	2	144	+0.67	+0.33	−0.707	−68	−34	+72
L_2L_3	288	2	144	+0.33	+0.67	0	0	0	0
L_0U_1	408	2	204	−0.94	−0.47	0	0	0	0
U_1U_2	288	2	144	−0.33	−0.67	−0.707	+34	+68	+72
U_2L_3	408	2	204	−0.47	−0.94	0	0	0	0
U_1L_1	288	1	288	+1.00	0	−0.707	−204	0	+144
U_1L_2	408	1	408	−0.47	+0.47	+1.00	−192	+192	+408
L_1U_2	408	1	408	0	0	+1.00	0	0	+408
U_2L_2	288	1	288	+0.33	+0.67	−0.707	−34	−136	+144
Σ							$\dfrac{-498}{E}$	$\dfrac{+90}{E}$	$\dfrac{+1248}{E}$

Draw the influence line for the redundant, L_1U_2, and note that the force in any other member equals $F' + X\mu$.

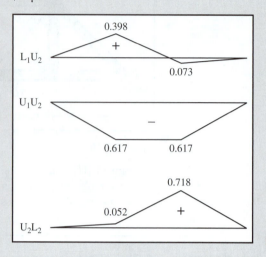

17.4 PROBLEMS FOR SOLUTION

Draw quantitative lines for the situations listed in Problems 17.1 to 17.6.

17.1 Reactions for all supports of the beam shown. Place unit load at 10-ft intervals. (*Ans.* Load 10 ft from left support: $V_A = +0.406 \uparrow$, $V_B = +0.688 \uparrow$, $V_C = -0.094 \downarrow$)

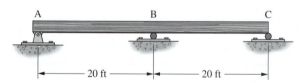

17.2 The left vertical reaction and the moment reaction at the fixed end for the beam shown. Place unit load at 5-ft intervals.

17.3 Shear immediately to the left of support B and moment at support B for the beam shown. Place unit load at 10-ft intervals. (*Ans.* Load 10 ft to right of left support: $V = -0.575$, $M_B = -1.50$)

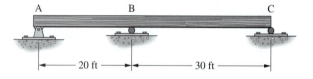

17.4 Shear and moment at a point 20 ft to the left of the fixed-end support C of the beam shown. Place unit load at 10-ft intervals.

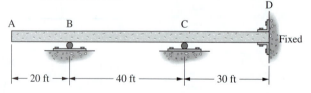

17.5 Vertical reactions and moments at A and B. Place unit load at 10-ft intervals. (*Ans.* Load @ point C: $V_A = -0.500$, $V_B = 1.50 \uparrow$, $M_A = 5.00$)

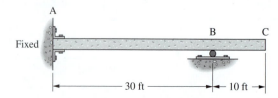

17.6 Vertical reactions at A and B, moment at A and x, and shear just to left of x. Place unit load at 10-ft intervals. (*Ans.* Load at x: $V_A = 0.722 \uparrow$, $V_B = 0.278 \uparrow$, $M_A = -8.88$, $M_x = 5.56$)

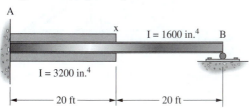

Using Müller-Breslau's principle, sketch influence lines qualitatively for the functions indicated in the structures of Problems 17.7 through 17.14.

17.7 (a) Reactions at A and C, (b) positive moment at x and y, and (c) positive shear at x.

17.8 (a) Reaction at A, (b) positive and negative moments at x, and (c) negative moment at B.

17.9 (a) Vertical reactions at A and C, (b) negative moment at A and positive moment at x, and (c) shear just to left of C.

17.10 (a) Vertical reactions at A and C, (b) negative moments at x and C, and (c) shear just to right of B.

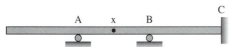

17.11 (a) Reaction at A, (b) positive moment at x, (c) positive shear at y, and (d) positive moment at z, assuming right side of column is bottom side.

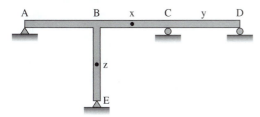

17.12 (a) Positive moment at x, (b) positive shear at x, and (c) negative moment just to the right of y.

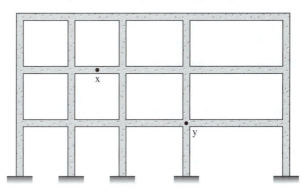

17.13 (a) Positive moment and shear at x, and (b) negative moment just to the right of y.

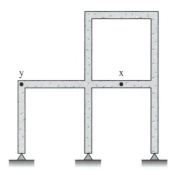

17.14 (a) Positive and negative moment at x and (b) positive shear just to right of y.

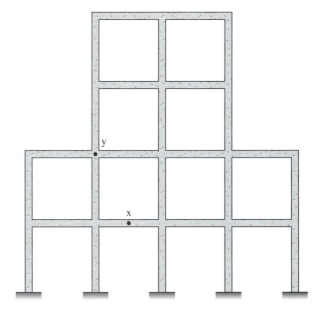

Draw quantitative influence lines for the situations listed in Problems 17.15 through 17.19. Solving these problems with a pocket calculator is quite tedious. As a result the author feels that the student may prefer to solve them with SABLE32 or SAP2000.

17.15 Reactions at all supports for the truss of Problem 16.2 (*Ans.* load @ L_3: $V_A = -0.15$, $V_B = +0.79$, $V_C = +0.35$)

17.16 Reactions for all supports for the truss. Member areas shown in circles on figure.

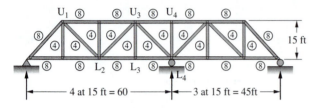

17.17 Forces in members U_1L_2, U_3U_4, and L_4L_5 of the truss of Problem 17.16. (*Ans.* load @ L_2: $U_1L_2 = +0.634$, $U_3U_4 = +0.203$, $L_4L_5 = -0.140$)

17.18 Forces in members U_1L_2, U_1U_2 and U_2L_2. All areas are equal.

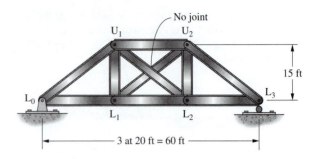

17.19 Force in member L_2U_3 and the center reaction of the truss. Member areas shown on figure. (*Ans.* Load at L_1: $V_B = +0.504 \uparrow$, $L_2U_3 = -0.0078$)

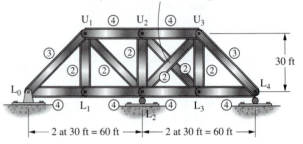

Chapter 18

Slope Deflection: A Displacement Method of Analysis

18.1 INTRODUCTION

George A. Maney introduced slope deflection in a 1915 University of Minnesota engineering publication.[1] His work was an extension of earlier studies of secondary stresses by Manderla[2] and Mohr.[3] For nearly 15 years, until the introduction of moment distribution, slope deflection was the popular "exact" method used for the analysis of continuous beams and frames in the United States.

Slope deflection is a method that takes into account the flexural deformations of beams and frames (rotations, settlements, etc.), but that neglects shear and axial deformations. Although this classical method is generally considered obsolete, its study can be useful for several reasons. These include:

1. Slope deflection is convenient for hand analysis of some small structures.
2. Knowledge of the method provides an excellent background for understanding the moment distribution method discussed in Chapters 20 and 21.
3. It is a special case of the displacement, or stiffness, method of analysis previously defined in Section 14.5 and provides a very effective introduction for the matrix formulation of structures described in Chapters 22 to 25.
4. The slopes and deflections determined by slope deflection enable the analyst to easily sketch the deformed shape of a particular structure. The result is that he or she has a better "feel" for the behavior of structures.

18.2 DERIVATION OF SLOPE-DEFLECTION EQUATIONS

The name *slope deflection* comes from the fact that the moments at the ends of the members in statically indeterminate structures are expressed in terms of the rotations (or slopes) and deflections of the joints. For developing the equations, members are assumed to have constant cross section between supports. Although it is possible to derive expressions for members of varying section, the results are so complex they are of little

---

[1]G. A. Maney, *Studies in Engineering,* No. 1 (Minneapolis: University of Minnesota, 1915).

[2]H. Manderla, "Die Berechnung der Sekundarspannungen," *Allg. Bautz* 45 (1880): 34.

[3]O. Mohr, "Die Berechnung der Fachwerke mit starren knotenverbingungen," *Zivilinginieur,* 1892.

practical value. It is further assumed that the joints in a structure may rotate or deflect, but the angles between the different members meeting at a joint remain unchanged.

Span AB of the continuous beam of Figure 18.1(a) is considered for the following discussion. If the span is completely fixed at each end, the slope of the elastic curve of the beam at the ends is zero. External loads produce fixed-end moments, and these moments cause the span to take the shape shown in Figure 18.1(b). Joints A and B, however, are not actually fixed. Under load, they will rotate slightly to positions such as those shown in Figure 18.1(c). In addition to the rotation of the joints, one or both of the supports may possibly settle. Settlement of the joints will cause a chord rotation of the member as shown in part (d), where support B is assumed to have settled an amount Δ.

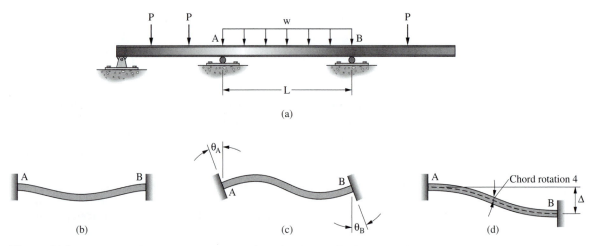

Figure 18.1 Illustration for development or equations for slope-deflection method

From Figure 18.1 we see that the values of the final end moments at A and B (M_{AB} and M_{BA}) are equal to the sum of the moments caused by the following:

1. The fixed-end moments (FEM_{AB} and FEM_{BA}) which can be determined with the moment-area theorems as illustrated in Example 11.9. A detailed list of fixed-end moment expressions is presented in Figure 20.4 of Chapter 20 and inside the front cover.
2. The moments caused by the rotations of joints A and B (θ_A and θ_B).
3. The moments caused by chord rotation ($\psi = \Delta/L$) if one or both of the joints settles or deflects.

When a joint in a structure rotates, the slopes of the tangents to the elastic curves of the members connected to that joint change. For a particular beam, the change in slope is equal to the end shear of the beam when it is loaded with the M/EI diagram. The beam is assumed to have the end moments FEM_{AB} and FEM_{BA}, shown in Figure 18.2(a). For convenience this diagram is broken down into two simple triangles in part(b). The moments are shown over EI values for convenience in the derivation to follow.

With the conjugate beam method the slope at the left end, θ_A, equals the shear (or reaction) at that end of the conjugate beam. This value is determined by taking moments about the right end of the beam. The slope at the right end, θ_B, is determined by taking moments about the left end.

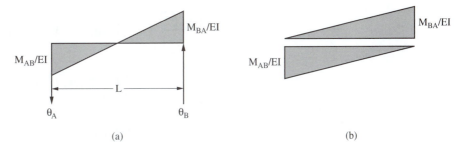

(a)   (b)

Figure 18.2

Poinsett Plaza, Greenville, South Carolina (Courtesy Britt, Peters and Associates)

$$\theta_A = \frac{\frac{1}{2}\left(\dfrac{M_{AB}}{EI}\right)(L)\left(\dfrac{2L}{3}\right) - \frac{1}{2}\left(\dfrac{M_{BA}}{EI}\right)(L)\left(\dfrac{L}{3}\right)}{L} = \frac{L}{6EI}(2M_{AB} - M_{BA})$$

$$\theta_B = \frac{\frac{1}{2}\left(\dfrac{M_{BA}}{EI}\right)(L)\left(\dfrac{2L}{3}\right) - \frac{1}{2}\left(\dfrac{M_{AB}}{EI}\right)(L)\left(\dfrac{L}{3}\right)}{L} = \frac{L}{6EI}(2M_{BA} - M_{AB})$$

If one of the supports of the beam settled or deflected an amount Δ, the angles θ_A and θ_B caused by joint rotation would be changed by Δ/L(or ψ), as was illustrated in Figure 18.1(d). When the chord rotation is added to the expressions above, the following equations are obtained for the slopes of the tangents to the elastic curves at the ends of the beams:

$$\theta_A = \frac{L}{6EI}(2M_{AB} - M_{BA}) + \psi$$

$$\theta_B = \frac{L}{6EI}(2M_{BA} - M_{AB}) + \psi$$

These equations can be solved simultaneously for M_{AB} and M_{BA}. The result are the following end moments in the beam due to the joint rotation and deflection:

$$M_{AB} = 2EK(2\theta_A + \theta_B - 3\psi)$$

$$M_{BA} = 2EK(\theta_A + 2\theta_B - 3\psi)$$

In these equations, the term I/L has been replaced with K. This term is called the *stiffness factor.* (We will see it again when we talk about moment distribution in Chapters 20 and 21.)

The final end moments are equal to the moments caused by rotation and deflection of the joints plus the fixed-end moments acting on the ends of the beam. The final slope-deflection equations are as follows:

$$M_{AB} = 2EK(2\theta_A + \theta_B - 3\psi) + FEM_{AB}$$

$$M_{BA} = 2EK(\theta_A + 2\theta_B - 3\psi) + FEM_{BA}$$

With these equations, we can express the end moments in a structure in terms of joint rotations and settlements. Using the methods previously studied, we have to write one equation for each redundant in the structure. The number of unknowns in these equations is equal to the number of redundants. The slope-deflection method appreciably reduces the amount of work involved in analyzing multiredundant structures, as compared to older analysis methods, because the unknown moments are expressed in terms of only a few unknown joint rotations and settlements. Even for multistory frames, the number of unknown slopes and chord rotations that appear in any one equation is rarely more than five or six, whereas the degree of indeterminacy of the structure is many times that figure.

18.3 APPLICATION OF SLOPE DEFLECTION TO CONTINUOUS BEAMS

Examples 18.1 to 18.4 illustrate the analysis of statically indeterminate beams using the slope-deflection equations. Each member in the beams is considered individually, its fixed-end moments are computed, and one equation is written for the moment at each end of the member. For span AB of Example 18.1, equations for M_{AB} and M_{BA} are written; for span BC, equations for M_{BC} and M_{CB} are written, and so on.

The moment equations are written in terms of the unknown values of θ at the supports. The two moments at an interior support must total zero, as $M_{BA} + M_{BC} = 0$ at support B in Example 18.1. Expressions are therefore written for the total moment at each support, which gives a set of simultaneous equations from which the unknown θ values may be obtained. Two conditions that will simplify the solution of the equations may exist. These are fixed ends for which the θ values must be zero and simple ends for which the moment is zero. A special simplifying expression is derived at the end of Example 18.1 for end spans that are simply supported.

The beams of Examples 18.1 and 18.2 have unyielding supports, and ψ is zero for all of the equations. Some support settlement occurs for the beams of Examples 18.3 and 18.4, and ψ is included in the equations. Chord rotation is considered positive when the chord of a beam is rotated clockwise by the settlements, meaning that the sign is the same no matter which end is being considered as the entire beam rotates in that direction.

When span lengths, moduli of elasticity, and moments of inertia are constant for the spans of a continuous beam, the 2EK values are constant and may be canceled from the equations. Should the values of K vary from span to span, as they often do, it is convenient to express them in terms of relative values, as is done in Example 18.6.

The major difficulty experienced in applying the slope-deflection equations lies in the use of correct signs. It is essential to understand these signs before attempting to apply the equations.

For previous work in this book a plus sign for the moment has indicated tension in the bottom fibers, whereas a negative sign has indicated tension in the top fibers. This sign convention was necessary for drawing moment diagrams.

In applying the slope-deflection equations it is simpler to use a convention in which signs are given to clockwise and counterclockwise moments at the member ends. Once these moments are determined, their signs can be easily converted to the usual convention for drawing moment diagrams (known as beam notation).

The following convention is used in this chapter for slope deflection and in Chapters 20 and 21 for moment distribution: Should a member cause a moment that tends to rotate a joint clockwise, the joint moment is considered negative; if counterclockwise, it is positive. In other words, a clockwise resisting moment on the member herein is considered positive, and a counterclockwise resisting moment is considered negative.

Figure 18.3 is presented to demonstrate this convention. The moment at the left end, M_{AB}, tends to rotate the joint clockwise and is considered negative. Note that the resisting moment would be counterclockwise. At the right end of the beam the loads have caused a moment that tends to rotate the joint counterclockwise, and M_{BA} is considered positive.

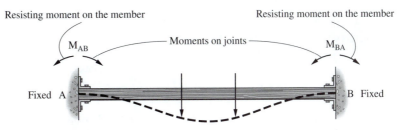

Figure 18.3

EXAMPLE 18.1

Determine moments at all supports of the beam of Figure 18.4 using slope deflection. E and I are constant for the entire member.

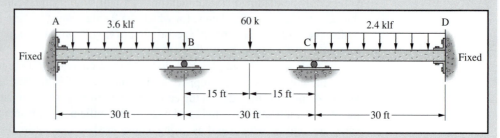

Figure 18.4

Solution.   Computing fixed end moments (see Figure 20.4 and inside the front cover).

$$\text{FEM}_{AB} = -\frac{(3.6)(30)^2}{12} = -270 \text{ ft-k}$$

$$\text{FEM}_{BA} = +\frac{(3.6)(30)^2}{12} = +270 \text{ ft-k}$$

$$\text{FEM}_{BC} = -\frac{(60)(15)(15)^2}{(30)^2} = -225 \text{ ft-k}$$

$$\text{FEM}_{CB} = +\frac{(60)(15)^2(15)}{(30)^2} = +225 \text{ ft-k}$$

$$\text{FEM}_{CD} = -\frac{(2.4)(30)^2}{12} = -180 \text{ ft-k}$$

$$\text{FEM}_{DC} = +\frac{(2.4)(30)^2}{12} = +180 \text{ ft-k}$$

Writing the moment equations noting that $\theta_A = \theta_D = \psi_{AB} = \psi_{BC} = \psi_{CD} = 0$ and 2EK is constant for all three spans.

$$M_{AB} = 2EK(\theta_B) - 270 = 2EK\theta_B - 270$$

$$M_{BA} = 2EK(2\theta_B) + 270 = 4EK\theta_B + 270$$

$$M_{BC} = 2EK(2\theta_B + \theta_C) - 225 = 4EK\theta_B + 2EK\theta_C - 225$$

$$M_{CB} = 2EK(\theta_B + 2\theta_C) + 225 = 2EK\theta_B + 4EK\theta_C + 225$$

$$M_{CD} = 2EK(2\theta_C) - 180 = 4EK\theta_C - 180$$

$$M_{DC} = 2EK(\theta_C) + 180 = 2EK\theta_C + 180$$

$$\underline{\sum M_B = M_{BA} + M_{BC} = 0}$$

$$(1) \quad 8EK\theta_B + 2EK\theta_C + 45 = 0$$

$$\underline{\sum M_C = M_{CB} + M_{CD} = 0}$$

$$(2) \quad 2EK\theta_B + 8EK\theta_C + 45 = 0$$

Solving Equations (1) and (2) simultaneously

$$EK\theta_B = -4.50$$
$$EK\theta_C = -4.50$$

Final Moments

$$M_{AB} = (2)(-4.50) - 270 = -279 \text{ ft-k}$$
$$M_{BA} = (4)(-4.50) + 270 = +252 \text{ ft-k}$$
$$M_{BC} = (4)(-4.50) + (2)(-4.50) - 225 = -252 \text{ ft-k}$$
$$M_{CB} = (2)(-4.50) + (4)(-4.50) + 225 = +198 \text{ ft-k}$$
$$M_{CD} = (4)(-4.50) - 180 = -198 \text{ ft-k}$$
$$M_{DC} = (2)(-4.50) + 180 = +171 \text{ ft-k} \quad \blacksquare$$

18.4 CONTINUOUS BEAMS WITH SIMPLE ENDS

To analyze the continuous beam of Figure 18.5 with its simply supported ends, the individual moment equations may be written exactly as they were in the last example. After this is done, the support equations ($\sum M_A = \sum M_B = \sum M_C = \sum M_D$) can be written and solved simultaneously for $EK\theta_A$, $EK\theta_B$, $EK\theta_C$ and $EK\theta_D$. From these values the numerical values of the moments may be determined.

Figure 18.5

The solution procedure just described for this beam involves four simultaneous equations. Such a procedure seems rather wasteful of time and effort when two of the final moments are already known ($M_{AD} = M_{DC} = 0$). When simple end supports are involved, however, the slope deflection equations may be appreciably simplified as described in the following paragraphs.

The usual slope deflection equations are:

$$M_{AB} = 2EK(2\theta_A + \theta_B - 3\psi_{AB}) + FEM_{AB} \tag{1}$$
$$M_{BA} = 2EK(\theta_A + 2\theta_B - 3\psi_{AB}) + FEM_{BA} \tag{2}$$

Assuming end A to be simply end supported, the value of M_{AB} is zero. Solving the two equations simultaneously by eliminating θ_A gives a simplified expression for M_{BA}, which has only one unknown, θ_B. The resulting simplified equation will expedite considerably the solution of continuous beams with simple ends.

Two times Equation (2)

$$2M_{BA} = 2EK(2\theta_A + 4\theta_B - 6\psi) + 2FEM_{BA}$$

Minus Equation (1)

$$0 = 2EK(2\theta_A + \theta_B - 3\psi) + FEM_{AB}$$

Gives

$$2M_{AB} = 2EK(3\theta_B - 3\psi) + 2FEM_{BA} - FEM_{AB}$$

$$M_{BA} = 3EK(\theta_B - \psi) + FEM_{BA} - \frac{1}{2}FEM_{AB}$$

Using this equation, the number of equations to be solved is reduced by one for each simple end support. Of course, when computers are used this is not particularly important.

EXAMPLE 18.2

Determine the end moments for each span of the beam shown in Figure 18.6 using the modified slope deflection expressions for the end spans. E and I are constant for all three spans.

Solution.

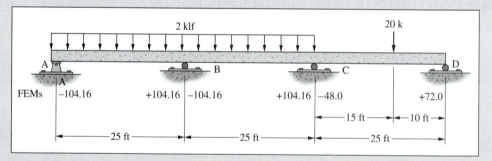

Figure 18.6

Noting $M_{AB} = M_{DA} = \psi_{AB} = \psi_{BC} = \psi_{CD} = 0$ and EK values are constant

$$M_{BA} = 3EK(\theta_B) + 104.16 - \left(\frac{1}{2}\right)(-104.16) = 3EK\theta_B + 156.24$$

$$M_{BC} = 2EK(2\theta_B + \theta_C) - 104.16 = 4EK\theta_B + 2EK\theta_C - 104.16$$

$$M_{CB} = 2EK(\theta_B + 2\theta_C) + 104.16 = 2EK\theta_B + 4EK\theta_C + 104.16$$

$$M_{CD} = 3EK(\theta_C) - 48 - \left(\frac{1}{2}\right)(72) = 3EK\theta_C - 84$$

$$\underline{\Sigma M_B = M_{BA} + M_{BC} = 0}$$

$$(1) \quad 7EK\theta_B + 2EK\theta_C + 52.08 = 0$$

$$\underline{\Sigma M_C = M_{CB} + M_{CD} = 0}$$

$$(2) \quad 2EK\theta_B + 7EK\theta_C + 20.16 = 0$$

$$\underline{\text{Solving Equations (1) and 2 simultaneously}}$$

$$EK\theta_B = -7.21$$

$$EK\theta_C = -0.82$$

Substituting these values into original moment expressions

$$M_{BA} = +134.6 \text{ ft-k} \qquad M_{CB} = +86.5 \text{ ft-k}$$
$$M_{BC} = -134.6 \text{ ft-k} \qquad M_{CD} = -86.5 \text{ ft-k} \quad \blacksquare$$

18.5 MISCELLANEOUS ITEMS CONCERNING CONTINUOUS BEAMS

Should stiffness factors vary for the different spans of a continuous beam their actual values or their relative values must be used in the slope deflection equations. Such a situation is illustrated in Figure 18.7

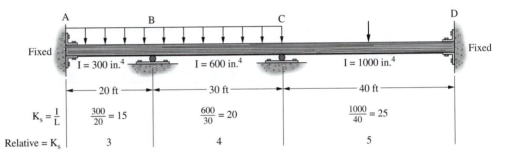

Figure 18.7

Sample equations noting $\theta_A = \theta_D = \psi_{AB} = \psi_{BC} = \psi_{CD} = 0$

$$M_{AB} = 2E3(\theta_B) + \text{FEM}_{AB}$$
$$M_{BC} = 2E4(2\theta_B + \theta_C) + \text{FEM}_{BC}$$

Sometimes a continuous member has an overhanging or cantilevered end perhaps supporting a balcony as illustrated in Figure 18.8. For such a member, the moment at the support end of the cantilever is obvious. For instance, the moment M_{BA} in this beam equals $+(3.6)(20)(10) = +720$ ft-kips. The other equations in the beam are written as before and solved for the support moments.

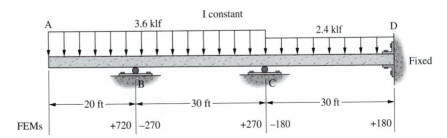

Figure 18.8

Writing slope deflection equations noting $\theta_D = \psi_{CD} = 0$

$$M_{BA} = +720 = -M_{BC}$$
$$M_{BC} = 2EK(2\theta_B) - 270$$
$$M_{CB} = 2EK(\theta_B + 270) \text{ and so on}$$

18.6 ANALYSIS OF BEAMS WITH SUPPORT SETTLEMENT

Examples 18.3 and 18.4 illustrate the analysis of continuous beams whose supports settle. Such settlements cause ψ values to occur in the equations. For cases such as these, the actual EK values are computed rather than just relative values.

EXAMPLE 18.3

Find the moment at support B in the beam of Figure 18.9, assuming B settles 0.25 in., or 0.0208 ft.

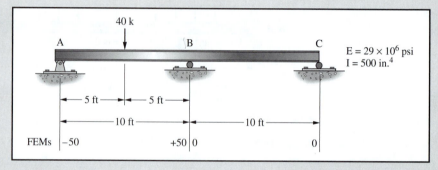

Figure 18.9

Solution. By writing the equations, and noting that $M_{AB} = M_{CB} = 0$,

$$M_{BA} = 3EK\left(\theta_B - \frac{+0.0208}{10}\right) + 50 - \left(\frac{1}{2}\right)(-50)$$

$$M_{BA} = 3EK\theta_B - 0.00624EK + 75$$

$$M_{BC} = 3EK\left(\theta_B - \frac{-0.0208}{10}\right)$$

$$M_{BC} = 3EK\theta_B + 0.00624EK$$

$$\sum M_B = 0 = M_{BA} + M_{BC}$$

$$3EK\theta_B - 0.0624EK + 75 + 3EK\theta_B + 0.0624EK = 0$$

$$6EK\theta_B + 75 = 0$$

$$EK\theta_B = -12.5$$

$$M_{BA} = (3)(-12.5) - 0.00624EK + 75$$

$$EK = \frac{(29 \times 10^6)(500)}{(12 \times 1000)(12 \times 10)} = 10,070$$

$$M_{BA} = -37.5 - 62.8 + 75 = -25.3' \text{ k}$$

$$M_{BC} = (3)(-12.5) + 62.8 = +25.3' \text{ k} \quad \blacksquare$$

EXAMPLE 18.4

Determine all moments for the beam of Figure 18.10 which is assumed to have the following support settlements: $A = 1.25$ in. $= 0.104$ ft, $B = 2.40$ in. $= 0.200$ ft, $C = 2.75$ in. $= 0.229$ ft, and $D = 1.10$ in. $= 0.0917$ ft.

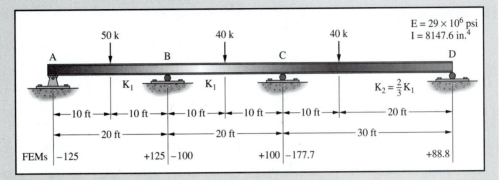

Figure 18.10

Solution.   For support B:

$$M_{BA} = 3EK_1\left(\theta_B - \frac{0.096}{20}\right) + 125 - \left(\frac{1}{2}\right)(-125)$$

$$M_{BA} = 3EK_1\theta_B - 0.0144EK_1 + 187.5$$

$$M_{BC} = 2EK_1\left[2\theta_B + \theta_C - 3\frac{0.029}{20}\right] - 100$$

$$M_{BC} = 4EK_1\theta_B + 2EK_1\theta_C - 0.0087EK_1 - 100$$

For support C:

$$M_{CB} = 2EK_1\left[\theta_B + 2\theta_C - (3)\left(\frac{0.029}{20}\right)\right] + 100$$

$$M_{CB} = 2EK_1\theta_B + 4EK_1\theta_C - 0.00087EK_1 + 100$$

$$M_{CD} = 3EK_2\left(\theta_C - \frac{-0.1373}{30}\right) - 177.7 - \left(\frac{1}{2}\right)(+88.8)$$

$$M_{CD} = 3EK_2\theta_C + 0.01373EK_2 - 222.1$$

$$M_{CD} = 2EK_1\theta_C + 0.00915EK_1 - 222.1$$

$$\sum M_B = 0 = M_{BA} + M_{BC}$$

$$3EK_1\theta_B - 0.0144EK_1 + 187.5 + 4EK_1\theta_B + 2EK_1\theta_C$$
$$- 0.0087EK_1 - 100 = 0$$

$$7EK_1\theta_B + 2EK_1\theta_C - 0.0231EK_1 + 87.5 = 0 \qquad (1)$$

$$\sum M_C = 0 = M_{CB} + M_{CD}$$

$$2EK_1\theta_B + 4EK_1\theta_C - 0.0087EK_1 + 100 + 2EK_1\theta_C$$
$$+ 0.00915EK_1 - 222.1 = 0$$

$$2EK_1\theta_B + 6EK_1\theta_C + 0.00045EK_1 - 122.1 = 0 \qquad (2)$$

By solving Equations (1) and (2) simultaneously for $EK_1\theta_B$ and $EK_1\theta_C$,

$$EK_1\theta_B = 0.00367EK_1 - 20.2$$

$$EK_1\theta_C = -0.00129EK_1 + 27.1$$

$$EK_1 = \frac{(29 \times 10^6)(8147.6)}{(12 \times 20)(12 \times 1000)} = 82{,}040'\text{ k}$$

$$EK_1\theta_B = (0.00367)(82{,}040) - 20.2 = +280.9$$

$$EK_1\theta_C = -(0.00129)(81{,}970) + 27.1 = -78.7$$

Final moments:

$$M_{BA} = (3)(280.9) - (0.0144)(82{,}040) + 187.5 = -151'\text{ k}$$

$$M_{BC} = (4)(280.9) + (2)(-78.7) - (0.0087)(82{,}040) - 100 = +152'\text{ k}$$

$$M_{CB} = (2)(280.9) + (4)(-78.7) - (0.0087)(82{,}040) + 100 = -367'\text{ k}$$

$$M_{CD} = (2)(-78.7) + (0.00915)(82{,}040) - 222.1 = +371'\text{ k}$$

Note: Some error occured due to rounding off of numbers. ∎

18.7 ANALYSIS OF FRAMES—NO SIDESWAY

The slope-deflection equations may be applied to statically indeterminate frames in the same manner as they were to continuous beams if there is no possibility for the frames to lean or deflect asymmetrically. A frame theoretically will not lean to one side or sway if it is symmetrical about the centerline as to dimensions, loads, and moments of inertia, or if it is prevented from swaying by other parts of the structure. The frame of Figure 18.11 cannot away because it is restrained against horizontal movement by member AB. Example 18.5 illustrates the analysis of a simple frame with no sidesway. Where sidesway occurs, the joints of the frame move, and that affects the θ values and the moments, as discussed in Section 18.8.

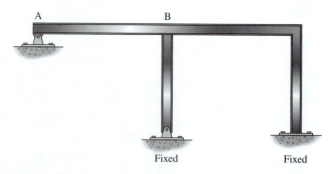

Figure 18.11

Lehigh Shop Center. (Courtesy Bethlehem Steel Corporation.)

EXAMPLE 18.5

Find all moments for the frame shown in Figure 18.12 which has no sidesway because of symmetry.

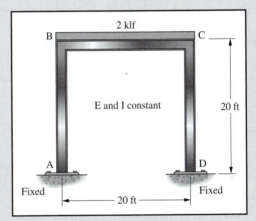

Figure 18.12

Solution.   By computing fixed-end moments,

$$\text{FEM}_{BC} = -\frac{(2)(20)^2}{12} = -66.7' \text{ k}$$

$$\text{FEM}_{CB} = +\frac{(2)(20)^2}{12} = +66.7' \text{ k}$$

By writing the equations, and noting $\theta_A = \theta_D = \psi_{AB} = \psi_{BC} = \psi_{CD} = 0$ and $2EK$ is constant for all members, which permits it to be neglected,

$$M_{AB} = \theta_B$$
$$M_{BA} = 2\theta_B$$
$$M_{BC} = 2\theta_B + \theta_C - 66.7$$
$$M_{CB} = \theta_B + 2\theta_C + 66.7$$
$$M_{CD} = 2\theta_C$$
$$M_{DC} = \theta_C$$

$$\sum M_B = 0 = M_{BA} + M_{BC}$$
$$2\theta_B + 2\theta_B + \theta_C - 66.7 = 0$$
$$4\theta_B + \theta_C = 66.7 \qquad\qquad (1)$$
$$\sum M_C = 0 = M_{CB} + M_{CD}$$
$$\theta_B + 2\theta_C + 66.7 + 2\theta_C = 0$$
$$\theta_B + 4\theta_C = -66.7 \qquad\qquad (2)$$

By solving Equations (1) and (2) simultaneously,

$$\theta_B = +22.2$$
$$\theta_C = -22.2$$

Final end moments

$$M_{AB} = +22.2' \text{ k} \qquad M_{CB} = +44.4' \text{ k}$$
$$M_{BA} = +44.4' \text{ k} \qquad M_{CD} = +44.4' \text{ k}$$
$$M_{BC} = -44.4' \text{ k} \qquad M_{DC} = -22.2' \text{ k} \quad \blacksquare$$

18.8 ANALYSIS OF FRAMES WITH SIDESWAY

The loads, moments of inertia, and dimensions of the frame of Figure 18.13 are not symmetrical about the centerline, and the frame will obviously sway to one side. Joints B and C deflect to the right, which causes chord rotations in members AB and CD, there being theoretically no rotation of BC if axial shortening (or lengthening) of AB and CD is neglected. Neglecting axial deformation of BC, each of the joints deflects the same horizontal distance Δ.

The chord rotations of members AB and CD, due to sidesway, can be seen to equal Δ/L_{AB} and Δ/L_{CD}, respectively. (Note that the shorter a member for the same Δ, the larger its chord rotation and thus the larger the effect on its moment.) For this frame ψ_{AB} is 1.5 times as large as ψ_{CD} because L_{AB} is only two-thirds of L_{CD}. It is convenient to work with only one unknown chord rotation, and in setting up the slope-deflection equations, relative values are used. The value ψ is used for the two equations for member AB, whereas $\frac{2}{3}\psi$ is used for the equations for member CD.

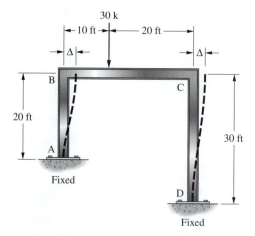

Figure 18.13

Example 18.6 presents the analysis of the frame of Figure 18.14. By noting $\theta_A = \theta_D = 0$, it will be seen that the six end-moment equations for the entire structure contain a total of three unknowns: θ_B, θ_C, and ψ. There are present, however, three conditions that permit their determination. They are

1. The sum of the moments at B is zero ($\Sigma M_B = 0 = M_{BA} + M_{BC}$).
2. The sum of the moments at C is zero ($\Sigma M_C = 0 = M_{CB} + M_{CD}$).
3. The sum of the horizontal forces on the entire structure must be zero.

The only horizontal forces are the horizontal reactions at A and D, and they will be equal in magnitude and opposite in direction. Horizontal reactions may be computed for each of the columns by dividing the column moments by the column heights, or, that is, by taking moments at the top of each column. The sum of the two reactions must be zero.

$$H_A = \frac{M_{AB} + M_{BA}}{L_{AB}} \qquad H_D = \frac{M_{CD} + M_{DC}}{L_{DC}}$$

$$\Sigma H = 0 = \frac{M_{AB} + M_{BA}}{L_{AB}} + \frac{M_{CD} + M_{DC}}{L_{CD}}$$

EXAMPLE 18.6

Determine the end moments of the members of the frame shown in Figure 18.14 for which E and I are constant.

Solution.   By using relative ψ values.

$$\psi_{AB} = \frac{\Delta}{20} \qquad \psi_{CD} = \frac{\Delta}{30} \qquad \frac{\psi_{AB}}{\psi_{CD}} = \frac{3}{2}$$

By using relative 2EK values,

for AB $2EK = \dfrac{2E}{20}$ say a relative value of 3 with respect to 2EK values of BC and CD

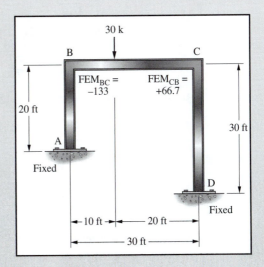

Figure 18.14

for BC $2EK = \dfrac{2E}{30}$ use relative value $= 2$

for CD $2EK = \dfrac{2E}{30}$ use relative value $= 2$

By writing the equations, and noting $\theta_A = \theta_D = \psi_{BC} = 0$,

$$M_{AB} = 3(\theta_B - 3\psi) = 3\theta_B - 9\psi$$

$$M_{BA} = 3(2\theta_B - 3\psi) = 6\theta_B - 9\psi$$

$$M_{BC} = 2(2\theta_B + \theta_C) - 133 = 4\theta_B + 2\theta_C - 133$$

$$M_{CB} = 2(\theta_B + 2\theta_C) + 66.7 = 2\theta_B + 4\theta_C + 66.7$$

$$M_{CD} = 2\left[(2\theta_C - (3)\left(\frac{2}{3}\psi\right)\right] = 4\theta_C - 4\psi$$

$$M_{DC} = 2\left[\theta_C - (3)\left(\frac{2}{3}\psi\right)\right] = 2\theta_C - 4\psi$$

$$\sum M_B = 0 = M_{BA} + M_{BC}$$

$$6\theta_B - 9\psi + 4\theta_B + 2\theta_C - 133 = 0$$

$$10\theta_B + 2\theta_C - 9\psi = 133 \tag{1}$$

$$\sum M_C = 0 = M_{CB} + M_{CD}$$

$$2\theta_B + 4\theta_C + 66.7 + 4\theta_C - 4\psi = 0$$

$$2\theta_B + 8\theta_C - 4\psi = -66.7 \tag{2}$$

$$\sum H = 0 = H_A + H_D$$

$$\frac{M_{AB} + M_{BA}}{L_{AB}} + \frac{M_{CD} + M_{DC}}{L_{CD}} = 0$$

$$\frac{3\theta_B - 9\psi + 6\theta_B - 9\psi}{20} + \frac{4\theta_C - 4\psi + 2\theta_C - 4\psi}{30} = 0$$

$$27\theta_B + 12\theta_C - 70\psi = 0 \tag{3}$$

By solving Equations (1), (2), and (3) simultaneously,

$$\theta_B = +21.2$$

$$\theta_C = -10.5$$

$$\psi = +6.4$$

Final moments:

$$M_{AB} = +6.0' \text{ k} \qquad M_{CB} = +67.4' \text{ k}$$

$$M_{BA} = +69.4' \text{ k} \qquad M_{CD} = -67.4' \text{ k}$$

$$M_{BC} = -69.4' \text{ k} \qquad M_{DC} = -46.6' \text{ k} \quad \blacksquare$$

Slope deflection can be applied to frames with more than one condition of sidesway, such as the two-story frame of Figure 18.15. Hand analysis of frames of this type usually is handled more conveniently by the moment-distribution method, but a knowledge of the slope-deflection solution is valuable in understanding the moment-distribution solution.

The horizontal loads cause the structure to lean to the right; joints B and E deflect horizontally a distance Δ_1; and joints C and D deflect horizontally $\Delta_1 + \Delta_2$, as shown in Figure 18.15. The chord rotations for the columns ψ_1 and ψ_2 will therefore equal Δ_1/L_{AB} for the lower level and Δ_2/L_{BC} for the upper level.

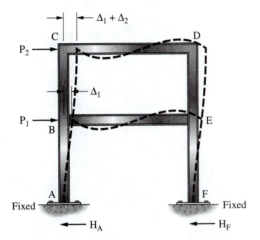

Figure 18.15

The slope-deflection equations may be written for the moment at each end of the six members, and the usual joint-condition equations are available as follows:

$$\sum M_B = 0 = M_{BA} + M_{BC} + M_{BE} \tag{1}$$

$$\sum M_C = 0 = M_{CB} + M_{CD} \tag{2}$$

$$\sum M_D = 0 = M_{DC} + M_{DE} \tag{3}$$

$$\sum M_E = 0 = M_{ED} + M_{EB} + M_{EF} \tag{4}$$

These equations involve six unknowns (θ_B, θ_C, θ_D, θ_E, ψ_1, and ψ_2), noting that θ_A and θ_F are equal to zero. Two more equations are necessary for determining the unknowns, and they are found by considering the horizontal forces or shears on the frame. It is obvious that the sum of the horizontal resisting forces on any level must be equal and opposite to the external horizontal shear on that level. For the bottom level the horizontal shear equals $P_1 + P_2$, and the reactions at the base of each column equal the end moments divided by the column heights. Therefore

$$P_1 + P_2 - H_A - H_F = 0$$

$$H_A = \frac{M_{AB} + M_{BA}}{L_{AB}}$$

$$H_F = \frac{M_{EF} + M_{EF}}{L_{EF}}$$

$$\frac{M_{AB} + M_{BA}}{L_{AB}} + \frac{M_{EF} + M_{FE}}{L_{EF}} + P_1 + P_2 = 0 \tag{5}$$

A similar condition equation may be written for the top level, which has an external shear of P_2. The moments in the columns produce shears equal and opposite to P_2, which permits the writing of the following equation for the level:

$$\frac{M_{BC} + M_{CB}}{L_{BC}} + \frac{M_{DE} + M_{ED}}{L_{DE}} + P_2 = 0 \tag{6}$$

Alcoa Building, San Francisco. Dominant feature of the topped-out steel framework of the "tiara" of San Francisco's Golden Gateway project, the Alcoa Building, is the diamond-patterned seismic-resisting shear walls. (Courtesy Bethlehem Steel Corporation.)

Six condition equations are available for determining the six unknowns in the end-moment equations, and the problem may be solved as illustrated in Example 18.7. It is desirable to assume all θ and ψ values to be positive in setting up the equations. In this way the signs of the answers will take care of themselves.

No matter how many floors the building has, one shear-condition equation is available for each floor. The slope-deflection procedure is not very practical for multistory buildings. For a six-story building four bays wide there will be 6 unknown ψ values and 30 unknown θ values, or a total of 36 simultaneous equations to solve. (Their solution is not quite as bad as it may seem because each equation would contain only a few unknowns, not 36.)

EXAMPLE 18.7

Find the moments for the frame of Figure 18.16 by the method of slope deflection.

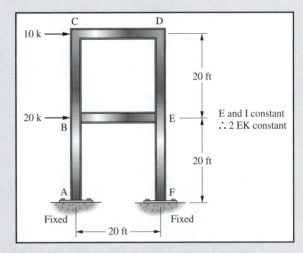

Figure 18.16

Solution.

By noting that chords BE and CD do not rotate,

$$\psi_{AB} = \psi_{EF} = \psi_1 \qquad \psi_{BC} = \psi_{DE} = \psi_2$$

By writing the equations, and noting that $\theta_A = \theta_F = 0$. As the 2EK values are all equal they are omitted.

$$M_{AB} = \theta_B - 3\psi_1 \qquad\qquad M_{DC} = \theta_C + 2\theta_D$$
$$M_{BA} = 2\theta_B - 3\psi_1 \qquad\qquad M_{DE} = 2\theta_D + \theta_E - 3\psi_2$$
$$M_{BC} = 2\theta_B + \theta_C - 3\psi_2 \qquad M_{ED} = \theta_D + 2\theta_E - 3\psi_2$$
$$M_{BE} = 2\theta_B + \theta_E \qquad\qquad M_{EB} = \theta_B + 2\theta_E$$
$$M_{CB} = \theta_B + 2\theta_C - 3\psi_2 \qquad M_{EF} = 2\theta_E - 3\psi_1$$
$$M_{CD} = 2\theta_C + \theta_D \qquad\qquad M_{FE} = \theta_E - 3\psi_1$$

$$\sum M_B = 0 = M_{BA} + M_{BC} + M_{BE}$$
$$2\theta_B - 3\psi_1 + 2\theta_B + \theta_C - 3\psi_2 + 2\theta_B + \theta_E = 0$$
$$6\theta_B + \theta_C + \theta_E - 3\psi_1 - 3\psi_2 = 0$$

$$\sum M_C = 0 = M_{CB} + M_{CD}$$
$$\theta_B + 2\theta_C - 3\psi_2 + 2\theta_2 + \theta_D = 0$$
$$\theta_B + 4\theta_C + \theta_D - 3\psi_2 = 0$$

$$\sum M_D = 0 = M_{DC} + M_{DE}$$
$$\theta_C + 2\theta_D + 2\theta_D + \theta_E - 3\psi_2 = 0$$
$$\theta_C + 4\theta_D + \theta_E - 3\psi_2 = 0$$

$$\sum M_E = 0 = M_{ED} + M_{EB} + M_{EF}$$
$$\theta_D + 2\theta_E - 3\psi_2 + \theta_B + 2\theta_E + 2\theta_E - 3\psi_1 = 0$$
$$\theta_B + \theta_D + 6\theta_E - 3\psi_1 - 3\psi_2 = 0$$

$\Sigma H = -10$, top level:

$$\frac{M_{BC} + M_{CB}}{20} + \frac{M_{DE} + M_{ED}}{20} = -10$$

$$\frac{2\theta_B + \theta_C - 3\psi_2 + \theta_B + 2\theta_C - 3\psi_2}{20}$$

$$+ \frac{2\theta_D + \theta_E - 3\psi_2 + \theta_D + 2\theta_E - 3\psi_2}{20} = -10$$

$$3\theta_B + 3\theta_C + 3\theta_D + 3\theta_E - 12\psi_2 = -200 \qquad (5)$$

$\Sigma H = 30$, bottom level:

$$\frac{M_{AB} + M_{BA}}{20} + \frac{M_{EF} + M_{FE}}{20} = -30$$

$$\frac{\theta_B - 3\psi_1 + 2\theta_B - 3\psi_1}{20} + \frac{2\theta_E - 3\psi_1 + \theta_E - 3\psi_1}{20} = -30$$

$$3\theta_B + 3\theta_E - 12\psi_1 = -600 \qquad (6)$$

By solving equations simultaneously,

$$\theta_B = \theta_E = +52.75 \qquad \psi_1 = +76.37$$
$$\theta_C = \theta_D = +21.82 \qquad \psi_2 = +53.95$$

Final moments:

$$M_{AB} = M_{FE} = -176.4' \text{ k} \qquad M_{BE} = M_{EB} = +158.2' \text{ k}$$
$$M_{BA} = M_{EF} = -123.6' \text{ k} \qquad M_{CB} = M_{DE} = -65.5' \text{ k}$$
$$M_{BC} = M_{ED} = -34.6' \text{ k} \qquad M_{CD} = M_{DC} = +65.5' \text{ k} \quad \blacksquare$$

18.9 ANALYSIS OF FRAMES WITH SLOPING LEGS

Space is not taken in this chapter to discuss the analysis of frames with sloping legs. A slope-deflection analysis of such a frame with hand calculations is tedious, and is therefore not likely to be used when other methods are available. Computer solutions using matrix methods and the moment-distribution procedure presented in Chapters 20 to 25 are more commonly used today. Nevertheless, once the analysis of frames with sloping legs using moment distribution is understood, the analyst will be able to return to this chapter and use slope deflection for the analysis of such structures.

18.10 PROBLEMS FOR SOLUTION

For Problems 18.1 through 18.15 compute the end moments of the beams with the slope-deflection equations. Draw shearing force and bending moment diagrams. E and I are constant for each member unless otherwise noted.

18.1   (*Ans.* $M_{AB} = -50$ ft-k, $M_{BC} = -140$ ft-k)

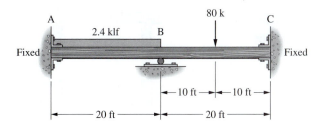

18.2

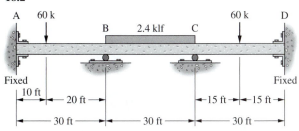

18.3   (*Ans.* $M_{BA} = -160$ ft-k, $M_{CD} = -250$ ft-k)

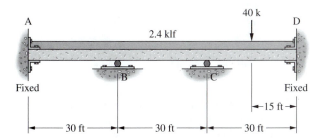

18.4

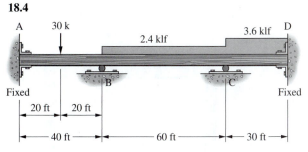

18.5   (*Ans.* $M_{AB} = -97.90$ ft-k, $M_{DC} = +67.3$ ft-k)

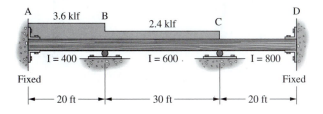

18.6

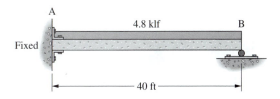

18.7   Use the regular slope deflection equations, that is, the ones that have not been modified for simple end supports. (*Ans.* $M_{BA} = -328.1$ ft-k, $M_{CB} = -173.9$ ft-k)

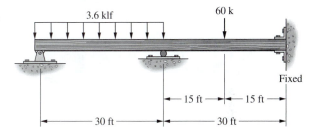

18.8   Repeat Problem 18.7 using the modified equations.

18.9   (*Ans.* $M_{CB} = -210$ ft-k, $V_C = 39.00$ k ↑)

18.10

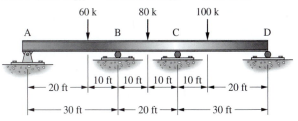

18.11   (*Ans.* $M_{BA} = -239.1$ ft-k, $M_{CD} = -405.7$ ft-k)

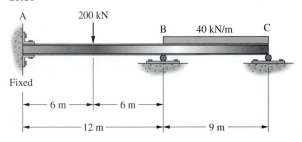

18.12

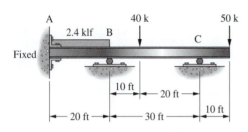

18.13 See inside front cover for fixed-end moments.
(*Ans.* $M_{AB} = -70.91$ ft-k, $M_{BA} = -42.44$ ft-k)

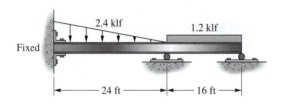

18.14 The beam of Problem 18.2 if both ends are simply supported.

18.15 The beam of Problem 18.4 if the right end is simply supported. (*Ans.* $M_{BA} = -545.8$ ft-k, $M_{CB} = -672.7$ ft-k)

For Problems 18.16 through 18.18 determine the end moments for the members of all of the structures by using the slope deflection equations. Notice sidesway is not permissible for these structures. I and E are constant.

18.16

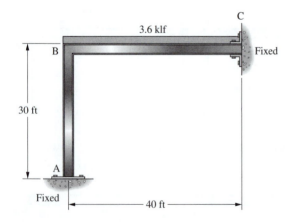

18.17 (*Ans.* $M_{BC} = -274.3$ ft-k, $M_{CB} = +122.0$ ft-k)

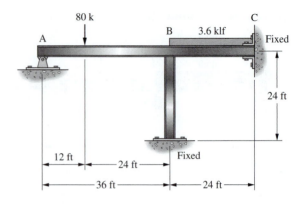

18.18

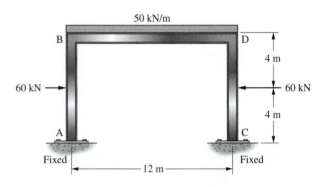

For Problems 18.19 through 18.22 determine the end moments for the members of all the structures by using the slope deflection equations. These structures are subject to sidesway. I and E are constant.

18.19 (*Ans.* $M_{BD} = +79.60$ ft-k, $M_{DC} = -220.53$ ft-k)

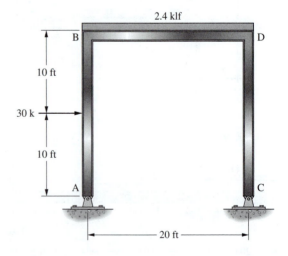

18.20 The frame of Problem 18.19 if the column bases are fixed.

18.21 (*Ans.* $M_{DF} = -166.1$ ft-k, $M_{FE} = -130.7$ ft-k)

18.22

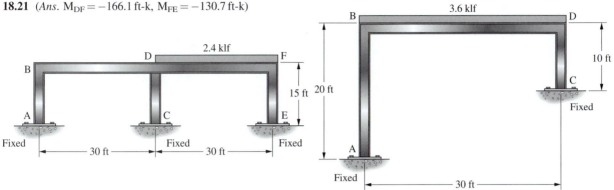

PART THREE

STATICALLY INDETERMINATE STRUCTURES

COMMON METHODS IN CURRENT PRACTICE

Approximate Analysis of Statically Indeterminate Structures

19.1 INTRODUCTION

The approximate methods presented in this chapter for analyzing statically indeterminate structures could very well be designated as *classical methods*. The same designation could be made for the moment distribution method presented in Chapters 20 and 21. The methods discussed in this and the next several chapters will occasionally be seen and used by an engineer in the course of everyday design—they are the methods of analysis sometimes used in current engineering practice.

Statically indeterminate structures may be analyzed "exactly" or "approximately." Several "exact" methods, which are based on elastic distortions, were discussed in Chapters 15 through 18. Approximate methods, which involve the use of simplifying assumptions, are presented in this chapter. These methods have many practical applications such as the following:

1. When costs are being estimated for alternative structural configurations and design concepts, approximate analyses are often very helpful. Approximate analyses and approximate designs of the various alternatives can be made quickly and used for initial cost estimates.
2. To analyze a statically indeterminate structure, an estimate of the member sizes must be made before the structure can be analyzed using an "exact" method. This is necessary because the analysis of a statically indeterminate structure is based on the elastic properties of the members. An approximate analysis of the structure will yield forces from which reasonably good initial estimates can be made of member sizes.
3. Today, computers are available with which "exact" analyses and designs of highly indeterminate structures can be made quickly and economically. To make use of computer programs, preliminary estimates of the size of the members should be made. If an approximate analysis of the structure has been done, very reasonable estimates of member sizes are possible. The result will be appreciable savings of both computer time and design hours.
4. Approximate analyses are quite useful for checking computer solutions, which is a very important matter.

5. An "exact" analysis may be too expensive for small noncritical systems, particularly when preliminary designs are being made. An acceptable and applicable approximate method is very appropriate for such a situation.

6. An additional advantage of approximate methods is that they provide the analyst with an understanding of the actual behavior of structures under various loading conditions. This important ability probably will not be developed from computer solutions.

To make an "exact" analysis of a complicated statically indeterminate structure, a qualified analyst must model the structure, that is, the analyst must make certain assumptions about the behavior of the structure. For instance, the joints are assumed to be simple, rigid or semi-rigid. Characteristics of material behavior and loading conditions must be assumed, and so on. The result of these assumptions is that all analyses are approximate. *We could say that we apply an "exact" analysis method to a structure that does not really exist.* Furthermore, all analysis methods are approximate in the sense that every structure is constructed within certain tolerances—no structure is perfect—and its behavior cannot be determined "exactly."

Many different methods are available for making approximate analyses. A few of the more common ones are presented here, with consideration being given to trusses, continuous beams, and building frames. The approximate methods described in this chapter hopefully will provide you with a general knowledge about a wide range of statically indeterminate structures. Not all types of statically indeterminate structures are considered in this chapter. However, based on the ideas presented, you should be able to make reasonable assumptions when other types of statically indeterminate structures are encountered.

To be able to analyze a structure using the equations of static equilibrium, there must be no more unknowns than there are available equations of static equilibrium. If a truss or frame has 10 more unknowns than equations of equilibrium, it is statically indeterminate to the 10th degree. To analyze it by an approximate method, one assumption for each degree of indeterminacy, a total of 10 assumptions must be made. Each assumption effectively provides another equation of equilibrium to use in the calculations.

19.2 TRUSSES WITH TWO DIAGONALS IN EACH PANEL

Diagonals Having Little Stiffness

The truss shown in Figure 19.1 has two diagonals in each panel. If one of the diagonal members were removed from each of the six panels, the truss would become statically determinate. Therefore the original truss is statically indeterminate to the sixth degree.

Frequently the diagonals in a truss are relatively long and slender, often being made of a pair of small steel angles. They can carry reasonably large tensile forces but have negligible capacity in compression. For this situation, it is logical to assume that the shearing force in each panel is carried entirely by the diagonal that would be in tension for that sense of the shearing force (positive or negative). The other diagonal is assumed to have no force. Making this assumption in each panel effectively provides six "equations" with which to evaluate the six redundants. The forces in the remaining members can be evaluated with the equations of static equilibrium cast as the method of joints or the method of sections. The forces in Figure 19.1 were obtained on this basis.

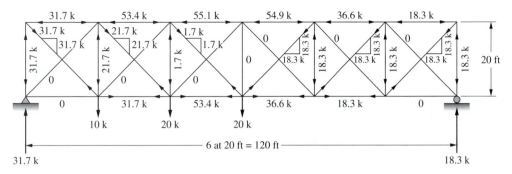

Figure 19.1   A truss analyzed assuming diagonals act only in tension

Diagonals Having Considerable Stiffness

In some trusses, the diagonals are constructed with sufficient stiffness to resist significant compressive loads. In panels with two substantial diagonals, the shearing force is carried by both diagonals. The division of shear causes one diagonal to be in tension and the other to be in compression. The usual approximation made is that each diagonal carries 50 % of the shearing force in the panel: other divisions of the shearing force are also possible. Another typical division is that one-third of shearing force is carried by the diagonal acting in compression and two-thirds is carried by the diagonal in tension.

The forces calculated for the truss in Figure 19.2 are based on a 50 % division of the shearing force in each panel.

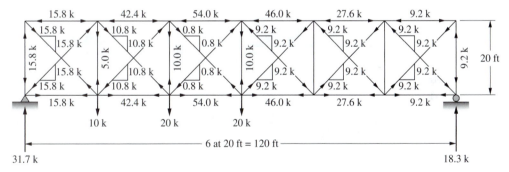

Figure 19.2   Approximate analysis of a truss assuming diagonals carry 50% of panel shearing force

19.3 CONTINUOUS BEAMS

Before beginning an "exact" analysis of a building frame, the sizes of the members in the frame must be estimated. Preliminary beam sizes can be obtained by considering their approximate moments. Frequently a portion of the building can be removed and analyzed separately from the rest of the structure. For instance, one or more beam spans may be taken out as a free body and assumptions made as to the moments in those spans. To facilitate such an analysis, moment diagrams are shown in Figure 19.3 for several different uniformly loaded beams.

It is obvious from the figure that the assumed types of supports can have a tremendous effect on the magnitude of the calculated moments. For instance, the uniformly loaded

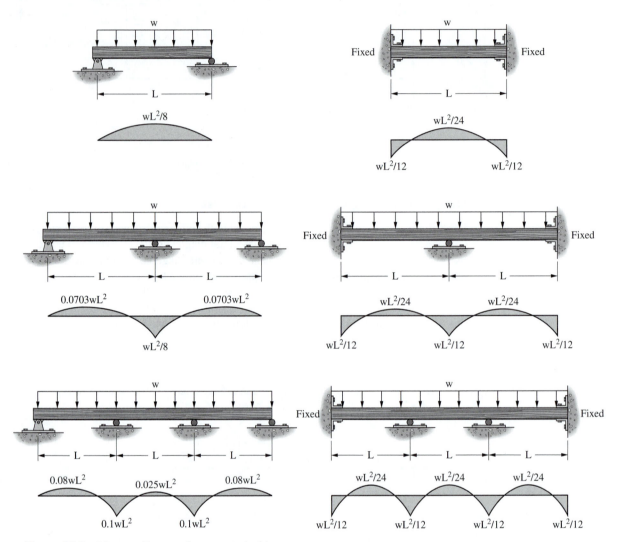

Figure 19.3   Moment diagrams for some typical beams

simple beam in Figure 19.3 will have a maximum moment equal to $wL^2/8$. On the other hand, the uniformly loaded single-span fixed-ended beam will have a maximum moment equal to $wL^2/12$. For a continuous uniformly loaded beam, the engineer may very well decide to estimate a maximum moment somewhere between the preceding values, at perhaps $wL^2/10$, and use that value for approximating the member size.

A very common method used for the approximate analysis of continuous reinforced-concrete structures involves the use of the American Concrete Institute bending moment and shearing force coefficients.[1] These coefficients, which are reproduced in Table 19.1 provide estimated maximum shearing forces and bending moments for buildings of normal proportions. The values calculated in this manner

[1]*Building Code Requirements for Reinforced Concrete*, ACI 318-05 (Detroit: American Concrete Institute) Section 8.3.3, pp. 96–97.

TABLE 19.1   ACI MOMENT COEFFICIENTS[*]

Positive moment	
End spans	
If discontinuous end is restrained	$\dfrac{1}{11}wL_n^2$
If discontinuous end is integral with the support	$\dfrac{1}{14}wL_n^2$
Interior Spans	$\dfrac{1}{16}wL_n^2$
Negative moment at the exterior face of the first interior support	
Two spans	$\dfrac{1}{9}wL_n^2$
More than two spans	$\dfrac{1}{10}wL_n^2$
Negative moment at other faces of interior supports	$\dfrac{1}{11}wL_n^2$
Negative moment at face of all supports for (a) slabs with spans not exceeding 10 ft and (b) beams and girders where ratio of sum of column stiffness to beam stiffness exceeds 8 at each end of the span	$\dfrac{1}{12}wL_n^2$
Negative moment at interior faces of exterior supports for members built integrally with their supports	
Where the support is a spandrel beam or girder	$\dfrac{1}{24}wL_n^2$
Where the support is a column	$\dfrac{1}{16}wL_n^2$
Shear in end members at face of first interior support	$\dfrac{1.15wL_n}{2}$
Shear at face of all other supports	$\dfrac{wL_n}{2}$

[*]American Concrete Institute ACI 318–05

usually will be somewhat larger than those that would be obtained with an exact analysis. Consequently, appreciable economy can normally be obtained by taking the time or effort to make an ''exact'' analysis. In this regard, the engineer should realize that the ACI coefficients are considered to apply best to continuous frames having more than three or four continuous spans.

In developing the coefficients, the negative moment values were reduced to take into account the usual support widths and some moment redistribution before collapse. In addition, the positive moment values have been increased somewhat to account for the moment redistribution. It will also be noted that the coefficients account for the fact that in monolithic construction, the supports are not simple and moments are present at end supports, where those supports are beams or columns.

In applying the coefficients, w is the design load per unit of length, while L_n is the clear span for calculating positive bending moments and the average of the adjacent clear spans for calculating negative bending moments. These values were developed for members with approximately equal spans—the larger of two adjacent spans does not exceed the smaller by more than 20%—and for cases where the ratio of the uniform service live load to the uniform service dead load is not greater than 3. In addition, the values are not applicable to prestressed-concrete members. Should these limitations not be met, a more precise method of analysis must be used.

For the design of a continuous beam or slab, the bending moment coefficients in effect provide two sets of moment diagrams for each span of the structure. One diagram is the result of placing the live loads so that they will cause maximum positive moment out in the span. The other is the result of placing the live loads to cause maximum negative moments at the supports. Actually, it is not possible to produce maximum negative moments at both ends of a span simultaneously. It takes one placement of the live loads to produce maximum negative moment at one end of the span and another placement to produce maximum negative moment at the other end. The assumption of both maximums occurring at the same time is on the safe side, however, because the resulting diagram will have greater critical values than are produced by either one of the two separate loading conditions.

The ACI coefficients give maximum values for a bending moment envelope for each span of a continuous frame. Typical envelopes are shown in Figure 19.4 for a continuous slab that is constructed integrally with its exterior supports, which are spandrel girders.

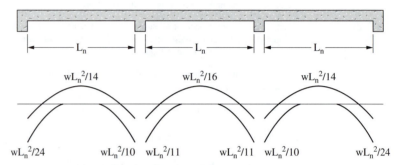

Figure 19.4 Moment envelopes for continuous slab constructed integrally with exterior supports that are spandrel girders

On some occasions the analyst will take out a portion of a structure that includes not only the beams but also the columns for the floor above and the floor below, as shown in Figure 19.5. This procedure, usually called the *equivalent frame method*, is applicable only for gravity loads. The sizes of the members are estimated and an analysis is made using one of the "exact" methods of analysis we have discussed.

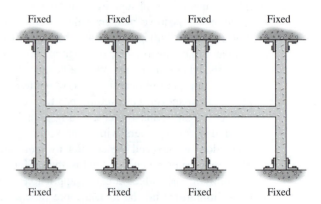

Figure 19.5 A portion of a building frame to be analyzed by the equivalent frame method

Multistory steel-frame building, showing structural skeleton;
Hotel Sheraton, Philadelphia, Pennsylvania (Courtesy
Bethlehem Steel Corporation)

19.4 ANALYSIS OF BUILDING FRAMES FOR VERTICAL LOADS

One approximate method for analyzing building frames for vertical loads involves the estimation of the location of the points of zero moment in the girders. These points, which occur where the moment changes from one sign to the other, are commonly called *points of inflection* (PIs) or points of contraflexure. One common practice is to assume the PIs are located at the one-tenth points from each girder end. In addition, the axial forces in the girders are assumed to be zero.[2]

These assumptions have the effect of creating a simple beam between the points of inflection, and the positive moments in the beam can be determined using simple equations of static equilibrium. Negative moments occur in the girders between their ends and the points of inflection. The negative moments are computed by treating the portion of the beam from the end to the point of inflection as a cantilever beam.

[2]C. H. Norris, J. B. Wilbur, and S. Utku, *Elementary Structural Analysis*, 3rd ed. (New York: McGraw-Hill, 1976), 200–201.

The shearing force at the end of the girders contributes to the axial forces in the columns. Similarly, the negative moments at the ends of the girders are transferred to the columns. At interior columns, the moments in the girders on each side oppose each other and may cancel. Exterior columns have moments caused by the girders framing into them on only one side; these moments need to be considered in design.

EXAMPLE 19.1

Determine the forces in beam AB of the building frame of Figure 19.6 by assuming the points of inflection are located at one-tenth points and the beam ends are fixed supports. The distributed loads acting on the frame are 3 klf each. The concentrated loads are acting at the third points on the beam.

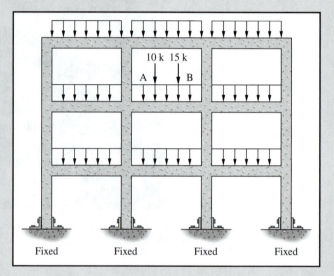

Figure 19.6

Solution. A detailed drawing of the beam and the location of the assumed points of inflection are shown in the next figure.

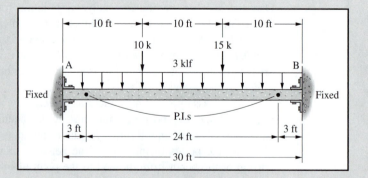

We can now split the beam into the "simple span" and the two "cantilever spans" and calculate the forces acting on each. After the forces are calculated the bending moment diagram can be drawn.

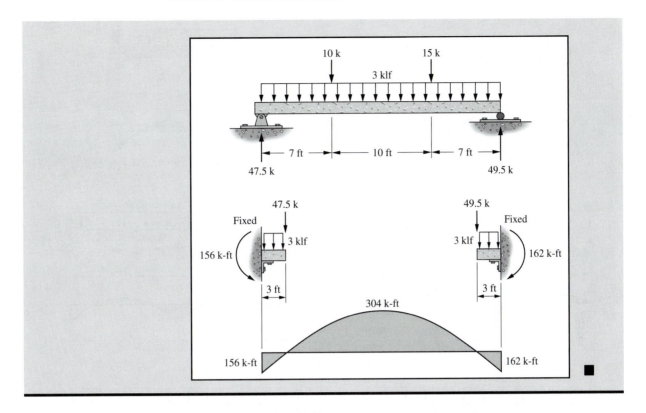

Sketching the deflected shape often helps to make reasonable estimates for the locations of the PIs. For illustration, a continuous beam is shown in Figure 19.7(a) and a sketch of its estimated deflected shape for the loads shown is given in part (b). From the sketch, an approximate location of the PIs is estimated.

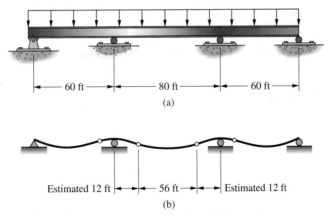

Figure 19.7   Estimated points of inflection in a continuous beam

It might be useful for the reader to see where PIs occur for a few types of statically indeterminate beams. These may be helpful in estimating where PIs will occur in other structures. Moment diagrams are shown for several beams in Figure 19.8. The PIs obviously occur where the moment diagrams change from positive to negative moments or vice versa.

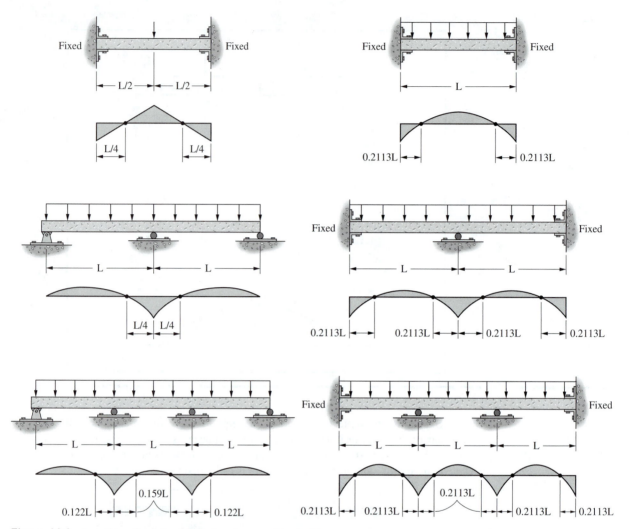

Figure 19.8 Location of points of inflection in several typical beams

19.5 ANALYSIS OF PORTAL FRAMES

Portal frames of the type shown in Figure 19.9 and Figure 19.10 may be fixed at their column bases, may be simply supported, or may be partially fixed. The columns of the frame of Figure 19.9(a) are assumed to have their bases fixed. Consequently, there will be three unknown forces of reaction at each support, for a total of six unknowns. The structure is statically indeterminate to the third degree, and to analyze it by an approximate method three assumptions must be made.

When a column is rigidly attached to its foundation, there can be no rotation at the base. Although the frame is subjected to wind loads causing the columns to bend laterally, a tangent to the column at the base will remain vertical. If the beam at the top of the columns is very stiff and rigidly fastened to the columns, the tangents to the columns at the junctions will remain vertical. A column rigidly fixed at the top and bottom will assume the shape of an S-curve when subjected to lateral loads, as shown in Figure 19.9(a).

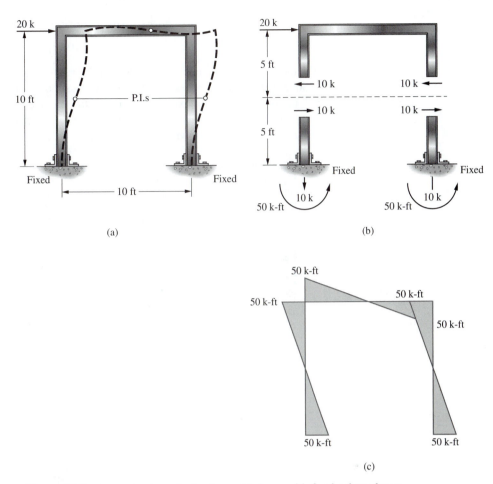

(a)

(b)

(c)

Figure 19.9 Approximate analysis of a portal frame with fixed column bases

At a point midway from the column base to the beam, the moment will be zero because it changes from a moment causing tension on one side of the column to a moment causing tension on the other side. If points of inflection are assumed in each column, two of the necessary three assumptions will have been made, that is, two equations for summation of moments are made available.

The third assumption usually made is that the horizontal shear is divided equally between the two columns at the plane of contraflexure. An "exact" analysis proves this is a very reasonable assumption as long as the columns are approximately the same size. If they are not similar in size, an assumption may be made that the shear is divided between them in a slightly different proportion, with the stiffer column carrying the larger proportional amount of the shear. A good assumption for such cases is to distribute the shear to the columns in proportion to their I/L^3 values.

To analyze the frame in Figure 19.9 the 20 kip lateral load is divided into 10 kip shearing forces at the column PIs as shown in Figure 19.9(b). The bending moment at the top and bottom of each column can be computed and is found to be equal to $10(5) = 50$ k-ft. Moments then are taken about the left column PI to determine the right-hand vertical reaction as follows:

$$20(5) - 10V_R = 0$$
$$V_R = 10 \text{ k} \uparrow$$

Finally, the moment diagrams are drawn with the results shown in Figure 19.9(c).

Should the column bases be assumed to be simply supported, that is, pinned or hinged, the points of contraflexure will occur at those hinges. Assuming the columns are the same size, the horizontal shear is split equally between the columns and the other values are determined as shown in Figure 19.10.

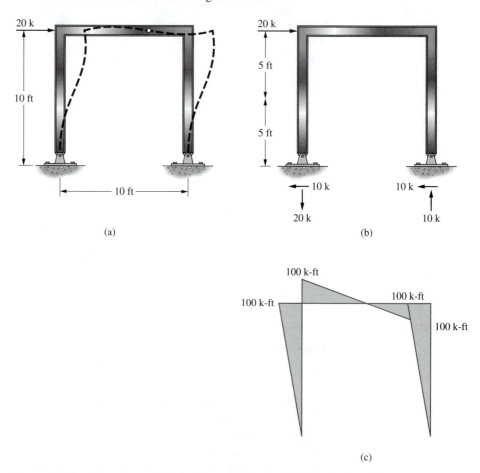

Figure 19.10 Approximate analysis of a portal frame with pinned column bases

The principles that we have discussed here can be applied to other framed structures regardless of the number of bays or the number of stories. Except for preliminary analysis, most structures are not analyzed using approximate methods. The ready availability of structural analysis software and the rapidity with which it can be used have now caused most analysis to be conducted with computers.

19.6 ANALYSIS OF BUILDING FRAMES FOR LATERAL LOADS

Building frames are subjected to lateral loads as well as to vertical loads. The necessity for careful attention to these forces increases as buildings become taller. Not only must a building have sufficient lateral resistance to prevent failure, it also must have sufficient resistance to deflections to prevent injury to its various parts.

Another item of importance is the provision of sufficient lateral rigidity to give the occupants a feeling of safety. They might not have this feeling in tall buildings that have a great deal of lateral movement in times of high winds. There have been actual instances of occupants of the upper floors of tall buildings complaining of motion sickness on very windy days.

Lateral loads can be taken care of by means of X or other types of bracing, by shear walls, or by moment-resisting wind connections. Only the last of these three methods is considered in this chapter.

Large, rigid-frame buildings are highly indeterminate and until the 1960s their analysis by the so-called "exact" methods usually was impractical because of the multitude of equations involved. The total degree of indeterminacy of a rigid-frame building (internal and external) can be determined by considering it to consist of separate portals. One level of the rigid frame of Figure 19.11 is broken down into a set of portals in Figure 19.12. Each of the portals is statically indeterminate to the third degree, and the total degree of indeterminancy of a rigid-frame building equals three times the number of individual portals in the frame.

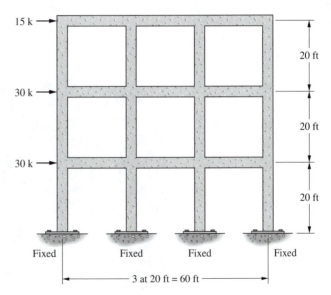

Figure 19.11

Figure 19.12   One level of the frame of Figure 19.11

Another method of obtaining the degree of indeterminacy is to assume that each of the girders is cut by an imaginary section. If these values—shear, axial force, and moment—are known in each girder, the free bodies produced can be analyzed by statics. The total degree of indeterminacy equals three times the number of girders.

Today such structures usually are analyzed "exactly" with modern computers, though approximate methods are used occasionally for preliminary analysis and sizing of members. Approximate methods also provide checks on computer solutions and give the analyst a better "feel" for and understanding of a structure's behavior than he or she can obtain by examining the seemingly endless printout of a computer.

The building frame of Figure 19.11 is analyzed by two approximate methods in the pages to follow, and the results are compared with those obtained with the computer program SABLE32. The dimensions and loading of the frame are selected to illustrate the methods involved while keeping the computations as simple as possible. There are nine girders in the frame, giving a total degree of indeterminacy of 27, and at least 27 assumptions will be needed to permit an approximate solution.

The reader should be aware that with the availability of digital computers today it is feasible to make "exact" analyses in appreciably less time than that required to make approximate analyses without the use of computers. The more-accurate values obtained permit the use of smaller members. The results of computer usage are money saving in analysis time and in the use of smaller members. It is possible to analyze in minutes structures (such as tall buildings) that are statically indeterminate with hundreds or even thousands of redundants, via the displacement method. The results for these highly indeterminate structures are far more accurate and more economically obtained than those obtained with approximate analyses.

Wachovia Plaza, Charlotte, North Carolina (Courtesy Bethlehem Steel Corporation)

The two approximate methods considered here are the portal and cantilever methods. These methods were used in so many successful building designs that they were almost the unofficial standard procedure for the design profession before the advent of modern computers. No consideration is given in either of these methods to the elastic properties of the members. These omissions can be very serious in asymmetrical frames and in very tall buildings. To illustrate the seriousness of the matter, the changes in member sizes are considered in a very tall building. In such a building probably there will not be a great deal of change in beam sizes from the top floor to the bottom floor. For the same loadings and spans the changed sizes would be due to the large wind moments in the lower floors. The change, however, in column sizes from top to bottom would be tremendous. The result is that the relative sizes of columns and beams on the top floors are entirely different from the relative sizes on the lower floors. When this fact is not considered, it causes large errors in the analysis.

In both the portal and cantilever methods, the entire wind loads are assumed to be resisted by the building frames, with no stiffening assistance from the floors, walls, and partitions. Changes in length of girders and columns are assumed to be negligible. However, they are not negligible in tall slender buildings whose height is five or more times the least horizontal dimension.

If the height of a building is roughly five or more times its least lateral dimension, it is generally felt that a more precise method of analysis should be used than the portal or cantilever methods. There are several excellent approximate methods that make use of the elastic properties of the structures and that give values closely approaching the results of the "exact" methods. These include the Factor method,[3] the Witmer method of K percentages,[4] and the Spurr method.[5] Should an "exact" hand-calculation method be desired, the moment-distribution procedure of Chapters 20 and 21 is convenient.

The Portal Method

The most common approximate method of analyzing building frames for lateral loads was at one time the portal method. Due to its simplicity, it probably was used more than any other approximate method for determining wind forces in building frames. This method, which was presented by Albert Smith in the *Journal of the Western Society of Engineers* in April 1915, was supposedly satisfactory for most buildings up to 25 stories in height.[6]

At least three assumptions must be made for each individual portal or for each girder. In the portal method, the frame is theoretically divided into independent portals (Figure 19.12) and the following three assumptions are made:

1. The columns bend in such a manner that there is a point of inflection at mid height Figure 19.9.
2. The girders bend in such a manner that there is a point of inflection at their centerlines.

---

[3]C. H. Norris, J. B. Wilbur, and S. Utku, *Elementary Structural Analysis*, 3rd ed. (New York: McGraw-Hill, 1976), 207–212.

[4]"Wind Bracing in Steel Buildings," *Transactions of the American Society of Civil Engineers* 105 (1940): 1725–1727.

[5]*Ibid*, pp. 1723–1725.

[6]*Ibid*, p. 1723.

3. The horizontal shears on each level are arbitrarily distributed between the columns. One commonly used distribution (and the one illustrated here) is to assume the shear divides among the columns in the ratio of one part to exterior columns and two parts to interior columns. The reason for this ratio can be seen in Figure 19.12. Each of the interior columns is serving two bents, whereas the exterior columns are serving only one. Another common distribution is to assume that the shear V taken by each column is in proportion to the floor area it supports. The shear distribution by the two procedures would be the same for a building with equal bays, but for one with unequal bays the results would differ with the floor area method, probably giving more realistic results.

For this frame (Figure 19.13) there are 27 redundants; to obtain their values, one assumption as to the location of the point of inflection has been made for each of the 21 columns and girders. Three assumptions are made on each level as to the shear split in each individual portal, or the number of shear assumptions equals one less than the number of columns on each level. For the frame, 9 shear assumptions are made, giving a total of 30 assumptions and only 27 redundants. More assumptions are made than necessary, but they are consistent with the solution (that is, if only 27 of the assumptions were used and the remaining values were obtained by statics, the results would be identical).

Frame Analysis

The frame of Figure 19.13 is analyzed on the basis of these assumptions. The arrows shown on the figure give the direction of the girder shears and the column axial forces. The reader can visualize the stress condition of the frame if he or she assumes the wind is tending to push it over from the left to right, stretching the left exterior columns and compressing the right exterior columns. Briefly, the calculations were made as follows.

I. Column Shears

The shears in each column on the various levels were first obtained. The total shear on the top level is 15 kips. Because there are two exterior and two interior columns the following expression may be written:

$$x + 2x + 2x + x = 15 \text{ k}$$
$$x = 2.5 \text{ k}$$
$$2x = 5.0 \text{ k}$$

The shear in column CD is 2.5 kips; in GH it is 5.0 kips, and so on. Similarly, the shears were determined for the columns on the first and second levels, where the total shears are 75 and 45 kips, respectively.

2. Column Moments

The columns are assumed to have points of inflection at their mid heights; therefore, their moments, top and bottom, equal the column shears times half the column heights.

3. Girder Moments and Shears

At any joint in the frame the sum of the moments in the girders equals the sum of the moments in the columns. The column moments have been previously determined. By

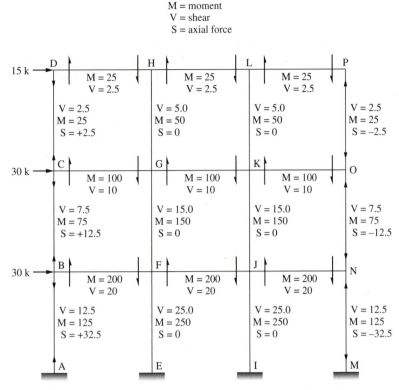

Figure 19.13   Building frame analyzed by portal method for lateral loads

beginning at the upper left-hand corner of the frame and working across from left to right, adding or subtracting the moments as the case may be, the girder moments were found in this order: DH, HL, LP, CG, GK, and so on. It follows that with points of inflection at girder centerlines, the girder shears equal the girder moments divided by half-girder lengths.

4. Column Axial Forces

The axial forces in the columns may be directly obtained from the girder shears. Starting at the upper left-hand corner, the column axial force in CD is numerically equal to the shear in girder DH. The axial force in column GH is equal to the difference between the two girder shears DH and HL, which equals zero in this case. (If the width of each of the portals is the same, the shears in the girder on one level will be equal, and the interior columns will have no axial force, since only lateral loads are considered.)

The Cantilever Method

Another simple method of analyzing building frames for lateral forces is the cantilever method presented by A. C. Wilson in *Engineering Record*, September 5, 1908. This method is said to be a little more desirable for high narrow buildings than the portal method and was used satisfactorily for buildings with heights not in excess of 25 to 35 stories.[7] It was not as popular as the portal method.

[7]*Ibid*, p. 1723.

Mr. Wilson's method makes use of the assumptions that the portal method uses, as to locations of points of inflection in columns and girders, but the third assumption differs somewhat. Rather than assume the shear on a particular level to divide between the columns in some ratio, the axial stress in each column is considered to be proportional to its distance from the center of gravity of all the columns on that level. If the columns on each level are assumed to have equal cross-sectional areas (as is done for the cantilever problems of this chapter) then their forces will vary in proportion to their distances from the center of gravity. The wind loads are tending to overturn the building, and the columns on the leeward side will be compressed, whereas those on the windward side will be put in tension. The greater the distance a column is from the center of gravity of its group of columns, the greater will be its axial stress.

This assumption is equivalent to making a number of axial-force assumptions equal to one less than the number of columns on each level. Again, the structure has 27 redundants, and 30 assumptions are made (21 column and girder point-of-inflection assumptions and 9 column axial-force assumptions), but the extra assumptions are consistent with the solution.

Frame Analysis

The frame previously analyzed by the portal method is analyzed by the cantilever method in Figure 19.14. Briefly, the calculations are made as follows.

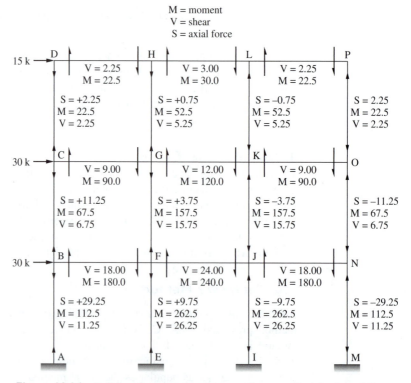

Figure 19.14 Building frame analyzed by cantilever method for lateral forces

I. Column Axial Forces

Considering first the top level, moments are taken about the point of contraflexure in column CD of the forces above the plane of contraflexure through the columns on that level. According to the third assumption, the axial force in GH will be only one-third of that in CD, and these forces in GH and CD will be tensile, whereas those in KL and OP will be compressive. The following expression is written, with respect to Figure 19.15 to determine the values of the column axial forces on the top level.

$$(15)(10) + (1S)(20) - (1S)(40) - (3S)(60) = 0$$
$$S = 0.75 \text{ k}$$
$$3S = 2.25 \text{ k}$$

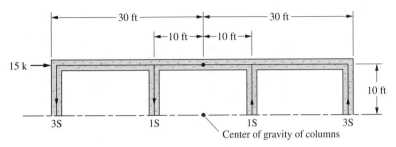

Figure 19.15

The axial force in CD is 2.25 kips and that in GH is 0.75 kip, and so on. Similar moment calculations are made for each level to obtain the column axial forces.

2. Girder Shears

The next step is to obtain the girder shears from the column axial forces. These shears are obtained by starting at the top left-hand corner and working across the top level, adding or subtracting the axial forces in the columns according to their signs. This procedure is similar to the method of joints used for finding truss forces.

3. Column and Girder Moments and the Column Shears

The final steps can be quickly summarized. The girder moments, as before, are equal to the girder shears times the girder half-lengths. The column moments are obtained by starting at the top left-hand corner and working across each level in succession, adding or subtracting the previously obtained column and girder moments as indicated. The column shears are equal to the column moments divided by half the column heights.

19.7 APPROXIMATE ANALYSES OF FRAME COMPARED TO "EXACT" ANALYSIS BY SABLE32

Table 19.2 compares the moments in the members of this frame as determined by the two approximate methods and by SABLE32

TABLE 19.2   MEMBER MOMENTS

Member	Portal	Cantilever	SABLE32	Member	Portal	Cantilever	SABLE32
AB	125	112.5	258.3	IJ	250	262.5	222.6
BA	125	112.5	148.9	JI	250	262.5	161.6
BC	75	67.5	57.4	JF	200	240	162.6
BF	200	180	206.3	JK	150	157.5	140.6
CB	75	67.5	100.7	JN	200	180	139.6
CD	25	22.5	7.1	KJ	150	157.5	146.4
CG	100	90	93.6	KG	100	120	114.5
DC	25	22.5	20.8	KL	50	52.5	54.2
DH	25	22.5	20.8	KO	100	90	86.1
EF	250	262.5	250.3	LK	50	52.5	71.9
FE	250	262.5	185.6	LH	25	30	47.4
FB	200	180	161.6	LP	25	22.5	24.5
FG	150	157.5	142.4	MN	125	112.5	179.8
FJ	200	240	166.3	NM	125	112.5	92.9
GF	150	157.5	152.4	NJ	200	180	165.5
GC	100	90	82.2	NO	75	67.5	72.7
GH	50	52.5	43.9	ON	75	67.5	87.4
GK	100	120	114.1	OK	100	90	103.1
HG	50	52.5	63.2	OP	25	22.5	15.6
HD	25	22.5	18.0	PO	25	22.5	37.4
HL	25	30	45.2	PL	25	22.5	37.4

Note that for several members the approximate results vary considerably from the results obtained by the "exact" method. Experience with the "exact" methods for handling indeterminate building frames will show that the points of inflection will not occur exactly at the midpoints. Using more realistic locations for the assumed inflection points will greatly improve results. The Bowman method[8] involves the location of the points of inflection in the columns and girders according to a specified set of rules depending on the number of stories in the building. In addition, the shear is divided between the columns on each level according to a set of rules that are based on the moments of inertia of the columns as well as the bay widths. Application of the Bowman method gives much better results than the portal and cantilever procedures.

19.8 MOMENT DISTRIBUTION

The moment distribution method described in Chapters 20 and 21 for the analysis of statically indeterminate beams and frames involves successive cycles of computation, each cycle drawing closer to the "exact" answers. When the computations are carried out until the changes in the numbers are very small, it is considered an "exact" method of analysis. Should the number of cycles be limited, the method becomes a splendid approximate method. As such it is discussed in Section 20.5 of the next chapter.

19.9 ANALYSIS OF VIERENDEEL "TRUSSES"

In previous chapters a truss has been defined as a structure assembled with a group of ties and struts connected at their joints with frictionless pins so the members are subjected

[8]H. Sutherland and H. L. Bowman, *Structural Theory* (New York: Wiley, 1950), 295–301.

only to axial tension or axial compression. A special type of truss (although it is not really a truss by the preceding definition) is the Vierendeel "truss." A typical example of a Vierendeel truss is the pedestrian bridge shown in Figure 19.16. You can see that these trusses are actually rigid frames or, as some people say, they are girders with big holes in them. The Vierendeel truss was developed in 1893 by the Belgian engineer and builder. Arthur Vierendeel. They have been used frequently in Europe, particularly in Belgium for highway and railroad bridges, but only occasionally in the United States.

Figure 19.16 An Example of a Vierendeel truss (Clinic Inn Pedestrian Bridge Cleveland, Ohio, Courtesy of the American Institute of Steel Construction, Inc.)

A Vierendeel truss is usually constructed with reinforced concrete, but also may be fabricated with structural steel. The external loads are supported by means of the flexural resistance of the short heavy members. The continuous moment-resisting joints cause the "trusses" to be highly statically indeterminate. Vierendeel trusses are rather inefficient, but because of their aesthetics, they are used in situations in which their large clear openings are desirable. In addition, they are particularly convenient in buildings where they can be constructed with depths equal to the story heights.

These highly statically indeterminate structures can be analyzed approximately by the portal and cantilever methods described in the preceding section. Figure 19.17 illustrates a Vierendeel truss (which is very similar to the one shown in Figure 19.16 and the results obtained by applying the portal method. For this symmetrical "truss", the cantilever method will yield the same results. To follow the calculations, you might like to turn the structure on its end because the shear being considered for the Vierendeel is vertical, whereas it was horizontal for the building frames previously considered. Because the "truss" is symmetric, the results are only shown for one half of the "truss".

For many Vierendeel trusses—particularly those that are several stories in height—the lower horizontal members may be much larger and stiffer than the other horizontal members. To obtain better results with the portal or cantilever methods, the non-uniformity of sizes should be taken into account by assuming that more shearing force is carried by the stiffer members.

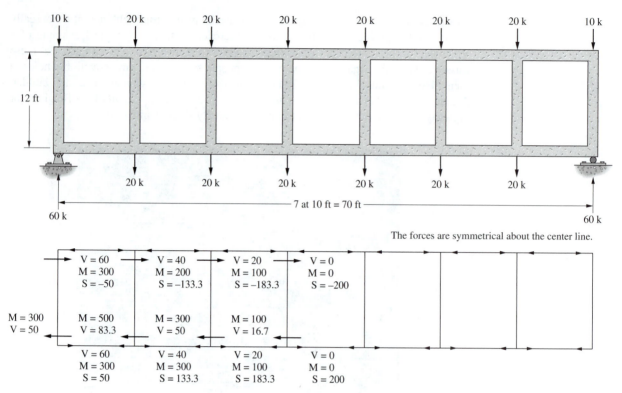

The forces are symmetrical about the center line.

Figure 19.17 Analysis of a Vierendeel truss using the portal method

19.10 PROBLEMS FOR SOLUTION

19.1 Compute the forces in the members of the truss shown in the accompanying illustration if the diagonals are unable to carry compression. (*Ans.* $U_0U_1 = -81.67$ k, $L_2L_3 = +133.34$ k, $U_3L_3 = -21.67$ k)

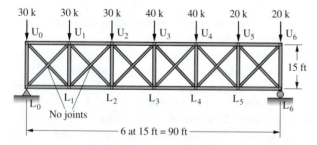

19.2 Repeat Problem 19.1 if the diagonals that theoretically are in compression can resist half of the shear in each panel.

19.3 Repeat Problem 19.1 if the diagonals that theoretically are in compression can resist one-third of the shear in each panel. (*Ans.* $U_0U_1 = -54.55$ k, $L_1L_2 = +98.89$ k)

19.4 Compute the forces in the members of the cantilever truss shown if the diagonals are unable to carry any compression.

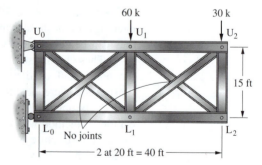

19.5 Repeat Problem 19.4 if the diagonals that would normally be in compression can resist only one-third of the shear in each panel. (*Ans.* $U_1U_2 = +13.33$ k, $U_1L_1 = -50$ k)

19.6 Compute the forces in the members of the truss shown if the interior diagonals that would normally be in compression can resist no compression.

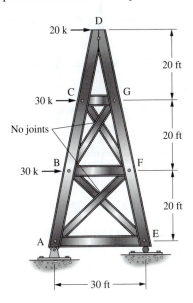

19.7 Prepare the shear and moment diagrams for the continuous beam shown using the ACI coefficients of Table 19.1. Assume the beam is constructed integrally with girders at all of its supports. Draw the moment diagrams as envelopes. (*Ans.* $M_{AB} = -2535$ ft lbs, $M_{CD} = -6084$ ft lbs)

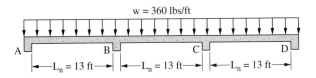

19.8 Analyze the portal shown if PIs are assumed to be located at column mid heights and at mid span of beam.

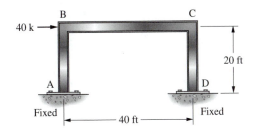

19.9 Repeat Problem 19.8 if the column PIs are assumed to be located 8 ft above the column bases. (*Ans.* $M_{AB} = 160$ ft-k, $M_{CB} = 240$ ft-k)

19.10 Repeat Problem 19.8 if the column bases are assumed to be pinned and the PIs located there.

For Problems 19.11 through 19.14 compute the bending moments, shear forces, and axial forces for all of the members of the frames shown with the portal method. All joints are assumed to be rigid.

19.11 (*Ans.* $M_{CD} = 25$ ft-k, $M_{GF} = 150$ ft-k, $M_{JI} = 250$ ft-k)

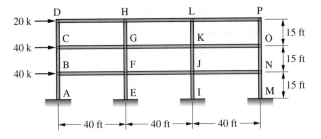

19.12

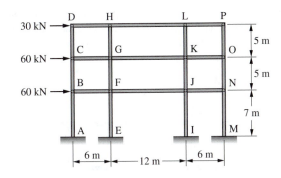

19.13 (*Ans.* $M_{AB} = 175$ ft-k, $S_{FG} = +21.67$ k)

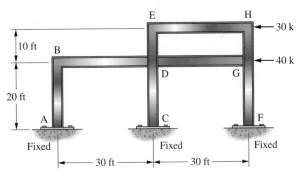

19.14

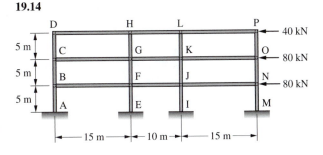

For Problems 19.15 to 19.17 analyze the frames with rigid joints using the cantilever method.

19.15 (*Ans.* $M_{BC} = 67.5$ ft-k, $V_{NM} = 15.00$ k)

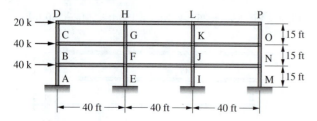

19.16 Rework Problem 19.12 if the column bases are assumed to be pinned.

19.17 Repeat Problem 19.13. (*Ans.* $M_{AB} = 250$ ft-k, $M_{FG} = 100$ ft-k)

For Problems 19.18 and 19.19 compute the bending moments, shearing forces, and axial forces for all of the members of the rigid jointed Vierendeel trusses using the portal method.

19.18

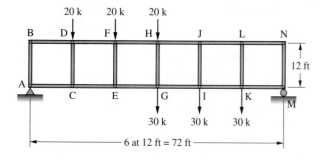

19.19 (*Ans.* $M_{BC} = 75$ ft-k, $M_{KL} = 60$ ft-k)

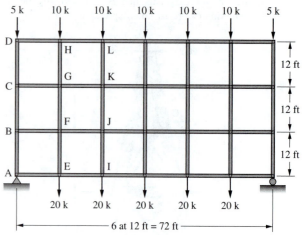

19.20 Repeat Problem 19.11 using SABLE32. Assume the frame members have constant areas, moduli of elasticity and moments of inertia.

Chapter 20

Moment Distribution for Beams

20.1 INTRODUCTION

The late Hardy Cross wrote papers about the moment-distribution method in 1929[1,2] and 1930[3] after having taught the subject to students at the University of Illinois since 1924. His papers began a new era in the analysis of statically indeterminate frames and gave added impetus to their use. The moment-distribution method of analyzing continuous beams and frames involves little more labor than the approximate methods but yields accuracy equivalent to that obtained from the infinitely more laborious classical "exact" methods.

The analysis of statically indeterminate structures in the preceding chapters frequently involved the solution of simultaneous equations. Simultaneous equations are not necessary in solutions by moment distribution, except in a few rare situations for complicated frames. The method that Cross developed involves successive cycles of computation, with each cycle drawing closer to the "exact" answers. The calculations may be stopped after two or three cycles, giving a very good approximate analysis, or they may be carried on to whatever degree of accuracy is desired. When these advantages are considered in light of the fact that the accuracy obtained by the lengthy "classical" methods is often questionable, the true worth of this quick and practical method is understood.

From the 1930s until the 1960s, moment distribution was the dominant method used for the analysis of continuous beams and frames. Since the 1960s, however, there has been a continually increasing use of computers for analyzing all types of structures. Computers are extremely efficient for solving the simultaneous equations that are generated by other methods of analysis. Generally, the software is developed from the matrix-analysis procedures described in Chapters 22 to 25 of this book.

Even with the computer software available, moment distribution continues to be an important hand-calculation method for analysis of continuous beams and frames. Structural engineers can use moment distribution to quickly make approximate analyses

---

[1]Hardy Cross, "Continuity as a Factor in Reinforced Concrete Design," *Proceedings of the American Concrete Institute*, Vol. 25 (1929), pp. 669–708.

[2]Hardy Cross, "Simplified Rigid Frame Design," Report of Committee 301, *Proceedings of the American Concrete Institute*, Vol. 26, (1929), pp. 170–183.

[3]Hardy Cross, "Analysis of Continuous Frames by Distributing Fixed-End Moments," *Proceedings of the American Society of Civil Engineers*, Vol. 56, No. 5 (May 1930): pp. 919–928. Also, *Transactions of American Society of Civil Engineers*, Vol. 96 (1932), pp. 1–10.

for preliminary designs and check computer results, which is very important. In addition, moment distribution may be solely used for the analysis of small structures.

The beauty of moment distribution lies in the simplicity of its theory and application. Readers will be able to grasp quickly the principles involved and will clearly understand what they are doing and why they are doing it. In the discussion that follows, certain assumptions have been made. These assumptions are:

1. The structures have members of constant cross section throughout their respective lengths. That is, the members are prismatic.
2. The joints at which two or more members frame together do not translate.
3. The joints to which members are connected can rotate, but the ends of all members connected to a joint rotate the same amount as the joint. At a joint, there is no rotation of the ends of members relative to each other or to the joint.
4. Axial deformation of members is neglected.

Largest curved "horizontal skyscraper" in the United States, Boston, Massachusetts (Courtesy Bethlehem Steel Corporation)

Considering the frame of Figure 20.1(a), joints A to D are seen to be fixed. Joint E, however, is not fixed, and the loads on the structure will cause it to rotate slightly, as represented by the angle θ_E in Figure 20.1(b).

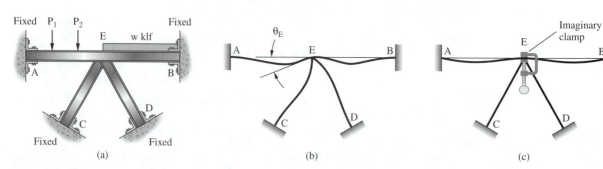

Figure 20.1

If an imaginary clamp is placed at E, fixing it so that it cannot be displaced, the structure will under load take the shape of Figure 20.1(c). For this situation, with all ends fixed, the fixed-end moments can be calculated with little difficulty by the usual expressions ($wL^2/12$ for uniform loads and Pab^2/L^2 or Pa^2b/L^2 for concentrated loads) as shown in Figure 20.4 and on the inside of the book cover.

If the clamp at E is removed, the joint will rotate slightly, rotating the ends of the members meeting there and causing a redistribution of the moments in the member ends. The changes in the moments or rotations at the E ends of members AE, BE, CE, and DE cause some effect at their other ends. When a moment is applied to one end of a member, the other end being fixed, there is some effect or *carryover* to the fixed end.

After the fixed-end moments are computed, the problem to be handled may be briefly stated as consisting of the calculation of (1) the moments caused in the E ends of the members by the rotation of joint E, (2) the magnitude of the moments carried over to the other ends of the members, and (3) the addition or subtraction of these latter moments to the original fixed-end moments.

These steps can be simply written as being the fixed-end moments plus the moments due to the rotation of joint E.

$$M = M_{fixed} + M_{\theta_E}$$

20.2 BASIC RELATIONS

Two questions must be answered in order to apply the moment-distribution method to actual structures. They are

1. What is the moment developed or carried over to a fixed end of a member when the other end is subjected to a certain moment?
2. When a joint is unclamped and rotates, what is the distribution of the unbalanced moment to the members meeting at the joint, or how much resisting moment is supplied by each member?

Carryover Moment

To determine the carryover moment, the unloaded beam of constant cross section in Figure 20.2(a) is considered. If a moment M_1 is applied to the left end of the beam, a

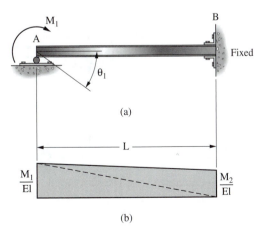

(a)

(b)

Figure 20.2   Concept of carryover moment

moment M_2 will be developed at the right end. The left end is at a joint that has been released and the moment M_1 causes it to rotate an amount θ_1. There will be, however, no deflection or translation of the left end with respect to the right end.

The second moment-area theorem may be used to determine the magnitude of M_2. The deflection of the tangent to the elastic curve of the beam at the left end with respect to the tangent at the right end (which remains horizontal) is equal to the moment of the area of the M/EI diagram taken about the left end and is also equal to zero. By drawing the M/EI diagram in Figure 20.2(b) and dividing it into two triangles to facilitate the area computations, the following expression may be written and solved for M_2:

$$\delta_A = \frac{\frac{1}{2}M_1L\left(\frac{1}{3}L\right) + \frac{1}{2}M_2L\left(\frac{2}{3}L\right)}{EI} = 0$$

$$\frac{M_1L^2}{6EI} + \frac{M_2L^2}{3EI} = 0$$

$$M_2 = -\frac{1}{2}M_1$$

A moment applied at one end of a prismatic beam, the other end being fixed, will cause a moment half as large and of opposite sign at the fixed end. The carryover factor is $-\frac{1}{2}$. The minus sign refers to strength-of-materials sign convention: A distributed moment on one end causing tension in bottom fibers must be carried over so that it will cause tension in the top fibers of the other end. A study of Figures 20.2 and 20.3 shows that carrying over with a $\frac{1}{2}$ value with the moment-distribution sign convention (which is discussed in Section 20.4) automatically takes care of the situation, and it is unnecessary to change signs with each carryover.

Distribution Factors

Usually members framed together at a joint have different flexural stiffnesses. When a joint is unclamped and begins to rotate under the unbalanced moment, the resistance to rotation varies from member to member. The problem is to determine how much of the unbalanced moment will be taken up by each of the members. A reasonable assumption is that the unbalance will be resisted in direct relation to the respective resistance to end rotation of each member.

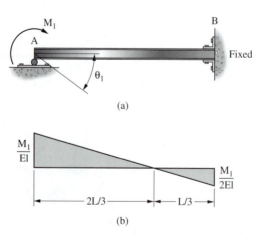

(a)

(b)

Figure 20.3 Moment Carry-over

The beam and M/EI diagram of Figure 20.2 are redrawn in Figure 20.3 with the proper relationship between M_1 and M_2, and an expression is written for the amount of rotation caused by moment M_1.

Using the first moment-area theorem, the angle θ_1 may be represented by the area of the M/EI diagram between A and B, the tangent at B remaining horizontal:

$$\theta_1 = \frac{\frac{1}{2} M_1 \left(\frac{2}{3}\right) L - \frac{1}{2} \left(\frac{1}{2}\right) M_1 \left(\frac{1}{3}\right) L}{EI}$$

$$\theta_1 = \frac{M_1 L}{4EI}$$

Assuming that all the members consist of the same material, having the same E values, the only variables in the foregoing equation affecting the amount of end rotation are the L and I values. The amount of rotation occurring at the end of a member obviously varies directly as the L/I value for the member. The larger the rotation of the member, the less moment it will carry. The moment resisted varies inversely as the amount of rotation or directly as the I/L value. This latter value is referred to as the *stiffness factor K*.

$$K = \frac{I}{L}$$

To determine the unbalanced moment taken by each of the members at a joint, the stiffness factors at the joint are totaled, and each member is assumed to carry a proportion of the unbalanced moment equal to its K value divided by the sum of all the K values at the joint. These proportions of the total unbalanced moment carried by each of the members are the *distribution factors*.

20.3 DEFINITIONS

The following terms are constantly used in discussing moment distribution.

Fixed-end Moments

When all of the joints of a structure are clamped to prevent any joint rotation, the external loads produce certain moments at the ends of the members to which they are applied. These moments are referred to as fixed-end moments. Figure 20.4 presents fixed-end moments for various loading conditions.

Unbalanced Moments

Initially the joints in a structure are considered to be clamped. When a joint is released, it rotates if the sum of the fixed-end moments at the joint is not zero. The difference between zero and the actual sum of the end moments is the unbalanced moment.

Distributed Moments

After the clamp at a joint is released, the unbalanced moment causes the joint to rotate. The rotation twists the ends of the members at the joint and changes their moments. In other words, rotation of the joint is resisted by the members and resisting moments are built up in the members as they are twisted. Rotation continues until equilibrium is reached—when the resisting moments equal the unbalanced moment—at which time the sum of the moments at the joint is equal to zero. The moments developed in the members resisting rotation are the distributed moments.

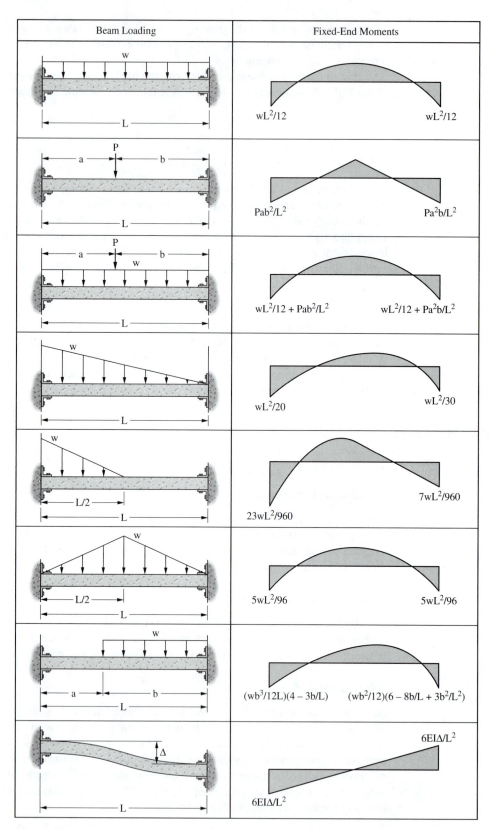

Figure 20.4 Fixed-end moment expressions

Carryover Moments

The distributed moments in the ends of the members cause moments in the other ends, which are assumed fixed. These are the carryover moments.

20.4 SIGN CONVENTION

The moments at the end of a member are assumed to be negative when they tend to rotate the member end clockwise about the joint (the resisting moment of the joint would be counterclockwise). The continuous beam of Figure 20.5 with all joints assumed to be clamped, has clockwise (or $-$) moments on the left end of each span and counterclockwise (or $+$) moments on the right end of each span. The usual sign convention used in strength of materials shows fixed-ended beams to have negative moments on both ends for downward loads, because tension is caused in the top fibers of the beams at those points. It should be noted that this sign convention, to be used for moment distribution, is the same one used in Chapter 18 for slope deflection.

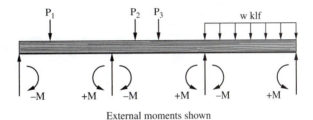

External moments shown

Figure 20.5 Sign convention for moment distribution

20.5 APPLICATION OF MOMENT DISTRIBUTION

The very few tools needed for applying moment distribution are now available, and the method of applying them is described.

A continuous beam with several loads applied to it is shown in Figure 20.6(a). In Figure 20.6(b), the interior joints B and C are assumed to be clamped, and the fixed-end moments are computed. At joint B, the unbalanced moment is computed and the clamp is removed, as seen in Figure 20.6(c). The joint rotates, thus distributing the unbalanced moment to the B ends of BA and BC in proportion to their distribution factors. The values of these distributed moments are carried over at the one-half rate to the other ends of the members AB and BC. When equilibrium is reached, joint B is clamped in its new rotated position and joint C is released, as shown in Figure 20.6(d). Joint C rotates under its unbalanced moment until it reaches equilibrium; the rotation causes distributed moments in the C ends of members CB and CD and their resulting carryover moments. Joint C is now clamped and joint B is released, in Figure 20.6(e).

The same procedure is repeated again and again for joints B and C. The amount of unbalanced moment quickly diminishes until the release of a joint causes only negligible rotation. This process, in brief, is moment distribution.

Examples 20.1 to 20.3 illustrate the analysis of three continuous beams. For Example 20.1, the moment of inertia I is constant for both spans and is assumed to equal 1.0 for calculating the relative stiffnesses or $\frac{I}{L}$ values of those members. With these values the distribution factors are determined as follows:

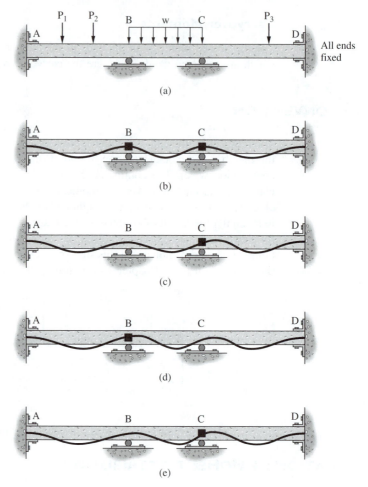

Figure 20.6   Procedure for moment distribution

$$DF_{BA} = \frac{K_{BA}}{\sum K} = \frac{\dfrac{1}{20}}{\dfrac{1}{20} + \dfrac{1}{25}} = 0.43$$

$$DF_{BC} = \frac{K_{BC}}{\sum K} = \frac{\dfrac{1}{15}}{\dfrac{1}{20} + \dfrac{1}{25}} = 0.57$$

A simple tabular form is used for Examples 20.1 and 20.2 to introduce the reader to moment distribution. For subsequent examples, a slightly varying but quicker solution much preferred by the author is used. The tabular procedure may be summarized as follows:

1. The fixed-end moments are computed and recorded on one line (line FEM's in Examples 20.1 and 20.2).
2. The unbalanced moments at each joint are balanced in the next line (Dist 1).
3. The carryovers are made from each of the joints on the next line (CO 1).
4. The new unbalanced moments at each joint are balanced (Dist 2), and so on. (As the beam of Example 20.1 has only one joint to be balanced, only one balancing cycle is necessary.

EXAMPLE 20.1

Determine the end moments for the members of the structure shown in Figure 20.7 by moment distribution.

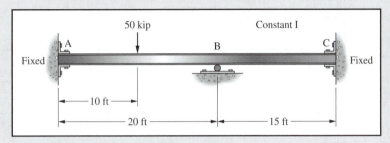

Figure 20.7

Solution.

0.0		0.43	0.57		0.0	Distribution Factors
−125		+125	0.0		0.0	Fixed-End Moments
0.0		−53.8	−71.2		0.0	Distribution 1
−26.9					−35.6	Carry Over 1
−151.9		+71.2	−71.2		−35.6	Final Moments

Carry overs shown with dashed lines

■

When the distribution has reached the accuracy desired, a double line is drawn under each column of figures. The final moment in the end of a member equals the sum of the moments at its position in the table. Unless a joint is fixed, the sum of the final end moments in the ends of the members meeting at the joint must total zero.

EXAMPLE 20.2

Using moment distribution, determine the member end moments for the structure of Figure 20.8. As the members have different I values, the actual numbers (or the relative values of those numbers) are used to compute the stiffness and distribution factors.

The computed value of $\frac{I}{L}$ for member AB is $\frac{200}{25} = 8$, while for member BC it's $\frac{300}{30} = 10$. The total stiffness of the members meeting at B is $8 + 10 = 18$. Member AB has $\frac{8}{18}$ ths of the total $= 0.44$, while member BC has $\frac{10}{18}$ ths of the total $= 0.56$. These values are the distribution factors at joint B. Similar calculations are made at joint C for the distribution factors.

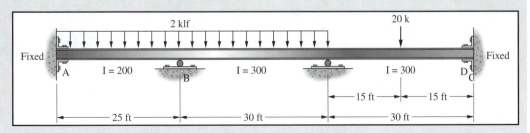

Figure 20.8

Solution.

0.0		0.44	0.56		0.5	0.5		0.0	D.F.s
−104.2		+104.2	−150.0		+150.0	−75.0		+75.0	FEM's
		+20.2	+25.6		−37.5	−37.5			Dist 1
+10.1			−18.8		+12.8			−18.8	CO 1
		+8.3	+10.5		−6.4	−6.4			Dist 2
+4.2			−3.2		+5.3			−3.2	CO 2
		+1.4	+1.8		−2.7	−2.7			Dist 3
+0.7			−1.3		+0.9			−1.3	CO 3
		+0.6	+0.7		−0.4	−0.4			Dist 4
+0.3			−0.2		+0.4			−0.2	CO 4
		+0.1	+0.1		−0.2	−0.2			Dist 5
−88.9		+134.8	−134.8		+122.2	−122.2		+51.5	Final

■

Beginning with Example 20.3, a slightly different procedure is used for distributing the moments. Only one joint at a time is balanced and the required carryovers are made from that joint. Generally speaking it is desirable (but not necessary) to balance the joint that has the largest imbalance; make the carryovers; balance the next joint with the largest imbalance, and so on, because such a process will result in the quickest convergence. This procedure is quicker than the tabular method used for Examples 20.1 and 20.2, and it follows along exactly with the description of the behavior of a continuous beam (with imaginary clamps) as pictured by the author in Figure 20.6.

EXAMPLE 20.3

Compute the end moments in the beam shown in the beam of Figure 20.9.

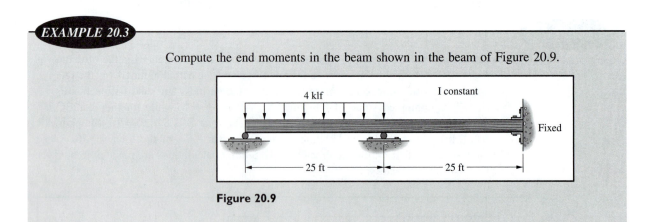

Figure 20.9

Solution.

0.0		0.5	0.5		0.0
−208.3		+208.3			
+208.3		−104.2			
−78.1		−156.2	−156.2		−78.1
+78.1		+39.0			
−9.8		−19.5	−19.5		−9.8
+9.8		+4.9			
−1.2		−2.5	−2.5		−1.2
+1.2		+0.6			
−0.2		−0.3	−0.3		−0.2
+0.2		+178.5	−178.5		−89.3
0.0					

■

In Chapter 19 several methods for approximately analyzing statically indeterminate structures were introduced. Moment distribution is one of the "exact" methods of analysis if it's carried out until the moments to be distributed and carried over become quite small. However, it can be used as a superb approximate method for statically indeterminate structures if only a few cycles of distribution are made.

Consider the beam of Example 20.2. In this problem, each cycle of distribution is said to end when the unbalanced moments are balanced. Table 20.1 shows the ratios of the total moments up through each cycle at joints A and C to the final moments at those joints after all the cycles are completed. These ratios should give you an idea of how good partial moment distribution can be as an approximate method.

TABLE 20.1 ACCURACY OF MOMENT DISTRIBUTION AFTER VARIOUS CYCLES OF BALANCING FOR THE BEAM OF EXAMPLE 20.2

Values given after cycle no.	Moments at support A	Ratio of approximate moment to "exact" moment support A	Moment at support C	Ratio of approximate moment to "exact" moment support C
1	104.2	1.172	112.5	0.921
2	94.1	1.058	118.9	0.973
3	89.9	1.011	121.5	0.994
4	89.2	1.003	122.0	0.998
5	88.9	1.000	122.2	1.000

For many cases where the structure and/or the loads are very unsymmetrical there will be a few cycles of major adjustment of the moments throughout the structure. Until these adjustments are substantially made, moment distribution will not serve as a very good approximate method. The analyst will easily be able to see when the distribution process can be stopped with good approximate results. This will occur when the unbalanced moments and carryover values become rather small as compared to the initial values.

20.6 MODIFICATION OF STIFFNESS FOR SIMPLE ENDS

The carryover factor was developed for carrying over to fixed ends, but it is applicable to simply supported ends, which must have final moments of zero. The simple end of Example 20.3 was considered to be clamped; the carryover was made to the end; and the joint was freed and balanced back to zero. Then one half of that balancing moment was carried over to the adjacent support. This procedure when repeated over and over is absolutely correct, but it involves a little unnecessary work, which may be eliminated by studying the stiffnesses of members with simply supported ends.

Shown in Figure 20.10(a) and (b) is a comparison of the relative stiffness of a member subjected to a moment M_1 when the far end is fixed and also when it is simply supported. In part (a), using principles of the moment-area method that were discussed in Chapter 11, the slope at the left end, which is represented by θ_1, is equal to $M_1L/4EI$.

For the simple end-supported beam shown in Figure 20.10(b), again using moment-area principles, the slope θ_1 is found to be $M_1L/3EI$. Therefore, the slope caused by the moment M_1 when the far end is fixed is only three-fourths as large $(M_1L/4EI \div M_1L/3EI = \frac{3}{4})$ when the far end is simply supported. The beam simply supported at the far end is only three-fourths as stiff as the one that is fixed. If the stiffness factors for end spans that are simply supported are multiplied by three-fourths, the simple end is initially balanced to zero, carryovers made to adjacent supports, no carryovers made to that end afterward, and the same results will be obtained. In Example 20.4 the stiffness modification is used for the beam that was previously evaluated in Example 20.3.

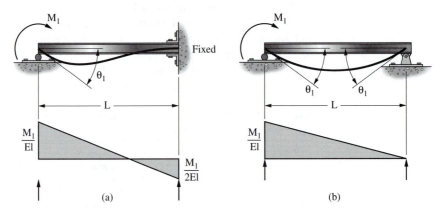

Figure 20.10 Comparison of relative stiffness for two support conditions

EXAMPLE 20.4

Determine the end moments of the structure shown in Figure 20.11 by using the simple endstiffness modification for the left end.

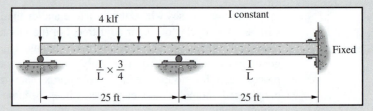

Figure 20.11

Solution.

0.0		0.43	0.57		0.0
−208.3		+208.3			
+208.3		+104.2			
0.0		−134.4	−178.1		−89.1
		+178.1	−178.1		−89.1

■

20.7 SHEARING FORCE AND BENDING MOMENT DIAGRAMS

Drawing shear and moment diagrams is an excellent way to check the final moments computed by moment distribution and to obtain an overall picture of the stress condition in the structure. Before preparing the diagrams, it is necessary to consider a few points relating the shear and the moment diagram sign convention to the one used for moment distribution. The usual conventions for drawing the diagrams will be used—tension in bottom fibers of beam is positive moment and upward shear to the left is positive shear.

The relationship between the signs of the moments for the two conventions is shown with the beams of Figure 20.12. Part (a) of the figure illustrates a fixed-end beam for which the result of moment distribution is a negative moment. The clockwise moment bends the beam as shown, causing tension in the top fibers or a negative moment for the shear and moment diagram convention. In Figure 20.12(b) the result of moment distribution is a positive moment, but again the top beam fibers are in tension, indicating a negative moment for the moment diagram.

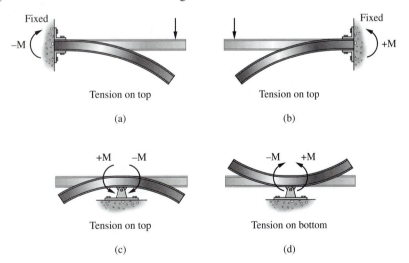

Figure 20.12   Interpreting sense of resulting moments

An interior simple support is represented by Figure 20.12(c) and (d). In part (c) moment distribution gives a negative moment to the right and a positive moment to the left, which causes tension in the top fibers. Part (d) shows the effect of moments of opposite sign at the same support considered in part (c).

From the preceding discussion it can be seen that the sign convention used herein for moment distribution for continuous beams agrees with the one used for drawing moment diagrams on the right-hand sides of supports, but disagrees on the left-hand sides.

To draw the diagrams for a vertical member, the right side is often considered to be the bottom side. Moments are distributed in the continuous beams in Examples 20.5 and 20.6 and the shearing force and bending moment diagrams are drawn.

The reactions shown in the solution of these problems were obtained by computing the reactions as though each span was simply supported and adding them to the reactions due to the moments at the beam supports.

EXAMPLE 20.5

Distribute moments and draw shear and moment diagrams for the structure shown in Figure 20.13.

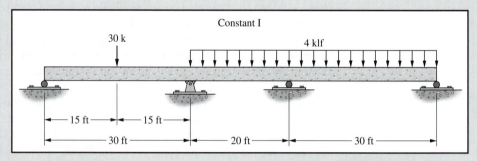

Figure 20.13

Solution.

0.0		0.33	0.67		0.67	0.33		0.0
−112.5		+112.5	−133.3		+133.3	−300.0		+300.0
+112.5		+56.2				−150.0		−300.0
			+105.6		+211.1	+105.6		
		−47.0	−94.0		−47.0			
			+15.7		+31.3	+15.7		
		−5.2	−10.5		−5.2			
			+1.7		+3.5	+1.7		
		−0.6	−1.1		−0.6			
			+0.2		+0.4	+0.2		
		−0.1	−0.1					
0.0		+115.8	−115.8		+326.8	−326.8		0.0
Reaction Computations								
"Simple beam" reactions	+15.00	+15.00	+40.00		+40.00	+60.00		+60.00
Moment reactions	−3.86	+3.86	−10.55		+10.55	+10.89		−10.89
Final reactions	+11.14	+18.86	+29.45		+50.55	+70.89		+49.11

Shearing Force and Bending Moment Diagrams:

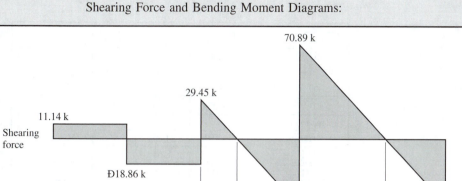

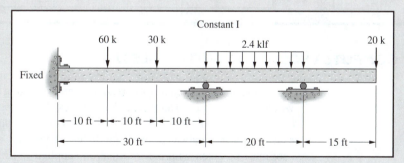

EXAMPLE 20.6

Distribute moments and draw shear and moment diagrams for the continuous beam shown in Figure 20.14.

Figure 20.14

Solution.

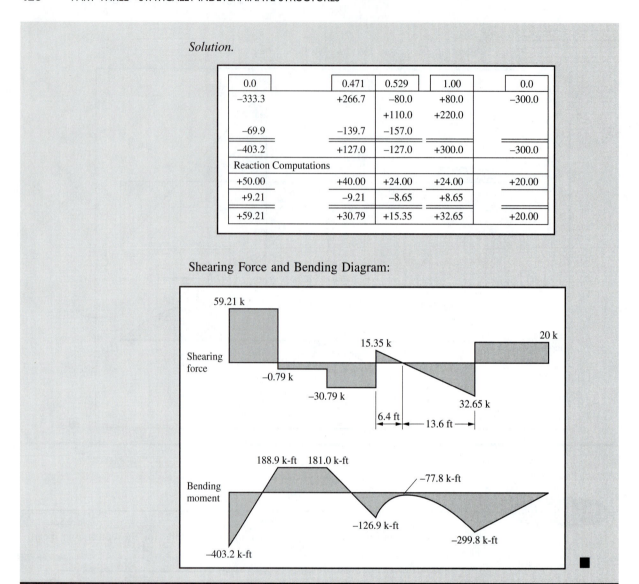

0.0		0.471	0.529	1.00		0.0
−333.3		+266.7	−80.0	+80.0		−300.0
			+110.0	+220.0		
−69.9		−139.7	−157.0			
−403.2		+127.0	−127.0	+300.0		−300.0
Reaction Computations						
+50.00		+40.00	+24.00	+24.00		+20.00
+9.21		−9.21	−8.65	+8.65		
+59.21		+30.79	+15.35	+32.65		+20.00

Shearing Force and Bending Diagram:

20.8 COMPUTER SOLUTION WITH SABLE32

The Sable software enables the reader to analyze quite quickly the problems contained within this chapter. Example 20.7 illustrates the analysis of the three-span continuous beam of Figure 20.15(a) with SABLE32. Although the data is input to the computer much as it was for the analysis of two- and three-dimensional trusses in earlier chapters, special care must be taken with the information supplied for the ends of the beam spans.

The reader will remember that the types of support for the beam are entered with the joint data. For this beam, the left fixed end support supplies resistance to movement in the x, y, and z directions, while the other supports only supply resistance to movement in the y direction.

When the beam data is supplied, the user should think of the ends of the spans as nodes and must decide whether the end of one span can move independently with respect to the adjoining span. For the beam of Figure 20.15, the reader can see that this is not the case at the fixed end nor is it possible at the interior nodes where the right ends of the left-hand spans cannot move with respect to the left-hand ends of the right-hand spans. As a result, the end conditions of the spans are as shown in Figure 20.15(b) with the letters F-F, which stand for fixed-fixed.

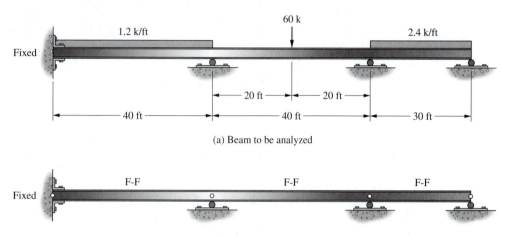

(a) Beam to be analyzed

(b) Support and node conditions for beam

Figure 20.15

All of the input data is shown with this solution as the author feels it may be helpful to the student in specifying the end conditions for his or her problems.

EXAMPLE 20.7

Analyze the beam of Figure 20.15 using SABLE32.

Solution. The member is resketched in Figure 20.16 and its joints and spans numbered.

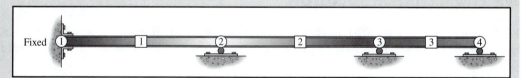

Figure 20.16

Data: Joint Location and Restraint Data

	Coordinates			Restraints		
Jnt	X	Y	Z	X	Y	Rot
1	0.000E+00	0.000E+00	0.000E+00	Y	Y	Y
2	4.000E+01	0.000E+00	0.000E+00	N	Y	N
3	8.000E+01	0.000E+00	0.000E+00	N	Y	N
4	1.100E+02	0.000E+00	0.000E+00	N	Y	N

Data: Beam Location and Property Data

Beam	i	j	Type	Stat	Beam Properties		
					Area	Izz	E
1	1	2	F-F	A	1.000E+00	1.000E+00	1.000E+00
2	2	3	F-F	A	1.000E+00	1.000E+00	1.000E+00
3	3	4	F-F	A	1.000E+00	1.000E+00	1.000E+00

Data: Applied Beam Loads

Beam	Case	P	a	W
1	1	0.000E+00	0.000E+00	-1.200E+00
2	1	-6.000E+01	2.000E+01	0.000E+00
3	1	0.000E+00	0.000E+00	-2.400E+00

Data: Calculated Beam End Forces

Beam	Case	End	Axial	Shear-Y	Moment-Z
1	1	i	0.000E+00	2.105E+01	1.207E+02
		j	0.000E+00	2.695E+01	-2.387E+02
2	1	i	0.000E+00	2.835E+01	2.387E+02
		j	0.000E+00	3.165E+01	-3.047E+02
3	1	i	0.000E+00	4.616E+01	3.047E+02
		j	0.000E+00	2.584E+01	6.148E-06

■

20.9 PROBLEMS FOR SOLUTION

For Problems 20.1 through 20.21 analyze the structures by the moment-distribution method and draw shear and moment diagrams. For answers given, the signs are based on the shear and moment diagram convention.

20.1   (*Ans.* $M_A = -192.5$ ft-k, $M_B = -215$ ft-k)

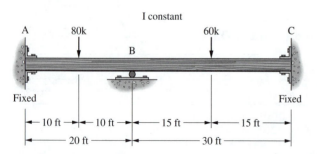

20.2

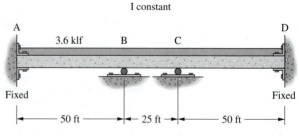

20.3 (*Ans.* $M_A = -414.5$ ft-k, $M_C = -269.2$ ft-k)

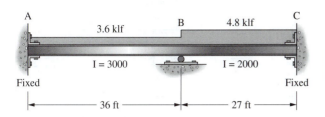

20.4 Rework Problem 20.3 if a 60 k load is added at the centerline of the right hand span.

20.5 (*Ans.* $M_B = -580.5$ ft-k, $M_C = -646.8$ ft-k)

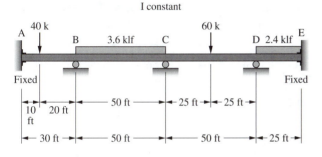

20.6

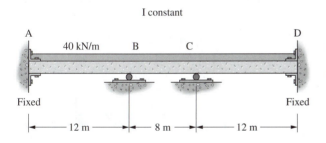

20.7 (*Ans.* $M_B = -146.28$ ft-k, $V_C = 73.75$ k ↑)

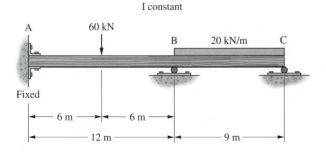

20.8

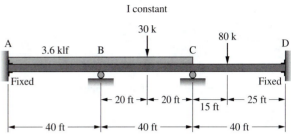

20.9 (*Ans.* $M_B = -600$ ft-k, $M_C = +203$ ft-k)

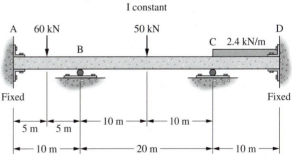

20.10

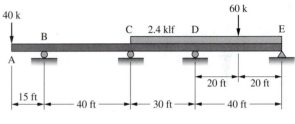

20.11 (*Ans.* $M_B = -171.9$ ft-k, $M_D = -75.9$ ft-k)

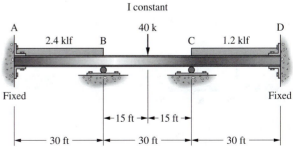

20.12

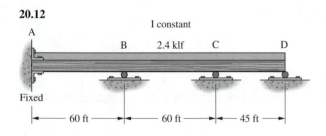

I constant

A B 2.4 klf C D

Fixed

| 60 ft | 60 ft | 45 ft |

20.13 (*Ans.* $M_A = -165.9$ ft-k, $M_C = -199.7$ ft-k)

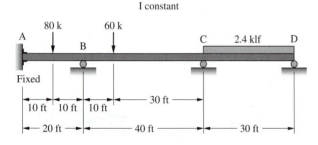

I constant

80 k   60 k

A B   C 2.4 klf D

Fixed

10 ft | 10 ft | 10 ft   30 ft

20 ft   40 ft   30 ft

20.14

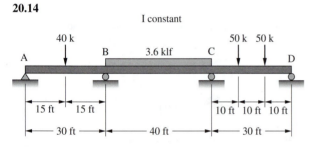

I constant

40 k   50 k 50 k

A   B 3.6 klf C   D

15 ft | 15 ft   10 ft | 10 ft | 10 ft

30 ft   40 ft   30 ft

20.15 Repeat Problem 20.2 if the end supports A and D are made simple supports. (*Ans.* $M_B = -723.3$ ft-k $= M_C$)

20.16

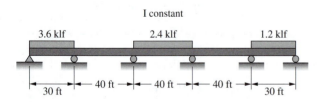

I constant

3.6 klf   2.4 klf   1.2 klf

30 ft | 40 ft | 40 ft | 40 ft | 30 ft

20.17 (*Ans.* $M_B = -538.8$ ft-k, $M_C = -715$ ft-k)

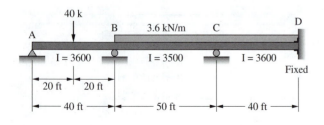

40 k

A   B 3.6 kN/m C   D

I = 3600   I = 3500   I = 3600

Fixed

20 ft | 20 ft

40 ft   50 ft   40 ft

20.18

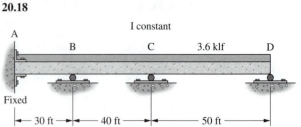

I constant

A B   C 3.6 klf D

Fixed

30 ft | 40 ft | 50 ft

20.19 (*Ans.* $M_A = -48.2$ ft-k, $M_B = -262.0$ ft-k)

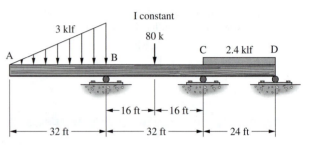

I constant

3 klf   80 k

A   B   C 2.4 klf D

16 ft | 16 ft

32 ft   32 ft   24 ft

20.20

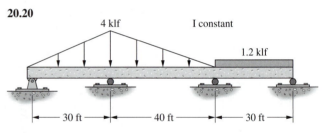

4 klf   I constant   1.2 klf

30 ft   40 ft   30 ft

20.21 (*Ans.* $M_A = -88.5$ ft-k, $M_C = -439.7$ ft-k)

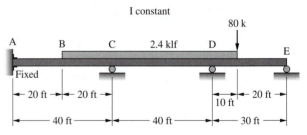

I constant

80 k

A B   C 2.4 klf D   E

Fixed

20 ft | 20 ft   20 ft
                    10 ft

40 ft   40 ft   30 ft

For Problems 20.22 to 20.26 use SABLE32.

20.22 Problem 20.1

20.23 Problem 20.3 (*Ans.* $M_A = -414.5$ ft-k, $M_C = -268.7$ ft-k)

20.24 Problem 20.6

20.25 Go to the library and read the paper written by Hardy Cross about the moment distribution method. See the footnotes on the first page of this chapter for the citation.

Chapter 21

Moment Distribution for Frames

21.1 FRAMES WITH SIDESWAY PREVENTED

Moment distribution for frames is handled in the same manner as it is for beams when sidesway or lateral movement cannot occur. Analysis of frames without sidesway is illustrated by Examples 21.1 and 21.2. Where sidesway is possible, however, it must be taken into account because joint displacements cause rotations at the ends of members connected to the joints, and therefore affect the moments in all the members.

As the structures being analyzed become more complex, a method of recording the calculations so they will not run into each other is necessary. In this chapter, a system is used for frames whereby the moments are recorded below beams on their left ends and above them on their right ends. For columns, the same system is used, the right sides being considered the bottom sides.

EXAMPLE 21.1

Determine the end moments for the frame shown in Figure 21.1. The relative value of I is shown for each member of the frame.

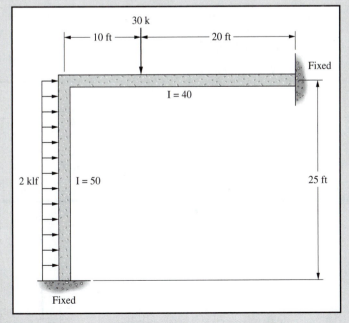

Figure 21.1

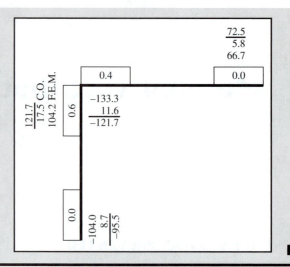

EXAMPLE 21.2

Compute the end moments of the structure shown in Figure 21.2. I values are constant for all members in the structure.

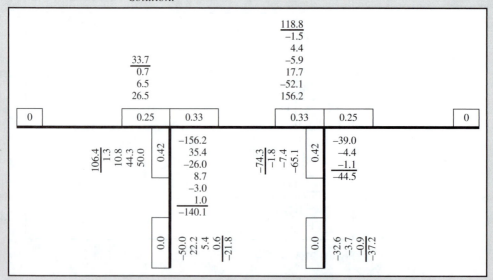

Figure 21.2

Solution.

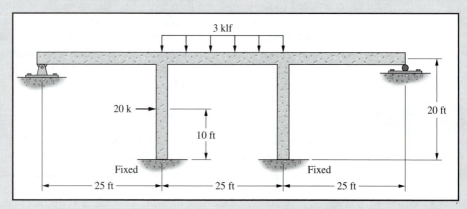

21.2 FRAMES WITH SIDESWAY

Structural frames, similar to the one shown in Figure 21.3, are usually so constructed that they may possibly sway to one side or the other under load. The frame in this figure is symmetrical, but it will tend to sway because the load P is not centered.

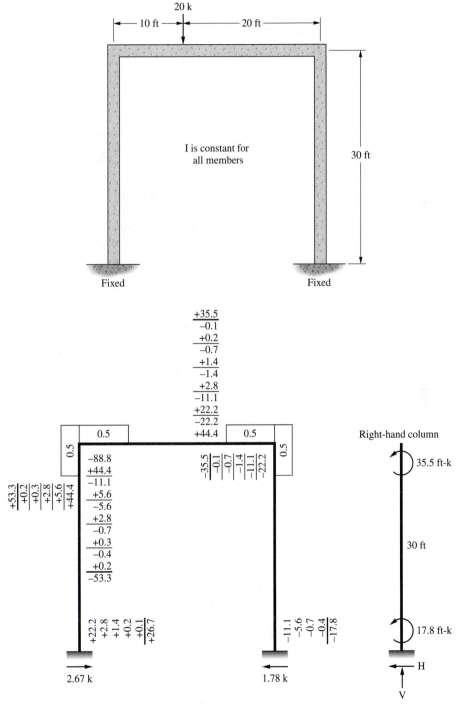

Figure 21.3

Analysis of the frame by the usual procedure, illustrated in the figure, gives inconsistent results.

The fixed-end moments are balanced and the horizontal reaction components are calculated at the supports. Each column is taken out as a free body. Moments are taken at the top of a particular column of the forces and moments applied to that column and the horizontal reaction component is determined. With reference to the sketch of the right-hand column in Figure 21.3, the following expression can be written noting that the vertical reaction component passes through the point where moments are being taken if the column is itself vertical.

$$\underline{\sum M_{\text{top of column}} = 0}$$
$$-35.5 - 17.8 + 30H = 0$$
$$H = +1.78 \text{ k} \leftarrow$$

The reaction at the base of the left-hand column is determined in a similar manner. From the results obtained it can be seen that the sum of the horizontal forces on the structure is not equal to zero. The sum of the forces to the right is 0.89 kip more than the sum of the forces to the left. If this symmetrical structure were subjected to an unbalanced force system such as this one, it would not be in equilibrium.

The usual analysis does not yield consistent results because the structure actually sways or deflects to one side, and the resulting deflections affect the moments. One possible solution is to compute the deflections caused by applying a force of 0.89 kips acting to the right at the top of the bent. The moments caused could be obtained for the computed deflections and added to the originally distributed fixed-end moments, but the method is rather difficult to apply.

A much more convenient method is to assume the existence of an imaginary support, sometimes called a virtual support, that prevents the structure from swaying, as shown in Figure 21.4. The fixed-end moments are distributed, and the force the imaginary support must supply to hold the frame in place is computed. For the frame of Figure 21.3 the fictitious support must supply 0.89 kip pushing to the left.

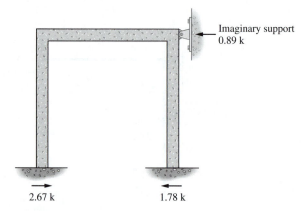

Figure 21.4

The support is imaginary and if removed will allow the frame to sway to the right. As the structure sways to the right the joints are assumed to be locked against rotation. The ends of the columns rotate in a clockwise direction and produce clockwise or negative moments at the joints (Figure 21.5). Assumed values of these sidesway moments can be placed in the columns. The necessary relations between these moments are discussed in Section 21.3.

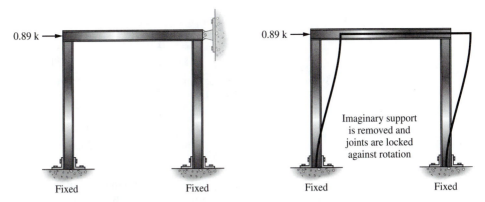

Figure 21.5

21.3 SIDESWAY MOMENTS

Should all of the columns in a frame be the same length and have the same moments of inertia, the assumed sidesway moments would be the same for each column. However, should the columns have different lengths and/or moments of inertia, this will not be the case. You will see in the following paragraphs that the assumed sidesway moments will vary from column to column in proportion to their respective ratios of I/L^2.

If the frame of Figure 21.6(a) is pushed laterally an amount Δ by the lateral load P, it will have the deflected shape shown in Figure 21.6(b). Theoretically, both columns will become perfect S-curves, if the beam is considered to be rigid. At their mid-height, the deflection for both columns will equal $\Delta/2$. Mid-height of the columns may be considered points of contra-flexure; the bottom half of each column may be dealt with as though it was a simple cantilevered beam. The deflection of a cantilevered beam with a concentrated load at its end is

$$\Delta = \frac{PL^3}{3EI}$$

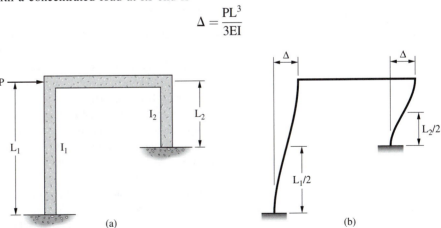

Figure 21.6   Simple frame for discussion of analysis considering sidesway

Since the deflections are the same for both columns, the following expressions may be written:

$$\frac{\Delta}{2} = \frac{P_1\left(\dfrac{L_1}{2}\right)^3}{3EI_1} = \frac{P_1L_1^3}{24EI_1}$$

$$\frac{\Delta}{2} = \frac{P_2 \left(\frac{L_2}{2}\right)^3}{3EI_2} = \frac{P_2 L_2^3}{24EI_2}$$

By solving these deflection expressions for P_1 and P_2, the forces pushing on the cantilevers, we have:

$$P_1 = \frac{12EI_1\Delta}{L_1^3}$$

$$P_2 = \frac{12EI_2\Delta}{L_2^3}$$

The moments caused by the two forces at the ends of their respective cantilevers are equal to the force times the cantilever length. These moments are written and the values of P_1 and P_2 are substituted in them:

$$M_1 = P_1 \frac{L_1}{2} = \left(\frac{12EI_1\Delta}{L_1^3}\right) \frac{L_1}{2} = \frac{6EI_1\Delta}{L_1^2}$$

$$M_2 = P_2 \frac{L_2}{2} = \left(\frac{12EI_2\Delta}{L_2^3}\right) \frac{L_2}{2} = \frac{6EI_2\Delta}{L_2^2}$$

From these expressions, a ratio of the moments may be written as follows because Δ is the same in each column:

$$\frac{M_1}{M_2} = \frac{6EI_1\Delta/L_1^2}{6EI_2\Delta/L_2^2} = \frac{I_1/L_1^2}{I_2/L_2^2}$$

This relationship must be used for assuming sidesway moments for the columns of a frame. Any convenient moments may be assumed, but they must be in proportion to each other as their I/L^2 values. Should their I and L values be equal, the assumed moments would be equal.

The procedure for applying the moment distribution method when sidesway is involved can be summarized as follows:

1. Determine the distribution factors for each member in the frame.
2. Calculate the fixed-end moments caused by the applied loads.
3. Distribute the fixed-end moments until convergence is achieved.
4. Compute the force imbalance, which is the force in the imaginary support that is preventing sidesway from occurring.
5. Compute the value of I/L^2 for each of the columns.
6. Select assumed sidesway moments in proportion to the I/L^2 values of each of the columns and distribute these moments until convergence is achieved. Any convenient moments may be selected as long as they are in proportion to the I/L^2 values of the columns.
7. Compute the lateral force caused by these moments.
8. Add these latter final moments, times the ratio of lateral forces, to the moments obtained in the original distribution. These are the final member-end moments in the frame.

The analysis of the frame of Figure 21.3 is completed in Figure 21.7 with the sidesway method. First, I/L^2 values are computed for each of the columns (equal in this case) in part (a) of the figure. Then, in part (b) sidesway moments in proportion to the I/L^2 values are assumed and distributed throughout the frame, after which the horizontal reactions at the column bases are calculated. The assumed sidesway moments of -10 ft-k each are found to produce horizontal reactions to the left totaling 0.92 kip. Only 0.89 kip was needed, and if 0.89/0.92 times the values of the distributed moments are added to the

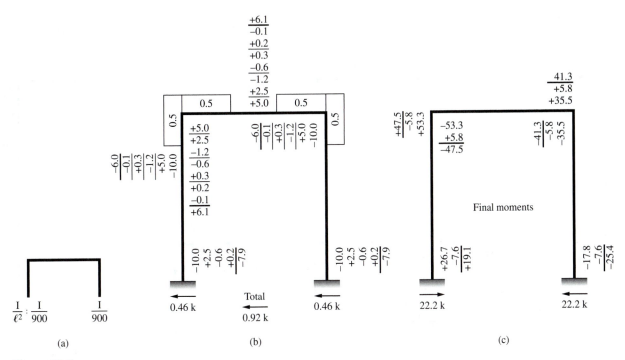

Figure 21.7

originally distributed fixed-end moments, the final moments will be obtained. The results are shown in Figure 21.7(c). The horizontal reactions at the column bases also are calculated and shown.

    Examples 21.3 to 21.5 present the solutions for additional sidesway problems. It will be noted in Example 21.4 that the column bases are pinned. Such a situation does not alter the method of solution. For the balancing of fixed-end moments or for balancing assumed sidesway moments, the bases are balanced to their correct zero values as was done in continuous beams. The three-quarter factor was used in computing the column stiffnesses for this example.

EXAMPLE 21.3

Using moment distribution find all of the member-end moments in the structure shown in Figure 21.8.

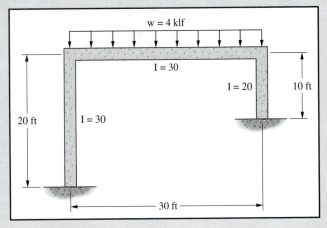

Figure 21.8

Solution. Distribute fixed-end moments and compute horizontal reactions at column bases.

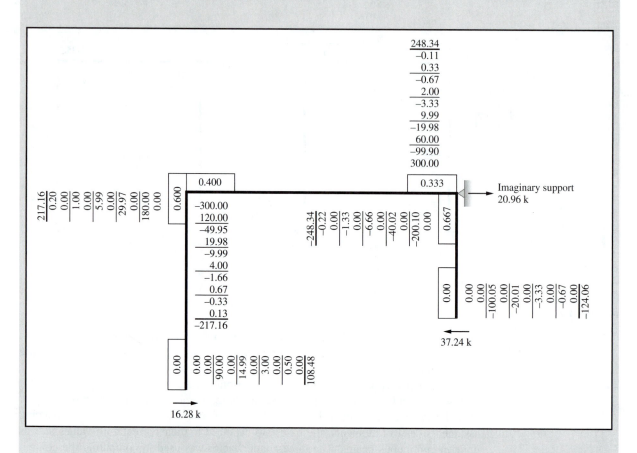

From these results we see that the imaginary support must exert a force of 20.96 kips acting to the right. That is the force necessary for static equilibrium in the horizontal direction. When the structure sways the assumed sidesway moments are negative and in the following proportion:

$$\frac{M_1}{M_2} = \frac{\dfrac{I_1}{L_1^2}}{\dfrac{I_2}{L_2^2}} = \frac{\dfrac{30}{20^2}}{\dfrac{20}{10^2}} = \frac{3}{8} \qquad \text{Let} \begin{cases} M_1 = 30 \text{ k-ft} \\ M_2 = 80 \text{ k-ft} \end{cases}$$

Next, distribute these assumed sidesway moments.

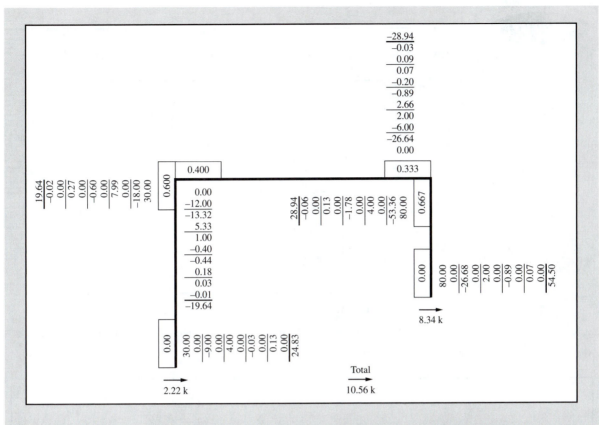

The factor by which the assumed sidesway moments are multiplied by is computed from

$$\frac{20.96}{10.56} = 1.985$$

The final member-end moments in the frame are then equal to 1.985 times the distributed assumed sidesway moments plus the distributed fixed end moments.

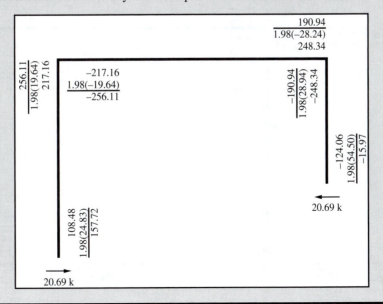

EXAMPLE 21.4

Determine the final member-end moments for the frame shown in Figure 21.9.

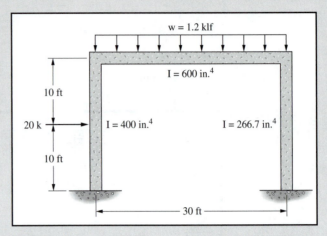

Figure 21.9

Solution. Distribute fixed-end moments and compute horizontal reactions at column bases.

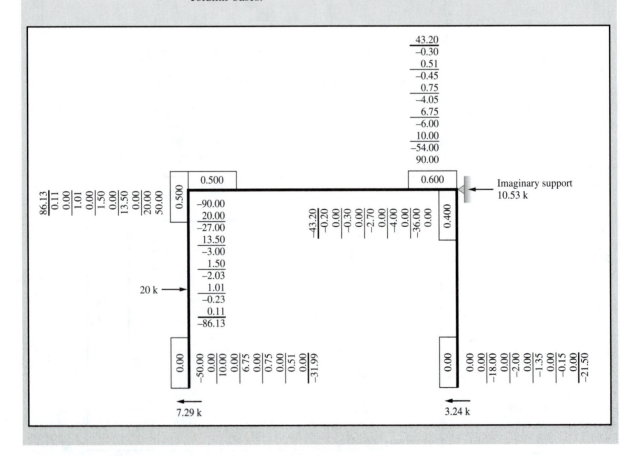

When calculating the shearing forces at the ends of the members, we must include any forces that are acting on the members. The left column has a concentrated load acting at midheight. The shearing force at the bottom of the left column, then, is

$$H_{Left} = \frac{+86.13 - 31.99 - (20)(10)}{20} = 7.29 \text{ k} \leftarrow$$

After calculating the reaction at the base of the right-hand column, we see that the imaginary support must exert a force of 10.53 kips acting to the left. That is the force necessary for static equilibrium in the horizontal direction, including the 20 kip lateral load. The assumed sidesway moments are determined as follows:

$$\frac{M_1}{M_2} = \frac{\dfrac{I_1}{L_1^2}}{\dfrac{I_2}{L_2^2}} = \frac{\dfrac{400}{20^2}}{\dfrac{266.7}{20^2}} = \frac{1.0}{0.667} \qquad \text{Let} \begin{cases} M_1 = 100 \text{ k-ft} \\ M_2 = 66.7 \text{ k-ft} \end{cases}$$

Next, distribute these assumed sidesway moments:

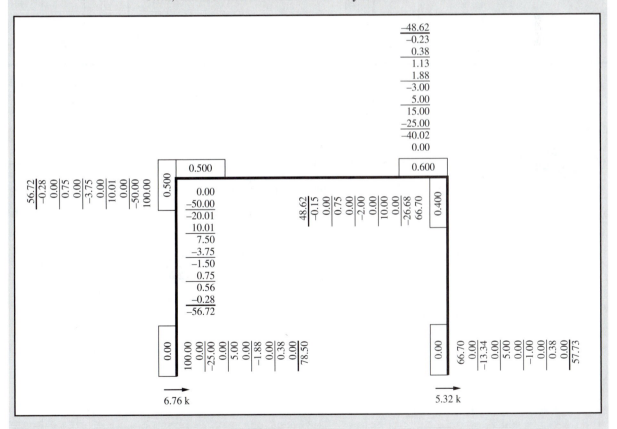

The factor by which the assumed sidesway moments are multiplied by is computed from the following noting the assumed sidesway moments should have been minus.

$$S = -\frac{20 - 7.29 - 3.24}{6.76 + 5.32} = -0.784$$

The final member-end moments are equal to the distributed fixed-end moments minus 0.784 times the distributed assumed sidesway moments.

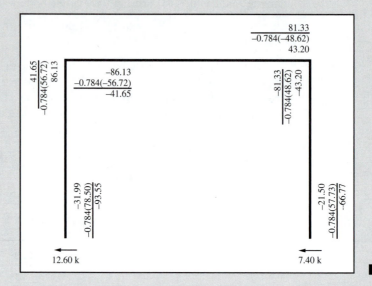

A two-bay or three-column frame is analyzed in Example 21.5 to show the same sidesway method may be used to analyze frames subject to sidesway no matter how many columns are present.

EXAMPLE 21.5

Compute the final end moments for the frame shown in Figure 21.10.

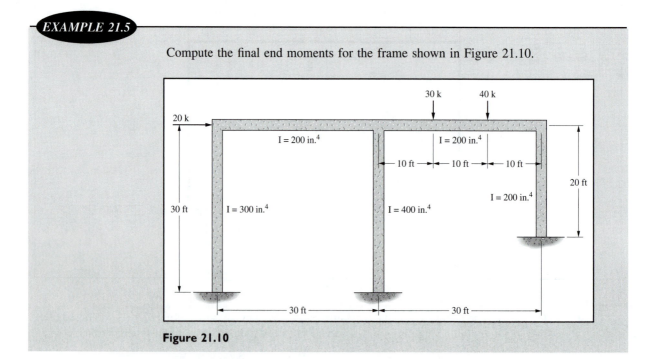

Figure 21.10

Solution.   Distribute fixed-end moments and compute horizontal reactions at column bases.

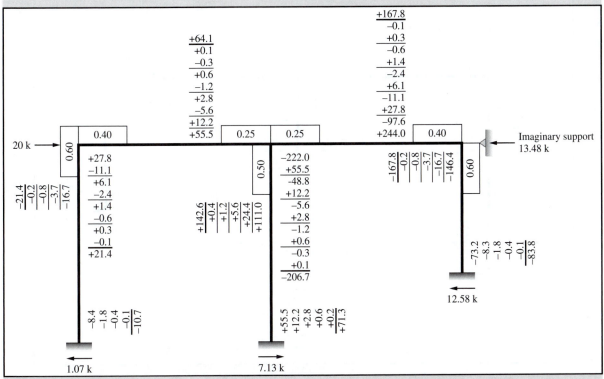

The structure sways to the right; therefore negative moments are assumed in the columns in proportion to the $\frac{I}{L^2}$ values.

$$M_1 : M_2 : M_3$$

$$\frac{30}{30^2} : \frac{40}{30^2} : \frac{20}{20^2}$$

$$30 : 40 : 45$$

Distribute assumed sidesway moments.

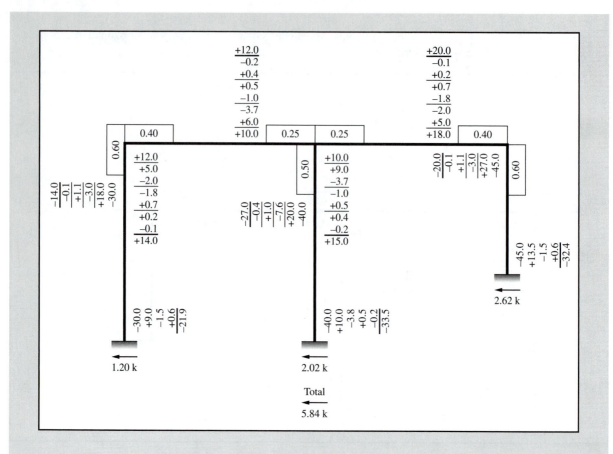

Final moments = distributed fixed end moments + $\frac{13.48}{5.84}$ times the distributed assumed sidesway moments.

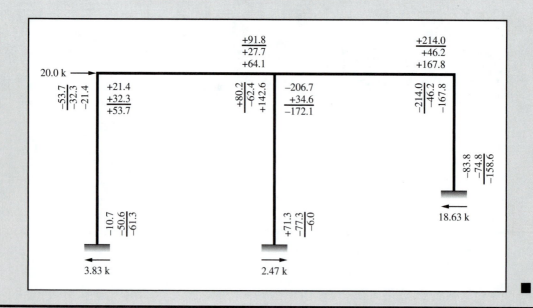

21.4 FRAMES WITH SLOPING LEGS

The frames considered up to this point have been made up of vertical and horizontal members. It was proved earlier in this chapter that when sidesway occurs in such frames, it causes fixed-end moments in the columns proportional to their $6EI\Delta/L^2$ values. (As $6E\Delta$ was a constant for these frames, moments were assumed proportional to their I/L^2 values.) Furthermore, lateral swaying did not produce fixed-end moments in the beams.

I-91 bridge, Lyndon, Vermont. (Courtesy of the Vermont Agency of Transportation.)

The sloping-leg frame of Figure 21.11 can be analyzed in much the same manner as the vertical-leg frames previously considered. The fixed-end moments due to the external loads are calculated and distributed; the horizontal reactions are computed; and the horizontal force needed at the imaginary support is determined.

Once again sidesway causes moments in the frame members proportional to their $\frac{6EI\Delta}{L^2}$ values. For sloping leg frames the Δ values often will be unequal for the various members, as will be the I and L values. Therefore, sidesway moments are assumed in proportion to the $\frac{I\Delta}{L^2}$ values. These assumed moments are distributed, the horizontal reactions are computed, and the necessary moments needed for balancing are calculated and superimposed on the distributed fixed-end moments.

For this discussion the frame of Figure 21.11(a) is considered and is assumed to sway to the right as shown in part (b) of the figure. It can be seen that lateral movement of

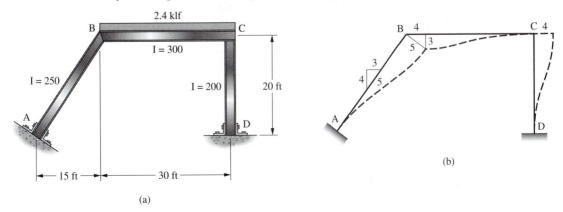

(a)

(b)

Figure 21.11

the frame will cause Δs, and thus moments in the beam as well as in the columns. To determine the $I\Delta/L^2$ value for each member it is necessary to determine the relative Δ values for the various members.

As the frame sways to the right, joint B moves in an arc about joint A and joint C moves in an arc about joint D. As these movements or arcs are very short, they are considered to consist of straight lines perpendicular to the respective members. Rather than attempting to develop complex trigonometric formulas for the relative Δs, the author has merely drawn deformation triangles on the figure.

Column AB has a slope of four vertically, three horizontally, or five inclined. If the relative movement of joint B perpendicular to AB is assumed to be five, then its vertical movement will be three and its horizontal movement will be four.

Joint C moves in a horizontal direction to the right perpendicular to member CD. If the change in length of member BC is neglected, joint C must move horizontally the same distance as joint B, or a distance of four. The relative Δ values are now available as follows: $\Delta_{AB} = 5$, clockwise; $\Delta_{BC} = 3$, counterclockwise; and $\Delta_{CD} = 4$, clockwise. The clockwise rotation produces a counterclockwise or negative resisting moment. These values are given at the end of this paragraph together with the $I\Delta/L^2$ values.

Relative Δ Values	Relative sidesway moments $I\Delta/L^2$	
$\Delta_{AB} = -5$	$M_{AB} = \dfrac{(250)(-5)}{(25)^2} = -2$	say -100 ft-k
$\Delta_{BC} = +3$		
$\Delta_{CD} = -4$	$M_{BC} = \dfrac{(300)(+3)}{(30)^2} = +1$	say $+50$ ft-k
	$M_{CD} = \dfrac{(200)(-4)}{(20)^2} = -2$	say -100 ft-k

Example 21.6 illustrates the analysis of the frame of Figure 21.11. The same procedure used for determining the horizontal reaction components at the bases of the sloping columns is used for the vertical columns. (That is, each column is considered to be a free body, and moments are taken at its top to determine the horizontal reaction at its base.) For vertical columns, the vertical reactions pass through the points where moments are taken and thus may be neglected. *This is not the case for sloping columns, and the vertical reactions will appear in the moment equations.*

The left column AB of the frame of Figure 21.11 is considered as a free body after the fixed-end moments are distributed and shown in Figure 21.12. Taking moments at B to determine H_A, the following equation results:

$$+59.9 + 120 + 15V_A - 20H_A = 0$$

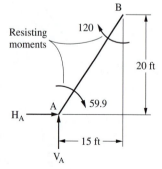

Figure 21.12

Note that it is necessary to compute V_A before the equation can be solved for H_A. The author finds it convenient to remove the beam BC as a free body and compute the vertical reaction applied at each end by the columns. Once the value at B is determined, the sum of the vertical forces on column AB can be equated to zero and V_A can be determined. The same procedure is followed after the assumed sidesway moments are distributed.

EXAMPLE 21.6

Determine the final end moments for the frame shown in Figure 21.11.

Solution. Distribute fixed-end moments and determine support reactions.

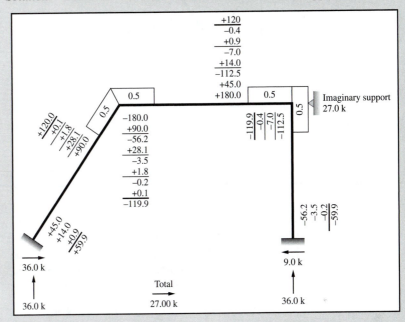

Distribute assumed sidesway moments.

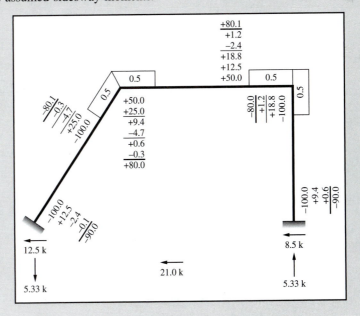

Final moments $= +\frac{27.0}{21.0}$ times the distributed assumed sidesway moments plus the distributed fixed end moments.

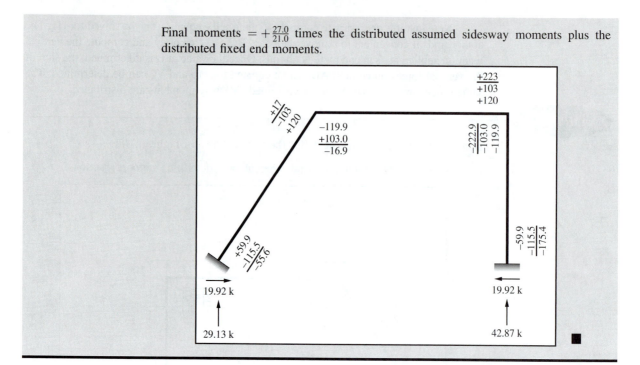

In Figure 21.13 another illustration of the determination of relative Δ values for a sloping leg frame is presented. With the numbers shown initially on the deformation triangles, joint B moves three units to the right and joint C moves four units to the right, but the horizontal movement of joints B and C must be equal. For this reason the initial values at C are marked through and multiplied by three quarters so the horizontal values will be equal. The resulting values are shown below.

Relative Δ values

$\Delta_{AB} = -3.605$
$\Delta_{BC} = +4.25$
$\Delta_{CD} = -3.75$

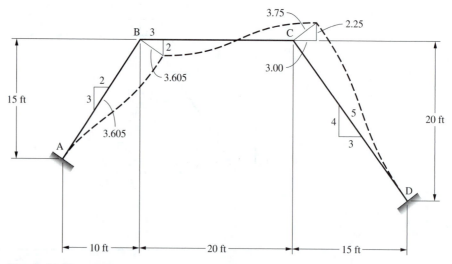

Figure 21.13

From the analyses considered in this chapter it is quite obvious that frame members are subject to axial forces (and thus axial deformations) as well as moments and shears. *The reader should clearly understand that the effects of axial deformations (which usually are negligible) are neglected in the moment distribution procedures described herein.*

21.5 MULTISTORY FRAMES

These are two possible ways in which the frame of Figure 21.14 may sway. The loads P_1 and P_2 obviously will cause both floors of the structure to sway to the right, but it is not known how much of the swaying is going to occur in the top floor (x condition) or how much will occur in the bottom floor (y condition). There are two sidesway conditions that need to be considered.

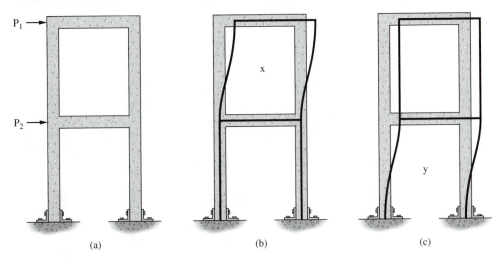

Figure 21.14   Story sway in a two story frame

To analyze the frame by the usual sidesway procedure would involve (1) an assumption of moments in the top floor for the x condition and the distribution of the moments throughout the frame, and (2) an assumption of moments in the lower floor for the y condition and the distribution of those moments throughout the frame. One equation could be written for the top floor by equating x times the horizontal forces caused by the x moments plus y times the horizontal forces caused by the y moments to the actual total shear on the floor, P_1. A similar equation could be written for the lower floor by equating the horizontal forces caused by the assumed moments to the shear on that floor, $P_1 + P_2$. Simultaneous solution of the two equations would yield the values of x and y. The final moments in the frame equal x times the x distributed moments plus y times the y distributed moments.

The sidesway method is not difficult to apply for a two-story frame, but for multistory frames it becomes unwieldy because each additional floor introduces another sidesway condition and another simultaneous equation.

Professor C. T. Morris of Ohio State University introduced a much simpler method for handling multistory frames, which involves a series of successive corrections.[1] His method also is based on the total horizontal shear along each level of a building. In

[1]C. T. Morris, "Morris on Analysis of Continuous Frames," *Transactions of the American Society of Civil Engineers* 96 (1932): 66–69.

considering the frame of Figure 21.15(a), which is being deflected laterally by the loads P_1 and P_2, each column is assumed to take the approximate S shape shown in (b).

P_1

P_2

V

h

$\frac{h}{2}$

$\frac{h}{2}$

(a)

(b)

(c)

Figure 21.15

At the middepth of the columns there is assumed to be a point of contraflexure. The column may be considered to consist of a pair of cantilevered beams, one above the point and one below the point, as shown in (c). The moment in each cantilever will equal the shear times $h/2 = Vh/2$, and the total moment top and bottom is equal to $Vh/2 + Vh/2 = Vh$. As such, the total moment in a column is equal to the shear carried by the column times the column height.

Office building, 99 Park Avenue, New York City. (Courtesy of the American Institute of Steel Construction, Inc.)

Similarly, on any one level the total moments top and bottom of all the columns will equal the total shear on that level multiplied by the column height. The method consists of initially assuming this total for the column moments on a floor and distributing it between the columns in proportion to their I/L^2 values.

The moment taken by each column is distributed half to top end and half to bottom end. The joints are balanced, including the fixed-end moments, making no carryovers until all the joints are balanced. At this time the sum of the column moments doesn't equal the correct final value—thus, they are corrected to their initial and final total, and the joints are balanced again. Successive corrections works exceptionally well for single or multistory frames, as illustrated by Examples 21.7 and 21.8. The procedure used is as follows:

1. Compute fixed-end moments.
2. Compute total moments in columns (equal to shear on the level times column height) for each level and distribute between the columns in proportion to their I/L^2 values, then divide each by half—one half to top of column and one half to bottom.
3. Balance all joints throughout the structure, making no carryovers.
4. Make carryovers for the entire frame.
5. The total of the column moments on each level has been changed and will not equal the shear times the column height. Determine the difference and add or subtract the amount back to the columns in proportion to their I/L^2 values.
6. Steps 3 to 5 are repeated over and over until the amount of the corrections to be made is negligible.

EXAMPLE 21.7

Determine final moments for the structure in Figure 21.16. Use the successive correction method developed by Professor Morris.

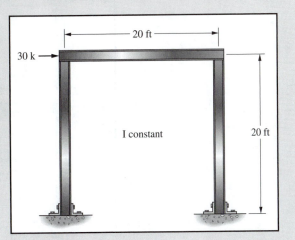

Figure 21.16

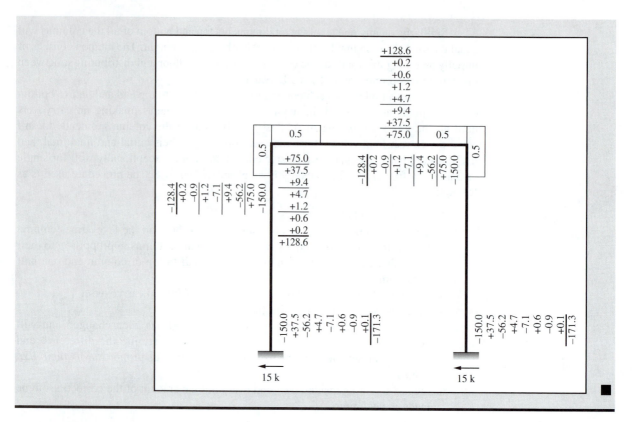

EXAMPLE 21.8

Determine the final moments for the frame of Figure 21.17 by successive corrections.

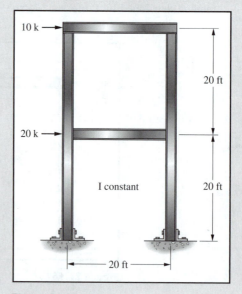

Figure 21.17

Solution.

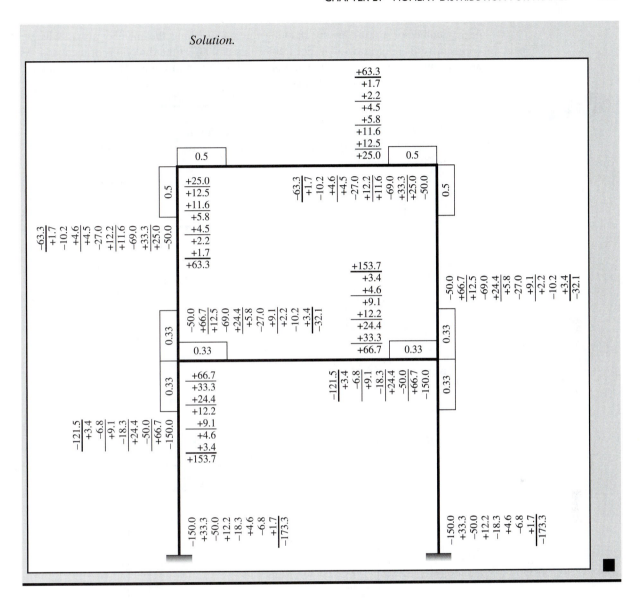

21.6 COMPUTER ANALYSIS OF FRAME

Example 21.9 presents the analysis of a one-story, one-bay frame with SABLE32. With one exception, the procedure is the same as the one previously used for beams. The one exception pertains to coordinate systems.

Two coordinate systems are used in SABLE32. These are the global coordinate system and the element coordinate system, which is also referred to as the local coordinate system. The global system is used to specify the data for the joints or nodes in a structure and to interpret the computed joint displacements and forces. The element or local system is used to specify the data concerning the elements or members in the structure and to interpret the computed element forces.

The two coordinate systems are discussed in detail in Chapter 24. The reader will learn in that matrix chapter that the local or element coordinate system is established for each member of a structure by drawing the x axis along the member's axis whether inclined, vertical, or horizontal. For frames this will affect the directions of the loads. As an illustration, the 50 kip horizontal member load shown in Figure 21.18 will be in the y direction perpendicular to the left column.

EXAMPLE 21.9

Determine the moments at the ends of the members of the frame shown in Figure 21.18 using SABLE32. The member areas, moments of inertia, and moduli of elasticity are constant for all of the members.

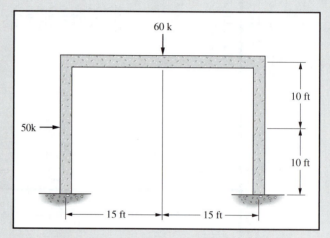

Figure 21.18

Solution. Numbering the joints and members of the frame.

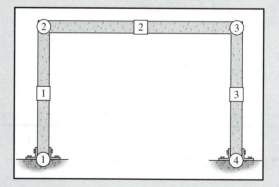

INPUT

Joint locations and restraint

	Coordinates			Restraints		
Jnt	X	Y	Z	X	Y	Rot
1	0.000E+00	0.000E+00	0.000E+00	Y	Y	Y
2	0.000E+00	2.000E+01	0.000E+00	N	N	N
3	3.000E+01	2.000E+01	0.000E+00	N	N	N
4	3.000E+01	0.000E+00	0.000E+00	Y	Y	Y

Beam locations and property data

Beam	i	j	Type	Stat	Area	Izz	E
1	1	2	F-F	A	1.000E+00	1.000E+00	1.000E+00
2	2	3	F-F	A	1.000E+00	1.000E+00	1.000E+00
3	3	4	F-F	A	1.000E+00	1.000E+00	1.000E+00

Applied beam loads

Beam	Case	P	a	W
1	1	−5.000E+01	1.000E+01	0.000E+00
2	1	−6.000E+01	1.500E+01	0.000E+00
3	1	0.000E+00	0.000E+00	0.000E+00

OUTPUT

Beam End Forces

Beam	Case	End	Axial	Shear-Y	Moment-Z
1	1	i	2.668E+01	2.857E+01	2.048E+02
		j	−2.668E+01	2.143E+01	−1.333E+02
2	1	i	2.143E+01	2.668E+01	1.333E+02
		j	−2.143E+01	3.332E+01	−2.330E+02
3	1	i	3.332E+01	2.143E+01	2.330E+02
		j	−3.332E+01	−2.143E+01	1.956E+02

21.7 PROBLEMS FOR SOLUTION

Balance moments and calculate horizontal reactions at bases of columns for Problems 21.1 through 21.6.

21.1   (*Ans.* $M_{BA} = 107.5$ ft-k, $M_{CB} = 123.1$ ft-k)

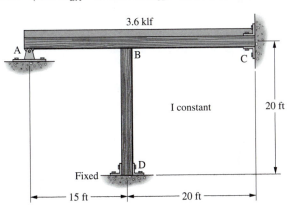

21.2

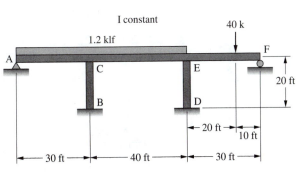

21.3   (*Ans.* $M_{BA} = 533.6$ kN, $M_{DB} = 533.2$ kN·m)

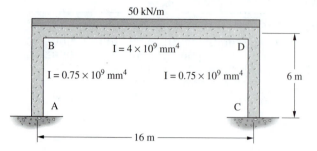

21.4

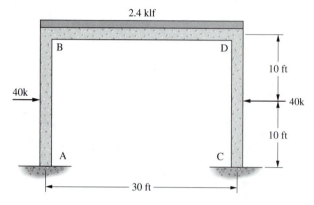

21.5   (*Ans.* $M_{AB} = 23.7$ ft-k, $M_{DF} = 94.7$ ft-k)

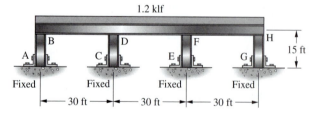

21.6

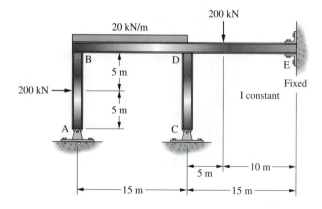

For problems 21.7 to 21.19 determine final moments with the sidesway method.

21.7   (*Ans.* $M_{BA} = 77.2$ ft-k, $M_{DC} = 437.2$ ft-k)

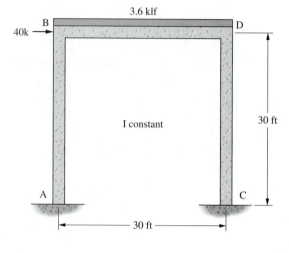

21.8

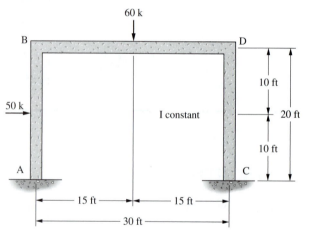

21.9   (*Ans.* $M_{BD} = 130.9$ ft-k, $M_{CD} = 85.7$ ft-k)

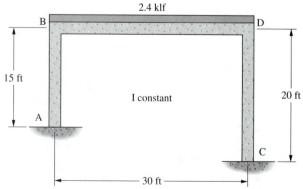

21.10 Rework Problem 21.7 if column bases are pinned.

21.11 Rework Problem 21.8 if column CD is pinned at its base. (*Ans.* $M_{BA} = 99.1$ ft-k, $M_{DC} = 219.5$ ft-k)

21.12

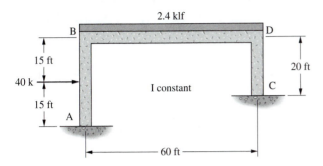

21.13 (*Ans.* $M_{AB} = 264.2$ ft-k, $M_{BD} = 135.7$ ft-k)

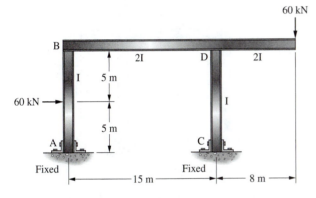

21.14

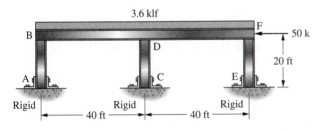

21.15 (*Ans.* $M_{AB} = 4.7$ ft-k, $M_{FE} = 256.9$ ft-k)

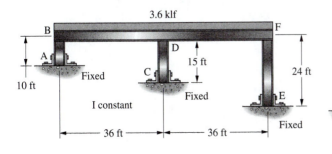

21.16

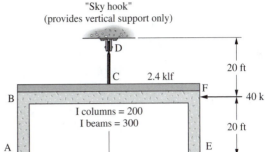

21.17 (*Ans.* $M_{AB} = 57.8$ ft-k, $M_{DC} = 2.1$ ft-k)

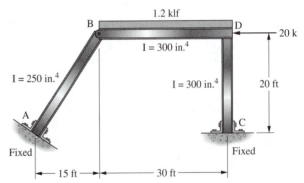

21.18

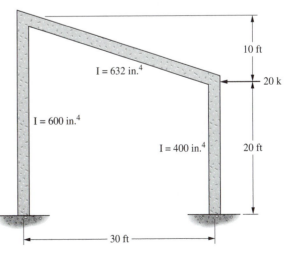

21.19   (*Ans.* $M_A = 18.9'$ k, $M_B = 21.1'$ k, $M_C = 24.4'$ k,
$M_D = 23.9'$ k)

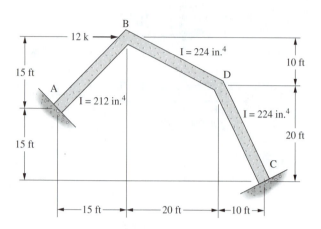

For Problems 21.20 through 21.23 analyze these structures using the successive correction method.

21.20   Rework Problem 21.7

21.21   (*Ans.* $M_{AB} = 1.6$ ft-k, $M_{BA} = 7.5$ ft-k, $M_{DF} = 106.4$ ft-k, $M_{FE} = 67.5$ ft-k answers after 3 cycles)

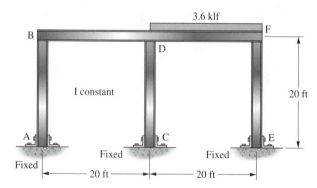

21.22

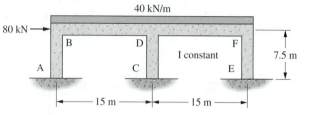

21.23   (*Ans.* $M_{AB} = 189.6$ ft-k, $M_{BE} = 222.1$ ft-k, $M_{EF} = 3.7$ ft-k answers after 5 cycles)

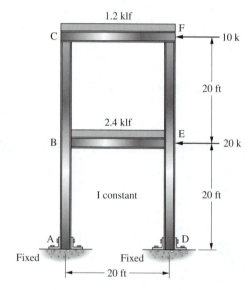

For Problems 21.24 to 21.26 rework the problems given using SABLE32 or SAP2000. Assume areas and moduli of elasticity are constant.

21.24   Problem 21.8

21.25   Problem 21.9 (*Ans.* $M_{AB} = 39.93$ ft-k, $M_{DC} = 143.8$ ft-k)

21.26   Problem 21.21

Chapter 22

Introduction to Matrix Methods

22.1 STRUCTURAL ANALYSIS USING THE COMPUTER

During the past several decades there have been tremendous changes in the structural analysis methods used in engineering practice. These changes have primarily occurred because of the great developments made with high-speed digital computers and the increasing use of very complex structures. Matrix methods of analysis provide a convenient mathematical language for describing complex structures and the necessary matrix manipulations can easily be handled with computers.

The decreasing cost of personal computers has escalated this trend and has made impressive computing power available in almost every design office. It is therefore important for all students of structural engineering to understand the fundamentals of structural analysis as performed on the computer, and to appreciate both the strengths and the weaknesses of this type of analysis.

Structural analysis, as performed on the computer, involves no new concepts of structural engineering. However, the organization of the work must be both versatile and precise. The computer is capable of extraordinary feats of arithmetic, but it can do only those tasks that can be described with simple, precise, and unambiguous instructions. In the following chapters the author has attempted to explain how structural analysis problems are organized for use by the computer, and how instructions can be written that permit the computer to solve a large variety of problems at the bidding of the analyst.

22.2 MATRIX METHODS

Structural engineers have been attempting to handle analysis problems for a good many years by applying the methods used by mathematicians in linear algebra.

Although many structures could be analyzed with the resulting equations, the work was extremely tedious, at least until large-scale computers became available. In fact, the usual matrix equations are not manageable with hand-held calculators unless the most elementary structures are involved.

Today matrix analysis (using computers) has almost completely replaced the classical methods of analysis in engineering offices. As a result, engineering educators and writers of structural analysis textbooks are faced with a difficult decision. Should they require a thorough study of the classical methods followed by a study of modern matrix methods; should they require the students to study both at the same time in an integrated approach; or should they just present a study of the modern methods? The reader can see

Cold Springs Interchange Bridge (U.S. 395), North of Reno, Nevada. (Courtesy of the Nevada Department of Transportation.)

from the preceding chapters that the author feels that an initial study of some of the classical methods followed by a study of the matrix methods will result in an engineer who has a better understanding of structural behavior.

Any method of analysis involving linear algebraic equations can be put into matrix notation and matrix operations used to obtain their solution. The possibility of the application of matrix methods by the structural engineer is very important because all linearly elastic, statically determinate and indeterminate structures are governed by systems of linear equations.

The simple numerical examples presented in the next few chapters could be solved more quickly by classical methods using a pocket calculator rather than with a matrix approach. However, as structures become more complex and as more loading patterns are considered, matrix methods using computers become increasingly useful.

22.3 REVIEW OF MATRIX ALGEBRA

Engineering students have some background in matrix algebra; however, the author's students seem to need a little review of the subject before they study the material presented in this and the next 3 chapters. If you should fall into this class of students, you might very well like to look over the matrix algebra review sections presented in Appendix B of this book.

22.4 FORCE AND DISPLACEMENT METHODS OF ANALYSIS

The methods presented in earlier chapters for analyzing statically indeterminate structures can be placed in two general classes. These are the force methods and the displacement methods. Both of these methods have been developed to a stage where they can be applied to almost any structure—trusses, beams, frames, plates, shells, and so on. The displacement procedures, however, are much more commonly used today since they can be more easily programmed for solution by computers (as described in Chapter 25. These two methods of analysis, which were previously discussed in Section 14.5 are redefined here, as it is felt that the material presented in the last few chapters will enable the reader to better understand the definitions.

Force Method of Analysis

With the force method, also called the *flexibility* or *compatibility method*, redundants are selected and removed from the structure so that a stable and statically determinate structure remains. An equation of deformation compatibility is written at each location where a redundant has been removed. These equations are written in terms of the redundants and the resulting equations are solved for the numerical values of the redundants. After the redundants are determined statics can be used to compute all other desired internal forces, moments, and so on. The Method of Consistent Distortions (see Chapter 15) is a force method.

Displacement Method of Analysis

With the displacement method of analysis, also called the *stiffness* or *equilibrium method*, the displacements of the joints necessary to describe fully the deformed shape of the structure are used in a set of simultaneous equations. When the equations are solved for these displacements, they are substituted into the force-deformation relations of each member to determine the various internal forces. Slope Deflection (see Chapter 18) is a displacement method.

Note that the number of unknowns in the displacement method is generally much greater than the number of unknowns in the force method. Despite this fact, the displacement method is of the greatest importance because, as will be shown it is the matrix method that can be computerized most easily for general usage. As a result, only Section 22.5 is devoted to the force method, while the remainder of this book is devoted to the displacement method.

22.5 INTRODUCTION TO THE FORCE OR FLEXIBILITY METHOD

This method is actually the method of consistent distortions (previously described in Chapter 15) cast in matrix form. The steps involved in analyzing a statically indeterminate structure by this method are outlined as follows:

1. A sufficient number of redundants is chosen and removed from the structure so as to make it statically determinate. The remaining structure, which is often called the *primary structure* or the *released structure*, must be stable.
2. The primary structure is analyzed to determine the deformations at and in the direction of the redundants that were removed.
3. A unit value is applied to the primary structure at the point and in the direction of one of the redundants and the deformation at that redundant and each of the other redundants is determined. For instance, the deflection due to a unit load at point 1 is computed and labelled $\delta_{1.1}$ herein: the deflection at point 2 due to a

unit load at point 1 is labeled $\delta_{2.1}$, and so on. This same procedure is followed with a unit value of a redundant applied at each of the other redundant locations.

The displacements due to the unit load are called *flexibility coefficients*. The actual displacement at joint 1 due to redundant R_1 is R_1 times the deflection caused by a unit load acting there, i.e., $R_1 * \delta_{1.1}$; the displacement at joint 2 due to R_1 is $R_1 * \delta_{2.1}$; and so on.

4. Finally, simultaneous equations of deformation compatibility are written at each of the redundant locations. The unknowns in these equations are the redundant forces. The equations are expressed in matrix form and solved for the redundants.

To illustrate this procedure the four-span beam of Figure 22.1 is considered This beam is statically indeterminate to the third degree and the three support reactions R_1, R_2, and R_3, have been selected as the redundants and considered to be removed from the structure as shown in part (b) of the figure. The external loads will cause the beam to deflect downward by the values δ_1, δ_2, and δ_3, as shown.

In Figure 22.1(c) a unit load is applied at point 1 acting upward. It causes upward deflections at points 1, 2, and 3, respectively, equal to $\delta_{1.1}$, $\delta_{2.1}$, and $\delta_{3.1}$. Similarly, in parts (d) and (e), upward unit loads are applied at points 2 and 3 respectively, and the deformations at the three redundant points determined.

Equations are then written for the total deformation at each of the points. It can be seen that these equations will be expressed in terms of all the redundants. This means that each redundant affects the displacement associated with each of the other redundants.

Now we can write an expression for the deformation at each of the joints Here δ_{R1} is the total deflection at point 1, δ_{R2} is the total deflection at joint 2, etc.

Since this beam is assumed to have unyielding supports these values are each equal to zero.

$$\delta_1 + \delta_{1.1}R_1 + \delta_{1.2}R_2 + \delta_{1.3}R_3 = \delta_{R_1}$$
$$\delta_2 + \delta_{2.1}R_1 + \delta_{2.2}R_2 + \delta_{2.3}R_3 = \delta_{R_2}$$
$$\delta_3 + \delta_{3.1}R_1 + \delta_{3.2}R_2 + \delta_{3.3}R_3 = \delta_{R_3}$$

(22.1)

These equations can be put into matrix form as follows:

$$\begin{Bmatrix} \delta_1 \\ \delta_2 \\ \delta_3 \end{Bmatrix} + \begin{bmatrix} \delta_{1.1} & \delta_{1.2} & \delta_{1.3} \\ \delta_{2.1} & \delta_{2.2} & \delta_{2.3} \\ \delta_{3.1} & \delta_{3.2} & \delta_{3.3} \end{bmatrix} \begin{Bmatrix} R_1 \\ R_2 \\ R_3 \end{Bmatrix} = \begin{Bmatrix} \delta_{R1} \\ \delta_{R2} \\ \delta_{R3} \end{Bmatrix}$$

(22.2)

This equation actually says that the displacements due to the loads plus the flexibility matrix times the redundants equals the final deformation at the supports. It can be written in compact symbolic form as follows:

$$\{\delta_L\} + [F]\{R\} = \{\delta_R\}$$

(22.3)

where

$\{\delta_L\}$ is a vector of displacements due to the imposed loads

$[F]$ is a matrix of flexibility coefficients

$\{R\}$ is a vector of redundant forces

$\{\delta_R\}$ is a vector of final deformations at the supports, (all equal to zero here)

This equation may be written in a somewhat altered form and solved for the redundants:

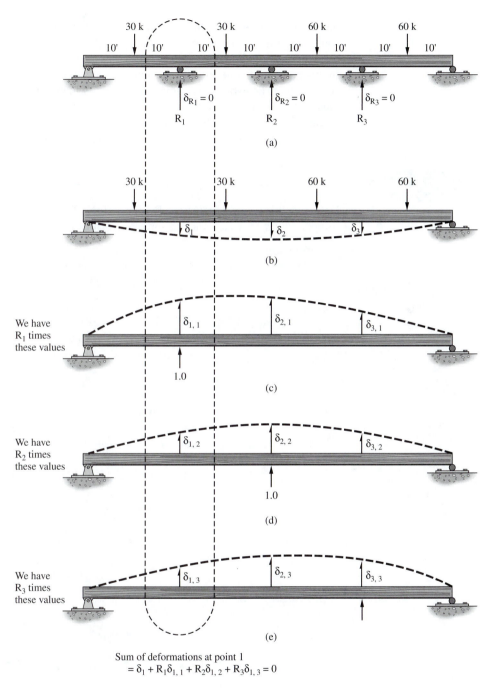

Figure 22.1

$$\{R\} = [F]^{-1}(\{\delta_R\} - \{\delta_L\}) \tag{4}$$

The symbol $[F]^{-1}$ represents the inverse of matrix $[F]$. This matrix may be found in a formal way through a mathematical procedure described in Appendix B. However, in most practical problems, the inverse is not found explicitly, but the set of algebraic equations is solved simultaneously by some other procedure, e.g., the Gauss method. The inverse notation will continue to be used in this book to represent symbolically the solution to the set of algebraic equations.

In Example 22.1, the redundant reactions are determined for the beam of Figure 22.1. In this problem it is necessary to compute three support deformations and nine flexibility coefficients. These rather lengthy calculations, which can be handled by methods previously presented in this book, are not shown here. Application of Maxwell's Law of Reciprocal Deflections shortens the work somewhat in that $\delta_{1.2} = \delta_{2.1}$, $\delta_{1.3} = \delta_{3.1}$, and $\delta_{2.3} = \delta_{3.2}$. In addition, because of symmetry for this particular beam, $\delta_{1.1} = \delta_{3.3}$.

EXAMPLE 22.1

Determine the values of reactions R_1, R_2, and R_3 of the continuous beam of Figure 22.1 using a matrix approach and the force or flexibility method.

Solution.

The following deflections are obtained using either the conjugate beam procedure of Chapter 12 or the virtual work method of Chapter 13.

$$\delta_1 = -\frac{850,000}{EI} \text{ ft}^3\text{-k}$$

$$\delta_2 = -\frac{1,230,000}{EI} \text{ ft}^3\text{-k}$$

$$\delta_3 = -\frac{905,000}{EI} \text{ ft}^3\text{-k}$$

$$\delta_{1.1} = \frac{6000}{EI} \text{ ft}^3\text{-k}$$

$$\delta_{1.2} = \delta_{2.1} = \frac{7333}{EI} \text{ ft}^3\text{-k}$$

$$\delta_{1.3} = \delta_{3.1} = \frac{4667}{EI} \text{ ft}^3\text{-k}$$

$$\delta_{2.2} = \frac{10,667}{EI} \text{ ft}^3\text{-k}$$

$$\delta_{2.3} = \delta_{3.2} = \frac{7333}{EI} \text{ ft}^3\text{-k}$$

$$\delta_{3.3} = \frac{6000}{EI} \text{ ft}^3\text{-k}$$

Writing the equations:

$$\frac{1}{EI}\left[-850,000 + 6000R_1 + 7333R_2 + 4667R_3\right] = 0$$

$$\frac{1}{EI}\left[-1,230,000 + 7333R_1 + 10,667R_2 + 7333R_3\right] = 0$$

$$\frac{1}{EI}\left[-905,000 + 4667R_1 + 7333R_2 + 6000R_3\right] = 0$$

Putting them in matrix form:

$$\frac{1}{EI}\begin{bmatrix} 6000 & 7333 & 4667 \\ 7333 & 10,667 & 7333 \\ 4667 & 7333 & 6000 \end{bmatrix} \begin{Bmatrix} R_1 \\ R_2 \\ R_3 \end{Bmatrix} = \frac{1}{EI}\begin{Bmatrix} 850,000 \\ 1,230,000 \\ 905,000 \end{Bmatrix}$$

Solving the equations:

$$R_1 = 34.0 \text{ k} \uparrow$$
$$R_2 = 40.2 \text{ k} \uparrow$$
$$R_3 = 75.3 \text{ k} \uparrow \quad \blacksquare$$

In Example 22.2, only axial deformations are considered in order to simplify the numerical calculations required while still illustrating the principles involved in the method. Similar problems are worked by the Stiffness Method in Chapter 23.

EXAMPLE 22.2

The axial force member shown in Figure 22.2(a) is supported at each end and loaded with axial forces at intermediate locations. Determine the reactions at the supports.

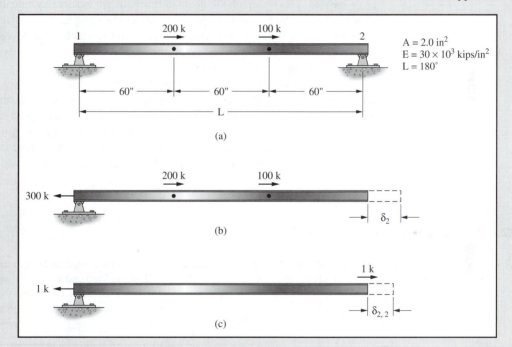

Figure 22.2

Solution.
The reaction at end 2 is considered the redundant for this problem. When this reaction is removed the remaining (primary) structure is shown in Figure 22.2(b). A unit load applied to the primary structure is shown in Figure 22.2(c). A member subject to an axial load changes in length by $\frac{PL}{AE}$ as described in Section 13.4 of this text. Thus

$$\delta_2 = \frac{300}{AE}\left(\frac{L}{3}\right) + \frac{100}{AE}\left(\frac{L}{3}\right) = \frac{(300)(60)}{(2.0)(30 \times 10^3)} + \frac{(100)(60)}{(2.0)(30 \times 10^3)}$$
$$= 0.40 \text{ in.}$$

$$\delta_{2.2} = 1\left(\frac{L}{AE}\right) = \frac{1(180)}{2.0(30 \times 10^3)} = 0.003 \text{ in.}$$

Reaction at end 2 is then computed from an equation similar to Eq. 22.1:

$$\delta_2 + \delta_{2.2}R_2 = \delta_{R_2} = 0$$

$$R_2 = -\frac{0.40}{0.003} = -133 \text{ k}$$

Reaction at end 1 is then computed from an equilibrium relationship.

$$\sum F_x = R_1 + 200 + 100 + R_2 = 0$$

$$R_1 = -167 \text{ k} \quad \blacksquare$$

An inherent difficulty with the use of the force method is that the analyst must know at the outset how many redundants there are in the structure to be analyzed. Furthermore he or she must choose which members or reactions are to serve as redundants. The choice of redundants, which is not unique, may affect the complexity involved in solving the problem at hand. With the stiffness method, however, which is introduced in Chapter 23, it is much easier to prepare a general program applicable to all types of statically determinate or indeterminate structures.

A full discussion of the force method is not included in this book because current structural analysis trends are decidedly in the direction of the displacement or stiffness method. It would be possible to expand the force method described herein to include other items such as the development of the flexibility coefficients using matrix methods, a consideration of different types of structures, and so on. The reader interested in such topics might like to study a textbook such as Weaver and Gere's[1] which describes the method in detail.

22.6 PROBLEMS FOR SOLUTION

22.1. Find the flexibility coefficients associated with joint loads applied at joint 2, i.e., find the coefficients of matrix [F] in the equation:

$$\begin{Bmatrix} v_2 \\ \theta_2 \end{Bmatrix} = [F] \begin{Bmatrix} Y_2 \\ M_2 \end{Bmatrix}$$

$$\left(Ans. \begin{Bmatrix} v_2 \\ \theta_2 \end{Bmatrix} = \begin{bmatrix} \dfrac{L^3}{3EI} & \dfrac{L^2}{2EI} \\ \dfrac{L^2}{2EI} & \dfrac{L}{EI} \end{bmatrix} \begin{Bmatrix} Y_2 \\ M_2 \end{Bmatrix} \right)$$

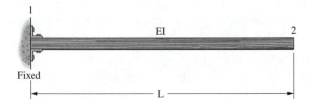

22.2. Use the flexibility method to determine the reactions at joint 2 of the fixed-end beam that supports a uniformly distributed load. Suggestion: Use the flexibility coefficients found in Prob. 22.1.

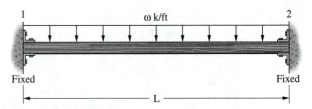

22.3 Determine the flexibility coefficients associated with the continuous beam shown for this problem. Use the moments at nodes 2 and 3 as redundants; the released structure will then appear as two simply supported beams, as shown adjacent to the figure. The flexibility coefficients will be part of matrix [F] in the matrix equation:

[1]W. Weaver and J. M. Gere, *Matrix Analysis of Framed Structures* 2nd ed. (New York: Van Nostrand Reinhold, 1980).

$$\left\{ \begin{array}{c} \theta_{2,1} - \theta_{2,3} \\ \theta_3 \end{array} \right\} = [F] \left\{ \begin{array}{c} M_2 \\ M_3 \end{array} \right\}$$

$$\left(Ans. \left\{ \begin{array}{c} \theta_{2,1} - \theta_{2,3} \\ \theta_{3,2} \end{array} \right\} = \begin{bmatrix} \dfrac{L}{3EI} + \dfrac{L}{3EI} & \dfrac{L}{6EI} \\ \dfrac{L}{6EI} & \dfrac{L}{3EI} \end{bmatrix} \left\{ \begin{array}{c} M_2 \\ M_3 \end{array} \right\} \right)$$

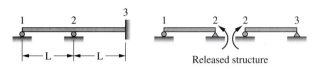

Released structure

22.4 Use the flexibility method to find the internal moments at joints 2 and 3 for the continuous beam shown. The beam is loaded with a uniformly distributed load over its entire length. (Suggestion: Use the flexibility coefficients determined in Prob. 22.3.)

22.5 Use the flexibility coefficients found in Prob. 22.3 to determine the moments at joints 2 and 3, for the continuous beam loaded with a concentrated load as shown.

$$\left(Ans. \left\{ \begin{array}{c} M_2 \\ M_3 \end{array} \right\} = \left\{ \begin{array}{c} \dfrac{2}{14} PL^2 \\ -\dfrac{1}{14} PL^2 \end{array} \right\} \right)$$

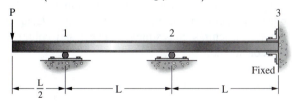

22.6 Determine the flexibility coefficients for the continuous beam shown. Choose reactions at joints 1 and 2 as redundants—the released configuration is shown adjacent to the figure. Find [F] in equation

$$\left\{ \begin{array}{c} v_1 \\ v_2 \end{array} \right\} = [F] \left\{ \begin{array}{c} Y_1 \\ Y_2 \end{array} \right\}$$

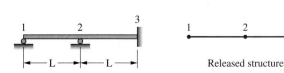

Released structure

22.7 Determine the reactions at joints 1 and 2 for the continuous beam shown in Prob. 22.6, assuming that a uniformly distributed load acts over the entire length of the beam, such as shown in Prob. 22.4.

$$\left(Ans. \left\{ \begin{array}{c} Y_1 \\ Y_2 \end{array} \right\} = \left\{ \begin{array}{c} \dfrac{11}{28} wL \\ \dfrac{32}{28} wL \end{array} \right\} \right)$$

22.8 Determine the reactions at nodes 1 and 2 for the continuous beam shown. Uniformly distributed load acts over the span between joints 2 and 3 only.

22.9 Determine the end moments and moment at joint 2 for the continuous beam shown. The loading carried by the beam consists of a uniformly distributed load over one span, as shown. Use the moments at the joints as redundants; the released structure is shown adjacent to the figure.

$$\left(Ans. \left\{ \begin{array}{c} M_1 \\ M_2 \\ M_3 \end{array} \right\} = \left\{ \begin{array}{c} -\dfrac{5}{48} wL^2 \\ -\dfrac{2}{48} wL^2 \\ \dfrac{1}{48} wL^2 \end{array} \right\} \right)$$

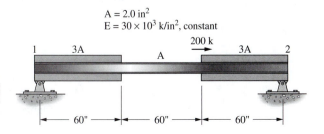

Released structure

22.10 An axial force member consisting of three segments having different cross-sectional areas, is supported at each end and loaded with an intermediate 200-kip load. Use the flexibility method to find the support reactions.

Chapter 23

Fundamentals of the Displacement or Stiffness Method

23.1 INTRODUCTION

When a structure is being analyzed with the displacement or stiffness method, the joint displacements (translations and rotations) are treated as unknowns. Equilibrium equations are written for each joint of the structure in terms of (1) the applied loads, (2) the properties of the members framing into the joint, and (3) the unknown joint displacements. The result is a set of linear, algebraic equations that can be solved simultaneously for the joint displacements. These displacements are then used to determine the internal member forces and support reactions.

The displacement method can be used with equal ease for the analysis of statically determinate or statically indeterminate structures. The analyst does not have to make a choice of redundants and does not have to specify or even know whether the structure is statically determinate or indeterminate. Furthermore, if the structure is unstable, no solution can be determined and the analyst is thereby warned of the instability.

In this chapter the author has developed the fundamentals of the stiffness method for continuous beams and struts. He has attempted to present the equations in a form that will enable the reader to grasp the physical significance of the terms involved. In Chapter 24 stiffness equations are developed for structures consisting of beams and columns. Inclined or sloping members are also considered.

23.2 GENERAL RELATIONSHIPS

The stiffness of a joint is usually defined as the force (or moment) required to produce a unit displacement (or rotation) at the joint if displacement is prevented at all other joints in the structure. For this initial discussion the linear spring shown in Figure 23.1 is considered.

Figure 23.1

From this figure the relationship between the applied force P_1 and the stretching of the spring δ_1 can be written as

$$P_1 = k\delta_1$$

In this expression k is the *spring constant* or the *force* required to produce a unit displacement:

$$k = P_1 \quad \text{if} \quad \delta_1 = 1.0$$

Therefore if the spring constant is known the displacement can be determined for any applied load P_1.

For most practical problems the displacement is needed at more than one joint or location. This is the case for the simple beam shown in Figure 23.2.

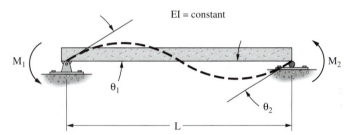

Figure 23.2

The end moments M_1 and M_2 produce the end rotations θ_1 and θ_2. Using the slope-deflection procedure of Section 18.2 the following equations may be written assuming no chord rotation occurs. In these expressions K is equal to I/L, the so-called stiffness factor.

$$M_1 = 2EK(2\theta_1 + \theta_2) = \frac{4EI}{L}\theta_1 + \frac{2EI}{L}\theta_2$$

$$M_2 = 2EK(\theta_1 + 2\theta_2) = \frac{2EI}{L}\theta_1 + \frac{4EI}{L}\theta_2$$

These equations are expressed in matrix form as follows:

$$\begin{Bmatrix} M_1 \\ M_2 \end{Bmatrix} = \begin{bmatrix} \dfrac{4EI}{L} & \dfrac{2EI}{L} \\ \dfrac{2EI}{L} & \dfrac{4EI}{L} \end{bmatrix} \begin{Bmatrix} \theta_1 \\ \theta_2 \end{Bmatrix}$$

The coefficients 4EI/L and 2EI/L may be written symbolically as $k_{i,j}$, where the subscripts define the row and column location of the coefficients in the stiffness matrix.

$$\begin{Bmatrix} M_1 \\ M_2 \end{Bmatrix} = \begin{bmatrix} k_{1,1} & k_{1,2} \\ k_{2,1} & k_{2,2} \end{bmatrix} \begin{Bmatrix} \theta_1 \\ \theta_2 \end{Bmatrix} \tag{23.1}$$

The stiffness coefficient $k_{1,1}$ may be interpreted as the moment that must be applied at end 1 of a beam in order to produce a unit rotation ($\theta_1 = 1$), while the opposite end of the beam is fixed ($\theta_2 = 0$), as shown in Figure 23.3. The coefficient $k_{2,1}$, is the resultant moment at end 2 of the beam for this situation. Similarly, the coefficients $k_{1,2}$ and $k_{2,2}$ may be interpreted to be the resultant moments at ends 1 and 2 of the beam, respectively, when $\theta_2 = 1$ and $\theta_1 = 0$. This situation is shown in Figure 23.4.

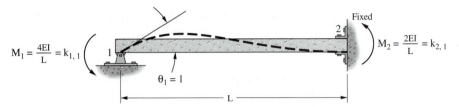

Figure 23.3

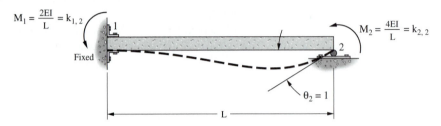

Figure 23.4

Matrix Equation 23.1 involves two algebraic equations and they may be solved simultaneously to obtain the end rotations θ_1 and θ_2. The result, written symbolically, is:

$$\left\{ \begin{array}{c} \theta_1 \\ \theta_2 \end{array} \right\} = \left[\begin{array}{cc} k_{1,1} & k_{1,2} \\ k_{2,1} & k_{2,2} \end{array} \right]^{-1} \left\{ \begin{array}{c} M_1 \\ M_2 \end{array} \right\} \tag{23.2}$$

A study of this simple beam with its two joints illustrates many of the key features of the stiffness method even though almost all practical structures analyzed by the stiffness method have far more than two joints. The remainder of this chapter is devoted to the development of the stiffness equation in a form applicable to beams or struts having any number of joints.

23.3 STIFFNESS EQUATIONS FOR AXIAL FORCE MEMBERS

For the discussion to follow the single axial force member or strut shown in Figure 23.5(a) is considered. This member may be thought of as a linear spring. Although such a spring has been considered in the last few pages the topic is reconsidered here so that additional characteristics of the stiffness method may be introduced.

The ends of the strut are identified as joints or node points (the two terms being used interchangeably). They are the locations where forces are applied and where displacements are determined. The forces acting at the node points are each given two subscripts. These represent the node numbers of the member on which the forces act. The first subscript is the node where the force is located while the second one represents the other end of the member. For instance $F_{1,2}$ is the force acting at node 1 of a member whose end nodes are 1 and 2 and force $F_{2,1}$ is the force acting at node 2 of the same member.

The x axis of the system is taken parallel to the axis of the member in Figure 23.5 and the positive direction is taken from left to right. Joint loads and joint displacements are assumed to be positive when they act in the positive sense of the member x axis. Thus in Figure 23.5, both $F_{1,2}$ and $F_{2,1}$ are positive. Similarly, displacements u_1 and u_2 have positive directions.

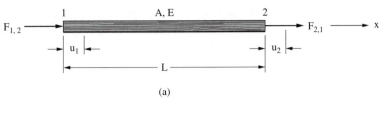

(a)

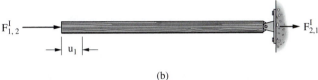

(b)

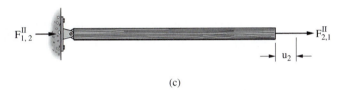

(c)

Figure 23.5

If end 2 of the strut is prevented from moving, as shown in part (b) of the figure, the following relationships (developed from Strength of Materials principles) will exist between the forces and displacements:

$$F^I_{1,2} = \left(\frac{AE}{L}\right)u_1$$

$$F^I_{2,1} = -F^I_{1,2} = -\left(\frac{AE}{L}\right)u_1$$

(23.3)

Written in matrix notation, these equations may be summarized as follows:

$$\left\{ \begin{array}{c} F_{1,2} \\ F_{2,1} \end{array} \right\}^I = \left\{ \begin{array}{c} \dfrac{AE}{L} \\ -\dfrac{AE}{L} \end{array} \right\} u_1$$

(23.4)

If joint 1 of the strut is now prevented from moving, as shown in Figure 23.5(c), the following relationships will exist between the forces and displacements:

$$F^{II}_{2,1} = \left(\frac{AE}{L}\right)u_2$$

$$F^{II}_{1,2} = -F^{II}_{2,1} = -\left(\frac{AE}{L}\right)u_2$$

(23.5)

Or in matrix form

$$\left\{ \begin{array}{c} F_{1,2} \\ F_{2,1} \end{array} \right\}^{II} = \left\{ \begin{array}{c} -\dfrac{AE}{L} \\ \dfrac{AE}{L} \end{array} \right\} u_2$$

(23.6)

Ends 1 and 2 of the strut can be given arbitrary displacements and, based on the principle of superposition, the following relationships written for the resulting forces at joints 1 and 2:

$$\left\{ \begin{array}{c} F_{1,2} \\ F_{2,1} \end{array} \right\}^{total} = \left\{ \begin{array}{c} F_{1,2} \\ F_{2,1} \end{array} \right\}^{I} + \left\{ \begin{array}{c} F_{1,2} \\ F_{2,1} \end{array} \right\}^{II} = \left\{ \begin{array}{c} \dfrac{AE}{L} \\ -\dfrac{AE}{L} \end{array} \right\} u_1 + \left\{ \begin{array}{c} -\dfrac{AE}{L} \\ \dfrac{AE}{L} \end{array} \right\} u_2 \qquad (23.7)$$

Or, in matrix form,

$$\left\{ \begin{array}{c} F_{1,2} \\ F_{2,1} \end{array} \right\}^{total} = \left[\begin{array}{cc} \dfrac{AE}{L} & -\dfrac{AE}{L} \\ -\dfrac{AE}{L} & \dfrac{AE}{L} \end{array} \right] \left\{ \begin{array}{c} u_1 \\ u_2 \end{array} \right\} \qquad (23.8)$$

Symbolically, Equation 23.8 may be written as

$$\{F\} = [K]\{u\} \qquad (23.9)$$

where

$\{F\}$ is a vector of joint forces,

$\{u\}$ is a vector of joint displacements, and

$[K]$ is a matrix of stiffness coefficients called a *stiffness matrix.*

It can be seen from Equations 23.7 and 23.8 that each column of the stiffness matrix represents the set of forces that corresponds to a *unit value of a single joint displacement.* An understanding of this characteristic of the stiffness matrix will enable the student to develop stiffness matrices for structures far more complex than simple struts.

Although matrix Equation 23.8 represents two distinct algebraic equations written in terms of two unknowns it is impossible to solve them for the displacements u_1 and u_2 in terms of the given forces $F_{1,2}$ and $F_{2,1}$. This fact may be verified by attempting to obtain a solution using the formal matrix inverse procedure, as follows:

$$\left\{ \begin{array}{c} u_1 \\ u_2 \end{array} \right\} = \left[\begin{array}{cc} \dfrac{AE}{L} & -\dfrac{AE}{L} \\ -\dfrac{AE}{L} & \dfrac{AE}{L} \end{array} \right]^{-1} \left\{ \begin{array}{c} F_{1,2} \\ F_{2,1} \end{array} \right\} \qquad (23.10)$$

The solution outlined by Equation 23.10 has no meaning since the inverse of matrix $[K]$ does not exist, i.e., matrix $[K]$ is singular. The reason for this peculiar circumstance is that rigid-body motions have not been eliminated from Equation 23.8. If u_1 should equal u_2 the strut may be displaced any arbitrary distance without the benefit of any axial forces $F_{1,2}$ or $F_{2,1}$. However, if either end of the strut is given a specified displacement as say $u_2 = 0$ a well-defined relationship will exist between the force $F_{1,2}$ and the resulting displacement at node 1:

$$u_1 = \left(\dfrac{L}{AE} \right) F_{1,2} \qquad (23.11)$$

The analysis of the strut shown in Figure 23.6(a) is a little more difficult than was the analysis of the two-joint strut of Figure 23.5. A study of this member, however, will better illustrate the principles involved in the past few paragraphs.

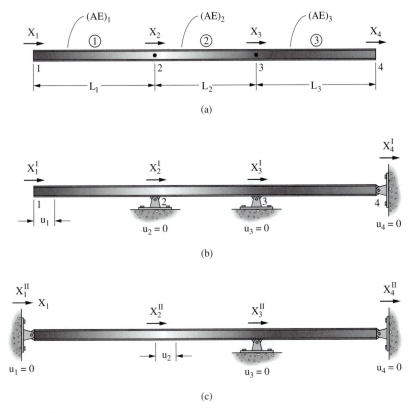

Figure 23.6

This strut has three segments, or elements, and four joints. The ends of the element are assigned node, or joint, numbers and each element is assigned an identifying number. These numbers are shown encircled in Figure 23.6(a). An external force acting at node j is given the symbol X_j.

If joint 1 of Figure 23.6(a) is given an arbitrary displacement and all other joints of the strut are prevented from displacing, as shown in Figure 23.6(b), the following relationships will exist between the joint forces and the joint displacements:

$$X_1^I = \left(\frac{AE}{L}\right)_1 u_1$$

$$X_2^I = -\left(\frac{AE}{L}\right)_1 u_1$$

$$X_3^I = 0$$

$$X_4^I = 0$$

The subscripts attached to the AE/L coefficients refer to the element numbers associated with the terms.

If joint 2 is given an arbitrary displacement while all other nodes are prevented from displacing as shown in Figure 23.6(c), the following relationships will exist between the joint forces and the joint displacements:

$$X_1^{II} = -\left(\frac{AE}{L}\right)_1 u_2$$

$$X_2^{II} = \left[\left(\frac{AE}{L}\right)_1 + \left(\frac{AE}{L}\right)_2\right] u_2$$

$$X_3^{II} = -\left(\frac{AE}{L}\right)_2 u_2$$

$$X_4^{II} = 0$$

This procedure can be repeated with each joint in turn given an arbitrary value while all other joints are prevented from moving. The superposition of all these relationships produces the following equation:

$$
\begin{Bmatrix} X_1 \\ X_2 \\ X_3 \\ X_4 \end{Bmatrix} = \begin{Bmatrix} \left(\frac{AE}{L}\right)_1 \\ -\left(\frac{AE}{L}\right)_1 \\ 0 \\ 0 \end{Bmatrix} u_1 + \begin{Bmatrix} -\left(\frac{AE}{L}\right)_1 \\ \left(\frac{AE}{L}\right)_1 + \left(\frac{AE}{L}\right)_2 \\ -\left(\frac{AE}{L}\right)_2 \\ 0 \end{Bmatrix} u_2
$$

$$
+ \begin{Bmatrix} 0 \\ -\left(\frac{AE}{L}\right)_2 \\ \left(\frac{AE}{L}\right)_2 + \left(\frac{AE}{L}\right)_3 \\ -\left(\frac{AE}{L}\right)_3 \end{Bmatrix} u_3 + \begin{Bmatrix} 0 \\ 0 \\ -\left(\frac{AE}{1}\right)_3 \\ \left(\frac{AE}{L}\right)_3 \end{Bmatrix} u_4
$$

(23.12)

Or in more compact form

$$
\begin{Bmatrix} X_1 \\ X_2 \\ X_3 \\ X_4 \end{Bmatrix} = \begin{bmatrix} \left(\frac{AE}{L}\right)_1 & -\left(\frac{AE}{L}\right)_1 & 0 & 0 \\ -\left(\frac{AE}{L}\right)_1 & \left(\frac{AE}{L}\right)_1 + \left(\frac{AE}{L}\right)_2 & -\left(\frac{AE}{L}\right)_2 & 0 \\ 0 & -\left(\frac{AE}{L}\right)_2 & \left(\frac{AE}{L}\right)_2 + \left(\frac{AE}{L}\right)_3 & -\left(\frac{AE}{L}\right)_3 \\ 0 & 0 & -\left(\frac{AE}{L}\right)_3 & \left(\frac{AE}{L}\right)_3 \end{bmatrix} \begin{Bmatrix} u_1 \\ u_2 \\ u_3 \\ u_4 \end{Bmatrix}
$$

(23.13)

This relationship can be written symbolically as

$$\{P\} = [K]\{\delta\}$$

The stiffness matrix $[K]$ shown in Equation 23.13 is singular and does not have an inverse. Therefore, as before, one cannot solve for the displacements u_1, u_2, u_3, and u_4 in terms of the nodal forces X_1, X_2, X_3, and X_4. However, if rigid-body motion is prevented for the strut by specifying one or more nodal displacements, a solution is possible. As an example, let $u_1 = 0$ and $u_4 = 0$. This will result in the configuration shown in Figure 23.7.

Figure 23.7

Since u_1 and u_4 are equal to zero, columns 1 and 4 of the stiffness matrix may be eliminated from the equation set. (The reader may like to look back earlier in this section to recall the physical meaning of the stiffness matrix.) In a similar manner equations 1 and 4 may also be eliminated from the set of equations. (The justification for eliminating these equations is presented in Chapter 25.) These manipulations can conveniently be handled when performing hand calculations by striking through the rows and columns associated with the displacement components whose values are specified to be zero. For this illustration the first and fourth columns and the first and fourth rows are eliminated from the equation set.

$$\left\{\begin{array}{c} X_1 \\ X_2 \\ X_3 \\ X_4 \end{array}\right\} = \left[\begin{array}{cccc} \left(\dfrac{AE}{L}\right)_1 & \left(\dfrac{AE}{L}\right)_1 & 0 & 0 \\ -\left(\dfrac{AE}{L}\right)_1 & \left(\dfrac{AE}{L}\right)_1 + \left(\dfrac{AE}{L}\right)_2 & -\left(\dfrac{AE}{L}\right)_2 & 0 \\ 0 & -\left(\dfrac{AE}{L}\right)_2 & \left(\dfrac{AE}{L}\right)_2 + \left(\dfrac{AE}{L}\right)_3 & -\left(\dfrac{AE}{L}\right)_3 \\ 0 & 0 & -\left(\dfrac{AE}{L}\right)_3 & \left(\dfrac{AE}{L}\right)_3 \end{array}\right] \left\{\begin{array}{c} u_1 = 0 \\ u_2 \\ u_3 \\ u_4 = 0 \end{array}\right\}$$

(23.14)

The remaining two equations may be solved simultaneously to determine the values of the free displacements of the nodal forces.

$$\left\{\begin{array}{c} X_2 \\ X_3 \end{array}\right\} = \left[\begin{array}{cc} \left(\dfrac{AE}{L}\right)_1 + \left(\dfrac{AE}{L}\right)_2 & -\left(\dfrac{AE}{L}\right)_2 \\ -\left(\dfrac{AE}{L}\right)_2 & \left(\dfrac{AE}{L}\right)_2 + \left(\dfrac{AE}{L}\right)_3 \end{array}\right] \left\{\begin{array}{c} u_1 \\ u_2 \end{array}\right\}$$

(23.15)

The following values are assumed so that numerical answers may be obtained for this illustration.

$$\left(\frac{AE}{L}\right)_1 = \left(\frac{AE}{L}\right)_2 = \left(\frac{AE}{L}\right)_3 = 1000.0 \text{ k/in.}$$

$$X_2 = 200 \text{ k and } X_3 = 100 \text{ k}$$

Then, Equation 23.15 may be solved with the following results:

$$\left\{\begin{array}{c} u_2 \\ u_3 \end{array}\right\} = \left[\begin{array}{cc} 2000 & -1000 \\ -1000 & 2000 \end{array}\right]^{-1} \left\{\begin{array}{c} 200 \\ 100 \end{array}\right\}$$

$$= 10^{-3} \left[\begin{array}{cc} \dfrac{2}{3} & \dfrac{1}{3} \\ \dfrac{1}{3} & \dfrac{2}{3} \end{array}\right] \left\{\begin{array}{c} 200 \\ 100 \end{array}\right\} = \left\{\begin{array}{c} 0.167 \text{ in.} \\ 0.133 \text{ in.} \end{array}\right\}$$

(23.16)

Now the reactions X_1 and X_4 may be determined by substituting the values of u_1 and u_2 into the equations previously eliminated from the equation set.

$$X_1 = -\left(\frac{AE}{L}\right)_1 u_2 = -1000(0.167) = -167 \text{ k}$$

$$X_4 = -\left(\frac{AE}{L}\right)_3 u_3 = -1000(0.1333) = -133 \text{ k}$$

$$(23.17)$$

The reader should carefully note that, although the continuous strut of Figure 23.7 is statically indeterminate, its solution by the stiffness method did not require the identification of the redundants or even the knowledge that the structure was statically indeterminate. *A rather peculiar characteristic of the stiffness method is that the number of algebraic equations that must be solved simultaneously decreases as the order of redundancy increases.* This is in direct contrast to the characteristics of the force or flexibility method.

23.4 STIFFNESS EQUATIONS FOR FLEXURAL MEMBERS

Stiffness equations for flexural members were derived by the slope-deflection method in Chapter 18. These derivations are repeated in this section, however, to present a uniform notation system and to introduce some topics that were not previously mentioned. For this discussion the single beam element shown in Figure 23.8 is first considered.

Figure 23.8

Node numbers are assigned to the ends of the member just as they were for struts. The x axis is taken parallel to the axis of the member and a positive sense designated by the analyst as shown in the figure. A positive y axis is taken perpendicular to the member x axis and a positive sense of rotation is determined using the standard right-hand rule.

As was the case for struts, the beam nodes are designated as the locations where forces are applied and where displacements are measured. For the beam element shown in Figure 23.8 the forces applied at the nodes consist of transverse shears and bending moments. In this figure these forces are designated with the symbols Y_i and M_i, respectively, and are shown acting in their positive sense. Displacements at the nodes consist of translations v_i parallel to the shear forces, and rotations θ_i. Axial forces, discussed in the previous section, may also act on the beam element, but for simplification are assumed to equal zero in this case.

The relations between joint forces and joint displacements are developed, as before, by assigning an arbitrary value to a single displacement component, while requiring all other displacement components to remain equal to zero. As an illustration, node 1 of the beam shown in Figure 23.8 is given an arbitrary value while $v_2 = \theta_1 = \theta_2 = 0$. A sketch of the resulting deformed beam is shown in Figure 23.9.

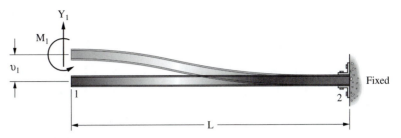

Figure 23.9

Quite a few force-displacement relations have been presented in the earlier chapters of this book. Because of their considerable importance to the stiffness method the most frequently used of these relations are shown in Figure 23.10.

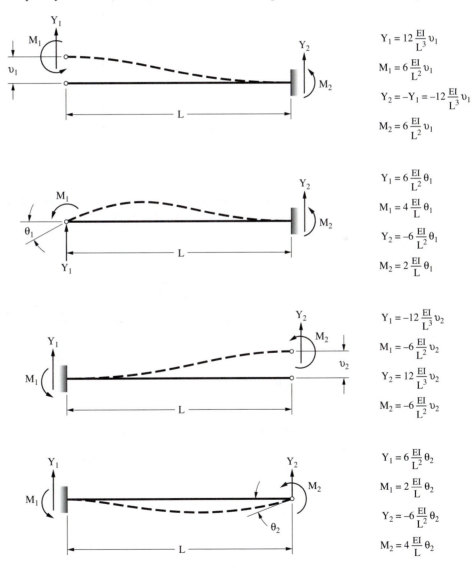

$$Y_1 = 12\frac{EI}{L^3}\upsilon_1$$

$$M_1 = 6\frac{EI}{L^2}\upsilon_1$$

$$Y_2 = -Y_1 = -12\frac{EI}{L^3}\upsilon_1$$

$$M_2 = 6\frac{EI}{L^2}\upsilon_1$$

$$Y_1 = 6\frac{EI}{L^2}\theta_1$$

$$M_1 = 4\frac{EI}{L}\theta_1$$

$$Y_2 = -6\frac{EI}{L^2}\theta_1$$

$$M_2 = 2\frac{EI}{L}\theta_1$$

$$Y_1 = -12\frac{EI}{L^3}\upsilon_2$$

$$M_1 = -6\frac{EI}{L^2}\upsilon_2$$

$$Y_2 = 12\frac{EI}{L^3}\upsilon_2$$

$$M_2 = -6\frac{EI}{L^2}\upsilon_2$$

$$Y_1 = 6\frac{EI}{L^2}\theta_2$$

$$M_1 = 2\frac{EI}{L}\theta_2$$

$$Y_2 = -6\frac{EI}{L^2}\theta_2$$

$$M_2 = 4\frac{EI}{L}\theta_2$$

Figure 23.10

For the deformed beam shown in Figure 23.9 and with reference to the force-displacement relations of Figure 23.10 the resulting nodal forces may be summarized as follows:

$$Y_1 = \left(\frac{12EI}{L^3}\right) v_1$$

$$M_1 = \left(\frac{6EI}{L^2}\right) v_1$$

$$Y_2 = -\left(\frac{12EI}{L^3}\right) v_1 \tag{23.18}$$

$$M_2 = \left(\frac{6EI}{L^2}\right) v_1$$

Or in matrix notation

$$\begin{Bmatrix} Y_1 \\ M_1 \\ Y_2 \\ M_2 \end{Bmatrix} = \begin{Bmatrix} \dfrac{12EI}{L^3} \\ \dfrac{6EI}{L^2} \\ -\dfrac{12EI}{L^3} \\ \dfrac{6EI}{L^2} \end{Bmatrix} v_1 \tag{23.19}$$

If θ_1 is given an arbitrary value and all other nodal displacement components are made equal to zero the beam element of Figure 23.8 will assume the shape shown in Figure 23.11.

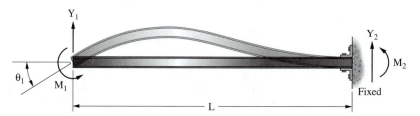

Figure 23.11

The resulting nodal forces may be summarized as follows:

$$Y_1 = \frac{6EI}{L^2}\theta_1$$

$$M_1 = \frac{4EI}{L}\theta_1$$

$$Y_2 = -\frac{6EI}{L^2}\theta_1 \tag{23.20}$$

$$M_2 = \frac{2EI}{L}\theta_1$$

Or in matrix notation

$$\begin{Bmatrix} Y_1 \\ M_1 \\ Y_2 \\ M_2 \end{Bmatrix} = \begin{Bmatrix} \dfrac{6EI}{L^2} \\[2mm] \dfrac{4EI}{L} \\[2mm] -\dfrac{6EI}{L^2} \\[2mm] \dfrac{2EI}{L} \end{Bmatrix} \theta_1 \tag{23.21}$$

Similar nodal displacements may also be imposed at node 2 of the beam element. Superposition of the nodal forces produced by each of the individual nodal displacements produces the following expression for the total nodal forces:

$$\begin{Bmatrix} Y_1 \\ M_1 \\ Y_2 \\ M_2 \end{Bmatrix}^{total} = \begin{Bmatrix} \dfrac{12EI}{L^3} \\[2mm] \dfrac{6EI}{L^2} \\[2mm] -\dfrac{12EI}{L^3} \\[2mm] \dfrac{6EI}{L^2} \end{Bmatrix} v_1 + \begin{Bmatrix} \dfrac{6EI}{L^2} \\[2mm] \dfrac{4EI}{L} \\[2mm] -\dfrac{6EI}{L^2} \\[2mm] \dfrac{2EI}{L} \end{Bmatrix} \theta_1 + \begin{Bmatrix} -\dfrac{12EI}{L^3} \\[2mm] -\dfrac{6EI}{L^2} \\[2mm] \dfrac{12EI}{L^3} \\[2mm] -\dfrac{6EI}{L^2} \end{Bmatrix} v_2 + \begin{Bmatrix} \dfrac{6EI}{L^2} \\[2mm] \dfrac{2EI}{L} \\[2mm] -\dfrac{6EI}{L^2} \\[2mm] \dfrac{4EI}{L} \end{Bmatrix} \theta_2 \tag{23.22}$$

Or in more compact matrix form

$$\begin{Bmatrix} Y_1 \\ M_1 \\ Y_2 \\ M_2 \end{Bmatrix} = \begin{bmatrix} \dfrac{12EI}{L^3} & \dfrac{6EI}{L^2} & -\dfrac{12EI}{L^3} & \dfrac{6EI}{L^2} \\[2mm] \dfrac{6EI}{L^2} & \dfrac{4EI}{L} & -\dfrac{6EI}{L^2} & \dfrac{2EI}{L} \\[2mm] -\dfrac{12EI}{L^3} & -\dfrac{6EI}{L^2} & \dfrac{12EI}{L^3} & -\dfrac{6EI}{L^2} \\[2mm] \dfrac{6EI}{L^2} & \dfrac{2EI}{L} & -\dfrac{6EI}{L^2} & \dfrac{4EI}{L} \end{bmatrix} \begin{Bmatrix} v_1 \\ \theta_1 \\ v_2 \\ \theta_2 \end{Bmatrix} \tag{23.23}$$

Equation 23.23 represents the stiffness equation for a single beam element.

For a beam that consists of two or more elements the stiffness equations may be determined in a similar manner. To illustrate this point the continuous two-span beam of Figure 23.12(a) is considered. Node numbers and member numbers are assigned to this structure in the same manner as they were for struts. If v_1 is given an arbitrary value (all other displacement components remaining equal to zero) the structure will deform as shown in Figure 23.12(b).

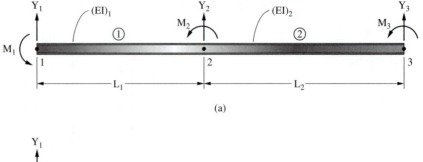

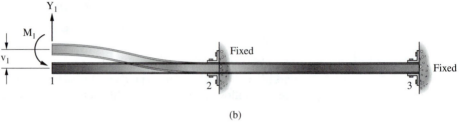

Figure 23.12

The resulting nodal forces may be summarized as follows:

$$\left\{ \begin{array}{c} Y_1 \\ M_1 \\ Y_2 \\ M_2 \\ Y_3 \\ M_3 \end{array} \right\} = \left\{ \begin{array}{c} 12\left(\dfrac{EI}{L^3}\right)_1 \\ 6\left(\dfrac{EI}{L^2}\right)_1 \\ -12\left(\dfrac{EI}{L^3}\right)_1 \\ 6\left(\dfrac{EI}{L^2}\right)_1 \\ 0 \\ 0 \end{array} \right\} v_1 \qquad (23.24)$$

It should be noted that there will be no forces imposed at node 3 of beam segment ② since neither node 2 nor 3 deforms in any way. Furthermore, the boundary conditions for a given structure do not have to be considered when the stiffness equations are being developed (they will be considered at a later stage in the analysis).

If displacement component v_2 is now given an arbitrary value, while all other displacements are kept equal to zero, the structure will take the deformed shape shown in Figure 23.13.

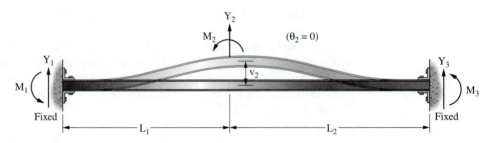

Figure 23.13

The resulting set of nodal forces that correspond to the deformed shape of Figure 23.13 may be summarized as follows:

$$
\begin{Bmatrix} Y_1 \\ M_1 \\ Y_2 \\ M_2 \\ Y_3 \\ M_3 \end{Bmatrix} =
\begin{Bmatrix}
-12\left(\dfrac{EI}{L^3}\right)_1 \\[2ex]
-6\left(\dfrac{EI}{L^2}\right)_1 \\[2ex]
12\left(\dfrac{EI}{L^3}\right)_1 + 12\left(\dfrac{EI}{L^3}\right)_2 \\[2ex]
-6\left(\dfrac{EI}{L^2}\right)_1 + 6\left(\dfrac{EI}{L^2}\right)_2 \\[2ex]
-12\left(\dfrac{EI}{L^3}\right)_2 \\[2ex]
6\left(\dfrac{EI}{L^2}\right)_2
\end{Bmatrix} v_2
\tag{23.25}
$$

When the process described here is repeated, with each displacement component in turn being given an arbitrary value while all other displacement components are held equal to zero, and the results superimposed, the following equation results:

$$
\begin{Bmatrix} Y_1 \\ M_1 \\ Y_2 \\ M_2 \\ Y_3 \\ M_3 \end{Bmatrix} =
\begin{Bmatrix}
12\left(\dfrac{EI}{L^3}\right)_1 & 6\left(\dfrac{EI}{L^2}\right)_1 & -12\left(\dfrac{EI}{L^3}\right)_1 & 6\left(\dfrac{EI}{L^2}\right)_1 & 0 & 0 \\[2ex]
6\left(\dfrac{EI}{L^2}\right)_1 & 4\left(\dfrac{EI}{L}\right)_1 & -6\left(\dfrac{EI}{L^2}\right)_1 & 2\left(\dfrac{EI}{L}\right)_1 & 0 & 0 \\[2ex]
-12\left(\dfrac{EI}{L^3}\right)_1 & -6\left(\dfrac{EI}{L^3}\right)_1 & 12\left(\dfrac{EI}{L^3}\right)_1 + 12\left(\dfrac{EI}{L^3}\right)_2 & -6\left(\dfrac{EI}{L^2}\right)_1 + 6\left(\dfrac{EI}{L^2}\right)_2 & -12\left(\dfrac{EI}{L^3}\right)_2 & 6\left(\dfrac{EI}{L^2}\right)_2 \\[2ex]
6\left(\dfrac{EI}{L^2}\right)_1 & 2\left(\dfrac{EI}{L}\right)_1 & -6\left(\dfrac{EI}{L^2}\right)_1 + 6\left(\dfrac{EI}{L^2}\right)_2 & 4\left(\dfrac{EI}{L}\right)_1 + 4\left(\dfrac{EI}{L}\right)_2 & -6\left(\dfrac{EI}{L^2}\right)_2 & 2\left(\dfrac{EI}{L}\right)_2 \\[2ex]
0 & 0 & -12\left(\dfrac{EI}{L^3}\right)_2 & -6\left(\dfrac{EI}{L^2}\right)_2 & 12\left(\dfrac{EI}{L^3}\right)_2 & -6\left(\dfrac{EI}{L^2}\right)_2 \\[2ex]
0 & 0 & 6\left(\dfrac{EI}{L^2}\right)_2 & 2\left(\dfrac{EI}{L}\right)_2 & -6\left(\dfrac{EI}{L^2}\right)_2 & 4\left(\dfrac{EI}{L}\right)_2
\end{Bmatrix}
\begin{Bmatrix} v_1 \\ \theta_1 \\ v_2 \\ \theta_2 \\ v_3 \\ \theta_3 \end{Bmatrix}
$$

$$\tag{23.26}$$

Although all the stiffness coefficients shown in Equation 23.26 are not listed explicitly in Figure 23.10, all coefficients can be found from this figure by properly adding (or subtracting) the coefficients associated with single beam segments. The reader is encourged to perform this operation for all coefficients of the stiffness matrix shown in Equation 23.26.

Example problems 23.1 and 23.2 are intended to illustrate many of the points made in the previous sections.

EXAMPLE 23.1

A continuous beam, shown in Figure 23.14(a), is subjected to a concentrated moment, M^*, applied at distance a from the left end. Determine the rotation and the reactions at the beam supports.

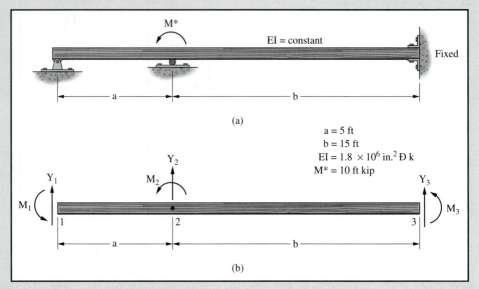

$$a = 5 \text{ ft}$$
$$b = 15 \text{ ft}$$
$$EI = 1.8 \times 10^6 \text{ in.}^2 \text{ Đ k}$$
$$M^* = 10 \text{ ft kip}$$

Figure 23.14

Solution.

The beam is divided into two segments so that a node point is located at each support point. Figure 23.14(b) shows the beam model that will be analyzed. The stiffness matrix for this structure was previously developed and is shown as Equation 23.26. The boundary conditions for this problem are expressed as

$$v_1 = 0$$
$$v_2 = 0$$
$$v_3 = 0$$
$$\theta_3 = 0$$

Initially the columns in the stiffness matrix that are associated with enforced zero displacements are deleted. Then the rows associated with the deleted columns are themselves deleted. The result of these operations is shown in Equation 23.27.

$$
\begin{Bmatrix} Y_1 \\ M_1 \\ Y_2 \\ M_2 \\ Y_3 \\ M_3 \end{Bmatrix} =
\begin{bmatrix}
\dfrac{12EI}{a^3} & \dfrac{6EI}{a^2} & \dfrac{-12EI}{a^3} & \dfrac{6EI}{a^2} & 0 & 0 \\[2mm]
\dfrac{6EI}{a^2} & \dfrac{4EI}{a} & \dfrac{-6EI}{a^2} & \dfrac{2EI}{a} & 0 & 0 \\[2mm]
-\dfrac{12EI}{a^3} & \dfrac{-6EI}{a^2} & \dfrac{12EI}{a^3}+\dfrac{12EI}{b^3} & \dfrac{-6EI}{a^2}+\dfrac{6EI}{b^2} & \dfrac{-12EI}{b^3} & \dfrac{6EI}{b^2} \\[2mm]
\dfrac{6EI}{a^2} & \dfrac{2EI}{a} & \dfrac{-6EI}{a^2}+\dfrac{6EI}{b^2} & \dfrac{4EI}{a}+\dfrac{4EI}{b} & \dfrac{-6EI}{b^2} & \dfrac{2EI}{b} \\[2mm]
0 & 0 & \dfrac{-12EI}{b^3} & \dfrac{-6EI}{b^2} & \dfrac{12EI}{b^3} & \dfrac{-6EI}{b^2} \\[2mm]
0 & 0 & \dfrac{6EI}{b^2} & \dfrac{2EI}{b} & \dfrac{-6EI}{b^2} & \dfrac{4EI}{b}
\end{bmatrix}
\begin{Bmatrix} v_1 = 0 \\ \theta_1 \\ v_2 = 0 \\ \theta_2 \\ v_3 = 0 \\ \theta_3 = 0 \end{Bmatrix}
$$

(23.27)

The remaining stiffness equations are written and solved for the unknown (free) displacements.

$$\begin{Bmatrix} M_1 = 0 \\ M_2 = M^* \end{Bmatrix} = \begin{bmatrix} \dfrac{4EI}{a} & \dfrac{2EI}{a} \\ \dfrac{2EI}{a} & \dfrac{4EI}{a} + \dfrac{4EI}{b} \end{bmatrix} \begin{Bmatrix} \theta_1 \\ \theta_2 \end{Bmatrix}$$

The numerical values shown in Figure 23.14 are substituted into the preceding equations and converted to inch units.

$$\begin{Bmatrix} 0 \\ 120 \end{Bmatrix} = \begin{bmatrix} 120,000 & 60,000 \\ 60,000 & 160,000 \end{bmatrix} \begin{Bmatrix} \theta_1 \\ \theta_2 \end{Bmatrix}$$

and

$$\begin{Bmatrix} \theta_1 \\ \theta_2 \end{Bmatrix} = \begin{bmatrix} 120,000 & 60,000 \\ 60,000 & 160,000 \end{bmatrix}^{-1} \begin{Bmatrix} 0 \\ 120 \end{Bmatrix} = \begin{Bmatrix} -4.62 \times 10^{-4} \\ 9.23 \times 10^{-4} \end{Bmatrix} \text{ rad}$$

The reaction components are now found by substituting the values of the free displacements into the equations that were shown deleted in Equation 23.27. The results follow:

$$Y_1 = \frac{6EI}{a^2}\theta_1 + \frac{6EI}{a^2}\theta_2 = 1.385 \text{ k}$$

$$Y_2 = \frac{-6EI}{a^2}\theta_1 + \left(\frac{-6EI}{a^2} + \frac{6EI}{b^2}\right)\theta_2 = -1.077 \text{ k}$$

$$Y_3 = \frac{-6EI}{b^2}\theta_2 = -0.308 \text{ k}$$

$$M_3 = \frac{2EI}{b}\theta_2 = 18.46 \text{ in.-k} \quad \blacksquare$$

EXAMPLE 23.2

The beam of Figure 23.15(a) has segments with different flexural stiffnesses. A concentrated load P acts at the midpoint of the beam. Determine the deflection under the load and the reactions at the support points.

Solution.
The beam is divided into two segments, with the load acting at one of the node points, as shown in Figure 23.15(b). The basic stiffness equation (23.26) is used for this problem, but now the boundary conditions are

$$v_1 = 0$$
$$\theta_1 = 0$$
$$v_2 = 0$$
$$\theta_2 = 0$$

When columns and rows are deleted from Equation 23.26 in order to reflect the boundary conditions, the result is

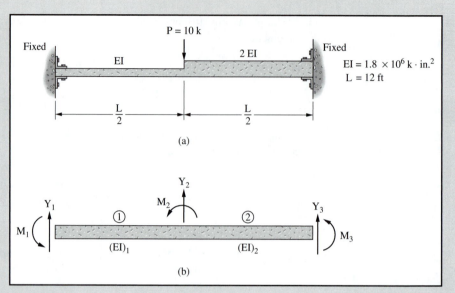

Figure 23.15

$$
\begin{Bmatrix} Y_1 \\ M_1 \\ Y_2 \\ M_2 \\ Y_3 \\ M_3 \end{Bmatrix} =
\begin{bmatrix}
x & x & \dfrac{-96EI}{L^3} & \dfrac{24EI}{L^2} & x & x \\[2mm]
x & x & \dfrac{-24EI}{L^2} & \dfrac{4EI}{L} & x & x \\[2mm]
x & x & \dfrac{288EI}{L^3} & \dfrac{24EI}{L^2} & x & x \\[2mm]
x & x & \dfrac{24EI}{L^2} & \dfrac{24EI}{L^2} & x & x \\[2mm]
x & x & \dfrac{-192EI}{L^3} & \dfrac{-48EI}{L^2} & x & x \\[2mm]
x & x & \dfrac{48EI}{L^2} & \dfrac{8EI}{L} & x & x
\end{bmatrix}
\begin{Bmatrix} v_1 = 0 \\ \theta_1 = 0 \\ v_2 \\ \theta_2 \\ v_3 = 0 \\ \theta_3 = 0 \end{Bmatrix}
\qquad \text{(23.28)}
$$

Two equations remain from the original set and these are solved for the displacements of the free nodes.

$$
\begin{Bmatrix} Y_2 = -10 \\ M_2 = 0 \end{Bmatrix} =
\begin{bmatrix}
\dfrac{288EI}{L^3} & \dfrac{24EI}{L^2} \\[2mm]
\dfrac{24EI}{L^2} & \dfrac{24EI}{L}
\end{bmatrix}
\begin{Bmatrix} v_2 \\ \theta_2 \end{Bmatrix}
$$

The numerical values shown in Figure 23.15 are substituted into the preceding expressions with the following results:

$$
\begin{Bmatrix} -10 \\ 0 \end{Bmatrix} =
\begin{bmatrix} 173.6 & 2083 \\ 2083 & 300,000 \end{bmatrix}
\begin{Bmatrix} v_2 \\ \theta_2 \end{Bmatrix}
$$

and

$$
\begin{Bmatrix} v_2 \\ \theta_2 \end{Bmatrix} =
\begin{bmatrix} 173.6 & 2083 \\ 2083 & 300,000 \end{bmatrix}^{-1}
\begin{Bmatrix} -10 \\ 0 \end{Bmatrix} =
\begin{Bmatrix} -0.063 \text{ in.} \\ 4.4 \times 10^{-4} \text{ rad} \end{Bmatrix}
$$

The values obtained for v_2 and θ_2 are substituted into the equations that were deleted in order to produce the solutions for the reactions:

$$Y_1 = \frac{-96EI}{L^3}v_2 + \frac{24EI}{L^2}\theta_2 = 4.545 \text{ k}$$

$$M_1 = \frac{-24EI}{L^2}v_2 + \frac{4EI}{L}\theta_2 = 152.7 \text{ in.-k}$$

$$Y_3 = \frac{-192EI}{L^3}v_2 - \frac{48EI}{L^2}\theta_2 = 5.455 \text{ k}$$

$$M_3 = \frac{48EI}{L^2}v_2 + \frac{8EI}{L}\theta_2 = -218.2 \text{ in.-k} \quad \blacksquare$$

The reader should note that the same basic stiffness equation (23.26) can be used to solve a large number of problems when different loadings, stiffnesses, and boundary conditions are specified.

23.5 STIFFNESS MATRIX FOR COMBINED AXIAL AND FLEXURAL MEMBERS

If axial forces act on members that are subjected simultaneously to shears and bending moments, the stiffness matrices can be prepared in the same manner as they were in the last section for members subjected to shears and moments only. For this discussion the element shown in Figure 23.16 is considered.

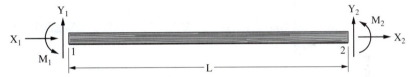

Figure 23.16

According to the small-displacement theory it is assumed that axial forces do not affect bending moments and transverse shears, and vice versa. The stiffness matrix, therefore, may be written as a superposition of individual loading cases.

$$
\begin{Bmatrix} X_1 \\ Y_1 \\ M_1 \\ X_2 \\ Y_2 \\ M_2 \end{Bmatrix} =
\begin{Bmatrix} \dfrac{AE}{L} \\ 0 \\ 0 \\ -\dfrac{AE}{L} \\ 0 \\ 0 \end{Bmatrix} u_1 +
\begin{Bmatrix} 0 \\ \dfrac{12EI}{L^3} \\ \dfrac{6EI}{L^2} \\ 0 \\ -\dfrac{12EI}{L^3} \\ \dfrac{6EI}{L^2} \end{Bmatrix} v_1 +
\begin{Bmatrix} 0 \\ \dfrac{6EI}{L^2} \\ \dfrac{4EI}{L} \\ 0 \\ -\dfrac{6EI}{L^2} \\ \dfrac{2EI}{L} \end{Bmatrix} \theta_1 +
\begin{Bmatrix} -\dfrac{AE}{L} \\ 0 \\ 0 \\ \dfrac{AE}{L} \\ 0 \\ 0 \end{Bmatrix} u_2 +
\begin{Bmatrix} 0 \\ -\dfrac{12EI}{L^3} \\ -\dfrac{6EI}{L^2} \\ 0 \\ \dfrac{12EI}{L^3} \\ -\dfrac{6EI}{L^2} \end{Bmatrix} v_2 +
\begin{Bmatrix} 0 \\ \dfrac{6EI}{L^2} \\ \dfrac{2EI}{L} \\ 0 \\ -\dfrac{6EI}{L^2} \\ \dfrac{4EI}{L} \end{Bmatrix} \theta_2
$$

$$(23.29)$$

In matrix notation the resulting stiffness equation is

$$
\begin{Bmatrix} X_1 \\ Y_1 \\ M_1 \\ X_2 \\ Y_2 \\ M_2 \end{Bmatrix} =
\begin{bmatrix}
\dfrac{AE}{L} & 0 & 0 & \dfrac{-AE}{L} & 0 & 0 \\[2mm]
0 & \dfrac{12EI}{L^3} & \dfrac{6EI}{L^2} & 0 & \dfrac{-12EI}{L^3} & \dfrac{6EI}{L^2} \\[2mm]
0 & \dfrac{6EI}{L^2} & \dfrac{4EI}{L} & 0 & \dfrac{-6EI}{L^2} & \dfrac{2EI}{L} \\[2mm]
-\dfrac{AE}{L} & 0 & 0 & \dfrac{AE}{L} & 0 & 0 \\[2mm]
0 & \dfrac{-12EI}{L^3} & \dfrac{-6EI}{L^2} & 0 & \dfrac{12EI}{L^3} & \dfrac{-6EI}{L^2} \\[2mm]
0 & \dfrac{6EI}{L^2} & \dfrac{2EI}{L} & 0 & \dfrac{-6EI}{L^2} & \dfrac{4EI}{L}
\end{bmatrix}
\begin{Bmatrix} u_1 \\ v_1 \\ \theta_1 \\ u_2 \\ v_2 \\ \theta_2 \end{Bmatrix}
\qquad \textbf{(23.30)}
$$

Example 23.3 illustrates the analysis of a cantilevered beam using the stiffness method.

EXAMPLE 23.3

Determine the end displacements and the support reactions for the cantilevered beam of Figure 23.17.

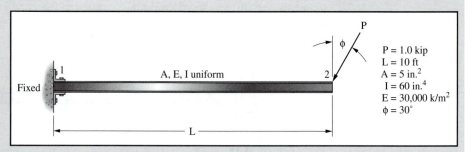

P = 1.0 kip
L = 10 ft
A = 5 in.^{2}
I = 60 in.^{4}
E = 30,000 k/m^2
$\phi = 30°$

Figure 23.17

Solution.
Equation 23.30 represents the total stiffness matrix for this structure since it consists of only one beam element. The appropriate rows and columns are eliminated from the equation to reflect the imposed boundary conditions.

$$
\begin{Bmatrix} X_1 \\ Y_1 \\ M_1 \\ X_2 = -P\sin\phi \\ Y_2 = -P\cos\phi \\ M_2 = 0 \end{Bmatrix} =
\begin{bmatrix}
x & x & x & \dfrac{AE}{L} & 0 & 0 \\[2mm]
x & x & x & 0 & \dfrac{-12EI}{L^3} & \dfrac{6EI}{L^2} \\[2mm]
x & x & x & 0 & \dfrac{-6EI}{L^2} & \dfrac{2EI}{L} \\[2mm]
x & x & x & \dfrac{AE}{L} & 0 & 0 \\[2mm]
x & x & x & 0 & \dfrac{12EI}{L^3} & \dfrac{-6EI}{L^2} \\[2mm]
x & x & x & 0 & \dfrac{-6EI}{L^2} & \dfrac{4EI}{L}
\end{bmatrix}
\begin{Bmatrix} u_1 = 0 \\ v_1 = 0 \\ \theta_1 = 0 \\ u_2 \\ v_2 \\ \theta_2 \end{Bmatrix}
\qquad \textbf{(23.31)}
$$

The remaining matrix equation is

$$\begin{Bmatrix} -P \sin \phi \\ -P \cos \phi \\ 0 \end{Bmatrix} = \begin{bmatrix} \dfrac{AE}{L} & 0 & 0 \\ 0 & \dfrac{12EI}{L^3} & \dfrac{-6EI}{L^2} \\ 0 & -6\dfrac{EI}{L^2} & \dfrac{4EI}{L} \end{bmatrix} \begin{Bmatrix} u_2 \\ v_2 \\ \theta_2 \end{Bmatrix} \tag{23.32}$$

After substitution of the numerical values given in Figure 23.17 the end displacements are determined by solving the equation set simultaneously.

$$\begin{Bmatrix} u_2 \\ v_2 \\ \theta_2 \end{Bmatrix} = \begin{bmatrix} 1250 & 0 & 0 \\ 0 & 12.5 & -750 \\ 0 & -750 & 60,000 \end{bmatrix}^{-1} \begin{Bmatrix} -0.50 \\ -0.866 \\ 0 \end{Bmatrix} = \begin{Bmatrix} -0.0004 \text{ in.} \\ -0.27713 \text{ in.} \\ -0.00346 \text{ rad} \end{Bmatrix}$$

Reactions are found by substituting the values of these displacements into the equations that were eliminated from Equation 23.31.

$$\begin{Bmatrix} X_1 \\ Y_1 \\ M_1 \end{Bmatrix} = \begin{bmatrix} -\dfrac{AE}{L} & 0 & 0 \\ 0 & \dfrac{-12EI}{L^3} & \dfrac{6EI}{L^2} \\ 0 & \dfrac{-6EI}{L^2} & \dfrac{2EI}{L} \end{bmatrix} \begin{Bmatrix} u_2 \\ v_2 \\ \theta_2 \end{Bmatrix}$$

$$= \begin{bmatrix} -1250 & 0 & 0 \\ 0 & -12.5 & 750 \\ 0 & 750 & 30,000 \end{bmatrix} \begin{Bmatrix} u_2 \\ v_2 \\ \theta_2 \end{Bmatrix} = \begin{Bmatrix} 0.50 \text{ k} \\ 0.866 \text{ k} \\ 103.9 \text{ in.-k} \end{Bmatrix} \quad \blacksquare$$

Although beam problems seldom involve appreciable axial forces, the stiffness equation (23.30) becomes a fundamental equation in the solution of problems that involve columns and diagonal members, as well as beam members. These problems, and their solutions, are discussed in Chapter 24.

23.6 CHARACTERISTICS OF STIFFNESS MATRICES

In the previous sections stiffness matrices have been developed for struts, for flexural members, and for members that have axial forces acting together with shears and bending moments. These stiffness matrices are all written in a special order, which should be carefully studied. The general stiffness equation is written symbolically as follows:

$$\{P\} = [K]\{\delta\}$$

The order of listing the nodal forces in the matrix $[K]$ should be the same as the order of listing of the corresponding displacements in the matrix $\{\delta\}$. Thus if the first listed nodal force in $\{P\}$ is X_1, then the first listed displacement in the matrix $\{\delta\}$ should be u_1, and so on. If this ordering of the equations is preserved throughout the analysis, all stiffness matrices will have the following characteristics:

1. They will be symmetric, i.e., $k_{i,j} = k_{j,i}$. A proof of this characteristic can be obtained from Maxwell's Law of Reciprocal Deformations.
2. The full stiffness matrix, for either a single element or for a general structure, is singular. However, if sufficient boundary conditions are specified so that the structure is stable (and the stiffness matrix modified to reflect these conditions), the resulting stiffness matrix is nonsingular.
3. The stiffness coefficients on the main diagonal are always positive. The reason for this characteristic stems from the fact that the stiffness coefficients on the main diagonal each represent the force at a node required to produce a corresponding displacement at that node. A negative force required to produce a positive displacement is contrary to observed physical behavior of structures.

23.7 RELATION BETWEEN STIFFNESS AND FLEXIBILITY MATRICES

Although the stiffness method and the flexibility method are two distinct methods of structural analysis the matrices developed in the two methods are closely related. It is the purpose of this section to point out these similarities and to show some of the differences between them as well.

As described in Chapter 22 flexibility coefficients represent displacements at specific locations due to unit loading at other specific locations. As an example, a set of flexibility coefficients are developed in this section for the simply supported beam of Figure 23.18(a).

A unit moment is applied at node 1 of the beam, as shown in Figure 23.18(b). The resultant rotations at nodes 1 and 2 are the flexibility coefficients, $\delta_{1,1}$ and $\delta_{2,1}$, respectively. Their values are shown in Figure 23.18(b). Similarly a unit moment is applied at node 2 of the beam, as shown in Figure 23.18(c), in order to determine the flexibility coefficients $\delta_{1,2}$ and $\delta_{2,2}$. The values of these coefficients are shown in Figure 23.18(c).

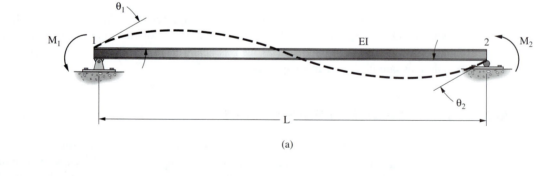

(a)

(b)   (c)

Figure 23.18

The total displacements at nodes 1 and 2 due to the application of the arbitrary moments M_1 and M_2 are then written using the superposition principle.

$$\theta_1 = \left(\frac{L}{3EI}\right)M_1 - \left(\frac{L}{6EI}\right)M_2$$

$$\theta_2 = -\left(\frac{L}{6EI}\right)M_1 + \left(\frac{L}{3EI}\right)M_2$$

(23.33)

Or in matrix form

$$\left\{\begin{array}{c} \theta_1 \\ \theta_2 \end{array}\right\} = \left[\begin{array}{cc} \dfrac{L}{3EI} & -\dfrac{L}{6EI} \\ -\dfrac{L}{6EI} & \dfrac{L}{3EI} \end{array}\right] \left\{\begin{array}{c} M_1 \\ M_2 \end{array}\right\}$$

(23.34)

Symbolically, Equation 23.34 is written as

$$\{\delta\} = [F]\{P\}$$

(23.35)

In Section 23.2, a comparable relation for this beam was derived using the stiffness method, and was shown in Equation 23.1. This equation is repeated as Equation 23.36.

$$\left\{\begin{array}{c} M_1 \\ M_2 \end{array}\right\} = \left[\begin{array}{cc} \dfrac{4EI}{L} & \dfrac{2EI}{L} \\ \dfrac{2EI}{L} & \dfrac{4EI}{L} \end{array}\right] \left\{\begin{array}{c} \theta_1 \\ \theta_2 \end{array}\right\}$$

(23.36)

Or, symbolically,

$$\{P\} = [K]\{\delta\}$$

(23.37)

A comparison of Equations 23.35 and 23.37 indicates that the stiffness matrix [K] is the inverse of the flexibility matrix [F].

$$[K] = [F]^{-1}$$

(23.38)

The reader may prove this relation by working out the following from Equations 23.34 and 23.36.

$$\left[\begin{array}{cc} \dfrac{L}{3EI} & -\dfrac{L}{6EI} \\ -\dfrac{L}{6EI} & -\dfrac{L}{3EI} \end{array}\right]^{-1} = \left[\begin{array}{cc} \dfrac{4EI}{L} & \dfrac{2EI}{L} \\ \dfrac{2EI}{L} & \dfrac{4EI}{L} \end{array}\right]$$

Although Equation 23.38 is correct for specific cases, there is a more fundamental difference between the stiffness and the flexibility matrices. The reader will recall from Equation 23.23 that the total stiffness matrix for a beam may be written as follows:

$$\left\{\begin{array}{c} Y_1 \\ M_1 \\ Y_2 \\ M_2 \end{array}\right\} = \left\{\begin{array}{cccc} \dfrac{12EI}{L^3} & \dfrac{6EI}{L^2} & \dfrac{-12EI}{L^3} & \dfrac{6EI}{L^2} \\ \dfrac{6EI}{L^2} & \dfrac{4EI}{L} & \dfrac{-6EI}{L^2} & \dfrac{2EI}{L} \\ \dfrac{-12EI}{L^3} & \dfrac{-6EI}{L^2} & \dfrac{12EI}{L^3} & \dfrac{-6EI}{L^2} \\ \dfrac{6EI}{L^2} & \dfrac{2EI}{L} & \dfrac{-6EI}{L^2} & \dfrac{4EI}{L} \end{array}\right\} \left\{\begin{array}{c} v_1 \\ \theta_1 \\ v_2 \\ \theta_2 \end{array}\right\}$$

(23.39)

Thus the stiffness matrix shown in Equation 23.36 represents only a portion of the total stiffness matrix for the beam. If different boundary conditions are specified for the beam, other reduced stiffness matrices apply. Thus the total stiffness matrix shown in Equation 23.39 encompasses reduced stiffness matrices for a variety of other boundary conditions, such as the ones shown in Figures 23.19(a) to 23.19(f). However, note that there is no flexibility matrix that is a counterpart to the total stiffness matrix, i.e., flexibility matrices exist only for stable structures.

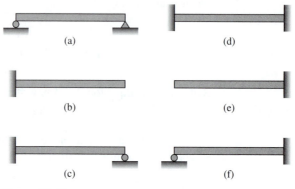

Figure 23.19

The preceding statements for a single beam element apply equally well to any structure, no matter how complex. For each individual support condition of the structure there is a distinct reduced stiffness matrix and a distinct flexibility matrix. They are related to each other through Equation 23.38. However, the structure possesses only one *total* stiffness matrix.

23.8 PROBLEMS FOR SOLUTION

23.1

$$\left(\frac{AE}{L}\right)_1 = \left(\frac{AE}{L}\right)_2 = \left(\frac{AE}{L}\right)_3 = 100.0 \text{ k/in.}$$

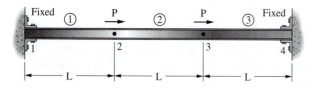

For the axial force structure shown, determine the following:

(a) Displacement of nodes 2 and 3;
(b) Reactions at nodes 1 and 4;
(c) Internal member forces in members 1, 2, and 3.

Ans. (a) $\left\{\begin{matrix} u_2 \\ u_3 \end{matrix}\right\} = \dfrac{P}{100}\left\{\begin{matrix} 1 \\ 1 \end{matrix}\right\}$

(b) $\left\{\begin{matrix} x_1 \\ x_4 \end{matrix}\right\} = \left\{\begin{matrix} -P \\ -P \end{matrix}\right\}$

(c) $\left\{\begin{matrix} F_{1,1} \\ F_{1,2} \end{matrix}\right\} = \left\{\begin{matrix} -P \\ P \end{matrix}\right\}$

23.2 Repeat Problem 23.1 using altered loads acting at nodes 2 and 3, as shown.

$$\frac{AE}{L} = 100.0 \text{ k/in.}$$

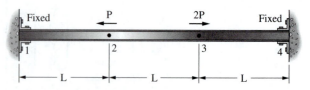

23.3 Repeat Problem 23.1 using altered stiffnesses for members 1 and 3, as shown.

$$\left(\frac{AE}{L}\right)_2 = 100.0 \text{ k/in.}$$

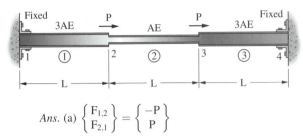

Ans. (a) $\left\{\begin{array}{c} F_{1,2} \\ F_{2,1} \end{array}\right\} = \left\{\begin{array}{c} -P \\ P \end{array}\right\}$

(b) $\left\{\begin{array}{c} F_{2,3} \\ F_{3,2} \end{array}\right\} = \left\{\begin{array}{c} 0 \\ 0 \end{array}\right\}$

(c) $\left\{\begin{array}{c} F_{3,4} \\ F_{4,3} \end{array}\right\} = \left\{\begin{array}{c} P \\ -P \end{array}\right\}$

23.4

$$\frac{AE}{L} = 100.0 \text{ k/in.}$$

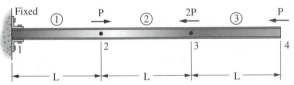

For the axial force structure shown determine the following:
(a) Displacements at nodes 2, 3, and 4;
(b) Reaction at node 1;
(c) Internal member forces in members 1, 2, and 3.

23.5

$$\frac{AE}{L} = 100.0 \text{ k/in.}$$
all members
$u_2 = 0.05$ in.

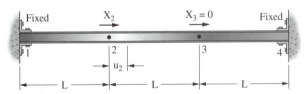

For the axial force structure shown assume that an external force X_2, acting at node 2, produces a displacement $u_2 = 0.02$ in. Determine the following:
(a) External force X_2;
(b) Resulting displacement at node 3;
(c) Reactions at nodes 1 and 4.

Ans. (a) and (b) $\left\{\begin{array}{c} x_3 \\ u_3 \end{array}\right\} = \left\{\begin{array}{c} 3 \text{ k} \\ 0.01 \text{ in.} \end{array}\right\}$

(c) $\left\{\begin{array}{c} R_1 \\ R_4 \end{array}\right\} = \left\{\begin{array}{c} -2 \\ -1 \end{array}\right\}$

23.6
$$\frac{AE}{L} = \text{constant} = 100.0 \text{ k/in.}$$
all members
$u_2 = 0.05$ in.

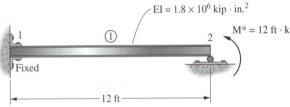

For the axial force member shown assume that a known force P acts at node 3, and an unknown force X_2 acts at node 2. The resulting displacement for the combined loading is measured and found to be $u_2 = 0.05$. Determine
(a) Nodal force X_2;
(b) Displacement at node 3;
(c) Reactions at nodes 1 and 4.

23.7

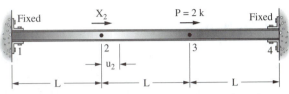

Using the stiffness matrix given by Equation 23.23, find the following:
(a) Rotation at node 2 due to externally applied moment M*; (*Ans.* $\theta_2 = 0.00288$ radians)
(b) Reaction at nodes 1 and 2. (*Ans.* $y_2 = -1.5$ k)

23.8

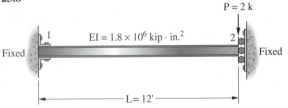

The beam shown can translate vertically at node 2 but cannot rotate. Assuming a transverse load P acting at node 2, determine the following:
(a) Displacement at node 2;
(b) Reactions at nodes 1 and 2.

Chapter 24

Stiffness Matrices for Inclined Members

24.1 GENERAL

In Chapter 23, stiffness equations were written for struts and beams that were oriented along a single horizontal axis. Most structures consist of a combination of beams and columns, and perhaps some inclined members. The development of stiffness equations for members whose axes are not all parallel is addressed in this chapter.

24.2 AXIAL FORCE MEMBERS

The pin-connected truss shown in Figure 24.1 consists of a number of members whose axes are oriented at various angles to a horizontal reference axis.

Figure 24.1

An inclined member of this truss is shown in Figure 24.2.

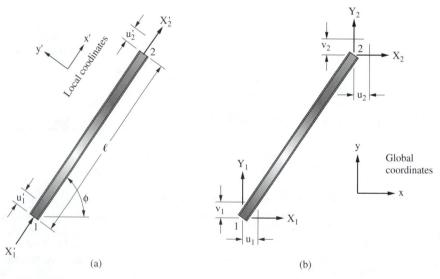

     (a)   (b)

Figure 24.2

Recreation Center, Lander College, Greenwood, South Carolina (Courtesy of Britt, Peters and Associates)

To describe the force-displacement relation for this member and all other inclined members, it is convenient to establish two coordinate systems—a *local coordinate system* and a *global coordinate system*. The local coordinate system is established by drawing the local x axis along the axis of the inclined member (see Figure 24.2a). This axis is labelled x′ to distinguish it from the global x axis. The local y axis, labelled y′, is drawn perpendicular to the local x′ axis. The global coordinate system is a single reference system established by the analyst for the entire structure. The global axes are labelled x and y for planar structures (see Figure 24.2b).

Forces and displacements referenced with respect to the local axes are designated with primes, e.g., X_1' represents a force parallel to the x′ axis, while u_1' represents a displacement parallel to this axis. Conversely, forces and displacements written without primes represent values referenced to the global axes. Thus X_1 and Y_1 represent forces parallel to the global x and y axes, respectively, while u_1 and v_1 represent displacement components parallel to these axes.

The force-displacement relations developed in Chapter 23 are valid only when forces and displacements are referenced to the local axes of the struct. Should several struts meet at a joint, such as shown in Figure 24.1, there will be several sets of local reference axes. Since the direction of the axial forces and the axial displacements of struts meeting at a joint are different, their individual stiffnesses cannot be added directly. Therefore in this section the force-displacement relationships for a strut are re-derived with respect to global axes, so that stiffnesses of struts meeting at a joint can be added.

To establish the force-displacement relation with respect to global axes for an inclined strut, one global component of displacement is given an arbitrary value. All other components of displacement are kept equal to zero. Figure 24.3 shows a strut that has

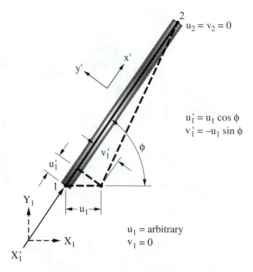

Figure 24.3

been given an arbitrary value of displacement, u_1. This displacement may be broken up, or decomposed, into components perpendicular and parallel to the local member axes as shown in this figure.

The displacement component v_1' does not change the length of the strut significantly and, in small-displacement theory, does not produce stresses in the strut. However, the displacement component u_1' is significant and does cause appreciable stresses in the strut. The force-displacement relation has previously been established as

$$X_1' = \left(\frac{AE}{L}\right) u_1' \tag{24.1}$$

The local member force X_1' can be decomposed into two global components X_1 and Y_1 as follows:

$$X_1 = X_1' \cos \phi$$
$$Y_1 = X_1' \sin \phi \tag{24.2}$$

Then, through appropriate substitution, all forces and displacement components written with respect to the local x' axis are written with respect to the global x and y axes:

$$X_1 = X_1' \cos \phi = \left(\frac{AE}{L}\right) \cos \phi \, u_1' = \left(\frac{AE}{L} \cos^2 \phi\right) u_1$$
$$Y_1 = X_1' \sin \phi = \left(\frac{AE}{L}\right) \sin \phi \, u_1' = \left(\frac{AE}{L} \sin \phi \, \cos \phi\right) u_1 \tag{24.3}$$

The forces at node 2, corresponding to the enforced displacement component u_1, can be established from equilibrium:

$$X_2 = -X_1 = -\left(\frac{AE}{L} \cos^2 \phi\right) u_1$$
$$Y_2 = -Y_1 = -\left(\frac{AE}{L} \sin \phi \cos \phi\right) u_1 \tag{24.4}$$

In a similar manner an arbitrary value of displacement v_1 may be given to node 1, while all other displacement components are held equal to zero. This configuration is shown in Figure 24.4.

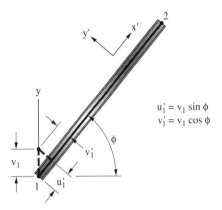

$$u_1' = v_1 \sin \phi$$
$$v_1' = v_1 \cos \phi$$

Figure 24.4

The displacement component v_1, may be resolved into the components u_1' and v_1', as shown in Figure 24.4. The force-displacement relation, written with respect to local coordinates is

$$X_1' = \left(\frac{AE}{L}\right) u_1' \tag{24.5}$$

East Belt Freeway (I-440). Little Rock, Arkansas. (Courtesy of the Arkansas State Highway and Transportation Department.)

When the force X_1' is resolved into the global components X_1 and Y_1 the following equations may be written, which relate global force components to the global displacement component v_1:

$$X_1 = X_1' \, \cos \phi = \left(\frac{AE}{L} \cos \phi\right) u_1' = \left(\frac{AE}{L} \sin \phi \, \cos \phi\right) v_1$$

$$Y_1 = X_1' \, \sin \phi = \left(\frac{AE}{L} \sin \phi\right) u_1' = \left(\frac{AE}{L} \sin^2 \phi\right) v_1$$

$$X_2 = -X_1 = -\left(\frac{AE}{L} \sin \phi \, \cos \phi\right) v_1 \qquad (24.6)$$

$$Y_2 = -Y_1 = -\left(\frac{AE}{L} \sin^2 \phi\right) v_1$$

A similar procedure may be used to establish the force-displacement relation associated with arbitrary displacement components u_2 and v_2. A superposition of all these relationships produces the following equation:

$$\begin{Bmatrix} X_1 \\ Y_1 \\ X_2 \\ Y_2 \end{Bmatrix} = \frac{AE}{L} \begin{Bmatrix} \cos^2 \phi \\ \sin \phi \, \cos \phi \\ -\cos^2 \phi \\ -\sin \phi \, \cos \phi \end{Bmatrix} u_1 + \frac{AE}{L} \begin{Bmatrix} \sin \phi \, \cos \phi \\ \sin^2 \phi \\ -\sin \phi \, \cos \phi \\ -\sin^2 \phi \end{Bmatrix} v_1$$

$$+ \frac{AE}{L} \begin{Bmatrix} -\cos^2 \phi \\ -\sin \phi \, \cos \phi \\ \cos^2 \phi \\ \sin \phi \, \cos \phi \end{Bmatrix} u_2 + \frac{AE}{L} \begin{Bmatrix} -\sin \phi \, \cos \phi \\ -\sin^2 \phi \\ \sin \phi \, \cos \phi \\ \sin^2 \phi \end{Bmatrix} v_2 \qquad (24.7)$$

In more compact matrix form, Equation 24.7 may be written as:

$$\begin{Bmatrix} X_1 \\ Y_1 \\ X_2 \\ Y_2 \end{Bmatrix} = \frac{AE}{L} \begin{bmatrix} c^2 & sc & -c^2 & -sc \\ sc & s^2 & -sc & -s^2 \\ -c^2 & -sc & c^2 & sc \\ -sc & -s^2 & sc & s^2 \end{bmatrix} \begin{Bmatrix} u_1 \\ v_1 \\ u_2 \\ v_2 \end{Bmatrix} \qquad (24.8)$$

where

$$s = \sin \phi$$

$$c = \cos \phi$$

Equation 24.8 represents a completely general relation between nodal forces and nodal displacements for a strut, all components of which are written with respect to the global axes.

A stiffness equation such as Equation 24.8 may be used to solve for the global components of displacement of a structure. For a given problem it may also be necessary to find the displacement components expressed in terms of local axes. The relations between global and local components of displacement are shown in Figures 24.3 and 24.4, and are summarized in matrix form as follows:

$$\begin{Bmatrix} u_1' \\ v_1' \end{Bmatrix} = \begin{bmatrix} \cos \phi & \sin \phi \\ -\sin \phi & \cos \phi \end{bmatrix} \begin{Bmatrix} u_1 \\ v_1 \end{Bmatrix} \qquad (24.9)$$

A similar relation can be written for end 2 of the strut by substituting subscript 2 in place of subscript 1 in Equation 24.9. The usefulness of expressions such as Equation 24.9 will be illustrated in the example to follow.

Example 24.1 illustrates the determination of the displacements at one point in a structure that consists of two inclined members joined together. The member forces are also calculated.

EXAMPLE 24.1

The truss shown in Figure 24.5 is subjected to a horizontal force P applied at node 3. Determine the resulting displacement components at the free node (node 3), and the resulting member forces.

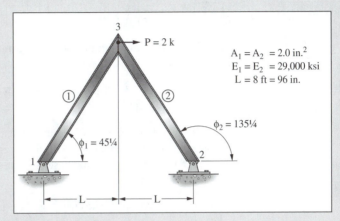

$A_1 = A_2 = 2.0 \text{ in.}^2$
$E_1 = E_2 = 29,000 \text{ ksi}$
$L = 8 \text{ ft} = 96 \text{ in.}$

Figure 24.5

Solution.
The stiffness equations for the structure may be written by enforcing, in turn, the joint displacements u_3 and v_3. When displacement u_3 is enforced, the configuration shown in Figure 24.6(a) results.

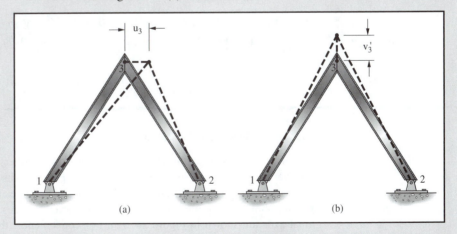

Figure 24.6

The force-displacement relations may be written with the help of Equation 24.7 as

$$\left\{ \begin{array}{c} X_3 \\ Y_3 \end{array} \right\} = \frac{AE}{\sqrt{2}1} \underbrace{\left\{ \begin{array}{c} c^2 \\ sc \end{array} \right\}_{\phi_1 = 45°}}_{\text{for member } \textcircled{1}} u_3 + \frac{AE}{\sqrt{2}1} \underbrace{\left\{ \begin{array}{c} c^2 \\ sc \end{array} \right\}_{\phi_2 = 135°}}_{\text{for member } \textcircled{2}} u_3 = 427.2 \left\{ \begin{array}{c} 1 \\ 0 \end{array} \right\} u_3$$

When v_3 is enforced, the configuration of Figure 24.6(b) results and the resultant force-displacement relations are

$$\begin{Bmatrix} X_3 \\ Y_3 \end{Bmatrix} = \frac{AE}{\sqrt{21}} \begin{Bmatrix} sc \\ s^2 \end{Bmatrix}_{\phi_1=45°} v_3 + \frac{AE}{\sqrt{21}} \begin{Bmatrix} sc \\ s^2 \end{Bmatrix}_{\phi_2=135°} v_3 = 427.2 \begin{Bmatrix} 0 \\ 1 \end{Bmatrix} v_3$$

for member ①   for member ②

When the two enforced displacement relations are superimposed, we have

$$\begin{Bmatrix} X_3 \\ Y_3 \end{Bmatrix} = 427.2 \begin{Bmatrix} 1 \\ 0 \end{Bmatrix} u_3 + 427.2 \begin{Bmatrix} 0 \\ 1 \end{Bmatrix} v_3 = 427.2 \begin{bmatrix} 1 & 0 \\ 0 & 1 \end{bmatrix} \begin{Bmatrix} u_3 \\ v_3 \end{Bmatrix} \tag{24.10}$$

The stiffness equation, given by Equation 24.10, may be solved for the nodal displacements u_3 and v_3:

$$\begin{Bmatrix} u_3 \\ v_3 \end{Bmatrix} = \frac{1}{427.2} \begin{bmatrix} 1 & 0 \\ 0 & 1 \end{bmatrix}^{-1} \begin{Bmatrix} X_3 = P = 2 \\ Y_3 = 0 \end{Bmatrix} = \begin{Bmatrix} 0.0047 \\ 0 \end{Bmatrix} \text{ in.}$$

The member forces may be found by using Equation 24.9 to resolve the global displacement components into local displacement components for each individual member, and then applying the basic stiffness relation of Equation 23.8:

For member 1 ($\phi_1 = 45°$):

$$\begin{Bmatrix} u_3' \\ v_3' \end{Bmatrix} = \begin{bmatrix} \cos \phi_1 & \sin \phi_1 \\ -\sin \phi_1 & \cos \phi_1 \end{bmatrix} \begin{Bmatrix} u_3 \\ v_3 \end{Bmatrix} = \begin{Bmatrix} 0.0047/\sqrt{2} \\ -0.0047/\sqrt{2} \end{Bmatrix} = \begin{Bmatrix} 0.0033 \\ -0.0033 \end{Bmatrix} \text{ in.}$$

and

$$X_{3,1}' = \frac{AE}{\sqrt{21}} u_3' = \frac{2(29000)}{\sqrt{2}(96)} (0.0033) = 1.414 \text{ k}$$

For member 2 ($\phi_2 = 135°$):

$$\begin{Bmatrix} u_3' \\ v_3' \end{Bmatrix} = \begin{bmatrix} \cos \phi_2 & \sin \phi_2 \\ -\sin \phi_2 & \cos \phi_2 \end{bmatrix} \begin{Bmatrix} u_3 \\ v_3 \end{Bmatrix} = \begin{Bmatrix} -0.0047/\sqrt{2} \\ -0.0047/\sqrt{2} \end{Bmatrix} = \begin{Bmatrix} -0.0033 \\ -0.0033 \end{Bmatrix} \text{ in.}$$

and

$$X_{3,2}' = \frac{AE}{\sqrt{21}} u_3' = \frac{2(29000)}{\sqrt{2}(96)} (-0.0033) = -1.414 \text{ k} \quad \blacksquare$$

24.3 FLEXURAL MEMBERS

Stiffness equations for arbitrarily oriented flexural members may be developed in the same general manner as they were for struts. For this discussion the single flexural member shown in Figure 24.7 is considered. Member forces expressed in terms of local axes are shown in Figure 24.7(a), while comparable forces expressed in terms of the global axes are shown in Figure 24.7(b). For convenience, the effects of axial force, as well as shears and moments, are treated together in this section.

In Figure 24.7(a) the local y axis (designated as y') is shown acting upward to the left, although this choice is somewhat arbitrary. For the purposes of this book, the author has chosen the sense of the y' axis so that it acts *upward* if the x' points to the viewer's *right*. Therefore once the analyst chooses the sense of the x' axis, the y' axis is determined automatically. The angle ϕ that describes the orientation of the member with respect to the global coordinates is measured in the positive right-hand rule sense from the positive x axis to the positive x' axis. Figures 24.8(a) and 24.8(b) illustrate two particular sets of orientations of local axes.

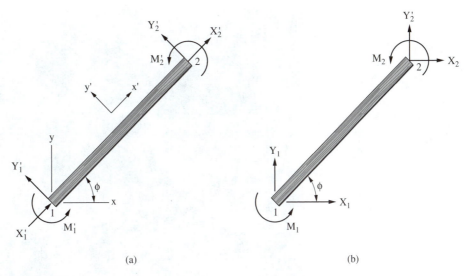

(a)   (b)

Figure 24.7

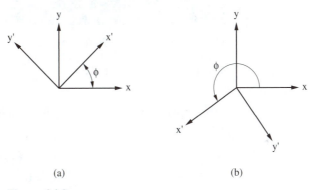

(a)   (b)

Figure 24.8

To establish force-displacement relations for the flexural member in terms of global components, a specific displacement component, u_1, is given an arbitrary value, while all other displacement components remain equal to zero. The displaced configuration is shown in Figure 24.9.

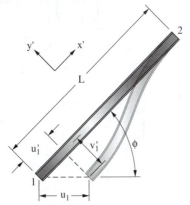

Figure 24.9

Westvaco Building, New York City. The most massive trusses and girders that have ever gone into a building foundation along Manhattan's Blue Chip Row. (Courtesy Bethlehem Steel Co.)

The displacement component, u_1, is decomposed into local displacement components:

$$
\begin{aligned}
u_1' &= u_1 \cos \phi \\
v_1' &= -u_1 \sin \phi \\
\theta_1' &= 0
\end{aligned}
\qquad (24.11)
$$

The force components acting at node 1 in response to the displacement u_1' and v_1' (which were previously summarized in Equation 23.29) are repeated here:

$$
\left\{ \begin{array}{c} X_1' \\ Y_1' \\ M_1' \end{array} \right\}
= \left\{ \begin{array}{c} \dfrac{AE}{L} \\ 0 \\ 0 \end{array} \right\} u_1'
+ \left\{ \begin{array}{c} 0 \\ \dfrac{12EI}{L^3} \\ \dfrac{6EI}{L^2} \end{array} \right\} v_1'
+ \left\{ \begin{array}{c} 0 \\ \dfrac{6EI}{L^2} \\ \dfrac{4EI}{L} \end{array} \right\} \theta_1'
\qquad (24.12)
$$

Substitution of Equation 24.11 into Equation 24.12 produces the following:

$$
\left\{ \begin{array}{c} X_1' \\ Y_1' \\ M_1' \end{array} \right\}
= \left[\begin{array}{c} \dfrac{AE}{L} \cos \phi \\ 0 \\ 0 \end{array} \right] u_1
+ \left\{ \begin{array}{c} 0 \\ \dfrac{-12EI}{L^3} \sin \phi \\ \dfrac{-6EI}{L^2} \sin \phi \end{array} \right\} u_1
= \left\{ \begin{array}{c} \dfrac{AE}{L} \cos \phi \\ \dfrac{-12EI}{L^3} \sin \phi \\ \dfrac{-6EI}{L^2} \sin \phi \end{array} \right\} u_1
$$

$$
(24.13)
$$

Similarly, a displacement v_1 is now enforced at node 1 and the resultant displaced configuration is shown in Figure 24.10.

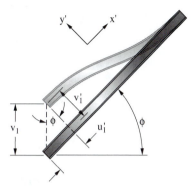

Figure 24.10

The displacement v_1 is decomposed into local components:

$$u'_1 = v_1 \, \sin \phi$$
$$v'_1 = v_1 \, \cos \phi \qquad\qquad \textbf{(24.14)}$$
$$\theta'_1 = 0$$

The force resultants at node 1 may be written by substituting these displacement values into Equation 24.12:

$$\begin{Bmatrix} X'_1 \\ Y'_1 \\ M'_1 \end{Bmatrix} = \begin{bmatrix} \dfrac{AE}{L}\sin \phi \\ 0 \\ 0 \end{bmatrix} v_1 + \begin{Bmatrix} 0 \\ \dfrac{12EI}{L^3}\cos \phi \\ \dfrac{6EI}{L^2}\cos \phi \end{Bmatrix} v_1 = \begin{Bmatrix} \dfrac{AE}{L}\sin \phi \\ \dfrac{12EI}{L^3}\cos \phi \\ \dfrac{6EI}{L^2}\cos \phi \end{Bmatrix} v_1 \qquad \textbf{(24.15)}$$

Rotation θ'_1 (in local coordinates) is exactly the same as θ_1. Therefore the stiffness relations due to an enforced joint rotation are written from Equation 24.12 as

$$\begin{Bmatrix} X'_1 \\ Y'_1 \\ M'_1 \end{Bmatrix} = \begin{Bmatrix} 0 \\ \dfrac{6EI}{L^2} \\ \dfrac{4EI}{L} \end{Bmatrix} \theta'_1 \qquad\qquad \textbf{(24.16)}$$

Forces X'_1 and Y'_1, expressed in local coordinates, are now resolved into the global components X_1 and Y_1:

$$X_1 = X'_1 \, \cos \phi - Y'_1 \, \sin \phi$$
$$Y_1 = X'_1 \, \sin \phi + Y'_1 \, \cos \phi \qquad\qquad \textbf{(24.17)}$$
$$M'_1 = M'_1$$

Substitution of the values of X_1' and Y_1' from Equations 24.13, 24.15, and 24.16 into Equation 24.17 produces the following:

$$X_1 = \cos \phi \left[\frac{AE}{L} \cos \phi \, u_1 + \frac{AE}{L} \sin \phi \, v_1 \right]$$

$$- \sin \phi \left[\frac{-12EI}{L^3} \sin \phi \, u_1 + \frac{12EI}{L^3} \cos \theta \, v_1 \right]$$

$$Y_1 = \sin \phi \left[\frac{AE}{L} \cos \phi \, u_1 + \frac{AE}{L} \sin \phi \, v_1 \right] \qquad \textbf{(24.18)}$$

$$+ \cos \phi \left[\frac{-12EI}{L^3} \sin \phi \, u_1 + \frac{12EI}{L^3} \cos \phi \, v_1 \right]$$

$$M_1 = \frac{-6EI}{L^2} \sin \phi \, u_1 + \frac{6EI}{L^2} \cos \phi \, v_1$$

Force components X_2, Y_2, and M_2, which result from enforced-displacement components at node 1 may be found in a similar manner, starting with the local force components at node 2 given in Equation 23.29. The entire process may be repeated using enforced-displacement components at node 2. A summary of the results of these operations is shown in matrix form as follows:

$$
\begin{Bmatrix} X_1 \\ Y_1 \\ M_1 \\ X_2 \\ Y_2 \\ M_2 \end{Bmatrix}
=
\begin{bmatrix}
c^2\left(\dfrac{AE}{L}\right) + s^2\left(\dfrac{12EI}{L^3}\right) & & & & & \\[2ex]
sc\left(\dfrac{AE}{L} - \dfrac{12EI}{L^3}\right) & s^2\left(\dfrac{AE}{L}\right) + c^2\left(\dfrac{12EI}{L^3}\right) & & \text{(Symmetric)} & & \\[2ex]
-s\left(\dfrac{6EI}{L^2}\right) & c\left(\dfrac{6EI}{L^2}\right) & \left(\dfrac{6EI}{L}\right) & & & \\[2ex]
-c^2\left(\dfrac{AE}{L}\right) - s^2\left(\dfrac{12EI}{L^3}\right) & -sc\left(\dfrac{AE}{L} - \dfrac{12EI}{L^3}\right) & s\left(\dfrac{6EI}{L^2}\right) & c^2\left(\dfrac{AE}{L}\right) + s^2\left(\dfrac{12EI}{L^3}\right) & & \\[2ex]
-sc\left(\dfrac{AE}{L} - \dfrac{12EI}{L^3}\right) & -s^2\left(\dfrac{AE}{L}\right) - c^2\left(\dfrac{12EI}{L^3}\right) & -c\left(\dfrac{6EI}{L^2}\right) & sc\left(\dfrac{AE}{L} - \dfrac{12EI}{L^3}\right) & s^2\left(\dfrac{AE}{L}\right) + c^2\left(\dfrac{12EI}{L^3}\right) & \\[2ex]
-s\left(\dfrac{6EI}{L^2}\right) & c\left(\dfrac{6EI}{L^2}\right) & \left(\dfrac{2EI}{L}\right) & s\left(\dfrac{6EI}{L}\right) & -c\left(\dfrac{6EI}{L^2}\right) & \left(\dfrac{4EI}{L}\right)
\end{bmatrix}
\begin{Bmatrix} u_1 \\ v_1 \\ \theta_1 \\ u_2 \\ v_2 \\ \theta_2 \end{Bmatrix}
$$

$$\textbf{(24.19)}$$

where

$$s = \sin \phi \quad \text{and} \quad c = \cos \phi.$$

Equation 24.19 represents a completely general stiffness equation for a two-dimensional axial-flexural member oriented at an arbitrary angle ϕ with respect to the global axes.

Examples 24.2 and 24.3 illustrate the analysis of structures whose members experience axial forces, as well as shears and bending moments.

EXAMPLE 24.2

Determine the displacements at node 2, the reactions at nodes 1 and 3, and the internal member forces in members 1 and 2 for the structure shown in Figure 24.11:

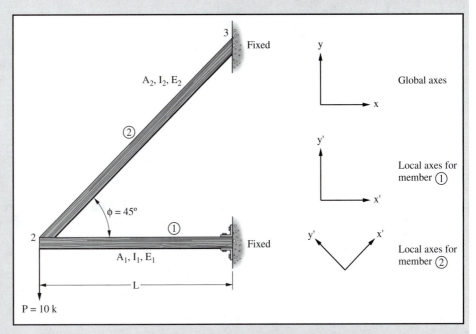

Figure 24.11

Solution.

The stiffness equation for member ①, including axial force contributions, is given by Equation 23.30, and rewritten here:

$$
\begin{Bmatrix} X_{2,1} \\ Y_{2,1} \\ M_{2,1} \\ X_{1,2} \\ Y_{1,2} \\ M_{1,2} \end{Bmatrix} =
\begin{bmatrix}
\left(\dfrac{AE}{L}\right)_1 & & & & & \\[6pt]
0 & \left(\dfrac{12EI}{L^3}\right)_1 & & & \text{(Symmetric)} & \\[6pt]
0 & \left(\dfrac{6EI}{L^2}\right)_1 & 4\left(\dfrac{EI}{L}\right)_1 & & & \\[6pt]
-\left(\dfrac{AE}{L}\right)_1 & 0 & 0 & \left(\dfrac{AE}{L}\right)_1 & & \\[6pt]
0 & -12\left(\dfrac{EI}{L^3}\right)_1 & -6\left(\dfrac{EI}{L^2}\right)_1 & 0 & 12\left(\dfrac{EI}{L^3}\right)_1 & \\[6pt]
0 & 6\left(\dfrac{EI}{L^2}\right)_1 & 2\left(\dfrac{EI}{L}\right)_1 & 0 & -6\left(\dfrac{EI}{L^2}\right)_1 & 4\left(\dfrac{EI}{L}\right)_1
\end{bmatrix}
\begin{Bmatrix} u_2 \\ v_2 \\ \theta_2 \\ u_1 \\ v_1 \\ \theta_1 \end{Bmatrix}
$$

(24.20)

The stiffness equations for member ② may be found directly from Equation 24.19, with an appropriate substitution of subscripts 2 and 3 (instead of 1 and 2) to reflect the node numbers at the ends of member ②:

$$
\begin{Bmatrix} X_{2,3} \\ Y_{2,3} \\ M_{2,3} \\ X_{3,2} \\ Y_{3,2} \\ M_{3,2} \end{Bmatrix} =
\begin{bmatrix}
\frac{1}{2}\left(\frac{AE}{L}+\frac{12EI}{L^3}\right)_2 & & & & & \\[6pt]
\frac{1}{2}\left(\frac{AE}{L}-\frac{12EI}{L^3}\right)_2 & \frac{1}{2}\left(\frac{AE}{L}+\frac{12EI}{L^3}\right)_2 & & \text{(Symmetric)} & & \\[6pt]
-\frac{1}{\sqrt{2}}\left(\frac{6EI}{L^2}\right)_2 & \frac{1}{\sqrt{2}}\left(\frac{6EI}{L^2}\right)_2 & 4\left(\frac{EI}{L}\right)_2 & & & \\[6pt]
-\frac{1}{2}\left(\frac{AE}{L}+\frac{12EI}{L^3}\right)_2 & -\frac{1}{2}\left(\frac{AE}{L}-\frac{12EI}{L^3}\right)_2 & \frac{1}{\sqrt{2}}\left(\frac{6EI}{L^2}\right)_2 & \frac{1}{2}\left(\frac{AE}{L}+\frac{12EI}{L^3}\right)_2 & & \\[6pt]
-\frac{1}{2}\left(\frac{AE}{L}-\frac{12EI}{L^3}\right)_2 & -\frac{1}{2}\left(\frac{AE}{L}+\frac{12EI}{L^3}\right)_2 & -\frac{1}{\sqrt{2}}\left(\frac{6EI}{L^2}\right)_2 & \frac{1}{2}\left(\frac{AE}{L}-\frac{12EI}{L^3}\right)_2 & \frac{1}{2}\left(\frac{AE}{L}+\frac{12EI}{L^3}\right)_2 & \\[6pt]
-\frac{1}{\sqrt{2}}\left(\frac{6EI}{L^2}\right)_2 & \frac{1}{\sqrt{2}}\left(\frac{EI}{L}\right)_2 & 2\left(\frac{EI}{L}\right)_2 & \frac{1}{\sqrt{2}}\left(\frac{6EI}{L^2}\right)_2 & -\frac{1}{\sqrt{2}}\left(\frac{6EI}{L^2}\right)_2 & 4\left(\frac{EI}{L}\right)_2
\end{bmatrix}
\begin{Bmatrix} u_2 \\ v_2 \\ \theta_2 \\ u_3 \\ v_3 \\ \theta_3 \end{Bmatrix}
$$

(24.21)

After accounting for the imposed boundary conditions:

$$u_1 = v_1 = \theta_1 = u_3 = v_3 = \theta_3 = 0$$

the stiffness equation for the total structure is written in terms of the three free-displacement components u_2, v_2, and θ_2 by adding appropriate stiffness terms of the individual members, i.e.,

$$
\begin{Bmatrix} X_2 \\ Y_2 \\ M_2 \end{Bmatrix} = \begin{Bmatrix} 0 \\ -P \\ 0 \end{Bmatrix} = \begin{Bmatrix} X_{2,1}+X_{2,3} \\ Y_{2,1}+Y_{2,3} \\ M_{2,1}+M_{2,3} \end{Bmatrix}
$$

(24.22)

The result of this operation produces the stiffness equation to follow:

$$
\begin{Bmatrix} 0 \\ -P \\ 0 \end{Bmatrix} =
\begin{bmatrix}
\left(\frac{AE}{L}\right)_1+\frac{1}{2}\left(\frac{AE}{L}+\frac{12EI}{L^3}\right)_2 & \frac{1}{2}\left(\frac{AE}{L}-\frac{12EI}{L^3}\right)_2 & -\frac{1}{\sqrt{2}}\left(\frac{6EI}{L^3}\right)_2 \\[6pt]
\frac{1}{2}\left(\frac{AE}{L}-\frac{12EI}{L^3}\right)_2 & \left(\frac{12EI}{L^3}\right)_1+\frac{1}{2}\left(\frac{AE}{L}+\frac{12EI}{L^3}\right)_2 & 6\left(\frac{AE}{L^2}\right)_1+\frac{1}{\sqrt{2}}\left(\frac{6EI}{L^2}\right)_2 \\[6pt]
-\frac{1}{\sqrt{2}}\left(\frac{6EI}{L^2}\right)_2 & \left(6\frac{EI}{L^2}\right)_1+\frac{1}{\sqrt{2}}\left(\frac{6EI}{L^2}\right)_2 & 4\left(\frac{EI}{L}\right)_1+4\left(\frac{EI}{L}\right)_2
\end{bmatrix}
\begin{Bmatrix} u_2 \\ v_2 \\ \theta_2 \end{Bmatrix}
$$

(24.23)

Solutions for the displacements at node 2, for the following specific values of member properties, are shown below:

$$\left(\frac{AE}{L}\right)_1 = \left(\frac{AE}{L}\right)_2 = \left(\frac{12EI}{L^3}\right)_1 = \left(\frac{12EI}{L^3}\right)_2 = 1000\,\frac{k}{in.}$$

$$\phi_2 = 45°$$
$$L = 6\text{ ft} = 72\text{ in.}$$
$$P = 10\text{ k}$$

Thus

$$
\begin{Bmatrix} 0 \\ -10 \\ 0 \end{Bmatrix} = 10^3
\begin{bmatrix} 2 & 0 & -36 \\ 0 & 2 & 72 \\ -36 & 72 & 5184 \end{bmatrix}
\begin{Bmatrix} u_2 \\ v_2 \\ \theta_2 \end{Bmatrix}
$$

Solution of the equation set above produces:

$$
\begin{Bmatrix} u_2 \\ v_2 \\ \theta_2 \end{Bmatrix} = \begin{Bmatrix} 3.333\text{ in.} \\ -11.667\text{ in.} \\ 0.185\text{ rad} \end{Bmatrix} \times 10^{-3}
$$

Substitution of these nodal displacements into Equations 24.20 and 24.21 provides the reactions at nodes 1 and 3:

$$\left\{ \begin{array}{c} X_1 \\ Y_1 \\ M_1 \end{array} \right\} = \left\{ \begin{array}{c} X_{1,2} \\ Y_{1,2} \\ M_{1,2} \end{array} \right\} = 10^3 \left[\begin{array}{ccc} -1 & 0 & 0 \\ 0 & -1 & -36 \\ 0 & 36 & 864 \end{array} \right] \left\{ \begin{array}{c} u_2 \\ v_2 \\ \theta_2 \end{array} \right\} = \left\{ \begin{array}{c} -3.33 \text{ k} \\ 5.00 \text{ k} \\ -260 \text{ in.-k} \end{array} \right\}$$

and

$$\left\{ \begin{array}{c} X_3 \\ Y_3 \\ M_3 \end{array} \right\} = \left\{ \begin{array}{c} X_{3,2} \\ Y_{3,2} \\ M_{3,2} \end{array} \right\} = 10^3 \left[\begin{array}{ccc} -1 & 0 & 36 \\ 0 & -1 & -36 \\ -36 & 36 & 1728 \end{array} \right] \left\{ \begin{array}{c} u_2 \\ v_2 \\ \theta_2 \end{array} \right\} = \left\{ \begin{array}{c} 3.33 \text{ k} \\ 5.00 \text{ k} \\ -220 \text{ in.-k} \end{array} \right\}$$

The reactions shown above are also member forces, expressed in global coordinates. In order to find the shears and axial force components, which are necessary to perform stress calculations, the forces expressed in global components must be resolved into local components through relations similar to Equation 24.17:

$$X' = X \cos \phi + Y \sin \phi$$

$$Y' = -X \sin \phi + Y \cos \phi$$

$$M' = M$$

Substitution into these equations produces:

Member ①
$(\phi = 0°)$
$$\left\{ \begin{array}{l} X'_{1,2} = -3.33 \text{ k} \\ Y'_{1,2} = 5.00 \text{ k} \\ M'_{1,2} = -260 \text{ in.-k} \end{array} \right.$$

Member ②
$(\phi = 45°)$
$$\left\{ \begin{array}{l} X'_{3,2} = 5.89 \text{ k} \\ Y'_{3,2} = 1.18 \text{ k} \\ M'_{3,2} = -220 \text{ in.-k} \end{array} \right.$$

Sketches showing the nodal reactions and member forces for this structure are shown in Figures 24.12(a) and 24.12(b):

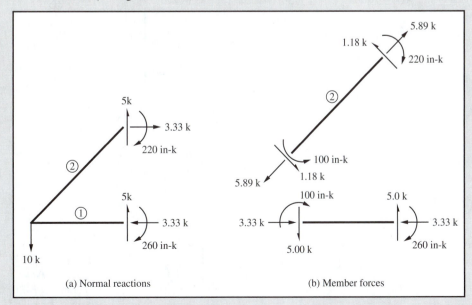

(a) Normal reactions   (b) Member forces

Figure 24.12

EXAMPLE 24.3

The frame of Figure 24.13, consisting of four equal-length members connected at one point, is subjected to a concentrated moment at its center node. Determine the displacements of this node.

Solution.

Node and member numbers are assigned to the structure as shown in Figure 24.13. The stiffness equation for the structure may be determined by considering the forces required to move, consecutively, the ends of each individual member the arbitrary distances u_5, v_5, and θ_5. These force coefficients may be determined from Equation 24.19 follows:

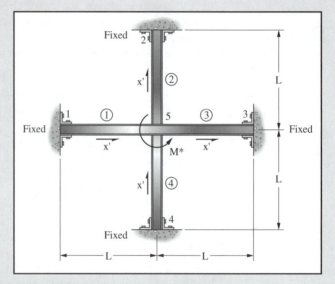

Figure 24.13

$$(\phi_1 = 0) \begin{Bmatrix} X_{5,1} \\ Y_{5,1} \\ M_{5,1} \end{Bmatrix} = \begin{bmatrix} \left(\dfrac{AE}{L}\right)_1 & 0 & 0 \\ 0 & \left(\dfrac{12EI}{L^3}\right)_1 & -\left(\dfrac{6EI}{L^2}\right)_1 \\ 0 & -\left(\dfrac{6EI}{L^2}\right)_1 & \left(\dfrac{4EI}{L}\right)_1 \end{bmatrix} \begin{Bmatrix} u_5 \\ v_5 \\ \theta_5 \end{Bmatrix}$$

$$(\phi_2 = 90°) \begin{Bmatrix} X_{5,2} \\ Y_{5,2} \\ M_{5,2} \end{Bmatrix} = \begin{bmatrix} \left(\dfrac{12EI}{L^3}\right)_2 & 0 & -\left(\dfrac{6EI}{L^2}\right)_2 \\ 0 & \dfrac{AE}{L} & 0 \\ \dfrac{-6EI}{L^2} & 0 & \dfrac{4EI}{L} \end{bmatrix} \begin{Bmatrix} u_5 \\ v_5 \\ \theta_5 \end{Bmatrix}$$

$$(\phi_3 = 0) \begin{Bmatrix} X_{5,3} \\ Y_{5,3} \\ M_{5,3} \end{Bmatrix} = \begin{bmatrix} \left(\dfrac{AE}{L}\right)_3 & 0 & 0 \\ 0 & \left(\dfrac{12EI}{L^3}\right)_3 & \left(\dfrac{6EI}{L^2}\right)_3 \\ 0 & \left(\dfrac{6EI}{L^2}\right)_3 & \left(\dfrac{4EI}{L}\right)_3 \end{bmatrix} \begin{Bmatrix} u_5 \\ v_5 \\ \theta_5 \end{Bmatrix}$$

$$(\phi_4 = 90°) \begin{Bmatrix} X_{5,4} \\ Y_{5,4} \\ M_{5,4} \end{Bmatrix} = \begin{bmatrix} \left(\dfrac{12EI}{L^3}\right)_4 & 0 & \left(\dfrac{6EI}{L^2}\right)_4 \\ 0 & \left(\dfrac{AE}{L}\right)_4 & 0 \\ \left(\dfrac{6EI}{L^2}\right)_4 & 0 & \left(\dfrac{4EI}{L}\right)_4 \end{bmatrix} \begin{Bmatrix} u_5 \\ v_5 \\ \theta_5 \end{Bmatrix}$$

The total resistance to movement of joint 5 is simply the sum of the individual force coefficients for the four members. Thus

$$\begin{Bmatrix} X_5 \\ Y_5 \\ M_5 \end{Bmatrix} = \begin{bmatrix} \left(\dfrac{AE}{L}\right)_1 + \left(\dfrac{12EI}{L^3}\right)_2 + \left(\dfrac{AE}{L}\right)_3 + \left(\dfrac{12EI}{L^3}\right)_4 & 0 & -\left(\dfrac{6EI}{L^2}\right)_2 + \left(\dfrac{6EI}{L^2}\right)_4 \\ 0 & \left(\dfrac{12EI}{L^3}\right)_1 + \left(\dfrac{AE}{L}\right)_2 + \left(\dfrac{12EI}{L^3}\right)_3 + \left(\dfrac{AE}{L}\right)_4 & -\left(\dfrac{6EI}{L^2}\right)_1 + \left(\dfrac{6EI}{L^2}\right)_3 \\ -\left(\dfrac{6EI}{L^2}\right)_2 + \left(\dfrac{6EI}{L^2}\right)_4 & -\left(\dfrac{6EI}{L^2}\right)_1 + \left(\dfrac{6EI}{L^2}\right)_3 & -\left(\dfrac{4EI}{L}\right)_1 + \left(\dfrac{4EI}{L}\right)_2 + \left(\dfrac{4EI}{L}\right)_3 + \left(\dfrac{4EI}{L}\right)_4 \end{bmatrix} \begin{Bmatrix} u_5 \\ v_5 \\ \theta_5 \end{Bmatrix}$$

(24.24)

To obtain a numerical solution, let

$$\left(\frac{AE}{L}\right)_1 = \left(\frac{AE}{L}\right)_2 = \psi \qquad \left(\frac{EI}{L^3}\right)_1 = \left(\frac{EI}{L^3}\right)_2 = 0.05\psi$$

$$\left(\frac{AE}{L}\right)_3 = \left(\frac{AE}{L}\right)_4 = 2\psi \qquad \left(\frac{EI}{L^3}\right)_3 = \left(\frac{EI}{L^3}\right)_4 = 0.10\psi$$

The stiffness equation then becomes:

$$\begin{Bmatrix} X_5 = 0 \\ Y_5 = 0 \\ M_5 = M^* \end{Bmatrix} = \psi \begin{bmatrix} 4.8 & 0 & 0.3L \\ 0 & 4.8 & 0.3L \\ 0.3L & 0.3L & 24L^2 \end{bmatrix} \begin{Bmatrix} u_5 \\ v_5 \\ \theta_5 \end{Bmatrix} \qquad \textbf{(24.25)}$$

The solution for the displacement components at node 5 is obtained by solving simultaneously the three equations shown in matrix Equation 24.25:

$$\begin{Bmatrix} u_5 \\ v_5 \\ \theta_5 \end{Bmatrix} = \frac{1}{\psi} \begin{bmatrix} 4.8 & 0 & 0.3L \\ 0 & 4.8 & 0.3L \\ 0.3L & 0.3L & 24L^2 \end{bmatrix}^{-1} \begin{Bmatrix} 0 \\ 0 \\ M^* \end{Bmatrix}$$

$$= \begin{Bmatrix} -0.0026L \\ -0.0026L \\ +0.04173 \end{Bmatrix} \frac{M^*L}{(EI)_1}$$

(24.26)

A comparable solution performed using the Moment Distribution Method produces the slightly different results:

$$\begin{Bmatrix} u_5 \\ v_5 \\ \theta_5 \end{Bmatrix} = \begin{Bmatrix} 0 \\ 0 \\ 0.04167 \end{Bmatrix} \frac{M^*L}{(EI)_1}$$

The differences between the solution obtained using the stiffness equations and the Moment Distribution Method occur because the stiffness equations take into account *axial deformation*, whereas the Moment Distribution Method is based on the assumption of zero axial deformations. Neglecting the axial deformations for simple structures such as the one considered here does not introduce appreciable errors. However, for more involved structures, e.g., a high-rise frame building, the zero-axial-deformation assumption may introduce significant errors. ■

Chetco River Bridge, Brookings, Oregon. (Courtesy of the Oregon Department of Transportation.)

24.4 LOADING BETWEEN NODES

The problems that have been discussed in the previous sections have involved concentrated forces and moments applied to nodes only. Since almost all real structures are subjected to distributed or concentrated loading between nodes, the solution technique using the stiffness method must be expanded to permit the handling of such situations. Fortunately, the changes required are relatively minor.

The beam of Figure 24.14, which has a uniformly distributed load over segment 2, is considered for this discussion.

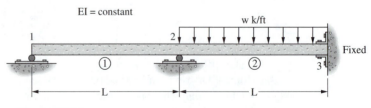

Figure 24.14

If all the degrees of freedom of the structure are restrained (i.e., $v_1 = v_2 = v_3 = \theta_1 = \theta_2 = \theta_3 = 0.0$), the resultant nodal reactions are termed "fixed-end forces." As used here, the term "forces" includes both transverse forces and moments. For the problem shown in Figure 24.14, the fixed end forces are

$$\{P\}^F = \begin{Bmatrix} Y_1^F \\ M_1^F \\ Y_2^F \\ M_2^F \\ Y_3^F \\ M_3^F \end{Bmatrix} = \begin{Bmatrix} 0 \\ 0 \\ \dfrac{wL}{2} \\ \dfrac{wL^2}{12} \\ \dfrac{wL}{2} \\ -\dfrac{wL^2}{12} \end{Bmatrix} \tag{24.27}$$

The additional nodal forces due to nodal displacement are given by Equation 23.26 (for a beam in which axial displacements are neglected):

$$\begin{Bmatrix} Y_1 \\ M_1 \\ Y_2 \\ M_2 \\ Y_3 \\ M_2 \end{Bmatrix} = \begin{bmatrix} \dfrac{12EI}{L^3} & & & & & \\ \dfrac{6EI}{L^2} & \dfrac{4EI}{L} & & \text{(Symmetric)} & & \\ -\dfrac{12EI}{L^3} & -\dfrac{6EI}{L^2} & \dfrac{24EI}{L^3} & & & \\ \dfrac{6EI}{L^2} & \dfrac{2EI}{L} & 0 & \dfrac{8EI}{L} & & \\ 0 & 0 & -\dfrac{12EI}{L^3} & -\dfrac{6EI}{L^2} & \dfrac{12EI}{L^3} & \\ 0 & 0 & \dfrac{6EI}{L^2} & \dfrac{2EI}{L} & -\dfrac{6EI}{L^2} & \dfrac{4EI}{L} \end{bmatrix} \begin{Bmatrix} v_1 \\ \theta_1 \\ v_2 \\ \theta_2 \\ v_3 \\ \theta_3 \end{Bmatrix} \tag{24.28}$$

or,

$$\{P\}^N = [K]\{\delta\}$$

Therefore the resulting nodal loading when distributed loads act on the structure is expressed, in general, as the sum of the right-hand sides of Equations 24.27 and 24.28:

$$\{P\}^{\text{total}} = \{P\}^F + [K]\{\delta\} \tag{24.29}$$

After deleting rows and columns associated with enforced-zero-boundary displacements (v_1, v_2, v_3, θ_3), the remaining stiffness equations are

$$\begin{Bmatrix} M_1 = 0 \\ M_2 = 0 \end{Bmatrix}^{\text{total}} = \begin{Bmatrix} 0 \\ \dfrac{wL^2}{12} \end{Bmatrix} + \begin{bmatrix} \dfrac{4EI}{L} & \dfrac{2EI}{L} \\ \dfrac{2EI}{L} & \dfrac{8EI}{L} \end{bmatrix} \begin{Bmatrix} \theta_1 \\ \theta_2 \end{Bmatrix} \tag{24.30}$$

The solution for the free displacements is found by solving the equation set simultaneously:

$$\left\{\begin{array}{c} \theta_1 \\ \theta_2 \end{array}\right\} = \begin{bmatrix} \dfrac{4EI}{L} & \dfrac{2EI}{L} \\ \dfrac{2EI}{L} & \dfrac{8EI}{L} \end{bmatrix}^{-1} \left(\left\{\begin{array}{c} 0 \\ 0 \end{array}\right\} - \left\{\begin{array}{c} 0 \\ \dfrac{wL^2}{12} \end{array}\right\} \right) = \left\{\begin{array}{c} 1 \\ -2 \end{array}\right\} \dfrac{wL^3}{12EI} \qquad \textbf{(24.31)}$$

Nodal forces are found by substituting the values of the free displacements into Equation 24.29:

$$\left\{\begin{array}{c} Y_1 \\ Y_2 \\ Y_3 \\ M_3 \end{array}\right\} = \left\{\begin{array}{c} 0 \\ \dfrac{wL^2}{2} \\ \dfrac{wL^2}{2} \\ \dfrac{-wL^2}{12} \end{array}\right\} + \begin{bmatrix} \dfrac{6EI}{L^2} & \dfrac{6EI}{L^2} \\ \dfrac{-6EI}{L^2} & 0 \\ 0 & \dfrac{-6EI}{L^2} \\ 0 & \dfrac{2EI}{L} \end{bmatrix} \left\{\begin{array}{c} \theta_1 \\ \theta_2 \end{array}\right\} = \left\{\begin{array}{c} -3 \\ 39 \\ 48 \\ -9L \end{array}\right\} \dfrac{wL}{84}$$

Internal member forces dues to nodal displacements can be found by substituting the derived values of θ_1 and θ_2 into the stiffness matrices for the individual beam elements. These internal moments are computed as follows:

$$\left\{\begin{array}{c} M_{1,2} \\ M_{2,1} \end{array}\right\} = \begin{bmatrix} \dfrac{4EI}{L} & \dfrac{2EI}{L} \\ \dfrac{2EI}{L} & \dfrac{4EI}{L} \end{bmatrix} \left\{\begin{array}{c} \theta_1 \\ \theta_2 \end{array}\right\} = \left\{\begin{array}{c} 0 \\ \dfrac{-6}{168}wL^2 \end{array}\right\}$$

and

$$\left\{\begin{array}{c} M_{2,3} \\ M_{3,2} \end{array}\right\} = \begin{bmatrix} \dfrac{4EI}{L} & \dfrac{2EI}{L} \\ \dfrac{2EI}{L} & \dfrac{4EI}{L} \end{bmatrix} \left\{\begin{array}{c} \theta_2 \\ \theta_3 = 0 \end{array}\right\} = \left\{\begin{array}{c} \dfrac{-8}{168}wL^2 \\ \dfrac{-4}{168}wL^2 \end{array}\right\}$$

The total internal member forces are those that result from "fixed-end" conditions *plus* those that result from nodal displacements. For this example problem, no internal fixed-end moments develop in member ① since there are no member loads. For member ②, the internal fixed-end moments due to the loading are

$$\left\{\begin{array}{c} M_{2,3} \\ M_{3,2} \end{array}\right\} = \left\{\begin{array}{c} \dfrac{wL^2}{12} \\ \dfrac{-wL^2}{12} \end{array}\right\}$$

Thus the final internal moments are

$$\left\{\begin{array}{c} M_{1,2} \\ M_{2,1} \end{array}\right\}^{\text{final}} = \left\{\begin{array}{c} 0 \\ 0 \end{array}\right\}^{F} + \left\{\begin{array}{c} 0 \\ \dfrac{-6}{168}wL^2 \end{array}\right\} = \left\{\begin{array}{c} 0 \\ \dfrac{-6}{168}wL^2 \end{array}\right\}$$

$$\left\{\begin{array}{c} M_{2,3} \\ M_{3,2} \end{array}\right\}^{\text{final}} = \left\{\begin{array}{c} \dfrac{wL^2}{12} \\ \dfrac{-wL^2}{12} \end{array}\right\}^{F} + \left\{\begin{array}{c} \dfrac{-8}{168}wL^2 \\ \dfrac{-4}{168}wL^2 \end{array}\right\} = \left\{\begin{array}{c} \dfrac{6}{168}wL^2 \\ \dfrac{-18}{168}wL^2 \end{array}\right\}$$

Example 24.4 illustrates the analysis of a frame that consists of a horizontal and an inclined member, both of which are subjected to loads located between nodes. The purpose of this example is to show the reader how to handle distributed loads applied to inclined members and concentrated loads acting between nodes, and how to deal with fixed-end forces on adjacent members.

EXAMPLE 24.4

The frame of Figure 24.15 supports a uniformly distributed load over member ① and a concentrated load at the centerline of member ②. Prepare the stiffness equations required to solve the problem.

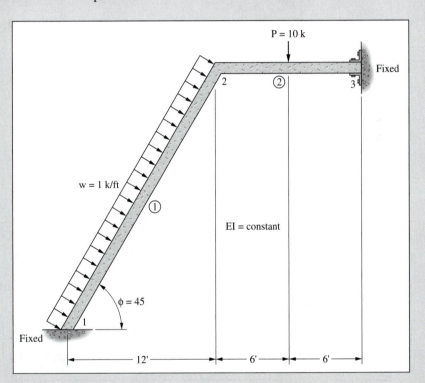

Figure 24.15

Solution.
The "fixed-end forces" acting on member ① are shown in Figure 24.16(a), with the shear forces shown acting parallel to the local y axis of member ①. These forces are resolved into forces acting in the global direction, and the "fixed-end forces" are redrawn in Figure 24.16(b). The appropriate "fixed-end forces" acting on member ② are shown in Figure 24.16(c).

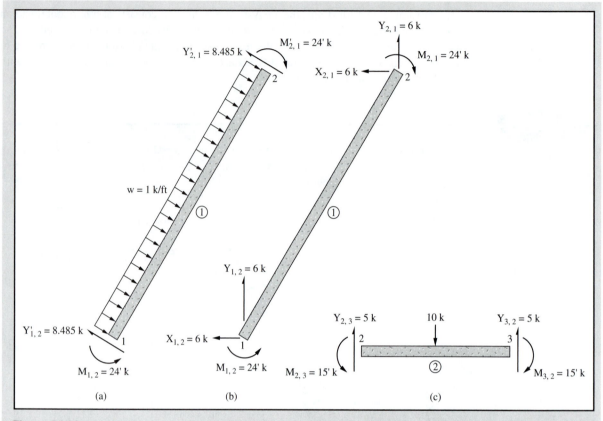

Figure 24.16

The addition of all the fixed-end forces produces the total fixed-end loading for the frame shown in Figure 24.17.

The stiffness equations associated with the free nodes of this structure may then be written as

$$\begin{Bmatrix} X_2 \\ Y_2 \\ M_2 \end{Bmatrix}^{total} = \begin{Bmatrix} 0 \\ 0 \\ 0 \end{Bmatrix} = \begin{bmatrix} k_{1,1} & k_{1,2} & k_{1,3} \\ k_{2,1} & k_{2,2} & k_{2,3} \\ k_{3,1} & k_{3,2} & k_{3,3} \end{bmatrix} \begin{Bmatrix} u_2 \\ v_2 \\ \theta_2 \end{Bmatrix} + \begin{Bmatrix} -6 \\ 11 \\ -9 \end{Bmatrix}^F$$

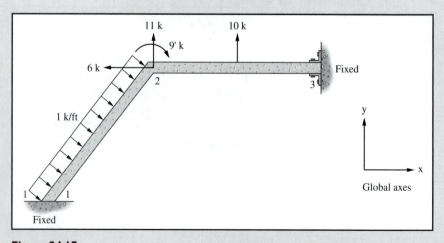

Figure 24.17

The free-node displacements are determined by using a suitable simultaneous equation solving technique.

$$\left\{ \begin{array}{c} u_2 \\ v_2 \\ \theta_2 \end{array} \right\} = \begin{bmatrix} k_{1,1} & k_{1,2} & k_{1,3} \\ k_{2,1} & k_{2,2} & k_{2,3} \\ k_{3,1} & k_{3,2} & k_{3,3} \end{bmatrix}^{-1} \left\{ \begin{array}{c} 6 \\ -11 \\ 9 \end{array} \right\}$$

The frame model shown in Figure 24.15 consists of two members and 3 nodes. As has been shown in the example, the concentrated force acting between nodes of member ② is treated in a manner similar to that used for distributed loading. Alternatively, the analyst can elect to add another node to the model at the location of the concentrated force. This choice eliminates the need for special treatment of the concentrated load, but does add three more displacement components to the set of unknowns in the problems.

Although the example problems dealt with very special cases of loading between nodes, the method illustrated in these examples can be used for any form of distributed or concentrated loading acting between nodes. To apply the stiffness method to these problems the analyst is faced only with the additional task of determining fixed-end forces. A condensed summary of fixed-end forces for commonly occurring loading is provided in Figure 20.4 and inside the cover. ■

24.5 PROBLEMS FOR SOLUTION

24.1 Determine the global stiffness matrix [K] for the vertical member shown. Assume that local axes are defined as shown on the sketch. (*Hint*: Use Equation 24.19, with $\theta = 90°$.)

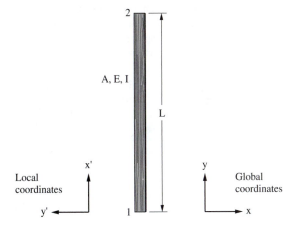

24.2 Repeat Problem 24.1, but use a newly defined set of local axes. Note the differences in the stiffness matrices determined in Problems 24.2 and 24.1.

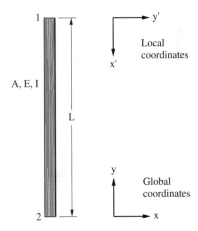

24.3 A cantilevered beam is oriented at 45° with the horizontal. A horizontal load P acts at the end of the strut. Perform the following:
(a) Determine the stiffness matrix for the strut;
(b) Determine the displacements at node 2 (expressed with respect to the global coordinate system);
(c) Resolve the global displacements at node 2 into a set of local displacements (u_2', v_2', θ_2')

(d) Use the answers to part (c) to obtain the member forces for the strut (expressed with respect to local axes).

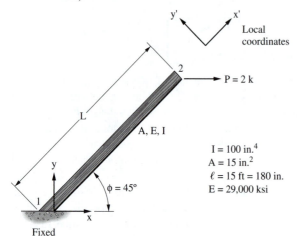

Local coordinates

$I = 100$ in.^{4}
$A = 15$ in.^{2}
$\ell = 15$ ft $= 180$ in.
$E = 29,000$ ksi

Fixed

Ans. (b) $u_2 = 0.6708$ (c) $X_2' = 1.41$ in.
   $v_2 = -0.6700$ $Y_2' = -1.41$ in.
   $\theta_2 = -0.0079$ $M_2' = 0$

24.4 A cruciform structure is made of four members joined together at node 5, as shown. A nodal force, P^*, acts at node 5. Perform the following:
(a) Determine the global stiffness matrix for the cruciform;
(b) Apply boundary conditions and determine the nodal displacements at node 5 due to P^*;
(c) Let the cross-sectional area A of all members go to zero (i.e., $A = 0$). Recompute the displacements at node 5.

For all members:
   $A = 4$ in.^{2}
   $I = 100$ in.^{4}
   $E = 29,000$ ksi
   $\ell = 12$ ft $= 144$ in.
   $P^* = 10$ k

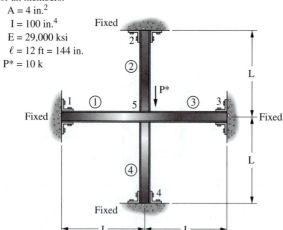

24.5 A two-member frame is rigidly supported at node 1 and pinned at node 3. Additionally, a roller support is provided at node 2.

(a) Determine the displacements at nodes 2 and 3.
(b) Let the cross-sectional area of the members go to zero (i.e., $A = 0$). Recompute the nodal displacements. Can you explain the nature of the solution for part (b)?

For both members:
   $A = 20$ in.^{2}
   $I = 100$ in.^{4}
   $E = 29,000$ ksi
   $L = 12$ ft $= 144$ in.

Ans. (a) $\begin{Bmatrix} v_2 \\ \theta_2 \\ \theta_3 \end{Bmatrix} = \begin{Bmatrix} -0.00745 \text{ in.} \\ 0.00002 \text{ rads.} \\ 0.00007 \text{ rads.} \end{Bmatrix}$

(b) $\begin{Bmatrix} v_2 \\ \theta_2 \\ \theta_3 \end{Bmatrix} = \begin{Bmatrix} -18.01 \text{ in.} \\ 0.536 \text{ rads.} \\ 0.161 \text{ rads.} \end{Bmatrix}$

24.6 A frame consisting of two members joined together at node 2 is rigidly supported at nodes 1 and 3. The frame is also guided by a roller support at node 2. A concentrated force P acts horizontally at node 2. Perform the following:
(a) Determine the global stiffness matrix for the frame;
(b) Apply boundary conditions and determine the resultant nodal displacements at node 2;
(c) Determine the reactions at nodes 1, 2, and 3.

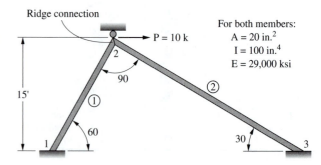

For both members:
   $A = 20$ in.^{2}
   $I = 100$ in.^{4}
   $E = 29,000$ ksi

24.7 Use the method described in Section 24.4 to determine the following for the beam shown, loaded with a uniformly distributed loading w:

(a) Rotation at node 2;

(b) Reactions at nodes 1 and 2.

$$EI = 1.8 \times 10^6 \, k \cdot in.^2$$

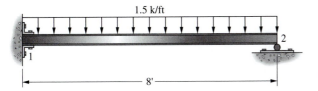

1.5 k/ft

8'

Ans. (a) $\theta_2 = 0.00128$ rads   (b) Y_1 Total $= 7.5$ k

$Y_1 = 1.5$ in. M_1 Total $= 144$ in.-k

$M_1 = 48.0$ in.-k Y_2 Total $= 4.5$ k

Y_2 (net) $= -1.5$ in. M_2 Total $= 0$

24.8 A continuous beam has a uniformly distributed load over its complete length. Determine the following:

(a) Rotations at nodes 2 and 3;

(b) Reactions at nodes 1, 2, and 3.

$$EI = 1.8 \times 10^6 \, k \cdot in.^2$$

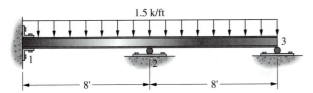

1.5 k/ft

8'   8'

24.9 A single-span beam has a uniformly distributed load over its complete length. The right end of the beam is free to translate vertically, but it cannot rotate. Determine the following:

(a) Transverse displacement at node 2;

(b) Reactions at nodes 1 and 2.

$$EI = 1.8 \times 10^6 \, k \cdot in.^2$$

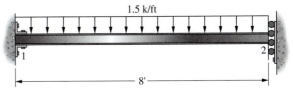

1.5 k/ft

8'

Ans. (a) $Y_1 = 6.00$ in.   (b) Y_1 Total $= 12.0$ in.

$M_1 = 288.0$ in.-k M_1 Total $= 384.0$ in.-k

$M_2 = 288.0$ in.-k M_2 Total $= 192.0$ in.-k

24.10. A beam is fixed at both ends, but has an internal hinge, as shown. A vertical force P acts at the hinge. Determine the following:[*]

(a) Transverse displacement at node 2;

(b) Rotations on either side of node 2, i.e., $\theta_{2,1}$ and $\theta_{2,3}$;

(c) Reactions at nodes 1 and 3.

$$EI = 1.8 \times 10^6 \, k \cdot in.^2$$

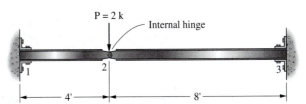

P = 2 k

Internal hinge

4'   8'

---

[*]In forming the stiffness matrix for the structure, treat rotations $\theta_{2,1}$ and $\theta_{2,3}$ as independent variables (i.e., $\theta_{2,1} \neq \theta_{2,3}$). After boundary conditions are imposed, the stiffness equations will contain three independent displacement variables: v_2, $\theta_{2,1}$, and $\theta_{2,3}$.

Chapter 25

Additional Matrix Procedures

25.1 GENERAL

The introduction to the stiffness method presented in Chapters 23 and 24 was written with the intention of providing an insight into the physical principles involved in the method. Although the implementation of the stiffness procedure is quite straightforward, the details involved in its application to all but the simplest structures are discouragingly tedious for anyone attempting a solution by hand. However, the usefulness of the method stems from its adaptability for use with digital computers. In this chapter some aspects of the stiffness method are recast in a form that will make the computer-programming steps more apparent. In addition, the method is applied to several structures that have not been previously discussed. Emphasis is placed on the manipulation of the algebraic equations through matrix methods, a form ideally suited for existing computer languages.

25.2 ADDITION OF STIFFNESS EQUATIONS

The derivation of the stiffness equations involves a new formulation for every different structure. However, as the reader may already have realized, the stiffness matrix for an entire structure can be found by adding, in a proper way, the stiffness matrices for the individual elements of the structure. To illustrate this fact, the axial force member shown in Figure 25.1 is considered.

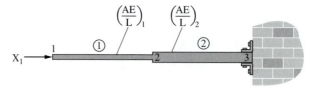

Figure 25.1

Clemson University Football Stadium, Clemson, South Carolina (Courtesy of Britt, Peters and Associates)

The member is divided into two elements and the stiffness equation (Equation 23.8) is written for each element. The subscripts i and j are used to denote node numbers and m is used for member numbers.

$$\begin{Bmatrix} F_{i,j} \\ F_{j,i} \end{Bmatrix} = \begin{bmatrix} \left(\dfrac{AE}{L}\right)_m & -\left(\dfrac{AE}{L}\right)_m \\ -\left(\dfrac{AE}{L}\right)_m & \left(\dfrac{AE}{L}\right)_m \end{bmatrix} \begin{Bmatrix} u_i \\ u_j \end{Bmatrix} \tag{25.1}$$

Since the total structure involves three nodes, there are three possible nodal displacements (also called degrees of freedom in this book), and a three-by-three total stiffness matrix will result. The stiffness matrix for each individual element is expanded, or *augmented*, to include all of the degrees of freedom of the structure. The rewritten member stiffness equations now appear as

$$\begin{Bmatrix} F_{1,2} \\ F_{2,1} \\ 0 \end{Bmatrix} = \begin{bmatrix} \left(\dfrac{AE}{L}\right)_1 & -\left(\dfrac{AE}{L}\right)_1 & 0 \\ -\left(\dfrac{AE}{L}\right)_1 & \left(\dfrac{AE}{L}\right)_1 & 0 \\ 0 & 0 & 0 \end{bmatrix} \begin{Bmatrix} u_1 \\ u_2 \\ u_3 \end{Bmatrix} \tag{25.2}$$

and

$$\begin{Bmatrix} 0 \\ F_{2,3} \\ F_{3,2} \end{Bmatrix} = \begin{bmatrix} 0 & 0 & 0 \\ 0 & \left(\dfrac{AE}{L}\right)_2 & -\left(\dfrac{AE}{L}\right)_2 \\ 0 & -\left(\dfrac{AE}{L}\right)_2 & \left(\dfrac{AE}{L}\right)_2 \end{bmatrix} \begin{Bmatrix} u_1 \\ u_2 \\ u_3 \end{Bmatrix} \tag{25.3}$$

The two matrix equations (25.2 and 25.3) are added directly to produce

$$
\begin{Bmatrix} F_{1,2} \\ F_{2,1} \\ 0 \end{Bmatrix} + \begin{Bmatrix} 0 \\ F_{2,3} \\ F_{3,2} \end{Bmatrix} = \left(\begin{bmatrix} \left(\dfrac{AE}{L}\right)_1 & -\left(\dfrac{AE}{L}\right)_1 & 0 \\ -\left(\dfrac{AE}{L}\right)_1 & \left(\dfrac{AE}{L}\right)_1 & 0 \\ 0 & 0 & 0 \end{bmatrix} + \begin{bmatrix} 0 & 0 & 0 \\ 0 & \left(\dfrac{AE}{L}\right)_2 & -\left(\dfrac{AE}{L}\right)_2 \\ 0 & -\left(\dfrac{AE}{L}\right)_2 & \left(\dfrac{AE}{L}\right)_2 \end{bmatrix} \right) \begin{Bmatrix} u_1 \\ u_2 \\ u_3 \end{Bmatrix}
$$

or

$$
\begin{Bmatrix} F_{1,2} = X_1 \\ F_{2,1} + F_{2,3} = X_2 \\ F_{3,2} = X_3 \end{Bmatrix} = \begin{bmatrix} \left(\dfrac{AE}{L}\right)_1 & -\left(\dfrac{AE}{L}\right)_1 & 0 \\ -\left(\dfrac{AE}{L}\right)_1 & \left(\dfrac{AE}{L}\right)_1 + \left(\dfrac{AE}{L}\right)_2 & -\left(\dfrac{AE}{L}\right)_2 \\ 0 & -\left(\dfrac{AE}{L}\right)_2 & \left(\dfrac{AE}{L}\right)_2 \end{bmatrix} \begin{Bmatrix} u_1 \\ u_2 \\ u_3 \end{Bmatrix}
$$

(25.4)

The internal member forces $F_{2,1}$ and $F_{2,3}$ are related to the external force acting at node 2 through equilibrium, as illustrated in Figure 25.2.

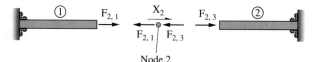

Figure 25.2

From this figure the following relation is written for node 2:

$$
\sum F_x = -F_{2,1} - F_{2,3} + X_2 = 0
$$
$$
\therefore X_2 = F_{2,1} + F_{2,3}
$$

(25.5)

Therefore adding the individual member stiffness equations produces the stiffness equation for the complete structure. In general, a computer program need contain only the instructions for forming a general member stiffness matrix and the instructions for adding properly the stiffness matrices for the individual members in order to form the stiffness matrix for the entire structure.

25.3 STIFFNESS MATRICES FOR INCLINED MEMBERS

In Chapter 24 the stiffness matrix for a general planar flexural member was derived in terms of the member's properties and its angle of inclination with a global reference system. A systematic procedure for manipulating equations written with respect to rotated reference systems is useful when writing computer programs to perform these tasks. An introduction to such methods is presented here.

The inclined flexural member of Figure 25.3 is now considered.

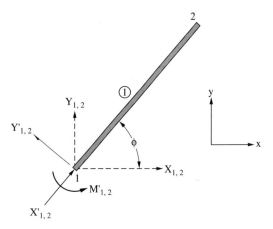

Figure 25.3

Nodal forces expressed in the local-coordinate system are shown with primes, while their equivalent components expressed with respect to the global-coordinate system are shown unprimed. The relation between the force components in the local-coordinate system and the equivalent components in the global-coordinate system has previously been shown as Equation 24.17 and is repeated here in slightly altered notation:

$$X_{1,2} = X'_{1,2} \cos \phi - Y'_{1,2} \sin \phi$$
$$Y_{1,2} = X'_{1,2} \sin \phi + Y'_{1,2} \cos \phi$$
$$M_{1,2} = M'_{1,2}$$

Written in matrix form, the relation is

$$\begin{Bmatrix} X_{1,2} \\ Y_{1,2} \\ M_{1,2} \end{Bmatrix} = \begin{bmatrix} \cos \phi & -\sin \phi & 0 \\ \sin \phi & \cos \phi & 0 \\ 0 & 0 & 1 \end{bmatrix} \begin{Bmatrix} X'_{1,2} \\ Y'_{1,2} \\ M'_{1,2} \end{Bmatrix} \qquad (25.6)$$

Symbolically, Equation 25.6 may be written as

$$\{P_i\} = [\gamma]\{P'_i\} \qquad (25.7)$$

where $[\gamma]$ is called a transformation matrix.

Equation 25.6 may also be used to express the local nodal forces in terms of the global force components:

$$\begin{Bmatrix} X'_{1,2} \\ Y'_{1,2} \\ M'_{1,2} \end{Bmatrix} = \begin{bmatrix} \cos \phi & -\sin \phi & 0 \\ \sin \phi & \cos \phi & 0 \\ 0 & 0 & 1 \end{bmatrix}^{-1} \begin{Bmatrix} X_{1,2} \\ Y_{1,2} \\ M_{1,2} \end{Bmatrix}$$
$$= \begin{bmatrix} \cos \phi & \sin \phi & 0 \\ -\sin \phi & \cos \phi & 0 \\ 0 & 0 & 1 \end{bmatrix} \begin{Bmatrix} X_{1,2} \\ Y_{1,2} \\ M_{1,2} \end{Bmatrix} \qquad (25.8)$$

The transformation matrix $[\gamma]$ has the property (not true for matrices in general) that $[\gamma]^{-1} = [\gamma]^T$. This property may easily be verified by multiplying $[\gamma]$ by its transpose to obtain an identity matrix. Therefore

$$\{P'\} = [\gamma]^{-1}\{P\} = [\gamma]^T\{P\} \qquad (25.9)$$

A similar transformation of nodal forces may be performed at end 2 of the member. The two transformations may be combined in a single matrix equation:

$$\left\{ \begin{matrix} \{P_1\} \\ \{P_2\} \end{matrix} \right\} = \left[\begin{matrix} [\gamma] & [0] \\ [0] & [\gamma] \end{matrix} \right] \left\{ \begin{matrix} \{P_1'\} \\ \{P_2'\} \end{matrix} \right\}$$

or

$$\{P\} = [\Gamma]\{P'\}$$

where

$$[\Gamma] = \left[\begin{matrix} [\gamma] & [0] \\ [0] & [\gamma] \end{matrix} \right] \tag{25.10}$$

and

$$\{P\} = \left\{ \begin{matrix} \{P_1\} \\ \{P_2\} \end{matrix} \right\}$$

A transformation similar to the one described in Equation 25.10 may be performed for any other vector acting at the node. In particular, nodal displacements may be transformed from local to global coordinates

$$\{\delta_i\} = [\gamma]\{\delta_i'\} \tag{25.11}$$

and

$$\left\{ \begin{matrix} \{\delta_1\} \\ \{\delta_2\} \end{matrix} \right\} = \{\delta\}_{\text{combined}} = [\Gamma]\{\delta'\}_{\text{combined}} \tag{25.12}$$

A stiffness equation such as the one given by Equation 25.30, referred to local axes, may now be transformed to global-coordinate axes with the help of Equations 25.10 and 25.12:

$$\{P'\} = [K']\{\delta'\}$$

where $[K']$ is called a local stiffness matrix.

Substitution of Equations 25.10 and 25.12 produces:

$$[\Gamma]^{-1}\{P\} = [K'][\Gamma]^{-1}\{\delta\} \tag{25.13}$$

If both sides of Equation 25.13 are premultiplied by $[\Gamma]$, the result is

$$\{P\} = ([\Gamma][K'][\Gamma]^{-1})\{\delta\} \tag{25.14}$$

or

$$\{P\} = [K]\{\delta\} \tag{25.15}$$

where

$$[K] = [\Gamma][K'][\Gamma]^{-1}, \text{ global stiffness matrix} \tag{25.16}$$

The matrix triple product shown in Equation 25.16 will produce exactly the same result for $[K]$ as has been shown previously in Equation 24.9.

Transformation equations such as the ones given by Equations 25.10 and 25.12 are very useful for handling inclined members since they permit the analyst to program equations for use in a computer in both local- and global-coordinate systems with relative ease. The typical solution of a structure that has inclined members requires several stages of transformations from one coordinate system to another.

25.4 STIFFNESS EQUATIONS FOR STRUCTURES WITH ENFORCED DISPLACEMENTS

Although, in general, structures deflect in response to applied loads, there are occasions when a structure is forced to displace a specified amount. A continuous beam whose supports are not aligned properly is one example. The misalignment may be corrected by jacking the structure into position. The analyst may want to know what forces will be required to force the beam through the necessary displacement in order to make it fit.

Node 1 of the continuous beam of Figure 25.4(a) is assumed to be misaligned by an amount v_1^*.

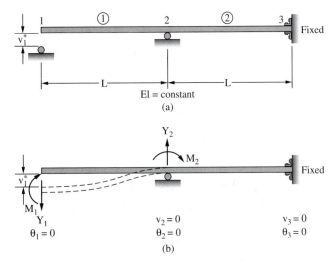

Figure 25.4

The forces needed to displace the beam through this distance while preventing any other nodal displacements can be determined from Figure 23.10. The deformed shape of the beam is sketched in Figure 25.4(b). A stiffness equation that includes these "fixed-end forces," as well as the forces developed by nodal displacements, can be written:

$$\begin{Bmatrix} Y_1 \\ M_1 \\ Y_2 \\ M_2 \\ Y_3 \\ M_3 \end{Bmatrix} = \begin{Bmatrix} \dfrac{-12EI}{L^3} \\[2mm] \dfrac{-6EI}{L^2} \\[2mm] \dfrac{12EI}{L^3} \\[2mm] \dfrac{-6EI}{L^2} \\[2mm] 0 \\[2mm] 0 \end{Bmatrix} v_1^* + [K] \begin{Bmatrix} \Delta v_1 = 0 \\ \theta_1 \\ v_2 = 0 \\ \theta_2 \\ v_3 = 0 \\ \theta_3 = 0 \end{Bmatrix} \qquad (25.17)$$

For the problem posed in Figure 25.4(a), the stiffness equation, after considering boundary conditions, is

$$\left\{ \begin{array}{c} M_1 \\ M_2 \end{array} \right\}^{total} = \left\{ \begin{array}{c} 0 \\ 0 \end{array} \right\} = \left\{ \begin{array}{c} \dfrac{-6EI}{L^2} \\ \dfrac{-6EI}{L^2} \end{array} \right\} v_1^* + \left[\begin{array}{cc} \dfrac{4EI}{L} & \dfrac{2EI}{L} \\ \dfrac{2EI}{L} & \dfrac{8EI}{L} \end{array} \right] \left\{ \begin{array}{c} \theta_1 \\ \theta_2 \end{array} \right\} \tag{25.18}$$

If no external nodal moments are applied, the solution for resultant nodal rotations, due to the enforced nodal displacement v_1^*, is found by solving the resultant equation set (Equation 25.18) simultaneously:

$$\left[\begin{array}{cc} \dfrac{4EI}{L} & \dfrac{2EI}{L} \\ \dfrac{2EI}{L} & \dfrac{8EI}{L} \end{array} \right] \left\{ \begin{array}{c} \theta_1 \\ \theta_2 \end{array} \right\} = \left\{ \begin{array}{c} \dfrac{6EI}{L^2} \\ \dfrac{6EI}{L^2} \end{array} \right\} v_1^*$$

and

$$\left\{ \begin{array}{c} \theta_1 \\ \theta_2 \end{array} \right\} = \left\{ \begin{array}{c} \dfrac{9}{7} \\ \dfrac{3}{7} \end{array} \right\} \dfrac{v_1^*}{L} \tag{25.19}$$

Internal member forces are found, as usual, by substituting the derived nodal-displacement components into the stiffness equations for the individual members, and adding the fixed-end forces to the result.

25.5 STIFFNESS EQUATIONS FOR STRUCTURES WITH MEMBERS EXPERIENCING TEMPERATURE CHANGES

Frequently, one or more members of a structure are subject to changes in temperature with resulting changes in length. For a statically indeterminate structure such length changes will generally result in the development of internal forces. The analyst may need to know the forces and displacements caused by temperature changes.

Member ② of the continuous strut of Figure 25.5(a) is assumed to experience an increase in temperature equal to ΔT.

If all the nodes of the structure are held immobile, forces will develop at both ends of the member. These are shown in Figure 25.5(b) as X_2^F and X_3^F, and their values are computed with the following expression, in which α is the thermal coefficient of expansion.

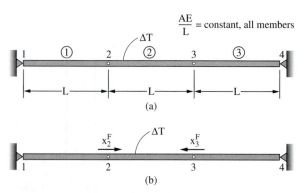

Figure 25.5

$$X_2^F = \alpha(\Delta T)EA$$

$$X_3^F = -\alpha(\Delta T)EA$$

The stiffness equation for the structure can be written as a superposition of "fixed-end forces" and the forces resulting from nodal displacements:

$$
\begin{Bmatrix} X_1 \\ X_2 \\ X_3 \\ X_4 \end{Bmatrix} = \frac{AE}{L} \begin{bmatrix} 1 & -1 & 0 & 0 \\ -1 & 2 & -1 & 0 \\ 0 & -1 & 2 & -1 \\ 0 & 0 & -1 & 1 \end{bmatrix} \begin{Bmatrix} u_1 \\ u_2 \\ u_3 \\ u_4 \end{Bmatrix} + \begin{Bmatrix} 0 \\ \alpha(\Delta T)EA \\ -\alpha(\Delta T)EA \\ 0 \end{Bmatrix} \tag{25.20}
$$

When boundary conditions are taken into account the resulting stiffness equation becomes

$$
\begin{Bmatrix} X_2 \\ X_3 \end{Bmatrix} = \begin{bmatrix} \dfrac{2AE}{L} & -\dfrac{AE}{L} \\ -\dfrac{AE}{L} & \dfrac{2AE}{L} \end{bmatrix} \begin{Bmatrix} u_2 \\ u_3 \end{Bmatrix} + \begin{Bmatrix} \alpha(\Delta T)EA \\ -\alpha(\Delta T)EA \end{Bmatrix} \tag{25.21}
$$

If there are no external forces applied at node 2 or 3, the solution for this thermal load problem is

$$
\begin{Bmatrix} u_2 \\ u_3 \end{Bmatrix} = \begin{bmatrix} \dfrac{2AE}{L} & -\dfrac{AE}{L} \\ -\dfrac{AE}{L} & \dfrac{2AE}{L} \end{bmatrix}^{-1} \begin{Bmatrix} -\alpha(\Delta T)EA \\ \alpha(\Delta T)EA \end{Bmatrix} = \begin{Bmatrix} \dfrac{-\alpha(\Delta T)L}{3} \\ \dfrac{\alpha(\Delta T)L}{3} \end{Bmatrix} \tag{25.22}
$$

Substitution of the nodal displacements into the individual-element stiffness equations, plus the addition of fixed-end forces, will provide the solution for internal member forces.

$$
\begin{Bmatrix} X_{1,2} \\ X_{2,1} \end{Bmatrix} = \begin{Bmatrix} 0 \\ 0 \end{Bmatrix}^F + \frac{AE}{L} \begin{bmatrix} 1 & -1 \\ -1 & 1 \end{bmatrix} \begin{Bmatrix} u_1 = 0 \\ u_2 \end{Bmatrix} = \begin{Bmatrix} \dfrac{\alpha(\Delta T)EA}{3} \\ -\dfrac{\alpha(\Delta T)EA}{3} \end{Bmatrix}
$$

$$
\begin{Bmatrix} X_{2,3} \\ X_{3,2} \end{Bmatrix} = \begin{Bmatrix} \alpha(\Delta T)EA \\ -\alpha(\Delta T)EA \end{Bmatrix}^F + \frac{AE}{L} \begin{bmatrix} 1 & -1 \\ -1 & 1 \end{bmatrix} \begin{Bmatrix} u_2 \\ u_3 \end{Bmatrix} = \begin{Bmatrix} \dfrac{\alpha(\Delta T)EA}{3} \\ -\dfrac{\alpha(\Delta T)EA}{3} \end{Bmatrix}
$$

$$
\begin{Bmatrix} X_{3,4} \\ X_{4,3} \end{Bmatrix} = \begin{Bmatrix} 0 \\ 0 \end{Bmatrix}^F + \frac{AE}{L} \begin{bmatrix} 1 & -1 \\ -1 & 1 \end{bmatrix} \begin{Bmatrix} u_3 \\ u_4 = 0 \end{Bmatrix} = \begin{Bmatrix} \dfrac{\alpha(\Delta T)EA}{3} \\ -\dfrac{\alpha(\Delta T)EA}{3} \end{Bmatrix}
$$

As seen from the solution, the structure experiences a uniform compression throughout as a result of a temperature increase in member ②.

25.6 STIFFNESS EQUATIONS FOR STRUCTURES WHOSE MEMBERS HAVE INCORRECT LENGTHS

Many structures are fabricated with one or more members having incorrect lengths. If the structure is "forced" together during assembly, forces may develop in various portions of the structure even when no external loads are applied. The procedure for determining these "locked-in" forces is essentially the same as has been described for members that experience temperature changes.

As an example, member ② of the structure shown in Figure 25.5(a) is assumed to be fabricated with an excessive length Δl. For this analysis it is assumed that the structure is forced together while keeping the nodes immobile. Nodal forces applied at nodes 2 and 3 will result, as shown in Figure 25.5(b). For this case, however, the values of the nodal forces are computed as

$$
\begin{aligned}
X_2^F &= AE\frac{\Delta L}{L} \\
X_3^F &= -AE\frac{\Delta L}{L}
\end{aligned}
\tag{25.23}
$$

The remainder of the procedure is exactly as was described for the thermal load problem, i.e., a stiffness equation similar to Equation 25.20 is written for the structure, boundary conditions are applied, and values determined for the displacement components at the free nodes. Internal member forces are then found by substituting the values of the free node displacements into the individual member stiffness equations.

25.7 APPLICATIONS OF MATRIX PARTITIONING

In the previous discussion of the stiffness method, the complete set of stiffness equations for a structure was generally derived first. The solution for the displacements of the free nodes was determined by first eliminating from the stiffness matrix the rows and columns corresponding to boundary displacements whose values were specified equal to zero assuming no moments at supports. The justification for this procedure is seen clearly by considering the full set of stiffness equations for a structure. Written in partitioned form, the equation is

$$
\left\{\frac{P_f}{P_s}\right\} = \left[\begin{array}{cc} K_{ff} & K_{fs} \\ K_{sf} & K_{ss} \end{array}\right] \left\{\frac{\delta_f}{\delta_s}\right\}
\tag{25.24}
$$

Equation 25.24 has been written so that all boundary displacements are grouped together and identified as $\{\delta_s\}$. The displacements of the free nodes are identified as $\{\delta_F\}$. In Equation 25.24, the underbar associated with the various terms indicates that the term is a matrix, and not just a scalar. The left-hand side of Equation 25.24 represents nodal forces, i.e.,

$$
\begin{aligned}
P_f &= \{P_f\}, && \text{a vector of applied nodal forces,} \\
P_s &= \{P_s\}, && \text{a vector of support reactions.}
\end{aligned}
$$

The matrix [K] is partitioned into four segments corresponding to the division between displacements of free and supported nodes.

Matrix Equation 25.24 can be written as two reduced matrix equations. The upper matrix equation is

$${P_f} = [K_{ff}]{\delta_f} + [K_{fs}]{\delta^0_s} = [K_{ff}]{\delta_f} \tag{25.25}$$

and the lower matrix equation is:

$${P_s} = [K_{sf}]{\delta_f} + [K_{ss}]{\delta^0_s} = [K_{sf}]{\delta_f} \tag{25.26}$$

A portion of the right-hand sides of both Equations 25.25 and 25.26 is eliminated since the vector ${\delta_s}$ is identically equal to zero.

Therefore Equation 25.25 may be solved for ${\delta_f}$:

$${\delta_f} = [K_{ff}]^{-1}{P_f} \tag{25.27}$$

and the values of ${\delta_f}$ may now be substituted in Equation 25.26 to produce the values of the reactions:

$${P_s} = [K_{sf}]{\delta_f} = [K_{sf}][K_{ff}]^{-1}{P_f} \tag{25.28}$$

This is exactly the procedure that has been used in solving the problems involving the stiffness method in the previous two chapters.

25.8 CONDENSATION

The discussion of solution procedures in the previous section is a special case of an important topic frequently referred to as "condensation". This procedure, which is discussed briefly here, permits an analyst to reduce the order of the set of stiffness equations to simplify the problem, or to accomplish some other objective.

To illustrate the procedure, the cantilevered beam of Figure 25.6 is considered. After accounting for boundary conditions, the structure has 4 degrees of freedom (v_1, θ_1, v_2, θ_2). The corresponding stiffness equations may be written symbolically (from Equation 23.26) as

$$
\begin{Bmatrix} Y_1 = 1\,k \\ Y_2 = 3\,k \\ M_1 = 0 \\ M_2 = 0 \end{Bmatrix} =
\begin{bmatrix}
\dfrac{12EI}{L^3} & \dfrac{-12EI}{L^3} & \dfrac{6EI}{L^2} & \dfrac{6EI}{L^2} \\[2mm]
\dfrac{-12EI}{L^3} & \dfrac{24EI}{L^3} & \dfrac{-6EI}{L^2} & 0 \\[2mm]
\dfrac{6EI}{L^2} & \dfrac{-6EI}{L^2} & \dfrac{4EI}{L} & \dfrac{2EI}{L} \\[2mm]
\dfrac{6EI}{L^2} & 0 & \dfrac{2EI}{L} & \dfrac{8EI}{L}
\end{bmatrix}
\begin{Bmatrix} v_1 \\ v_2 \\ \theta_1 \\ \theta_2 \end{Bmatrix} \tag{25.29}
$$

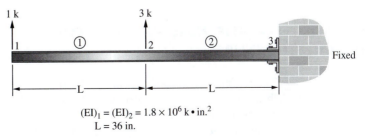

$(EI)_1 = (EI)_2 = 1.8 \times 10^6 \, k \cdot in.^2$
$L = 36$ in.

Figure 25.6

In Equation 25.29, the rotations θ_1 and θ_2 and the displacements v_1 and v_2 have been grouped together, and the stiffness matrix partitioned to reflect this division. For a particular problem the analyst may know that there are no external moments applied at nodes 1 and 2. Therefore the lower matrix equation shown in Equation 25.29 may be solved for the rotations θ_1 and θ_2 in terms of the displacements v_1 and v_2:

$$\begin{Bmatrix} 0 \\ 0 \end{Bmatrix} = \frac{6EI}{L^2}\begin{bmatrix} 1 & -1 \\ 1 & 0 \end{bmatrix}\begin{Bmatrix} v_1 \\ v_2 \end{Bmatrix} + \frac{2EI}{L}\begin{bmatrix} 2 & 1 \\ 1 & 4 \end{bmatrix}\begin{Bmatrix} \theta_1 \\ \theta_2 \end{Bmatrix}$$

and

$$\begin{Bmatrix} \theta_1 \\ \theta_2 \end{Bmatrix} = \frac{3}{7}\begin{bmatrix} -3 & 4 \\ -1 & -1 \end{bmatrix}\begin{Bmatrix} \dfrac{v_1}{L} \\ \dfrac{v_2}{L} \end{Bmatrix} \tag{25.30}$$

Equation 25.30 may now be substituted in the upper matrix equation shown in Equation 25.29 to produce the following result:

$$\begin{Bmatrix} 1 \\ 3 \end{Bmatrix} = \frac{12EI}{L^3}\begin{bmatrix} 1 & -1 \\ -1 & 2 \end{bmatrix}\begin{Bmatrix} v_1 \\ v_2 \end{Bmatrix} + \left(\frac{6EI}{L^2}\right)\left(\frac{3}{7}\right)\begin{bmatrix} 1 & 1 \\ -1 & 0 \end{bmatrix}\begin{bmatrix} -3 & 4 \\ -1 & -1 \end{bmatrix}\begin{Bmatrix} \dfrac{v_1}{L} \\ \dfrac{v_2}{L} \end{Bmatrix}$$

$$= \frac{EI}{L^3}\begin{bmatrix} \dfrac{12}{7} & \dfrac{-30}{7} \\ \dfrac{-30}{7} & \dfrac{96}{7} \end{bmatrix}\begin{Bmatrix} v_1 \\ v_2 \end{Bmatrix} \tag{25.31}$$

Equation 25.31 is solved for the displacements $\begin{Bmatrix} v_1 \\ v_2 \end{Bmatrix}$ in terms of the applied nodal forces:

$$\begin{Bmatrix} v_1 \\ v_2 \end{Bmatrix} = \frac{L^3}{EI}\begin{bmatrix} \dfrac{12}{7} & \dfrac{-30}{7} \\ \dfrac{-30}{7} & \dfrac{96}{7} \end{bmatrix}^{-1}\begin{Bmatrix} 1 \\ 3 \end{Bmatrix} = \frac{L^3}{EI}\begin{bmatrix} 2.667 & 0.833 \\ 0.833 & 0.333 \end{bmatrix}\begin{Bmatrix} 1 \\ 3 \end{Bmatrix} = \begin{Bmatrix} 0.134\,\text{in.} \\ 0.048\,\text{in.} \end{Bmatrix}$$

$$\tag{25.32}$$

Note that Equation 25.31 represents a set of only two equations that have to be solved simultaneously in order to find the displacements, whereas the original matrix equation (Equation 25.29) involves four equations that have to be solved simultaneously. Although the reduced order of the set of equations that must be solved simultaneously is obtained at the expense of additional matrix algebra, the net result may be a significant savings in computer costs for solving large problems.

25.9 BAND WIDTH OF STIFFNESS MATRICES FOR GENERAL STRUCTURES

The reader may have noted, in the illustrations of stiffness matrices, that almost all of these matrices are populated with a large number of zeroes. If the node-numbering scheme for a structure is performed optimally, the nonzero elements of the matrix will be grouped together in a "band." As an example, consider the stiffness matrix for the uniform axial force member consisting of four elements, shown in Figure 25.7.

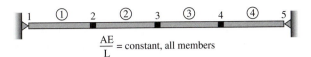

Figure 25.7

The stiffness matrix for this structure can be written, with the help of Equation 23.8, as

$$[K] = \frac{AE}{L} \begin{bmatrix} 1 & -1 & 0 & 0 & 0 \\ -1 & 2 & -1 & 0 & 0 \\ 0 & -1 & 2 & -1 & 0 \\ 0 & 0 & -1 & 2 & -1 \\ 0 & 0 & 0 & -1 & 1 \end{bmatrix} \tag{25.33}$$

Dashed lines in the stiffness matrix denote the limits of the band within which all nonzero elements of the matrix lie.

Manufactures & Traders Trust Company Building. (Courtesy Bethlehem Steel Co.)

Computer solutions of the stiffness equations utilize the banded characteristics of the stiffness matrix to save storage space in the computer. Algorithms can be written that require the storage of only those elements within the band of a stiffness matrix. For a large structural problem, the number of elements outside the band may be many times larger than the number of elements within the band of the matrix. By taking proper advantage of the banded characteristics of a stiffness matrix, dramatic savings can be achieved in computer storage space and resulting time for solution.

If a structure consists of a continuous beam or strut, optimal node numbering is achieved by numbering the nodes sequentially from left to right or from right to left. If the structure is a frame, the optimal node-numbering scheme is far less apparent. However, the following guidelines will permit the analyst to achieve a "near-minimum" bandwidth:

1. For each node of the structure, perform the following:
   (a) determine the largest node number associated with any member attached to the node in question; also determine the smallest node number.
   (b) compute the difference between the largest node number and the smallest node number.

2. Minimize this difference, when considering all nodes.

Though the use of these guidelines will not always result in absolutely the minimum bandwidth for the associated stiffness matrix, they will provide a value satisfactorily close to the minimum.

To illustrate these guidelines, a simple, multistory frame with two node-numbering schemes, shown in Figures 25.8(a) and (b), is considered. For scheme (a), the largest difference occurs simultaneously at nodes 2 to 7 and 9 to 13, and has a value of 8. For scheme (b) the largest difference occurs simultaneously at nodes 3 to 11 and 4 to 12, and has a value of 4. Thus node-numbering scheme (b) leads to a smaller bandwidth and will generally lead to a more economical solution in terms of required computer time.

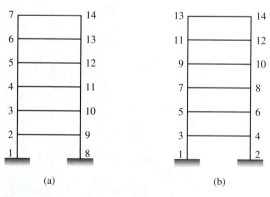

(a)   (b)

Figure 25.8

Most computer programs are written so as to take advantage of the symmetry in the stiffness matrix. Therefore only that portion of the matrix that is either above or below the main diagonal is stored and used in the computations. For these programs, the term bandwidth should be replaced by half-band width. However, the rules regarding attainment of economical node numbering schemes remain the same.

25.10 PROBLEMS FOR SOLUTION

25.1 An axial force member consists of three elements (members) attached to each other and to supports through four nodes. Perform the following:

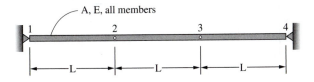

A, E, all members

(a) Determine the individual stiffness matrices for all three members;

(b) Add the element stiffness matrices together into a global stiffness matrix for the structure. The stiffness matrix is to be arranged so that it can be used in the following stiffness equation:

$$\begin{Bmatrix} X_1 \\ X_3 \\ X_4 \\ X_2 \end{Bmatrix} = [K_{total}] \begin{Bmatrix} u_1 \\ u_3 \\ u_4 \\ u_2 \end{Bmatrix}$$

(c) Compare the "bandwidth" of the stiffness matrix found in part (b) to that shown in Equation 25.33 [*Ans.* Bandwidth (above) = 4 vs Bandwidth (Eq. 25.33) = 3]

25.2 Use Equation 25.16 to verify that the stiffness matrix [K], expressed in global coordinates, is the same as given by Equation 24.19

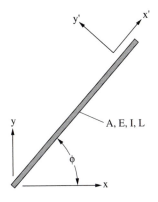

A, E, I, L

25.3 Determine the stiffness matrix for an inclined member, such as shown in the figure, but with the local x axis oriented as shown (pointing from upper right to lower left). Also, find the stiffness matrix for the same strut if the local x axis is oriented as shown in Problem 25.2. Point out the differences, if any, that exist between these two stiffness matrices.

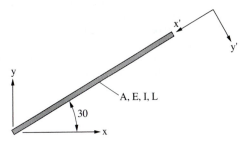

A, E, I, L

25.4 Verify Equation 25.9, i.e., show that $[\gamma]^{-1} = [\gamma]^T$

25.5 Show that the matrix $[\Gamma]$ in Equation 25.10 has the same properties as does $[\gamma]$, i.e., show that $[\Gamma]^{-1} = [\Gamma]^T$.

25.6 A fixed-ended beam has an intermediate support at node 2. Support 2 is a distance v_2^* below an axis drawn between nodes 1 and 3. Determine the nodal forces that develop at nodes 1, 2, and 3 when the beam is jacked into position (with node 2 attached to the beam).

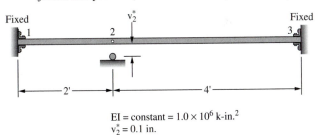

$$EI = \text{constant} = 1.0 \times 10^6 \text{ k-in.}^2$$
$$v_2^* = 0.1 \text{ in.}$$

Ans. $Y_1 = 54.2$ k, $Y_2 = -73.2$ k; $Y_3 = 19.0$ k
$M_1 = 780.9$ in.-k $M_3 = -390.0$ in.-k

25.7 A frame consists of three uniform members, as shown in the sketch. Member ① experiences an increase of temperature ΔT_1, while members ② and ③ remain at ambient temperature. Determine the resultant reactions at nodes 1, 2, and 3 due to the temperature increase. [*Ans.* Total reactions (Parts A and B) $x_1^{Total} = 2.17$ k]

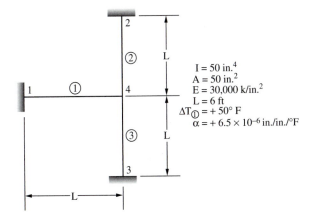

$$I = 50 \text{ in.}^4$$
$$A = 50 \text{ in.}^2$$
$$E = 30,000 \text{ k/in.}^2$$
$$L = 6 \text{ ft}$$
$$\Delta T_① = +50° \text{ F}$$
$$\alpha = +6.5 \times 10^{-6} \text{ in./in./°F}$$

25.8 For the frame shown in Problem 25.7, assume that all members are at ambient temperature, but member ② is fabricated too short a distance $\Delta L_2 = -0.10$ in. Determine the resulting reactions at nodes 1, 2, and 3 if member ② is forced into position (*Note*: assume that $u_4 = 0.0$, to simplify computation).

Ans.

$$\begin{Bmatrix} X_1 \\ Y_1 \\ M_1 \end{Bmatrix} = \begin{Bmatrix} 0 \\ -1.79\,k \\ -71.7\,\text{in.-k} \end{Bmatrix}$$

$$\begin{Bmatrix} X_2 \\ Y_2 \\ M_2 \end{Bmatrix} = \begin{Bmatrix} 0.598\,k \\ -103.3\,k \\ 14.34\,\text{in.-k} \end{Bmatrix} + \begin{Bmatrix} 0.0 \\ 208.3\,k \\ 0.0 \end{Bmatrix} = \begin{Bmatrix} 0.598\,k \\ 105.0\,k \\ 14.34\,\text{in.-k} \end{Bmatrix}$$

25.9. A two-bay, multistory structure is modeled with two different node-numbering schemes, shown in sketches (a) and (b). Which node-numbering scheme will result in a stiffness matrix having the smallest bandwidth? Explain your answer. [*Ans.* node numbering scheme (b) produces least band width with maximum difference = 6]

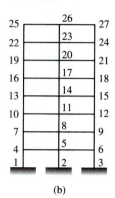

(a)   (b)

The Catenary Equation

Some civil engineering structures include cables as part of the load-carrying system. Among these are suspension bridges and offshore platform installations that include mooring lines, such as tension leg platforms. Because the cables are not weightless, as we often assume when designing guy-lines for small towers, the tension in the cable changes along its length and the cable sags along its length when in use. These cables assume the shape of a catenary. Developed in this appendix are the fundamental relationships necessary to evaluate the force in catenary cables.

The basic geometry of the catenary cable that we will be considering is shown in Figure A.1. Notice that the bottom of the cable is located at coordinate (0, 0) and that it is tangent to the horizontal coordinate axis at that point. The top of the cable is located at coordinate (X, Y).

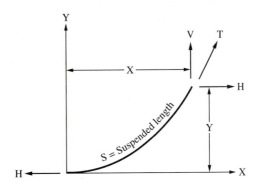

Figure A.1 Geometry of the cable

To develop the governing equations for a cable, consider the segment of length Δs that is shown in Figure A.2. The equations of equilibrium for this segment of the cable are

$$\sum F_x = (T + \Delta T)\cos(\theta + \Delta\theta) - T\cos\theta = 0$$
$$\sum F_y = (T + \Delta T)\sin(\theta + \Delta\theta) - T\sin\theta - w\Delta s = 0 \tag{A.1}$$

The term w in Equation A.1 is the weight of the cable per unit length. For convenience in mathematical manipulation, let us introduce the function:

$$H(T, \theta) = T\cos\theta \tag{A.2}$$

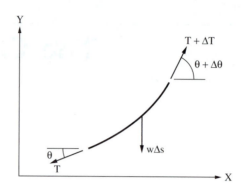

Figure A.2 A segment of the cable

This equation is the horizontal component of force in the cable. If we substitute Equation A.2 into the first of the equations in Equation A.1, that equation becomes

$$H(T + \Delta T, \theta + \Delta\theta) - H(T, \theta) = \Delta H = 0 \qquad (A.3)$$

Along the segment Δs, then, the rate of change of the horizontal force is

$$\frac{dH}{ds} = \lim_{\Delta s \to 0} \left(\frac{\Delta H}{\Delta s} \right) = \lim_{\Delta s \to 0} \left(\frac{0}{\Delta s} \right) = 0 \qquad (A.4)$$

Because the function H, as defined in Equation A.2, is the horizontal component of force in the cable, we can conclude from Equation A.4 that the horizontal component of force is constant along the length of the cable, namely:

$$H = T \cos \theta \qquad (A.5)$$

From this equation, we can further conclude that for H to be constant the force in the cable, T, must change along its length.

Now let us examine the second of the equations in Equation A.1. If we rearrange that equation and divide both sides by $\Delta\theta$, we can obtain:

$$\frac{(T + \Delta T) \sin(\theta + \Delta\theta) - T \sin \theta}{\Delta\theta} = \frac{w\Delta s}{\Delta\theta} \qquad (A.6)$$

Taking the limit of this equation as $\Delta\theta$ approaches zero, we find that:

$$\lim_{\Delta\theta \to 0} \left[\frac{(T + \Delta T) \sin(\theta + \Delta\theta) - T \sin \theta}{\Delta\theta} \right] = \lim_{\Delta\theta \to 0} \left[\frac{w\Delta s}{\Delta\theta} \right] \qquad (A.7)$$

which becomes

$$\frac{d}{d\theta} [T \sin \theta] = w \frac{ds}{d\theta} \qquad (A.8)$$

But from Equation A.5 it is known that

$$T = \frac{H}{\cos \theta} \qquad (A.9)$$

so Equation A.8 becomes

$$\frac{d}{d\theta} [H \tan \theta] = H \sec^2\theta = w \frac{ds}{d\theta} \qquad (A.10)$$

and can be rearranged to the form

$$\frac{ds}{d\theta} = \frac{H}{w} \sec^2\theta \qquad (A.11)$$

Upon integrating Equation A.11, the suspended length of the cable is found to be

$$S = \int_0^s ds = \frac{H}{w} \int_0^\theta \sec^2\theta \, d\theta = \frac{H}{w} \tan\theta \tag{A.12}$$

From calculus, it is known that

$$\cos\theta = \lim_{\Delta s \to 0} \left[\frac{\Delta x}{\Delta s}\right] = \frac{dx}{ds}$$

$$\sin\theta = \lim_{\Delta s \to 0} \left[\frac{\Delta y}{\Delta s}\right] = \frac{dy}{ds} \tag{A.13}$$

By applying the chain rule with Equations A.11 and A.13, we obtain the equations:

$$\frac{dx}{d\theta} = \left[\frac{H}{w}\right] \sec^2\theta \cos\theta = \left[\frac{H}{w}\right] \sec\theta$$

$$\frac{dy}{d\theta} = \left[\frac{H}{w}\right] \sec^2\theta \sin\theta \tag{A.14}$$

Integrating the first of these equations yields

$$X = \int_0^x dx = \frac{H}{w} \int_0^\theta \sec\theta \, d\theta \tag{A.15}$$

which reduces to

$$X = \left[\frac{H}{w}\right] \ln(\sec\theta + \tan\theta) \tag{A.16}$$

This is the X projection of the cable, the horizontal distance from one end to the other end. The Y projection can be found by working with the second equation in Equation A.14 and performing integration as was done to obtain the X projection. The result is

$$Y = \left[\frac{H}{w}\right] (\sec\theta - 1) \tag{A.17}$$

These last two equations are expressed in terms of the angle θ, which probably is not known. However, the angle can be expressed in terms of other parameters of cable.

From Equation A.16:

$$e^{wX/H} = \sec\theta + \tan\theta \tag{A.18}$$

and from trigonometry we know that

$$\sec^2\theta - \tan^2\theta = (\sec\theta + \tan\theta)(\sec\theta - \tan\theta) = 1 \tag{A.19}$$

If Equation A.18 is substituted into Equation A.19 the equation obtained is

$$\sec\theta - \tan\theta = e^{-wX/H} \tag{A.20}$$

Then, by adding Equations A.18 and A.20, $\sec\theta$ is found to be

$$\sec\theta = \tfrac{1}{2}[e^{wX/H} + e^{-wX/H}] = \cosh\left(\frac{wX}{H}\right) \tag{A.21}$$

and by subtracting Equation A.20 from Equation A.18 $\tan\theta$ is found to be

$$\tan\theta = \tfrac{1}{2}[e^{wX/H} - e^{-wX/H}] = \sinh\left(\frac{wX}{H}\right) \tag{A.22}$$

The length of the cable is found by substituting Equation A.22 into Equation A.12 The result is

$$S = \left[\frac{H}{w}\right] \sinh\left(\frac{wX}{H}\right) \tag{A.23}$$

By substituting Equation A.21 into Equation A.17, we find that the Y projection of the cable is

$$Y = \frac{H}{w}\left[\cosh\left(\frac{wX}{H}\right) - 1\right] \tag{A.24}$$

Lastly, using Equations A.21, A.22, and A.24, the X projection of the cable is found to be

$$X = \frac{H}{w}\cosh^{-1}\left(\frac{Y + \frac{H}{w}}{\frac{H}{w}}\right) \tag{A.25}$$

or if Equations A.21, A.22, and A.23 had been used, X would be of the form:

$$X = \frac{H}{w}\sinh^{-1}\left(\frac{Sw}{H}\right) \tag{A.26}$$

The last quantity needed is the tension in the cable. This can be obtained by using Equations A.5, A.13, and A.26. The result is

$$T = \sqrt{(wS)^2 + H^2} \tag{A.27}$$

From the form of this last equation, we recognize that T is computed using Pythagorean's theorem applied to the vertical and horizontal components of force in the cable. As such, the vertical component of force in the cable is

$$V = wS \tag{A.28}$$

The equations developed are sufficient to determine the forces in a cable and its profile. They can be used to evaluate a cable that does not pass though the origin, as was shown in Figure A.1. To do so, a cable passing through the origin, which contains cable AB, is determined as shown in Figure A.3. The forces and projections at points a and b are then computed. They represent the forces and profile in the cable ab.

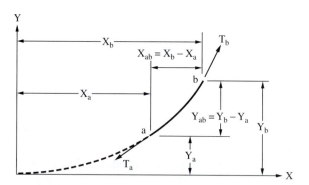

Figure A.3 Geometry of a cable segment not passing through the origin

EXAMPLE A.1

The anchor chain on a mobile offshore drilling unit is 2,000 ft long and has a submerged weight of 108 pounds per foot. It is tangent to the seafloor at the end with the anchor. The other end of the chain connects to the vessel at a point 800 ft above the seafloor. What is the tension in the chain and the distance from the vessel to the anchor?

By substituting Equation A.25 into Equation A.23 we obtain the equation:

$$S = \frac{H}{w} \sinh\left[\cosh^{-1}\left(\frac{Y + \frac{H}{w}}{\frac{H}{w}}\right)\right]$$

which can be solved for H, the horizontal component of force in the chain. The result is

$$H = \tfrac{1}{2}(S^2 + Y^2)\left(\frac{w}{Y}\right) = \tfrac{1}{2}(2,000^2 - 800^2)\left(\frac{108}{800}\right) = 226,800\,\text{lbs}$$

We then can substitute this value of H, and the other known values, into Equation A.25 to obtain X, the horizontal distance from the end of the chain to the anchor.

$$X = \frac{226,800}{108}\cosh^{-1}\left(\frac{800 + \dfrac{226,800}{108}}{\dfrac{226,800}{108}}\right) = 1,780\,\text{ft}$$

The tension in the chain can be computed using Equation A.27. The result is:

$$T = \sqrt{(108 \cdot 2000)^2 + (226,800)^2} = 313,200\,\text{lbs} \quad \blacksquare$$

Appendix B

Matrix Algebra

B.1 INTRODUCTION

An introduction to the basic rules of matrix algebra is presented here. The material is intended to provide introductory background of the subject. The focus of the material is on applications that may be of interest primarily to structural engineers. Other excellent references exist that will provide an interested reader with a more comprehensive treatment of the subject[1,2].

B.2 MATRIX DEFINITIONS AND PROPERTIES

A matrix is defined as an ordered arrangement of numbers in rows and columns, as:

$$[A] = \begin{bmatrix} a_{1,1} & a_{1,2} & \cdots & a_{1,n} \\ a_{2,1} & a_{2,2} & \cdots & a_{2,n} \\ \vdots & \vdots & \ddots & \vdots \\ a_{m,1} & a_{m,2} & \cdots & a_{m,n} \end{bmatrix} \qquad (B.1)$$

A representative matrix [A], such as shown in Equation B.1 consists of m rows and n columns of numbers enclosed in brackets. Matrix [A] is said to be of order $m \times n$. The elements $a_{i,j}$ in the array are identified by two subscripts. The first subscript designates the row in which the element is and the second subscript designates the column. Thus $a_{3,5}$ is the element located at the intersection of the third row and the fifth column of the matrix. Elements with repeated subscripts, for example, $a_{i,i}$ are located on the main diagonal of the matrix.

Although the number of rows and columns of a matrix may vary from problem to problem, two special cases deserve mention. When m is equal to 1, the matrix consists of only one *row* of elements and is called a row matrix. It is written as

$$\lfloor A \rfloor = \lfloor a_1 \quad a_2 \quad \cdots \quad a_n \rfloor \qquad (B.2)$$

[1]F. Ayres, *Theory and Problems of Matrices*, Schaum Outline Series (New York: McGraw-Hill, 1962).
[2]S. D. Conte and L. deBoor, *Elementary Numerical Analysis*, 2d ed. (New York: McGraw-Hill, 1972).

When $n = 1$, the matrix consists of only one *column* and is called a column matrix or a vector. It is written as

$$\{A\} = \begin{Bmatrix} a_1 \\ a_2 \\ \vdots \\ a_n \end{Bmatrix} \tag{B.3}$$

B.3 SPECIAL MATRIX TYPES

Square Matrix

When the number of rows, m, and the number of columns, n, are equal, the matrix is said to be *square*. As an example, if $m = n = 3$, the square matrix [A] may appear as

$$[A] = \begin{bmatrix} a_{11} & a_{12} & a_{13} \\ a_{21} & a_{22} & a_{23} \\ a_{31} & a_{32} & a_{33} \end{bmatrix} \tag{B.4}$$

Symmetric Matrix

A symmetric matrix is one in which the off-diagonal terms are reflected about the main diagonal. As such:

$$a_{ij} = a_{ji} \qquad \text{for } i \neq j \tag{B.5}$$

When the elements of a square matrix obey this rule, the matrix is said to be *symmetric* and the elements are arranged symmetrically about the main diagonal. An example of a symmetric 3×3 matrix is

$$[A] = \begin{bmatrix} 3 & -2 & 5 \\ -2 & 4 & 7 \\ 5 & 7 & 6 \end{bmatrix} \tag{B.6}$$

Symmetric matrices occur frequently in structural theory and play an important role in the matrix manipulations used to develop the theory.

Identity Matrix

When all the elements on the main diagonal of a square matrix are equal to unity and all of the other elements are equal to zero, the matrix is called an *identity matrix*. Sometimes it is also called a *unit matrix*. Matrices of this type are identified with the symbol [I]. An example of a 3×3 identity matrix is

$$[I] = \begin{bmatrix} 1 & 0 & 0 \\ 0 & 1 & 0 \\ 0 & 0 & 1 \end{bmatrix} \tag{B.7}$$

Transposed Matrix

When the elements of a given matrix are reordered so that the columns of the origin matrix become the corresponding rows of the new matrix, the new matrix is said to be the *transpose* of the original matrix. In this book, the transpose of matrix [A] is given the symbol $[A]^T$. An example of a specific matrix and its transpose is

$$[A] = \begin{bmatrix} 1 & 3 & -2 \\ 5 & 4 & 7 \\ 8 & 2 & 6 \end{bmatrix}$$

$$[A]^T = \begin{bmatrix} 1 & 5 & 8 \\ 3 & 4 & 2 \\ -2 & 7 & 6 \end{bmatrix}$$

(B.8)

The reader should note that if the matrix [A] is a symmetric matrix, then

$$[A]^T = [A] \tag{B.9}$$

This property is used frequently in the development of structural theory using matrices. As special cases, the transpose of a row matrix becomes a column matrix and vice versa. Thus:

$$\lfloor A \rfloor^T = \{A\}$$
$$\{A\}^T = \lfloor A \rfloor$$

(B.10)

B.4 DETERMINANT OF A SQUARE MATRIX

A determinant of a square matrix [A] is given by the symbol |A| and, in its expanded form, is written as:

$$|A| = \begin{vmatrix} a_{1,1} & a_{1,2} & \cdots & a_{1,n} \\ a_{2,1} & a_{2,2} & \cdots & a_{2,n} \\ \vdots & \vdots & \ddots & \vdots \\ a_{m,1} & a_{m,2} & \cdots & a_{m,n} \end{vmatrix} \tag{B.11}$$

This determinant is said to be of order m. Unlike matrix [A], which has no single value, the determinant |A| does have a single numerical value. The value of |A| is easily found for a 2×2 array of numbers, for example:

$$\begin{vmatrix} 2 & 1 \\ 4 & 5 \end{vmatrix} = 2 \cdot 5 - 1 \cdot 4 = 6 \tag{B.12}$$

The value of |A| in the example was determined by multiplying the numbers on the main diagonal and subtracting from this product the product of the numbers on the other diagonal. Unfortunately, this simple procedure does not work for determinants of order greater than two.

A general procedure for finding the value of a determinant sometimes is called "expansion by minors." The first *minor* of the matrix [A], corresponding to the element $a_{i,j}$ is defined as the determinant of a reduced matrix obtained by eliminating the ith row and the jth column from matrix [A]. The minor is a specific number, like any other determinant.

As an illustration, several first minors are shown for a matrix [A]:

$$[A] = \begin{bmatrix} 1 & 2 & 3 \\ -2 & 3 & 4 \\ 1 & 5 & 2 \end{bmatrix}$$

$$\text{minor of } a_{1,1} = \begin{vmatrix} 3 & 4 \\ 5 & 2 \end{vmatrix} = 6 - 20 = -14 \tag{B.13}$$

$$\text{minor of } a_{1,2} = \begin{vmatrix} -2 & 4 \\ 1 & 2 \end{vmatrix} = -4 - 4 = -8$$

$$\text{minor of } a_{2,3} = \begin{vmatrix} 1 & 2 \\ 1 & 5 \end{vmatrix} = 5 - 2 = 3$$

When the proper sign is attached to a minor, the result is called a *cofactor*, and is given the symbol A_{ij}. The sign of a minor is determined by multiplying the minor by $(-1)^{i+j}$. Several cofactors for the example matrix [A] are

$$A_{1,1} = (-1)^{1+1} \times \text{minor of } a_{1,1} = 1(-14) = -14$$
$$A_{1,2} = (-1)^{1+2} \times \text{minor of } a_{1,2} = (-1)(-8) = 8 \tag{B.14}$$
$$A_{2,3} = (-1)^{2+3} \times \text{minor of } a_{2,3} = -1(3) = -3$$

Now, to obtain the value of a general determinant, we can choose any arbitrary row i of matrix [A] and expand according to the relation:

$$|A| = \sum_{j=1}^{m} a_{i,j} A_{i,j} \tag{B.15}$$

The value of a determinant can also be found by choosing an arbitrary column j and expanding according to the relation:

$$|A| = \sum_{i=1}^{m} a_{i,j} A_{i,j} \tag{B.16}$$

If the order of the original determinant is large, the procedure described does not appear to produce a simple solution. For example, a 15th-order determinant will still have first minors that are of order 14. However, the 14th-order minors can be reduced to 13th-order minors by the expansion process. The process can be repeated until the resulting minors are of order 2. These minors can then be evaluated readily using the procedure described in the initial illustration of this section.

Although the procedure for evaluating determinants may appear long and tedious, computer algorithms can be written that will perform the necessary algebraic operations. Other simplifying procedures are available, which make use of special characteristics of determinants. These will not be discussed here, but interested readers may refer to books cited previously in footnotes 1 and 2 in this appendix.

B.5 ADJOINT MATRIX

A special matrix exists called an *adjoint* matrix and is given the symbol adj[A]. To find the adjoint matrix corresponding to an original matrix [A], first replace each element of [A] with its cofactor; the adjoint matrix is then the transpose of this resultant matrix.

Symbolically, adjoint matrix adj[A] is written as

$$\text{adj}[A] = \begin{vmatrix} A_{1,1} & A_{1,2} & \cdots & A_{1,m} \\ A_{2,1} & A_{2,2} & \cdots & A_{2,m} \\ \vdots & \vdots & \ddots & \vdots \\ A_{m,1} & A_{m,2} & \cdots & A_{m,m} \end{vmatrix}$$

(B.17)

A specific numerical example of a matrix [A] and its adjoint matrix is

$$[A] = \begin{bmatrix} 1 & 2 & 3 \\ 2 & 3 & 4 \\ 1 & 5 & 3 \end{bmatrix}$$

$$
\begin{array}{lll}
A_{1,1} = -11 & A_{1,2} = -2 & A_{1,3} = 7 \\
A_{2,1} = 9 & A_{2,2} = 0 & A_{2,3} = -3 \\
A_{3,1} = -1 & A_{3,2} = 2 & A_{3,3} = -1
\end{array}
$$

(B.18)

$$\text{adj}[A] = \begin{bmatrix} -11 & -2 & 7 \\ 9 & 0 & -3 \\ -1 & 2 & -1 \end{bmatrix}^{T} = \begin{bmatrix} -11 & 9 & -1 \\ -2 & 0 & 2 \\ 7 & -3 & -1 \end{bmatrix}$$

The adjoint matrix and the determinant both are used in the computation of the inverse of a matrix, a topic that is treated in a later section of this appendix.

B.6 MATRIX ARITHMETIC

Equality of Matrices

Two matrices are equal only if the corresponding elements of the two matrices are equal. Thus, equality of matrices can exist only between matrices of equal orders.

Addition and Subtraction of Matrices

Two matrices may be added or subtracted only if they have the same order. The addition of two matrices, [A] and [B], is performed as

$$[A] + [B] = [C]$$

(B.19)

where the elements of matrix [C] are the sum of corresponding elements of [A] and [B], that is

$$c_{i,j} = a_{i,j} + b_{i,j}$$

(B.20)

An example of the addition of two matrices is shown:

$$\begin{bmatrix} 1 & 2 \\ 3 & 4 \end{bmatrix} + \begin{bmatrix} 5 & 7 \\ 6 & 8 \end{bmatrix} = \begin{bmatrix} 6 & 8 \\ 9 & 12 \end{bmatrix}$$

(B.21)

Subtraction of two matrices is performed similarly by subtracting corresponding elements.

Both the commutative and the associative laws hold for the addition and subtraction of matrices. Thus:

$$[A] + [B] = [B] + [A]$$

(B.22)

and

$$[A] + ([B] + [C]) = ([A] + [B]) + [C]$$

(B.23)

Scalar Multiplication of Matrices

To multiply a matrix by a scalar, *each element* of the matrix is multiplied by the scalar. Thus:

$$\alpha[A] = \begin{bmatrix} \alpha a_{1,1} & \alpha a_{1,2} & \cdots & \alpha a_{1,n} \\ \alpha a_{2,1} & \alpha a_{2,2} & \cdots & \alpha a_{2,n} \\ \vdots & \vdots & \ddots & \vdots \\ \alpha a_{m,1} & \alpha a_{m,2} & \cdots & \alpha a_{m,n} \end{bmatrix} \tag{B.24}$$

Multiplication of Matrices

The product of two matrices exists only if the matrices are *conformable*. For the matrix product [A][B], conformability means that the number of columns of [A] equals the number of rows of [B]. The two matrices shown are conformable (in the order shown) and may be multiplied.

$$[A] = \begin{bmatrix} 1 & 3 \\ 2 & 5 \end{bmatrix} \quad [B] = \begin{bmatrix} 2 & 6 & 5 \\ 1 & 3 & -4 \end{bmatrix} \tag{B.25}$$

The number of columns in [A] is equal to the number of rows in [B]. However, if the order of the matrix multiplication is reversed, [B][A], the matrices are *not* conformable and the matrix product does not exist. Furthermore, even if the matrices [A] and [B] are square, and thus conformable in the order [A][B] and [B][A], the two matrix products are generally not the same. In general,

$$[A][B] \neq [B][A] \tag{B.26}$$

The formal definition of a matrix product between matrices that are conformable is given as follows:

$$[A]_{m \times 1}[B]_{1 \times n} = [C]_{m \times n} \tag{B.27}$$

where

$$c_{i,j} = \sum_{k=1}^{1} a_{i,k} b_{k,j} \tag{B.28}$$

Note that matrix [A] is not of the same order as [B], but the two matrices are conformable since the number of columns of [A] equals the number of rows of [B]. The order of the product matrix [C] is $m \times n$.

A simple illustration of a matrix product is shown:

$$\begin{bmatrix} 1 & 2 \\ -3 & 2 \end{bmatrix} \begin{bmatrix} 1 & 3 & 2 \\ 4 & 5 & 3 \end{bmatrix} = \begin{bmatrix} 9 & 13 & 8 \\ 5 & 1 & 0 \end{bmatrix} \tag{B.29}$$

In this matrix multiplication the terms in the matrix [C] are computed as follows:

$$\begin{aligned} c_{1,1} &= (1)(1) + (2)(4) = 9 \\ c_{1,2} &= (1)(3) + (2)(5) = 13 \\ c_{1,3} &= (1)(2) + (2)(3) = 8 \\ c_{2,1} &= (-3)(1) + (2)(4) = 5 \\ c_{2,2} &= (-3)(3) + (2)(5) = 1 \\ c_{2,3} &= (-3)(2) + (2)(3) = 0 \end{aligned} \tag{B.30}$$

Although the order in which two matrices are multiplied may not be reversed, in general, without obtaining different results, both the associative and the distributive laws are valid for matrix products. Thus

$$[A][B][C] = ([A][B])[C] = [A]([B][C]) \tag{B.31}$$

and

$$[A]([B] + [C]) = [A][B] + [A][C] \tag{B.32}$$

If a matrix [A] is multiplied by the identity matrix [I] (assuming that the matrices are conformable) the matrix [A] remains unchanged. Thus

$$\begin{bmatrix} 1 & 2 & 3 \\ -2 & 4 & 6 \\ 3 & 5 & 2 \end{bmatrix} \begin{bmatrix} 1 & 0 & 0 \\ 0 & 1 & 0 \\ 0 & 0 & 1 \end{bmatrix} = \begin{bmatrix} 1 & 2 & 3 \\ -2 & 4 & 6 \\ 3 & 5 & 2 \end{bmatrix} \tag{B.33}$$

In general,

$$[A][I] = [A]$$
$$[I][A] = [A] \tag{B.34}$$

Transpose of a Product

If a matrix product is transposed, the result is the reverse product of transpose of the individual matrices. The result is shown symbolically as:

$$([A][B])^T = [B]^T[A]^T \tag{B.35}$$

The transpose of a triple matrix product may be found by using Equation B.35 and the associative law in several stages, as shown

$$\begin{aligned} ([A][B][C])^T &= \{[A]([B][C])\}^T \\ &= ([B][C])^T[A]^T \\ &= [C]^T[B]^T[A]^T \end{aligned} \tag{B.36}$$

Note that in finding the transpose of a matrix product the order of multiplication changes.

Matrix Inverse

Although matrix addition, subtraction, and multiplication have been defined in the preceding sections, no mention has been made of matrix division. In fact, division of matrices in the form [A]/[B] does not exist. However, a matrix operation does exist that closely parallels algebraic division. This operation makes use of a matrix *inverse*.

The inverse of a square matrix [A] is given the symbol $[A]^{-1}$. It is defined such that:

$$[A][A]^{-1} = [I] \tag{B.37}$$

Many techniques exist with which a matrix inverse may be determined. One formal technique is described by the following relationship:

$$[A]^{-1} = \frac{\text{adj}[A]}{|A|} \tag{B.38}$$

As an example, consider the matrix [A] given by Equation B.18. The adjoint matrix adj[A] is also shown in Equation B.18. The determinant |A| may be found by using the elements and first minors of the first column of [A]:

$$|A| = a_{1,1}A_{1,1} + a_{2,1}A_{2,1} + a_{3,1}A_{3,1}$$
$$= 1(-11) + 2(9) + 1(-1) \qquad \text{(B.39)}$$
$$= 6$$

Thus:

$$[A]^{-1} = \frac{1}{6}\begin{bmatrix} -11 & 9 & -1 \\ -2 & 0 & 2 \\ 7 & -3 & -1 \end{bmatrix} \qquad \text{(B.40)}$$

The correctness of the values given for the coefficients of $[A]^{-1}$ may be verified by forming the matrix product $[A][A]^{-1}$ and checking to see if the result is the identity matrix. For the example given:

$$[A][A]^{-1} = \begin{bmatrix} 1 & 2 & 3 \\ 2 & 3 & 4 \\ 1 & 5 & 3 \end{bmatrix}\left(\frac{1}{6}\right)\begin{bmatrix} -11 & 9 & -1 \\ -2 & 0 & 2 \\ 7 & -3 & -1 \end{bmatrix} = \frac{1}{6}\begin{bmatrix} 6 & 0 & 0 \\ 0 & 6 & 0 \\ 0 & 0 & 6 \end{bmatrix} = [I] \qquad \text{(B.41)}$$

A special situation involving the inverse of a matrix deserves attention. If the determinant of a matrix [A] is equal to zero (|A|=0), then the division operation indicated by Equation B.38 cannot be performed. Under these circumstances, the inverse of matrix [A] does not exist and matrix [A] is said to be singular. Singular matrices occur frequently in structural theory and a reader should be aware of the meaning of this term. An example of a singular matrix is shown:

$$[A] = \begin{bmatrix} 1 & 2 & 4 \\ -2 & 3 & 2 \\ 3 & 6 & 12 \end{bmatrix} \qquad \text{(B.42)}$$

and

$$|A| = a_{1,1}A_{1,1} + a_{2,1}A_{2,1} + a_{3,1}A_{3,1} = 24 + 0 - 24 = 0 \qquad \text{(B.43)}$$

The inverse of a matrix product may be found using rules that are very similar to those used in finding the transpose of a matrix product as shown in Equation B.35. Specifically,

$$([A][B])^{-1} = [B]^{-1}[A]^{-1} \qquad \text{(B.44)}$$

and

$$([A][B][C])^{-1} = ([B][C])^{-1}[A]^{-1}$$
$$= [C]^{-1}[B]^{-1}[A]^{-1} \qquad \text{(B.45)}$$

Application of the Matrix Inverse

Consider a set of algebraic equations, each of which contains a number of unknown quantities, x_i, namely:

$$a_{1,1}x_1 + a_{1,2}x_2 + a_{1,3}x_3 = b_1$$
$$a_{2,1}x_1 + a_{2,2}x_2 + a_{2,3}x_3 = b_2 \qquad \text{(B.46)}$$
$$a_{3,1}x_1 + a_{3,2}x_2 + a_{3,3}x_3 = b_3$$

The set of algebraic equations may be cast in matrix form as follows:

$$\begin{bmatrix} a_{1,1} & a_{1,2} & a_{1,3} \\ a_{2,1} & a_{2,2} & a_{2,3} \\ a_{3,1} & a_{3,2} & a_{3,3} \end{bmatrix} \begin{Bmatrix} x_1 \\ x_2 \\ x_3 \end{Bmatrix} = \begin{Bmatrix} b_1 \\ b_2 \\ b_3 \end{Bmatrix} \qquad \textbf{(B.47)}$$

or, symbolically, as

$$[A]\{X\} = \{B\} \qquad \textbf{(B.48)}$$

The solution for the unknown quantities, x_1, x_2, and x_3, may be found by premultiplying both sides of Equation B.48 by $[A]^{-1}$, namely:

$$[A]^{-1}[A]\{X\} = [A]^{-1}\{B\}$$
$$[I]\{X\} = [A]^{-1}\{B\} \qquad \textbf{(B.49)}$$
$$\{X\} = [A]^{-1}\{B\}$$

Therefore, if $[A]^{-1}$ is known or can be computed, the values of x can be determined from a simple matrix product.

As a numerical example, consider the following algebraic equations. The coefficients in these equations are the same as those shown in matrix [A] in Equation B.18, namely:

$$2x_1 + 3x_2 + 4x_3 = 100$$
$$4x_1 + 3x_2 + 6x_3 = 140 \qquad \textbf{(B.50)}$$
$$3x_1 + 5x_2 + 3x_3 = 110$$

These are written in matrix form as follows

$$[A] = \begin{bmatrix} 2 & 3 & 4 \\ 4 & 3 & 6 \\ 3 & 5 & 3 \end{bmatrix}$$

The cofactor matrix of [A] is

$$\begin{bmatrix} \begin{vmatrix} 3 & 6 \\ 5 & 3 \end{vmatrix} & -\begin{vmatrix} 4 & 6 \\ 3 & 3 \end{vmatrix} & \begin{vmatrix} 4 & 3 \\ 3 & 5 \end{vmatrix} \\[2ex] -\begin{vmatrix} 3 & 4 \\ 5 & 3 \end{vmatrix} & \begin{vmatrix} 2 & 4 \\ 3 & 3 \end{vmatrix} & -\begin{vmatrix} 2 & 3 \\ 3 & 5 \end{vmatrix} \\[2ex] \begin{vmatrix} 3 & 4 \\ 3 & 6 \end{vmatrix} & -\begin{vmatrix} 2 & 4 \\ 4 & 6 \end{vmatrix} & \begin{vmatrix} 2 & 3 \\ 4 & 3 \end{vmatrix} \end{bmatrix}$$

Evaluating the determinants with the adjoint matrix

$$\text{adj}[A] = \begin{bmatrix} -21 & 11 & -6 \\ 6 & -6 & 4 \\ 11 & -1 & -6 \end{bmatrix}$$

$$[A] = \begin{bmatrix} 2 & 3 & 4 \\ 4 & 3 & 6 \\ 3 & 5 & 3 \end{bmatrix}$$

$$|A| = (2)(-21) + (3)(6) + (4)(11) = 20$$

The inverse of [A] is

$$A^{-1} = \frac{1}{20}\begin{bmatrix} -21 & 11 & -6 \\ 6 & -6 & 4 \\ 11 & -1 & -6 \end{bmatrix}\begin{bmatrix} 100 \\ 140 \\ 110 \end{bmatrix}$$

And the values of the unknowns are

$$\begin{aligned}
x_1 &= \tfrac{1}{20}[(-21)(100) + (11)(140) + (6)(110)] = 5 \\
x_2 &= \tfrac{1}{20}[(6)(100) + (-6)(140) + (4)(110)] = 10 \\
x_3 &= \tfrac{1}{20}[(11)(100) + (-1)(140) + (-6)(110)] = 15
\end{aligned}$$
(B.51)

Many other techniques exist for solving algebraic equations simultaneously. The use of the matrix inverse is a special technique that may be used at the option of the analyst.

B.7 GAUSS'S METHOD FOR SOLVING SIMULTANEOUS EQUATIONS

One of the most widely used methods for solving linear, algebraic equations simultaneously is the Gauss method. This method, or some variation of it, is used in many of the currently available computer programs that deal with structural problems. Interestingly, the method is also well adapted to hand calculations. The fundamentals of this method are illustrated next.

Consider a set of three algebraic equations written in terms of three unknowns x_1, x_2, and x_3:

$$\begin{aligned}
3x_1 + 1x_2 - 1x_3 &= 2 \\
1x_1 + 4x_2 + 1x_3 &= 12 \\
2x_1 + 1x_2 + 2x_3 &= 10
\end{aligned}$$
(B.52)

To solve for the unknown terms, first divide each equation of the set by its leading coefficient, so that the leading coefficient becomes unity, namely:

$$\begin{aligned}
x_1 + \tfrac{1}{3}x_2 - \tfrac{1}{3}x_3 &= \tfrac{2}{3} \\
x_1 + 4x_2 + x_3 &= 12 \\
x_1 + \tfrac{1}{2}x_2 + x_3 &= 5
\end{aligned}$$
(B.53)

Next, subtract the first equation of the resultant set from each of the other equations so that the leading coefficients of the second and third equations are equal to zero, namely:

$$\begin{aligned}
x_1 + 0.3333x_2 - 0.3333x_3 &= 0.6667 \\
3.6667x_2 + 1.3333x_3 &= 11.3333 \\
0.1667x_2 + 1.3333x_3 &= 4.3333
\end{aligned}$$
(B.54)

Repeat the two operations described, but now start with the second equation; divide the second and third equations by their leading coefficient so that the leading coefficients become unity:

$$\begin{aligned}
x_1 + 0.3333x_2 - 0.3333x_3 &= 0.6667 \\
x_2 + 0.3636x_3 &= 3.0909 \\
x_2 + 7.9996x_3 &= 25.9946
\end{aligned}$$
(B.55)

Subtract the second equation from the third equation so that the leading coefficient of the third equation becomes zero:

$$x_1 + 0.3333x_2 - 0.3333x_3 = 0.6667$$
$$x_2 + 0.3636x_3 = 3.0909 \tag{B.56}$$
$$7.6360x_3 = 22.9037$$

Divide the third equation by its leading coefficient. This operation yields a solution for x_3:

$$x_3 = \frac{22.9037}{7.6360} = 2.9994 \tag{B.57}$$

Now substitute the derived value of x_3 into the latest form of the second equation and solve for x_2:

$$x_2 + 0.3636x_3 = x_2 + 0.3636(2.9994) = 3.0909$$
$$x_2 = 2.0003 \tag{B.58}$$

Finally, substitute the derived value of x_2 and x_3 into the first equation to solve for x_1:

$$x_1 + 0.3333x_2 - 0.3333x_3 = x_1 + 0.3333(2.0003) - 0.3333(2.9994) = 0.6667$$
$$x_1 = 0.9997 \tag{B.59}$$

The solution, then, obtained using Gauss's method is

$$\begin{Bmatrix} x_1 \\ x_2 \\ x_3 \end{Bmatrix} = \begin{Bmatrix} 0.9997 \\ 2.0003 \\ 2.9994 \end{Bmatrix} \tag{B.60}$$

This solution compares closely to the exact solution, which is

$$\begin{Bmatrix} x_1 \\ x_2 \\ x_3 \end{Bmatrix} = \begin{Bmatrix} 1 \\ 2 \\ 3 \end{Bmatrix} \tag{B.61}$$

The inexactness of the solution shown is a function of round-off errors. Improved accuracy is attained by carrying a larger number of significant figures in the solution.

B.8 SPECIAL TOPICS

Matrix Partitioning

The manipulation of matrix equations is frequently made simpler by dividing the matrices into smaller matrices, called *partitions*. Partitioning is indicated in this book by horizontal and vertical lines between the rows and the columns of the matrices. Illustrations of matrices that have been partitioned are as follows:

$$[A] = \left[\begin{array}{cc|c} a_{1,1} & a_{1,2} & a_{1,3} \\ a_{2,1} & a_{2,2} & a_{2,3} \\ \hline a_{3,1} & a_{3,2} & a_{3,3} \end{array} \right]$$

$$\lfloor A \rfloor = \lfloor a_1 \; a_2 \mid a_3 \rfloor \tag{B.62}$$

$$\{A\} = \begin{Bmatrix} a_1 \\ a_2 \\ \hline a_3 \end{Bmatrix}$$

Partitioning of a matrix equation is

$$[A]\{X\} = \{B\}$$

$$\left[\begin{array}{cc|c} a_{1,1} & a_{1,2} & a_{1,3} \\ a_{2,1} & a_{2,2} & a_{2,3} \\ \hline a_{3,1} & a_{3,2} & a_{3,3} \end{array}\right] \left\{\begin{array}{c} x_1 \\ x_2 \\ x_3 \end{array}\right\} = \left\{\begin{array}{c} b_1 \\ b_2 \\ b_3 \end{array}\right\} \tag{B.63}$$

Note that the horizontal partition lines in Equation B.63 extend between the same two rows for each matrix in the equation. Furthermore, the column numbers that define the vertical partition line for the square matrix are the same as the row numbers that define the horizontal partition lines for the complete equation. Thus, if the horizontal partition lines run between the second and third rows, then the vertical partition line runs between the second and third column of the square matrix.

As an illustration of the use of partitioning, consider the same matrix equation as was described in Equation B.63:

$$\left[\begin{array}{cc|c} 1 & 2 & 3 \\ 2 & 3 & 4 \\ \hline 1 & 5 & 3 \end{array}\right] \left\{\begin{array}{c} x_1 \\ x_2 \\ x_3 \end{array}\right\} = \left\{\begin{array}{c} 13 \\ 19 \\ 22 \end{array}\right\} \tag{B.64}$$

In Equation B.64, partition lines have been drawn between the second and third rows of each matrix, and between the second and third columns of matrix [A]. When the matrix is partitioned, the partitions, are generally still are matrices and are manipulated as matrices. In this example, though, one of the partitions is a 1×1 matrix, which can be treated as a scalar. The original matrix equation may now be written as two matrix equations, namely:

$$\left[\begin{array}{cc} 1 & 2 \\ 2 & 3 \end{array}\right] \left\{\begin{array}{c} x_1 \\ x_2 \end{array}\right\} + \left\{\begin{array}{c} 3 \\ 4 \end{array}\right\} x_3 = \left\{\begin{array}{c} 13 \\ 19 \end{array}\right\} \tag{B.65}$$

and

$$\lfloor 1 \quad 5 \rfloor \left\{\begin{array}{c} x_1 \\ x_2 \end{array}\right\} + 3x_3 = 22 \tag{B.66}$$

Equation B.66 may be solved for x_3 in terms of x_1 and x_2 as

$$3x_3 = 22 - \lfloor 1 \quad 5 \rfloor \left\{\begin{array}{c} x_1 \\ x_2 \end{array}\right\}$$

$$x_3 = \frac{22}{3} - \frac{1}{3} \lfloor 1 \quad 5 \rfloor \left\{\begin{array}{c} x_1 \\ x_2 \end{array}\right\} \tag{B.67}$$

The value of x_3 from Equation B.67 is substituted into Equation B.65, and the resultant equation is solved for x_1 and x_2:

$$\left[\begin{array}{cc} 1 & 2 \\ 2 & 3 \end{array}\right] \left\{\begin{array}{c} x_1 \\ x_2 \end{array}\right\} + \left\{\begin{array}{c} 3 \\ 4 \end{array}\right\} \left(\frac{22}{3} - \frac{1}{3} \lfloor 1 \quad 5 \rfloor \left\{\begin{array}{c} x_1 \\ x_2 \end{array}\right\}\right) = \left\{\begin{array}{c} 13 \\ 19 \end{array}\right\}$$

$$\left[\begin{array}{cc} 1 & 2 \\ 2 & 3 \end{array}\right] \left\{\begin{array}{c} x_1 \\ x_2 \end{array}\right\} - \frac{1}{3} \left[\begin{array}{cc} 3 & 15 \\ 4 & 20 \end{array}\right] \left\{\begin{array}{c} x_1 \\ x_2 \end{array}\right\} = \left\{\begin{array}{c} 13 \\ 19 \end{array}\right\} - \frac{1}{3} \left\{\begin{array}{c} 66 \\ 88 \end{array}\right\} \tag{B.68}$$

$$\frac{1}{3} \left[\begin{array}{cc} 0 & -9 \\ 2 & -11 \end{array}\right] \left\{\begin{array}{c} x_1 \\ x_2 \end{array}\right\} = \frac{1}{3} \left\{\begin{array}{c} -27 \\ -31 \end{array}\right\}$$

$$\left\{\begin{array}{c} x_1 \\ x_2 \end{array}\right\} = \left[\begin{array}{cc} 0 & -9 \\ 2 & -11 \end{array}\right]^{-1} \left\{\begin{array}{c} -27 \\ -31 \end{array}\right\} = \left\{\begin{array}{c} 1 \\ 3 \end{array}\right\}$$

The values of x_1 and x_2 from Equation B.68 may now be substituted in Equation B.67 to obtain the value of x_3, namely:

$$x_3 = \frac{22}{3} - \frac{1}{3} \lfloor 1 \quad 5 \rfloor \left\{ \begin{array}{c} 1 \\ 3 \end{array} \right\} = \frac{1}{3}(22 - 16) = 2 \qquad \textbf{(B.69)}$$

One obvious advantage of partitioning is that the order of the matrices for which inverses must be found is reduced. However, this advantage is offset somewhat by the fact that additional algebraic manipulations are required when using partitioning schemes. Nonetheless, partitioning of matrix equations is used extensively in developing computer solutions for structural problems.

Differentiating and Integrating a Matrix

Previously in this appendix a matrix was defined as an ordered arrangement of numbers in rows and columns. Although the elements of a matrix generally are thought of as constants, they may also be variables, as shown by the following example:

$$[A] = \begin{bmatrix} 3x & -x^2 & 2x^4 \\ -x^2 & 5x^3 & 7x \\ 2x^4 & 7x & 2x^2 \end{bmatrix} \qquad \textbf{(B.70)}$$

Differentiating matrix [A] with respect to x is performed by differentiating each element in the matrix with respect to x. The result for the example is shown as follows:

$$\frac{d}{dx}[A] = \begin{bmatrix} 3 & -2x & 8x^3 \\ -2x & 15x^2 & 7 \\ 8x^3 & 7 & 4x \end{bmatrix} \qquad \textbf{(B.71)}$$

The elements of [A] are functions of more than one variable, say, x and y. Partial differentiation of matrix [A] with respect to either x or y is performed similarly, by partially differentiating each element of the matrix with respect to that variable.

In a similar way, integration of a matrix is performed by integrating each element of the matrix. Thus, for the matrix [A] given by Equation (1), integration produces the following results:

$$\int_a^b [A]\, dx = \begin{bmatrix} \dfrac{3x^2}{2} & \dfrac{-x^3}{3} & \dfrac{2x^5}{5} \\[2mm] \dfrac{-x^3}{3} & \dfrac{5x^4}{4} & \dfrac{7x^2}{2} \\[2mm] \dfrac{2x^5}{5} & \dfrac{7x^2}{2} & \dfrac{2x^3}{3} \end{bmatrix}_a^b$$

$$= \begin{bmatrix} \dfrac{3(b^2 - a^2)}{2} & \dfrac{-(b^3 - a^3)}{3} & \dfrac{2(b^5 - a^5)}{5} \\[2mm] \dfrac{-(b^3 - a^3)}{3} & \dfrac{5(b^4 - a^4)}{4} & \dfrac{7(b^2 - a^2)}{2} \\[2mm] \dfrac{2(b^5 - a^5)}{5} & \dfrac{7(b^2 - a^2)}{2} & \dfrac{2(b^3 - a^3)}{3} \end{bmatrix} \qquad \textbf{(B.72)}$$

A special application of matrix differentiation is worth noting because it frequently appears in the development of structural theory. Assume that a scalar variable U is defined in terms of a matrix triple product:

$$U = \tfrac{1}{2} \lfloor x \rfloor [A] \{x\} \tag{B.73}$$

where $\{x\}$ is a column matrix consisting of n variables $(x_1, x_2, \ldots, x_n)$. The square matrix [A] is a symmetric matrix. If U is differentiated successively with respect to $x_1, x_2, \ldots, x_n$, and the results arranged in a column matrix, the result is remarkably simple, namely:

$$\begin{Bmatrix} \dfrac{\partial U}{\partial x_1} \\[2mm] \dfrac{\partial U}{\partial x_2} \\[2mm] \vdots \\[2mm] \dfrac{\partial U}{\partial x_n} \end{Bmatrix} = [A]\{x\} \tag{B.74}$$

A further differentiation yields:

$$\frac{\partial^2 U}{\partial x_i \partial x_j} = a_{i,j} \tag{B.75}$$

where $a_{i,j}$ are the elements of the original matrix [A]. Although the proof of Equation B.74 is not given here, a simple example will verify their correctness. Consider the following matrix triple product:

$$U = \tfrac{1}{2} \lfloor x_1 \quad x_2 \rfloor \begin{bmatrix} 2 & 4 \\ 4 & 3 \end{bmatrix} \begin{Bmatrix} x_1 \\ x_2 \end{Bmatrix} \tag{B.76}$$

When U is expanded the result is

$$U = \tfrac{1}{2}(2x_1^2 + 8x_1 x_2 + 3x_2^2) \tag{B.77}$$

Differentiating U with respect to x_1 and with respect to x_2 yields

$$\frac{\partial U}{\partial x_1} = 2x_1 + 4x_2$$
$$\frac{\partial U}{\partial x_2} = 4x_1 + 3x_2 \tag{B.78}$$

Arrangement of these results in matrix form produces

$$\begin{Bmatrix} \dfrac{\partial U}{\partial x_1} \\[2mm] \dfrac{\partial U}{\partial x_2} \end{Bmatrix} = \begin{bmatrix} 2 & 4 \\ 4 & 3 \end{bmatrix} \begin{Bmatrix} x_1 \\ x_2 \end{Bmatrix} \tag{B.79}$$

This result corresponds to Equation B.74. Further differentiation produces

$$\frac{\partial^2 U}{\partial x_1^2} = 2 \qquad \frac{\partial^2 U}{\partial x_1 x_2} = 4$$
$$\frac{\partial^2 U}{\partial x_2 x_1} = 4 \qquad \frac{\partial^2 U}{\partial x_2^2} = 3 \tag{B.80}$$

These results agree with Equation B.75.

The resultant form of the matrix triple product given by Equation B.73 is sometimes called a *quadratic form* (because of the second-order appearance of the variables in the matrix product). Quadratic forms occur frequently in structural theory when strain energy is used to help derive the stiffness matrix. Although quadratic forms have not been used in this book, the topic is important in more advanced treatment of structural theory and is included here for future reference for interested readers.

Wind, Seismic and Snow Load Tables and Figures

TABLE C.I CLASSIFICATION OF BUILDINGS AND OTHER STRUCTURES FOR FLOOD, WIND, SNOW, EARTHQUAKE, AND ICE LOADS

Nature of Occupancy	Category
Buildings and other structures that represent a low hazard to human life in the event of failure including, but not limited to: • Agricultural facilities • Certain temporary facilities • Minor storage facilities	I
All buildings and other structures except those listed in Categories I, III, and IV	II
Buildings and other structures that represent a substantial hazard to human life in the event of failure including, but not limited to: • Buildings and other structures where more than 300 people congregate in one area • Buildings and other structures with day care facilities with capacity greater than 150 • Buildings and other structures with elementary school or secondary school facilities with capacity greater than 250 • Buildings and other structures with a capacity greater than 500 for colleges or adult education facilities • Health care facilities with a capacity of 50 or more resident patients but not having surgery or emergency treatment facilities • Jails and detention facilities • Power generating stations and other public utility facilities not included in Category IV Buildings and other structures not included in Category IV (including, but not limited to, facilities that manufacture, process, handle, store, use, or dispose of such substances as hazardous fuels, hazardous chemicals, hazardous waste, or explosives) containing sufficient quantities of hazardous materials to be dangerous to the public if released. Buildings and other structures containing hazardous materials shall be eligible for classification as Category II structures if it can be demonstrated to the satisfaction of the authority having jurisdiction by a hazard assessment as described in Section 1.5.2 that a release of the hazardous material does not pose a threat to the public.	III
Buildings and other structures designated as essential facilities including, but not limited to: • Hospitals and other health care facilities having surgery or emergency treatment facilities • Fire, rescue, ambulance, and police stations and emergency vehicle garages • Designated earthquake, hurricane, or other emergency shelters • Designated emergency preparedness, communication, and operation centers and other facilities required for emergency response • Power generating stations and other public utility facilities required in an emergency • Ancillary structures (including, but not limited to, communication towers, fuel storage tanks, cooling towers, electrical substation structures, fire water storage tanks or other structures housing or supporting water, or other fire-suppression material or equipment) required for operation of Category IV structures during an emergency • Aviation control towers, air traffic control centers, and emergency aircraft hangars • Water storage facilities and pump structures required to maintain water pressure for fire suppression • Buildings and other structures having critical national defense functions Buildings and other structures (including, but not limited to, facilities that manufacture, process, handle, store, use, or dispose of such substances as hazardous fuels, hazardous chemicals, hazardous waste, or explosives) containing extremely hazardous materials where the quantity of the material exceeds a threshold quantity established by the authority having jurisdiction. Buildings and other structures containing extremely hazardous materials shall be eligible for classification as Category II structures if it can be demonstrated to the satisfaction of the authority having jurisdiction by a hazard assessment as described in Section 1.5.2 that a release of the extremely hazardous material does not pose a threat to the public. This reduced classification shall not be permitted if the buildings or other structures also function as essential facilities.	IV

Source: American Society of Civil Engineers, <u>Minimum Design Loads for Buildings and Other Structures</u>, ASCE 7-02, 2002. Reproduced with permission of the American Society of Civil Engineers.

TABLE C.2   SURFACE ROUGHNESS CATEGORIES

Surface Roughness B: Urban and suburban areas wooded areas or other terrain with numerous closely spaced obstructions having the size of single-family dwellings or larger.

Surface Roughness C: Open terrain with scattered obstructions having heights generally less than 30 ft (9.1 m). This category includes flat open country, grasslands, and all water surfaces in hurricane-prone regions.

Surface Roughness D: Flat, unobstructed areas and water surfaces outside hurricane-prone regions. This category includes smooth mud flats, salt flats, and unbroken ice.

Note: Formerly there was a Category A for heavily built up urban areas but it has been removed by the ASCE.

TABLE C.3   ADJUSTMENT FACTOR FOR BUILDING HEIGHT AND EXPOSURE, λ

Mean roof height (ft)	Exposure		
	B	C	D
15	1.00	1.21	1.47
20	1.00	1.29	1.55
25	1.00	1.35	1.61
30	1.00	1.40	1.66
35	1.05	1.45	1.70
40	1.09	1.49	1.74
45	1.12	1.53	1.78
50	1.16	1.56	1.81
55	1.19	1.59	1.84
60	1.22	1.62	1.87

Unit Conversions — 1.0 ft = 0.3048 m.

TABLE C.4   IMPORTANCE FACTOR, I

Category	Non-Hurricane Prone Regions and Hurricane Prone Regions with V = 85–100 mph and Alaska	Hurricane Prone Regions with V > 100 mph
I	0.87	0.77
II	1.00	1.00
III	1.15	1.15
IV	1.15	1.15

Note: The building and structure classification categories are listed in Table C.1

Source: American Society of Civil Engineers, <u>Minimum Design Loads for Buildings and Other Structures</u>, ASCE 7-02, 2002.
Reproduced with permission of the American Society of Civil Engineers.

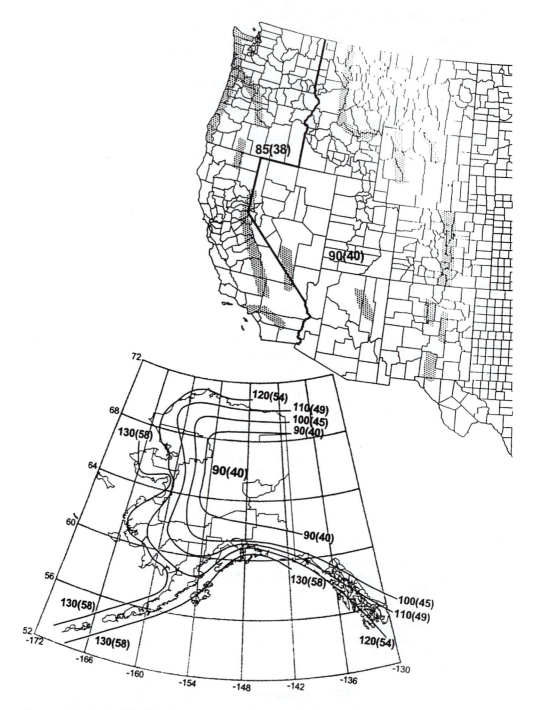

Figure C.1 Basic Wind Speed

Source: American Society of Civil Engineers, <u>Minimum Design Loads for Buildings and Other Structures,</u> ASCE 7-02, 2002.
Reproduced with permission of the American Society of Civil Engineers.

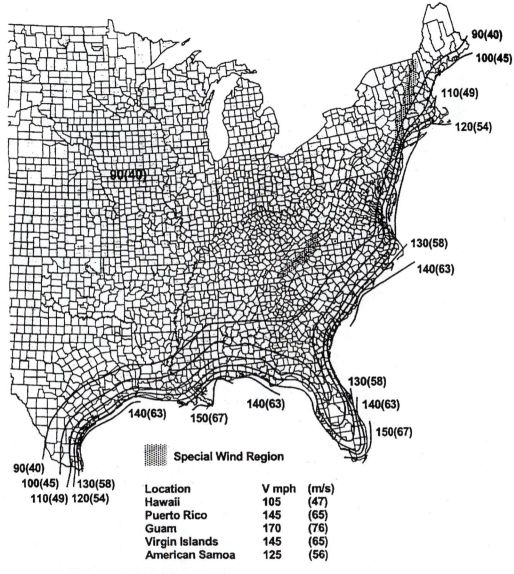

Location	V mph	(m/s)
Hawaii	105	(47)
Puerto Rico	145	(65)
Guam	170	(76)
Virgin Islands	145	(65)
American Samoa	125	(56)

Notes:
1. Values are nominal design 3-second gust wind speeds in miles per hour (m/s) at 33 ft (10 m) above ground for Exposure C category.
2. Linear interpolation between wind contours is permitted.
3. Islands and coastal areas outside the last contour shall use the last wind speed contour of the coastal area.
4. Mountainous terrain, gorges, ocean promontories, and special wind regions shall be examined for unusual wind conditions.

Figure C.1   *(Continued)*

Source: American Society of Civil Engineers, <u>Minimum Design Loads for Buildings and Other Structures</u>, ASCE 7-02, 2002. Reproduced with permission of the American Society of Civil Engineers.

Figure C.2 Main Wind Force Resisting System — Method 1

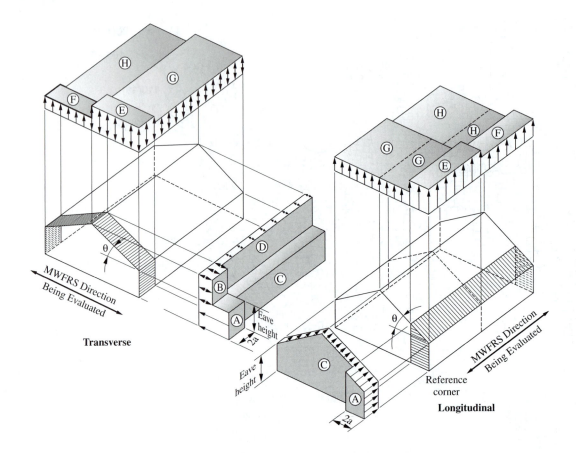

Simplified Design Wind Pressure, P$_{S30}$ (psf) (*Exposure B h = 30 ft. with I = 1.0*)

Basic Wind Speed (mph)	Roof Angle (degrees)	Load Case	Zones									
			Horizontal Pressures				Vertical Pressures				Overhangs	
			A	B	C	D	E	F	G	H	EOH	GOH
85	0 to 5°	1	11.5	−5.9	7.6	−3.5	−13.8	−7.8	−9.6	−6.1	−19.3	−15.1
	10°	1	12.9	−5.4	8.6	−3.1	−13.8	−8.4	−9.6	−6.5	−19.3	−15.1
	15°	1	14.4	−4.8	9.6	−2.7	−13.8	−9.0	−9.6	−6.9	−19.3	−15.1
	20°	1	15.9	−4.2	10.6	−2.3	−13.8	−9.6	−9.6	−7.3	−19.3	−15.1
	25°	1	14.4	2.3	10.4	2.4	−6.4	−8.7	−4.6	−7.0	−11.9	−10.1
		2	—	—	—	—	−2.4	−4.7	−0.7	−3.0	—	—
	30 to 45	1	12.9	8.8	10.2	7.0	1.0	−7.8	0.3	−6.7	−4.5	−5.2
		2	12.9	8.8	10.2	7.0	5.0	−3.9	4.3	−2.8	−4.5	−5.2
90	0 to 5°	1	12.8	−6.7	8.5	−4.0	−15.4	−8.8	−10.7	−6.8	−21.6	−16.9
	10°	1	14.5	−6.0	9.6	−3.5	−15.4	−9.4	−10.7	−7.2	−21.6	−16.9
	15°	1	16.1	−5.4	10.7	−3.0	−15.4	−10.1	−10.7	−7.7	−21.6	−16.9
	20°	1	17.8	−4.7	11.9	−2.6	−15.4	−10.7	−10.7	−8.1	−21.6	−16.9
	25°	1	16.1	2.6	11.7	2.7	−7.2	−9.8	−5.2	−7.8	−13.3	−11.4
		2	—	—	—	—	−2.7	−5.3	−0.7	−3.4	—	—
	30 to 45	1	14.4	9.9	11.5	7.9	1.1	−8.8	0.4	−7.5	−5.1	−5.8
		2	14.4	9.9	11.5	7.9	5.6	−4.3	4.8	−3.1	−5.1	−5.8

Figure C.2   (*Continued*)

Basic Wind Speed (mph)	Roof Angle (degrees)	Load Case	Zones									
			Horizontal Pressures				Vertical Pressures				Overhangs	
			A	B	C	D	E	F	G	H	EOH	GOH
100	0 to 5°	1	15.9	−8.2	10.5	−4.9	−19.1	−10.8	−13.3	−8.4	−26.7	−20.9
	10°	1	17.9	−7.4	11.9	−4.3	−19.1	−11.6	−13.3	−8.9	−26.7	−20.9
	15°	1	19.9	−6.6	13.3	−3.8	−19.1	−12.4	−13.3	−9.5	−26.7	−20.9
	20°	1	22.0	−5.8	14.6	−3.2	−19.1	−13.3	−13.3	−10.1	−26.7	−20.9
	25°	1	19.9	3.2	14.4	3.3	−8.8	−12.0	−6.4	−9.7	−16.5	−14.0
		2	—	—	—	—	−3.4	−6.6	−0.9	−4.2	—	—
	30 to 45	1	17.8	12.2	14.2	9.8	1.4	−10.8	0.5	−9.3	−6.3	−7.2
		2	17.8	12.2	14.2	9.8	6.9	−5.3	5.9	−3.8	−6.3	−7.2
110	0 to 5°	1	19.2	−10.0	12.7	−5.9	−23.1	−13.1	−16.0	−10.1	−32.3	−25.3
	10°	1	21.6	−9.0	14.4	−5.2	−23.1	−14.1	−16.0	−10.8	−32.3	−25.3
	15°	1	24.1	−8.0	16.0	−4.6	−23.1	−15.1	−16.0	−11.5	−32.3	−25.3
	20°	1	26.6	−7.0	17.7	−3.9	−23.1	−16.0	−16.0	−12.2	−32.3	−25.3
	25°	1	24.1	3.9	17.4	4.0	−10.7	−14.6	−7.7	−11.7	−19.9	−17.0
		2	—	—	—	—	−4.1	−7.9	−1.1	−5.1	—	—
	30 to 45	1	21.6	14.8	17.2	11.8	1.7	−13.1	0.6	−11.3	−7.6	−8.7
		2	21.6	14.8	17.2	11.8	8.3	−6.5	7.2	−4.6	−7.6	−8.7
120	0 to 5°	1	22.8	−11.9	15.1	−7.0	−27.4	−15.6	−19.1	−12.1	−38.4	−30.1
	10°	1	25.8	−10.7	17.1	−6.2	−27.4	−16.8	−19.1	−12.9	−38.4	−30.1
	15°	1	28.7	−9.5	19.1	−5.4	−27.4	−17.9	−19.1	−13.7	−38.4	−30.1
	20°	1	31.6	−8.3	21.1	−4.6	−27.4	−19.1	−19.1	−14.5	−38.4	−30.1
	25°	1	28.6	4.6	20.7	4.7	−12.7	−17.3	−9.2	−13.9	−23.7	−20.2
		2	—	—	—	—	−4.8	−9.4	−1.3	−6.0	—	—
	30 to 45	1	25.7	17.6	20.4	14.0	2.0	−15.6	0.7	−13.4	−9.0	−10.3
		2	25.7	17.6	20.4	14.0	9.9	−7.7	8.6	−5.5	−9.0	−10.3
130	0 to 5°	1	26.8	−13.9	17.8	−8.2	−32.2	−18.3	−22.4	−14.2	−45.1	−35.3
	10°	1	30.2	−12.5	20.1	−7.3	−32.2	−19.7	−22.4	−15.1	−45.1	−35.3
	15°	1	33.7	−11.2	22.4	−6.4	−32.2	−21.0	−22.4	−16.1	−45.1	−35.3
	20°	1	37.1	−9.8	24.7	−5.4	−32.2	−22.4	−22.4	−17.0	−45.1	−35.3
	25°	1	33.6	5.4	24.3	5.5	−14.9	−20.4	−10.8	−16.4	−27.8	−23.7
		2	—	—	—	—	−5.7	−11.1	−1.5	−7.1	—	—
	30 to 45	1	30.1	20.6	24.0	16.5	2.3	−18.3	0.8	−15.7	−10.6	−12.1
		2	30.1	20.6	24.0	16.5	11.6	−9.0	10.0	−6.4	−10.6	−12.1
140	0 to 5°	1	31.1	−16.1	20.6	−9.6	−37.3	−21.2	−26.0	−16.4	−52.3	−40.9
	10°	1	35.1	−14.5	23.3	−8.5	−37.3	−22.8	−26.0	−17.5	−52.3	−40.9
	15°	1	39.0	−12.9	26.0	−7.4	−37.3	−24.4	−26.0	−18.6	−52.3	−40.9
	20°	1	43.0	−11.4	28.7	−6.3	−37.3	−26.0	−26.0	−19.7	−52.3	−40.9
	25°	1	39.0	6.3	28.2	6.4	−17.3	−23.6	−12.5	−19.0	−32.3	−27.5
		2	—	—	—	—	−6.6	−12.8	−1.8	−8.2	—	—
	30 to 45	1	35.0	23.9	27.8	19.1	2.7	−21.2	0.9	−18.2	−12.3	−14.0
		2	35.0	23.9	27.8	19.1	13.4	−10.5	11.7	−7.5	−12.3	−14.0
150	0 to 5°	1	35.7	−18.5	23.7	−11.0	−42.9	−24.4	−29.8	−18.9	−60.0	−47.0
	10°	1	40.2	−16.7	26.8	−9.7	−42.9	−26.2	−29.8	−20.1	−60.0	−47.0
	15°	1	44.8	−14.9	29.8	−8.5	−42.9	−28.0	−29.8	−21.4	−60.0	−47.0
	20°	1	49.4	−13.0	32.9	−7.2	−42.9	−29.8	−29.8	−22.6	−60.0	−47.0
	25°	1	44.8	7.2	32.4	7.4	−19.9	−27.1	−14.4	−21.8	−37.0	−31.6
		2	—	—	—	—	−7.5	−14.7	−2.1	−9.4	—	—
	30 to 45	1	40.1	27.4	31.9	22.0	3.1	−24.4	1.0	−20.9	−14.1	−16.1
		2	40.1	27.4	31.9	22.0	15.4	−12.0	13.4	−8.6	−14.1	−16.1
170	0 to 5°	1	45.8	−23.8	30.4	−14.1	−55.1	−31.3	−38.3	−24.2	−77.1	−60.4
	10°	1	51.7	−21.4	34.4	−12.5	−55.1	−33.6	−38.3	−25.8	−77.1	−60.4
	15°	1	57.6	−19.1	38.3	−10.9	−55.1	−36.0	−38.3	−27.5	−77.1	−60.4
	20°	1	63.4	−16.7	42.3	−9.3	−55.1	−38.3	−38.3	−29.1	−77.1	−60.4
	25°	1	57.5	9.3	41.6	9.5	−25.6	−34.8	−18.5	−28.0	−47.6	−40.5
		2	—	—	—	—	−9.7	−18.9	−2.6	−12.1	—	—
	30 to 45	1	51.5	35.2	41.0	28.2	4.0	−31.3	1.3	−26.9	−18.1	−20.7
		2	51.5	35.2	41.0	28.2	19.8	−15.4	17.2	−11.0	−18.1	−20.7

Unit Conversions 1.0 psf = 0.0479 kN/m²

Source: American Society of Civil Engineers, <u>Minimum Design Loads for Buildings and Other Structures</u>, ASCE 7-02, 2002.
Reproduced with permission of the American Society of Civil Engineers.

Notes:

1. Pressures shown are applied to the horizontal and vertical projections, for exposure B, at $h = 30$ ft (9.1m), for $1 = 1.0$. Adjust to other exposures and heights with adjustment factor λ.

2. The load patterns shown shall be applied to each corner of the building in turn as the reference corner. (See Figure 6-10)

3. For the design of the longitudinal MWFRS use $\theta = 0°$, and locate the zone E/F, G/H boundary at the mid-length of the building.

4. Load cases 1 and 2 must be checked for $25° < \theta \leq 45°$. Load case 2 at 25° is provided only for interpolation between 25° to 30°.

5. Plus and minus signs signify pressures acting toward and away from the projected surfaces, respectively.

6. For roof slopes other than those shown, linear interpolation is permitted.

7. The total horizontal load shall not be less than that determined by assuming $p_S = 0$ in zones B & D.

8. The zone pressures represent the following:

   Horizontal pressure zones – Sum of the windward and leeward net (sum of internal and external) pressures on vertical projection of:

A - End zone of wall	C - Interior zone of wall
B - End zone of roof	D - Interior zone of roof

   Vertical zones – Net (sum of internal and external) pressures on horizontal projection of:

E - End zone of windward roof	G - Interior zone of windward roof
F - End zone of leeward roof	H - Interior zone of leeward roof

9. Where zone E or G falls on a roof overhang on the windward side of the building, use E_{OH} and G_{OH} for the pressure on the horizontal projection of the overhang. Overhangs on the leeward and side edges shall have the basic zone pressure applied.

10. Notation:

    a: 10 percent of least horizontal dimension or 0.4h, whichever is smaller, but not less than either 4% of least horizontal dimension or 3 ft (0.9 m).

    h: Mean roof height, in feet (meters), except that eave height shall be used for roof angles $<10°$.

    θ: Angle of plane of roof from horizontal, in degrees.

Source: American Society of Civil Engineers, <u>Minimum Design Loads for Buildings and Other Structures</u>, ASCE 7-02, 2002. Reproduced with permission of the American Society of Civil Engineers.

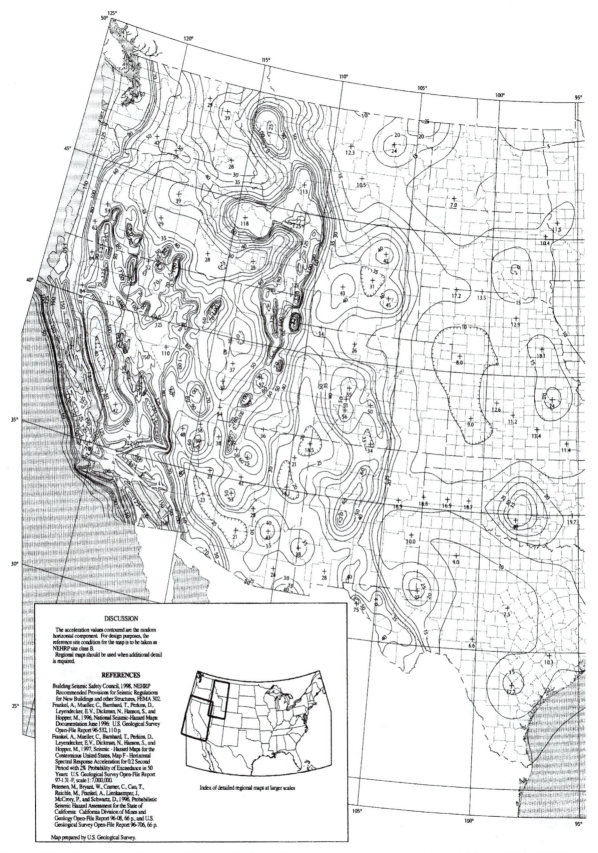

Figure C.3 Maximum considered earthquake ground motion for conterminous united states, of 0.2 s spectral response acceleration (5% of critical damping), site class B

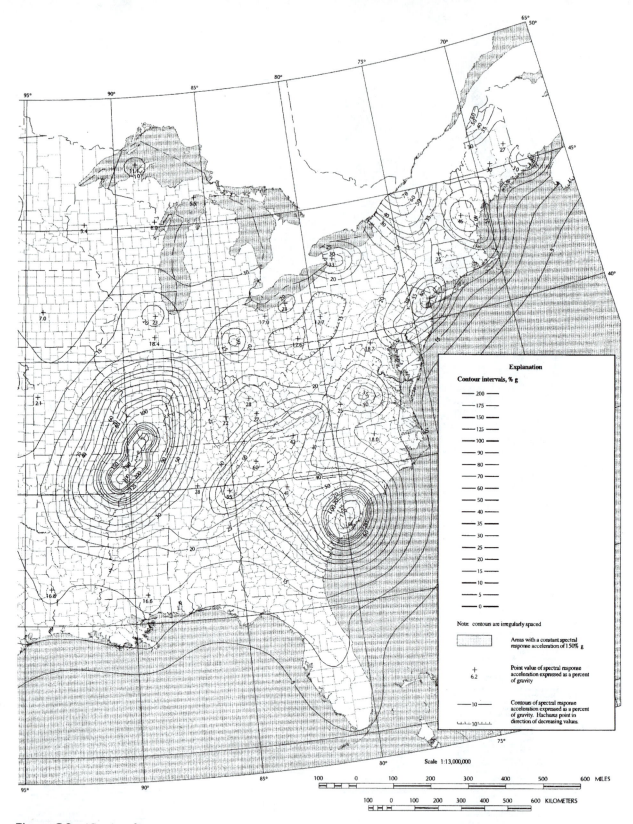

Figure C.3 *(Continued)*

Source: American Society of Civil Engineers, <u>Minimum Design Loads for Buildings and Other Structures</u>, ASCE 7-02, 2002. Reproduced with permission of the American Society of Civil Engineers.

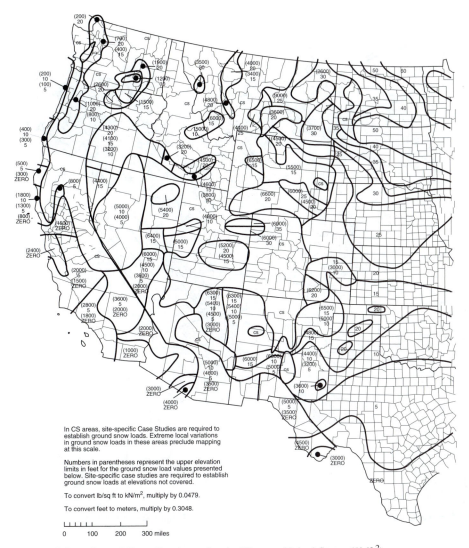

Figure C.4a   Ground Snow Loads, p_g for the Western United States (1b/ft^2)

Source: American Society of Civil Engineers, <u>Minimum Design Loads for Buildings and Other Structures</u>, ASCE 7-02, 2002.
Reproduced with permission of the American Society of Civil Engineers.

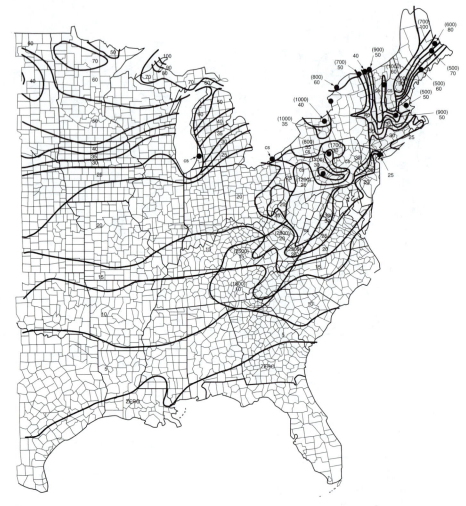

Figure C.4b Ground Snow Loads, p_g, for the Eastern United States (1b/ft^2).

Source: American Society of Civil Engineers, <u>Minimum Design Loads for Buildings and Other Structures</u>, ASCE 7-02, 2002. Reproduced with permission of the American Society of Civil Engineers.

Computer Analysis of Various Structures with SAP2000

D.I INTRODUCTION

As previously mentioned in Chapter 1, a student version of the widely used commercial program for structures, SAP2000, is available on the book's Web site. It is one of the general types of computer programs that the student will use if he or she works for a structural design firm after graduation.

In this appendix, six different types of structures are analyzed with the SAP2000 software. The author presents very little description of the data input for the problems. He assumes that if you read the HELP columns for each of the problem types, you will easily understand how to make the analyses. In addition, the user should note that the complete input to the computer for each of the examples is given in the example section of SAP2000 on the Web site.

D.2 ANALYSIS OF PLANE TRUSSES

In this section the author presents the analysis of a plane truss with Example D.1. Several remarks are presented in the Solutions part of the problem that should be helpful to the student in making the analysis.

EXAMPLE D.1

Using SAP2000, determine the forces in the members of the truss shown in Figure D.1. This same truss was previously analyzed in Examples 6.2 by joints and in Example 7.5 by sections and joints. For this analysis the units used are kips and inches. All members have cross sectional areas of $4\,\text{in.}^2$ and moduli of elasticity of 29,000 ksi. The moment of inertias are constant and can be taken as 1.

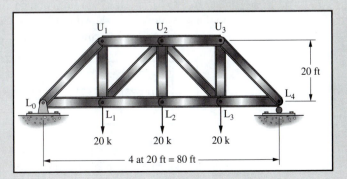

Figure D.1

Solution. To begin the analysis the user must associate a number with each of the members and nodes in the truss. The numbering scheme used in this analysis is shown in Figure D.2. These numbers are used to identify the joints and members when entering data into the computer and interpreting results of the analysis.

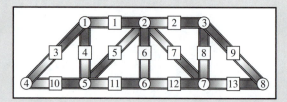

Figure D.2

As rotations are not considered in truss analysis, the rotational degree of freedom at each joint is restrained. Further, the X and Y translational degrees of freedom at joint 4 and the Y translational degree of freedom at joint 8 are restrained since these are the structural supports. A Pin-Pin member is assumed to be pin-connected at both ends. Results obtained from the computer for the first four members are shown in the table below.

Beam	Case	End	Axial	Shear Y	Moment Z
1	1	i	30.00	0.0	0.0
		j	−30.00	0.0	0.0
2	1	i	30.00	0.0	0.0
		j	−30.00	0.0	0.0
3	1	i	42.43	0.0	0.0
		j	−42.43	0.0	0.0
4	1	i	−30.00	0.0	0.0
		j	30.00	0.0	0.0

Observe that the shearing force and bending moment at each end of every member are equal to zero. This is as expected and is consistent with our assumption about behavior. Also observe from the calculated results that the axial force at one end of each member is positive whereas the axial force at the other end is negative. To determine whether a member is in axial tension or axial compression we must refer to the local coordinate system specified for the member. From the HELP information in SAP2000, we can see that the local coordinate system is as shown in Figure D.3. These are the directions of positive forces at each end of a member.

Figure D.3

From this sketch of the local coordinate system, it should be obvious that a positive axial force at the left end of the member indicates axial compression whereas a positive axial force at the right end indicates axial tension. The author uses the indicated result at the right end of the member, the j end, to interpret the magnitude and sense (tension or compression) of the axial forces in the members.   ∎

D.3 ANALYSIS OF SPACE TRUSSES

The analysis of space trusses with SAP2000 is handled very much in the same manner as was the plane truss of Example D.1. Example D.2 illustrates this fact.

EXAMPLE D.2

Analyze the space truss shown in Figure D.4 using SAP2000.

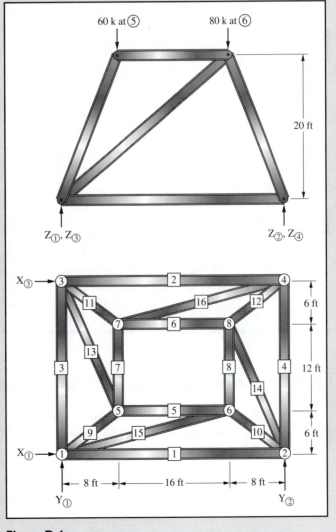

Figure D.4

Solution.   To analyze this truss using SAP2000, we must first establish the structural geometry. To do so, establish a grid that has four 8-ft spaces in the x directions and four 6-ft spaces in the y direction. There is one space in the z direction that is 20-ft high. Next, the element connectivity is designated as it was in Example D.1. The joint and frame member geometry is shown in the figure. Once again the complete input data file is contained in the example folder with the program as given on the book's Web site.

To analyze this truss, two modeling considerations need to be discussed. First, because truss members are assumed to only carry an axial force, each of the members in this structure needs to have the ends released against rotation in each direction. The members also need to have torsion released at the beginning of the member. These releases will result in a stable member that can only carry axial force.

Secondly, the three rotational degrees of freedom at each of the joints need to be restrained. Unless these are restrained, the computer will think that the structure is unstable, which it is not. Restraining these degrees of freedom has no affect on the forces in the members since rotational forces cannot be transferred to the members because of the manner in which we have released the ends. The forces in each of the members are shown in the table that follows.

Member	Axial Force (k)	Member	Axial Force (k)
1	22.00	9	−50.31
2	8.00	10	−83.85
3	13.50	11	0.00
4	6.00	12	−22.36
5	−24.00	13	−21.05
6	0.00	14	28.07
7	0.00	15	−7.95
8	−24.00	16	0.00

The user should display the deformed shape on the computer screen and then place it into motion to develop an understanding of how the structure deforms under load. This is accomplished by selecting the deformed shape from the display menu and clicking the animate button in the lower right-hand corner of the screen. ■

D.4 ANALYSIS OF STATICALLY INDETERMINATE PLANE TRUSSES

Statically indeterminate plane trusses are handled with the SAP2000 program just as are the plane statically determinate trusses. Example D.3 presents the analysis of a plane truss that is statically indeterminate internally.

EXAMPLE D.3

Determine the forces in the truss of Figure D.5 using SAP2000. This is the same truss that was previously analyzed in Example 16.2. In the figure the number shown by each member is the area of that member (in.^{2}).

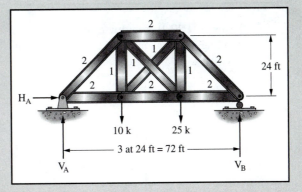

Figure D.5

Solution.   This truss was modeled in Figure D.6 in the same manner as the trusses in the previous chapters. The flexural degrees of freedom at the ends of each member were released as were the torsional degrees of freedom at the beginning of each member.

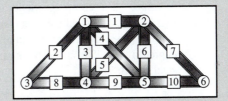

Figure D.6

The member forces that resulted from the analysis are shown in the table.

| Member | P | Beginning | | End | |
		M	V	M	V
1	−21.54	0	0	0	0
2	−21.21	0	0	0	0
3	8.46	0	0	0	0
4	9.25	0	0	0	0
5	2.18	0	0	0	0
6	18.46	0	0	0	0
7	−28.28	0	0	0	0
8	15.00	0	0	0	0
9	13.46	0	0	0	0
10	20.00	0	0	0	0

D.5 ANALYSIS OF COMPOSITE STRUCTURES

In this section the author uses the term composite structure to represent a structure that has different types of members. For instance, in Example D.4 the structure consists of one member that is a cable and another that is a beam.

EXAMPLE D.4

Using SAP2000, determine the total tension in the cable and the shearing forces and bending moments in the beam of the structure shown in Figure D.7. For this structure E is equal to 29,000 ksi.

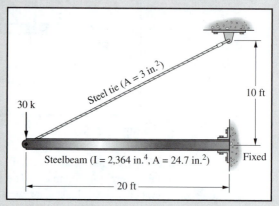

Figure D.7

Solution.   The computer model used in the solution of this problem is shown in Figure D.8. The steel cable is treated as a truss member and as such, the flexural degrees of freedom at each end of the member were released, as was the torsional degree of freedom at the beginning of the member. Notice that member ② is a steel beam.

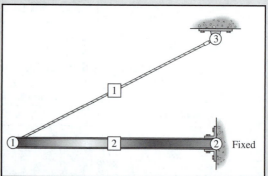

Figure D.8

Upon analysis, the forces in the members are found to be as follows:

Member	P	Beginning		End	
		M	V	M	V
1	43.42	0	0	0	0
2	−34.73	0	3.95	−947.8	3.95

D.6 ANALYSIS OF CONTINUOUS BEAMS AND FRAMES

In this section Examples D.5 and D.6 are presented. The first of these is for a continuous beam while the second one is for a rigid frame.

EXAMPLE D.5

Determine the final moments for the beam of Figure D.9 using SAP2000. This is the beam previously analyzed by moment distribution in Example 20.6.

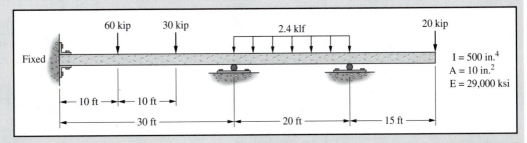

Figure D.9

Solution. The structural model that is used for the solution is shown in Figure D.10. Notice that the loads acting on the beam spans are a combination of point loads and uniformly distributed loads. The properties of the members are specified by creating a new section type called "general." The input data file is available with the program as given on the book's Web site.

Figure D.10

The results of the analysis for the beam members are given in the table to follow. Notice that these results are the same as were obtained with moment distribution.

Member	P	Beginning		End	
		M	V	M	V
1	0.0	−4835.9	−59.21	−1523.4	30.79
2	0.0	−1523.4	−15.35	−3600.0	32.65
3	0.0	−3600.0	−20.0	0.0	−20.0

In Example D.6 the sloping leg frame of Figure D.11 is analyzed. The reader will note that more information is provided with this computer solution than is directly obtained with moment distribution. With moment distribution, only the end moments are obtained directly and other forces such as shears, reactions, and axial forces have to be obtained with additional computations. All of this information is obtained at the same time with the computer solution.

EXAMPLE D.6

Determine the end moments, shearing forces, and axial forces in the frame of Figure D.11. This is the frame previously analyzed in Example 21.6.

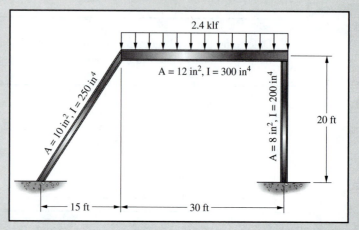

Figure D.11

Solution. The computer model that is used for the analysis is shown in Figure D.12. This is a rigid frame for which there are no releases at the ends of the members.

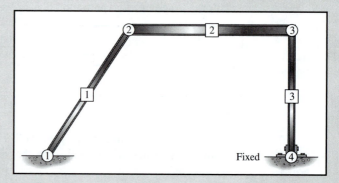

Figure D.12

The results of the analysis are summarized in the following table:

| Member | P | Beginning | | End | |
		M	V	M	V
1	−35.24	−667.07	−1.59	−201.41	−1.59
2	−19.88	−201.41	−29.14	−2670.04	42.86
3	−42.86	2099.98	19.88	−2067.04	19.88

■

GLOSSARY

Glossary

Approximate structural analysis Analysis of structures making use of certain simplifying assumptions or "reasonable approximations."

Beam A member that supports loads that are acting transverse to the member's axis.

Bending moment Algebraic sum of the moments of all of the external forces to one side or the other of a particular section in a member. The moments are taken about an axis through the centroid of the section.

Braced frame A frame that has resistance to lateral loads supplied by some type of auxiliary bracing.

Camber The construction of a member with an initial curvature which is opposite to that which is caused by the loads. The rusulting member deflections will as a result not appear to be large.

Cantilever A projecting or overhanging beam.

Cantilever construction Two simple beams, each with overhanging or cantilevered ends with another simple beam in between supported by the cantilevered ends.

Castigliano's theorems Energy methods for computing deformations and for analyzing statically indeterminate structures.

Cladding The exterior covering of the structural parts of a building.

Column A structural member whose primary function is to support compressive loads.

Concrete A mixture of sand, gravel, crushed rock, or other aggregates held together in a rocklike mass with a paste of cement and water.

Conjugate beam An imaginary beam that has the same length as a real beam being analyzed, and that has a set of boundary and internal continuity conditions such that the slopes and deflections in the real beam equal the shear and moment in the fictitious beam when it is loaded with the M/EI diagram.

Conservation of energy See Law of conservation of energy.

Dead loads Loads of constant magnitude that remain in one position. Examples are weights of walls, floors, roofs, fixtures, structural frames, and so on.

Diaphragms Structural components that are flat plates.

Effective length The distance between points of zero moments in a column, that is, the distance between its inflection points.

Elastic behavior When the external forces are removed, an elastic member will return to its original length.

Environmental loads The loads caused by the environment in which the structure is located. Included are snow, wind, rain, and earthquakes. Strictly speaking, these also are live loads.

"Exact" structural analysis Theoretical analysis of structures.

Fixed-end moments The moments at the ends of loaded members when the member joints are clamped to prevent rotation.

Floor beams The larger beams in many bridge floors that are perpendicular to the roadway of the bridge and that are used to transfer the floor loads from the stringers to the supporting girders or trusses.

Geometric instability A situation existing when a structure has a number of reaction components equal to or greater than the number of equilibrium equations available, and yet still is unstable.

Girder A rather loosely used term usually indicating a large beam and perhaps one into which smaller beams are framed.

Hooke's law A statement of the linear relationship existing between force and deformation in elastic members.

Impact loads The difference between the magnitudes of live loads actually caused and the magnitudes of those loads had they been dead loads.

Influence area The floor area of a building that *directly influences* the forces in a particular member.

Influence line A diagram whose ordinates show the magnitude and character of some function of a structure (shear, moment, deflection, etc.) as a unit load moves across the structure.

Joists The closely spaced beams supporting the floors and roofs of buildings.

Law of conservation of energy When a set of external loads is applied to a structure, the work performed by those loads equals the work performed in the elements of the structure by the internal forces.

Least work, principle of The internal work accomplished by each member or each portion of a structure subjected to a set of external loads is the least possible amount necessary to maintain equilibrium in supporting the loads.

Live loads Loads that change position and magnitude: They move or are moved. Examples are trucks, people, warehouse, materials, furniture, and so on.

Load and resistance factor design A method of design in which the loads are multiplied by certain load or over-capacity factors (larger than 1.0) and the members are designed to have design strengths sufficient to resist these so-called factored loads.

Matrix An ordered arrangement of numbers in rows and columns.

Maxwell's law of reciprocal deflections The deflection at one point A in a structure due to a load applied at another point B is exactly the same as the deflection at B if the same load is applied at A. The law is applicable to members consisting of materials that follow Hooke's law.

Moment distribution A successive correction or iteration method of analysis whereby fixed end and/or sidesway moments are balanced by a series of corrections.

Muller-Breslau's principle The deflected shape of a structure represents to some scale the influence line for a function of the structure such as shear, moment, deflection, among others, if the function in question is allowed to act through a unit displacement.

Nodes The locations in a structure where the elements are connected. In structures composed of beams and columns, the nodes usually are the joints.

Nominal strength Theoretical strength.

Open-web joist A small parallel chord truss whose members often are made from bars (hence the common name *bar joist*)

or small angles or other shapes. These joists are very commonly used to support floor and roof slabs.

Plane frame A frame that for purposes of analysis and design is assumed to lie in a single (or two-dimensional) plane.

Point of contraflexure *See* Point of inflection.

Point of inflection (PI) A point of zero moment. Also called *point of contraflexure*

Ponding A situation in which water accumulates on a roof faster than it runs off.

Principle of superposition If a structure is linearly elastic, the forces acting on the structure may be separated or divided in any convenient fashion and the structure analyzed for the separate cases. The final results can be obtained by adding together the individual parts.

Purlins Roof beams that span between trusses.

Qualitative influence line A sketch of an influence line in which no numerical values are given.

Quantitative influence line An influence line that shows numerical values.

Reinforced concrete A combination of concrete and steel reinforcing wherein the steel provides the tensile strength lacking in the concrete. The steel reinforcing also can be used to help the concrete resist compressive forces.

Scuppers Large holes or tubes in walls or parapets that enable water above a certain depth to quickly drain from roofs.

Seismic Of or having to do with an earthquake.

Service loads The actual loads that are assumed to be applied to a structure when it is in service (also called *working loads*).

Shear The algebraic summation of the external forces in a member to one side or the other of a particular section that are perpendicular to the axis of the member.

Sidesway The lateral movement of a structure caused by lateral or unsymmetrical loads and/or by an unsymmetrical arrangement of the members of the structure.

Skeleton construction Building construction in which the loads are transferred from each floor by beams to columns and thence to the foundation.

Slenderness ratio The ratio of the effective length of a member to its radius of gyration, both values pertaining to the same axis of bending.

Slope deflection A classical method of analyzing statically indeterminate structures in which the moments at the ends of the members are expressed in terms of the rotations (or slopes) and deflections of the joints.

Space truss A three-dimensional truss.

Statically determinate structures Structures for which the equations of equilibrium are sufficient to compute all of the external reactions and internal forces.

Statically indeterminate structures Structures for which the equations of equilibrium are insufficient for computing the external reactions and internal forces.

Steel An alloy consisting almost entirely of iron (usually over 98%). It also contains small quantities of carbon, silicon, manganese, sulfur, phosphorus, and other elements.

Stringers The beams in bridge floors that run parallel to the roadway.

Structural analysis The computation of the forces and deformations of structures under load.

Struts Structural members that are subjected only to axial compression forces.

Superposition principle *See* Principle of superposition.

Tension coefficient The force in a truss member divided by its length.

Three-moment theorem A classical theorem that presents the relationship between the moments in the different supports of a continuous beam.

Ties Structural members that are subjected only to axial tension forces.

Tributary area The loaded area of a structure that *directly contributes* to the load applied to a particular member.

Truss A structure formed by a group of members usually arranged in the shape of one or more triangles.

Unbraced frame A frame whose resistance to lateral forces is provided by its members and their connections.

Unstable equilibrium A support situation whereby a structure is stable under one arrangement of loads but is not stable under other load arrangements.

Vierendeel "truss" Though not really a truss by our usual definition, it is considered a special type of truss whose members are arranged in the shape of a set of rectangles. It requires moment-resisting joints.

Virtual displacement A fictitious displacement imposed on a structure.

Virtual work The work performed by a set of real forces during a virtual displacement.

Voussoirs The truncated wedge-shaped parts of a stone arch that are pushed together in compression.

Wichert truss A continuous statically determinate truss formerly patented by E. M. Wichert.

Working loads *See* Service loads.

Yield stress The stress at which there is a decided increase in the elongation or strain in a member without a corresponding increase in stress.

Zero-load test A procedure in which one member of a truss subjected to no external loads is given a force and the forces in the other members are computed. If all the joints balance or are in equilibrium, the structure is unstable.

INDEX

Index